Lecture Notes in Artificial Intelligence   3849

Edited by J. G. Carbonell and J. Siekmann

Subseries of Lecture Notes in Computer Science

W0112498

Isabelle Bloch   Alfredo Petrosino
Andrea G.B. Tettamanzi (Eds.)

Fuzzy Logic
and Applications

6th International Workshop, WILF 2005
Crema, Italy, September 15-17, 2005
Revised Selected Papers

 Springer

Series Editors

Jaime G. Carbonell, Carnegie Mellon University, Pittsburgh, PA, USA
Jörg Siekmann, University of Saarland, Saarbrücken, Germany

Volume Editors

Isabelle Bloch
Ecole Nationale Supérieure des Télécommunications CNRS
46 rue Barrault, 75013 Paris, France
E-mail: Isabelle.Bloch@enst.fr

Alfredo Petrosino
Universitá di Napoli "Parthenope", Dipartimento di Scienze Applicate
Via A. De Gasperi 5, 80131, Napoli, Italy
E-mail: alfredo.petrosino@unipartehnope.it

Andrea G.B. Tettamanzi
Università degli Studi di Milano
Dipartimento di Tecnologie dell'Informazione
Via Bramante 65, 26013 Crema (CR), Italy
E-mail: andrea.tettamanzi@unimi.it

Library of Congress Control Number: 2006920790

CR Subject Classification (1998): I.2.3, I.5, F.4.1, F.1, F.2, G.2, I.2, I.4

LNCS Sublibrary: SL 7 – Artificial Intelligence

ISSN   0302-9743
ISBN   3-540-32529-8 Springer Berlin Heidelberg New York
ISBN   978-3-540-32529-1 Springer Berlin Heidelberg New York

Springer is a part of Springer Science+Business Media

springer.com

© Springer-Verlag Berlin Heidelberg 2006

Typesetting: Camera-ready by author, data conversion by Scientific Publishing Services, Chennai, India
Printed on acid-free paper   SPIN: 11676935   06/3142   5 4 3 2 1 0

Preface

This volume contains the proceedings of the 6th International Workshop on Soft Computing and Applications (WILF 2005), which took place in Crema, Italy, on September 15–17, 2005, continuing an established tradition of biannual meetings among researchers and developers from both academia and industry to report on the latest scientific and theoretical advances, to discuss and debate major issues, and to demonstrate state-of-the-art systems.

This edition of the workshop included two special sessions, sort of subworkshops, focusing on the application of soft computing techniques (or computational intelligence) to image processing (SCIP) and bioinformatics (CIBB).

WILF began life in Naples in 1995. Subsequent editions of this event took place in 1997 in Bari, in 1999 in Genoa, in 2001 in Milan, and in 2003 back in Naples.

Soft computing, also known as computational intelligence, differs from conventional (hard) computing in that, unlike hard computing, it is tolerant of imprecision, uncertainty, partial truth, and approximation. The guiding principle of soft computing is to exploit the tolerance for imprecision, uncertainty, partial truth, and approximation to achieve tractability, robustness, and low solution cost. The main components of soft computing are fuzzy logic, neural computing, and evolutionary computation.

A rigorous peer-review selection process was applied to the 86 submitted papers. The Program Committee was carefully selected for their knowledge and expertise, and, as far as possible, papers were matched with the reviewer's particular interests and special expertise. The results of this process are seen here in the high quality of papers published within this volume.

Of the 50 published papers, 23 have an Italian provenance; the next strongest representation is from Korea, with 9 papers; all remaining papers are European, with the exception of 3 from the USA, 1 from Algeria and 1 from Iran. This distribution confirms the vocation of WILF to establish itself as a truly international event.

The success of this conference is to be credited to the contribution of many people. In the first place, we would like to thank the members of the Program Committee for their commitment to the task of providing high-quality reviews. We would also like to thank the Information Technology Department of the University of Milan, which hosted the workshop on its premises.

December 2005

Isabelle Bloch
Alfredo Petrosino
Andrea G.B. Tettamanzi

Organization

Organizing Committee

Chairs: Isabelle Bloch (Ecole Nationale Supérieure des
Télécommunications CNRS, France)
Alfredo Petrosino (University of Naples "Parthenope", Italy)
Andrea G. B. Tettamanzi (University of Milan, Italy)
Secretariat: Célia da Costa Pereira (University of Milan, Italy)
Raffaele Montella (University of Naples "Parthenope", Italy)

Steering Committee

Andrea Bonarini, Politecnico di Milano, Italy
Antonio Di Nola, University of Salerno, Italy
Gabriella Pasi, University of Milan "Bicocca", Italy
Vito Di Gesù, University of Palermo, Italy
Francesco Masulli, University of Pisa, Italy
Alfredo Petrosino, University of Naples "Parthenope", Italy

Program Committee

Jim Bezdek, University of West Florida, USA
Miguel Delgado, University of Granada, Spain
Didier Dubois, IRIT, Université Paul Sabatier, France
Marco Gori, University of Siena, Italy
Ugur Halici, METU – Ankara, Turkey
Jim Keller, University of Missouri-Columbia, USA
Etienne Kerre, Ghent University, Belgium
Rudolf Kruse, Universität Magdeburg, Germany
Jérôme Lang, IRIT, CNRS, France
Henrik Larsen, Roskilde University, Denmark
Sankar Pal, Indian Statistics Institute, India
Witold Pedrycz, University of Alberta, Canada
Élie Sanchez, Université de Marseille, France
Daniel Sánchez, University of Granada, Spain
Umberto Straccia, ISTI-CNR, Italy
Settimo Termini, University of Palermo and CNR, Italy
Enric Trillas, Universidad Politécnica de Madrid, Spain
Ronald Yager, Iona College, New York, USA
Hans-Jürgen Zimmermann, RWTH-Aachen, Germany

Sponsoring Institutions

Gruppo Italiano di Ricercatori in Pattern Recognition, Italy
IEEE Neural Networks Society – Italian RIG, Italy
INNS International Neural Network Society – SIG, Italy
Società Italiana Reti Neuroniche, Italy
Genetica s.r.l., Italy

Table of Contents

Pattern Recognition

Evolutionary Algorithms

Control

Special Session: CIBB

Special Session: SCIP

Knowledge Management

Miscellaneous Applications

A Bipolar Possibilistic Representation of Knowledge and Preferences and Its Applications

Didier Dubois and Henri Prade

IRIT, 118, route de Narbonne,
31062 Toulouse Cedex 9, France
{dubois, prade}@irit.fr

Abstract. When representing knowledge, it may be fruitful to distinguish between negative and positive information in the following sense. There are pieces of information ruling out what is known as impossible on the one hand, and pieces of evidence pointing out things that are guaranteed to be possible. But what is not impossible is not necessarily guaranteed to be possible. This applies as well to the modelling of the preferences of an agent when some potential choices are rejected since they are rather unacceptable, while others are indeed really satisfactory if they are available, leaving room for alternatives to which the agent is indifferent. The combination of negative information is basically conjunctive (as done classically in logic), while it is disjunctive in the case of positive information, which is cumulative by nature. This second type of information has been largely neglected by the logical tradition. Both types may be pervaded with uncertainty when modelling knowledge, or may be a matter of degree when handling preferences. The presentation will first describe how the two types of information can be accommodated in the framework of possibility theory. The existence of the two types of information can shed new light on the revision of a knowledge / preference base when receiving new information. It is also highly relevant when reasoning with (fuzzy) if-then rules, or for improving the expressivity of flexible queries.

1 Introduction

Generally speaking, bipolarity refers to the existence of positive and negative information. There are at least three different types of bipolarity. The simplest one, which may be termed *symmetric univariate bipolarity*, takes place when the negative and the positive parts are the exact images of each other. It is the situation in classical logic where negation is involutive. Graded versions of this type of bipolarity are provided in uncertainty modelling by probabilities (since $P(A) = 1 - P(\neg A)$), and in multiple criteria decision making, by bipolar univariate scales having a central element (e. g., the scale [0,1], where 1 (resp. 0) stands for fully satisfactory (resp. not satisfactory at all), and 1/2 models indifference). A second type of bipolarity, termed *dual bivariate*, refers to the use of two separate scales still pertaining to the same information, with generally a duality relation putting the scales in correspondence. An example of it is the representation of uncertainty by dual belief and plausibility functions in Shafer evidence theory, or dual necessity and possibility measures in possibility theory. A third type of *bipolarity*, called *"heterogeneous"*, which is

I. Bloch, A. Petrosino, and A.G.B. Tettamanzi (Eds.): WILF 2005, LNAI 3849, pp. 1 – 10, 2006.

addressed in this paper, takes place when dealing with two different kinds of information in parallel. For instance, incomplete information describing a subset of elements is naturally expressed in a bipolar way. On the one hand, we know that some elements of the referential do not belong to this subset: this is negative information and on the other hand, we know that some other elements belong to this subset: this is positive information. This applies to knowledge or preference representations.

Negative knowledge is usually given by pieces of generic knowledge, integrity constraints, laws, necessary conditions, which state what is impossible, forbidden. Observed cases, examples of solutions, sufficient conditions are positive pieces of information. Beware that positive knowledge may not just mirror what is not impossible. Indeed what is not impossible, not forbidden, does not coincide with what is explicitly possible or permitted. So, a situation that is not impossible (i.e., possible) is not necessarily guaranteed possible (i.e., positive) if it is not explicitly permitted, observed or given as an example. The bipolar view applies to preference representation as well if we have to identify positive desires among what is not more or less rejected. More precisely, negative preferences express solutions, which are rejected or unacceptable, while positive preferences express solutions that are desired, pursued. Bipolarity is supported by recent works on cognitive psychology showing that there are indeed two independent types of information, processed separately in the mind [7]. Indeed, positive and negative information require different representation models and reasoning techniques.

Bipolarity may also be present in other domains. For instance, learning processes bridge the gap between positive and negative information: situations that are often observed are eventually considered as normal and those never observed are considered as impossible. In inconsistency handling problems, argumentation frameworks compute arguments in favor and arguments against formulas. A conflict on a formula is then solved by evaluating the acceptability of arguments in favor of this formula with respect to arguments against this formula. Then, an argument may be accepted, rejected or in abeyance i.e., neither accepted nor rejected [2]. Multiple criteria decision-making can be addressed by evaluating the weight of criteria in favor of a decision and criteria against it.

Possibility theory is a suitable framework for modelling and reasoning about bipolar information. Then negative information and positive information are represented by two separate possibility distributions yielding possibility and guaranteed possibility measures respectively; see, e. g. [9, 13]. This paper provides an introduction on representation and reasoning about bipolar information in this framework. The next section first recalls the semantic representation of bipolar information, and then presents a more compact representation in terms of logical expressions. It also briefly discusses the fusion and the revision of bipolar information. Section 3 discusses bipolarity in the representation of if-then rules. Section 4 briefly discusses the handling of bipolar preferences in flexible querying.

2   Representation of Bipolar Information in Possibility Theory

Bipolar information is represented in possibility theory by *two* separate possibility distributions, denoted by δ and π, representing respectively positive and negative information. When δ and π are *binary* possibility distributions (i.e., $\forall\omega$, $\delta(\omega)$ (resp. $\pi(\omega)$) is 1 or 0), the set I of impossible situations is computed as follows:

$$I = \{\omega : \omega \in \Omega \text{ and } \pi(\omega) = 0\},$$

and the set GP of guaranteed possible situations is computed as follows:

$$GP = \{\omega: \omega \in \Omega \text{ and } \delta(\omega) = 1\}.$$

2.1 Semantic Representations: Combination and Revision

In uni-polar possibility theory, a possibility distribution [19] π, encodes a total pre-order on a set Ω of interpretations or possible states. It associates to each interpretation ω a real number $\pi(\omega) \in [0, 1]$, which represents the compatibility of the interpretation ω with the available knowledge on the real world (in case of uncertain knowledge), or how acceptable reaching state ω is, or equivalently to what extent ω is not rejected (in case of preferences). The larger $\pi(\omega)$, the more plausible or acceptable ω is, according to the problem under concern (knowledge or preference modelling respectively). The distribution π acts as a restriction on possible states. In particular, $\pi(\omega) = 0$ means that ω is totally impossible. The second possibility distribution δ should be understood differently. The degree $\delta(\omega) \in [0, 1]$ estimates to what extent the feasibility of ω is supported by evidence, or ω is really satisfactory, and $\delta(\omega) = 0$ just means that ω has not been observed yet, or when speaking of preference that ω is not especially satisfactory (but this does not mean rejection in any case).

A characteristic property of heterogeneous bipolarity is the fact that the sets of guaranteed possible (GP) and impossible (I) situations should be disjoint and not cover all the referential. This is expressed by the coherence condition GP $\subseteq$ NI, where NI = Ω − I is the set of non-impossible situations. This condition means that what is guaranteed to be possible should be not impossible. This applies as well to bipolar preference modelling where guaranteed possible and impossible situations are replaced by guaranteed satisfactory and non-rejected solutions respectively. When information is graded in presence of uncertainty (or priority), δ represents the fuzzy set of guaranteed possible (or satisfactory) elements and π represents the fuzzy set of not impossible (or not rejected) elements. The coherence condition now reads [13]:

$$\forall\omega \in \Omega, \delta(\omega) \leq \pi(\omega). \tag{1}$$

Example 1 - Bipolar knowledge: Assume for instance one has some information about the opening hours and prices of a museum M. We may know that museum M is open from 2 pm to 4 pm, and certainly closed at night (from 9 pm to 9 am). Note that nothing forbids museum M to be open in the morning although there is no positive evidence supporting it. Its ticket fare is neither less than 2 euros nor more than 8 euros (following legal regulations), prices between 4 and 5 euros are guaranteed to be possible (they are prices actually proposed by the museum).

Since cases referring to observations accumulate, while constraints induced by the knowledge of impossibilities eliminate possible worlds, positive information aggregate disjunctively, and negative information aggregate conjunctively. This can be understood in our setting in the following way. A constraint like « the value of X is restricted by A_i » is encoded by a possibility distribution π s. t. $\pi \le \mu_{Ai}$. Several constraints « X is A_i » for i = 1, …, n are thus equivalent to $\pi \le \min_i \mu_{Ai}$. By the principle of *minimal* commitment (anything not declared impossible is possible), it leads to choose the greatest possibility distribution $\pi = \min_i \mu_{Ai}$ compatible with the constraints. Hence a conjunctive combination. In the case of positive information « X is A_i » is equivalent to $\delta \ge \mu_{Ai}$, since it reports to what extent some states are feasible. Then several observations « X is A_i » for i = 1,…, n are equivalent to $\delta \ge \max_i \mu_{Ai}$. By closed world assumption (anything not observed as actually possible is not considered), one gets $\delta = \max_i \mu_{Ai}$. Hence a disjunctive combination.

Given a pair of possibility distributions π and δ, we can define: the *possibility* degree of a logical formula φ, $\Pi(\varphi) = \max\{\pi(\omega): \omega \in \Omega \text{ and } \omega \mid= \varphi\}$, the dual *necessity* degree $N(\varphi) = 1 - \Pi(\neg\varphi)$ on the one hand, and the *guaranteed possibility* degree of a formula φ, $\Delta(\varphi) = \min\{\delta(\omega): \omega \in \Omega \text{ and } \omega \mid= \varphi\}$ on the other hand (and if necessary the dual degree of potential necessity $\nabla(\varphi) = 1 - \Delta(\neg\varphi)$), where $\omega \mid= \varphi$ denotes that the interpretation ω makes φ true. Note that Π underlies an existential quantifier since $\Pi(\varphi)$ is high as soon as some ω satisfying φ is plausible enough. It agrees with the negative nature of information, since $\neg\varphi$ is impossible, i. e. $\Pi(\neg\varphi) = 0 \Leftrightarrow N(\varphi) = 1$, corresponds to the non-existence of an interpretation ω falsifying φ and having a non-zero degree of possibility $\pi(\omega)$. Δ underlies a universal quantifier since $\Delta(\varphi)$ is low as soon as some ω with a low plausibility satisfies φ. It agrees with the positive nature of information encoded by δ, since $\Delta(\varphi) = 1$ requires that all the interpretations of φ are fully feasible.

Merging bipolar information [5], by disjunctive (resp. conjunctive) combination of positive (resp. negative) information, may create inconsistency when the upper and lower possibility distributions, which represent the negative part and the positive part of the information respectively, fail to satisfy the consistency condition $\forall\omega, \pi(\omega) \ge \delta(\omega)$. Then, it is necessary to revise either π or δ (or their syntactic counterparts, see 2.2 below) for restoring consistency. When dealing with knowledge, the observations are generally regarded as more solid information than beliefs, and then the revision process is modelled by $\pi^{revised}(\omega) = \max(\pi(\omega), \delta(\omega))$, which describes the revision of π by δ once a new report has been fused with the current δ using max operation. Thus priority is given to reports on observed values and it is a belief revision process.

A dual type of revision, defined by $\delta^{revised}(\omega) = \min(\pi(\omega), \delta(\omega))$ would consist in changing δ into $\delta^{revised}$ when receiving information restricting π more (applying fusion based on min combination). This type of revision is natural when dealing with preferences, since then π is associated with a set of more or less imperative goals, while positive information corresponds to the expression of simple desires. Then goals have more priority than the wishes expressed as positive information.

2.2 Logical Representations of Qualitative Bipolar Possibilistic Information

It is generally difficult to specify a possibility distribution pointwisely on large discrete universes. The qualitative possibility theory framework offers three compact representations of a possibility distribution, namely a logical representation by means of a set of weighted formulas [12], a conditional representation by means of default rules, and a Bayesian-like directed acyclic graph representation. The representation of negative information has been widely investigated in these three formats [4]. These three formats can be adapted for representing positive information [5, 6]. We only consider the logical format here. For negative information, formulas refer to pieces of belief (with certainty levels) or to goals (with priority levels), while for positive information formulas report observations (with confidence levels), or describe classes of solutions (with their satisfaction levels).

The logical representation of bipolar information is given by means of two separate sets of weighted classical logic formulas of the form $\Sigma_N = \{(\varphi_i, a_i): i = 1, \ldots, n\}$ and $\Sigma_\Delta = \{[\psi_j, b_j] : j = 1, \ldots, n\}$ modelling respectively negative and positive information. The pair (φ_i, a_i) means that the necessity degree of φ_i is at least equal to a_i, i.e. $N(\varphi_i) \geq a_i$ (which implies $\Pi(\neg\varphi_i) \leq 1 - a_i$), and the pair $[\psi_j, b_j]$ means that the guaranteed possibility degree of ψ_j is at least equal to b_j i.e., $\Delta(\psi_j) \geq b_j$. Given the bases Σ_N and Σ_Δ, we can generate their associated possibility distributions as follows: $\forall \omega \in \Omega$,

$$\pi(\omega) = 1 - \max_i\{a_i : (\varphi_i, a_i) \in \Sigma_N, \omega \not\models \varphi_i\} ; \qquad \pi(\omega) = 1 \text{ if } \forall(\varphi_i, a_i) \in \Sigma_N, \omega \models \varphi_i$$
$$\delta(\omega) = \max_i\{b_j : [\psi_j, b_j] \in \Sigma_\Delta, \omega \models \psi_j\} ; \qquad \delta(\omega) = 0 \text{ if } \forall[\psi_j, b_j] \in \Sigma_\Delta, \omega \not\models \psi_i.$$

Note that adding new formulas to Σ_N leads to a more restrictive possibility distribution, which agrees with the fact that Π-measures model negative information. On the other hand, adding new formulas to Σ_Δ yields a larger possibility distribution, which fits with the fact that Δ-measures model positive information.

3 Bipolarity and If-Then Rules

In the tradition of expert systems, a rule is understood as a production rule, associated to a modus ponens-like deduction process. Such a rule is thus a kind of inference rule, however without a clear mathematical status. In more recent probabilistic expert systems, rules are encoded as conditional probabilities in a belief network. This view of a weighted rule, if mathematically sound, is at odds with the logical tradition, since the probability of a material implication describing a rule clearly differs from the corresponding conditional probability. This observation [17] has led to a vivid debate in philosophical circles since the late seventies [15] without fully settling the case.

A rule is not a two-valued entity, it is a three valued one [11]. To see it, consider a database containing descriptions of items in a set S. If a rule "if A then B" is to be evaluated in the face of this database, it clearly creates a 3-partition of S, namely:

-    the set of examples of the rule: $A \cap B$,
-    its set of counter-examples: $A \cap \neg B$,
-    the set of irrelevant items for the rule is: $\neg A$.

Each case should be encoded by means of a different truth-value. The two first cases only corresponding to the usual truth-values "true" and "false" respectively. The third case corresponds to a third truth-value that, according to the context, can be interpreted as "unknown", undetermined, irrelevant, etc. This idea of a rule as a "tri-event" actually goes back to De Finetti [8] in 1936. It is also the backbone of De Finetti's approach to conditional probability. Indeed it is obvious to see that the probability $P(B|A)$ is entirely defined by $P(A \cap B)$ and $P(A \cap \neg B)$. This framework for modeling a rule produces a precise mathematical model: a rule is modeled as a pair of disjoint sets representing the examples and the counter-examples of a rule, namely $(A \cap B, A \cap \neg B)$. This definition has several consequences. First, it justifies the claim made by De Finetti that a conditional probability $P(B|A)$ is the probability of a particular entity denoted by $B|A$ that can be called a conditional event.

Moreover it precisely shows that material implication only partially captures the intended meaning of an "if-then" rule. It is obvious that the set of items where the material implication $\neg A \cup B$ is true is the complement of the set of counter-examples of a rule. Hence the usual logical view does not single out the examples of the rule, only its counter-examples. This is clearly in agreement with the fact that propositions in classical logic represent negative information. On the other hand, the set of examples of a rule is $A \cap B$ and clearly represents positive information. Thus, the three-valued representation of an "if-then" rule also strongly suggests that a rule contains both positive and negative information. Note that in data mining [1] the merit of an association rule $A \Rightarrow B$ extracted from a database is evaluated by two indices: the support degree and the confidence degree, respectively corresponding to the probability $P(A \cap B)$ and the conditional probability $P(B|A) = P(A \cap B) / (P(A \cap B) + P(A \cap \neg B))$. This proposal may sound ad hoc. However the deep reason why two indices are necessary to evaluate the quality of a rule is because the rule generates a 3-partition of the database, and two evaluations are needed to picture their relative importance. In fact the primitive quality indices of an association rule are the proportion of its examples and the proportion of its counter-examples. All other indices derive from these basic evaluations.

This tri-valued representation also tolerates nonmonotonicity. It is intuitively satisfying to consider that a rule R1 = "if A then B" is safer than or entails a rule R2 = "if C then D", if R1 has more examples and less counterexamples than R2 (in the sense of inclusion). This entailment relation (denoted $\vdash\sim$) can be formally written as

$$B \,|\, A \;\vdash\sim\; D \,|\, C \text{ if and only if } A \cap B \subseteq C \cap D \text{ and } C \cap \neg D \subseteq A \cap \neg B.$$

Indeed, it has been shown [13] that the three-valued semantics of rules provide a representation for the calculus of conditional assertions of Kraus, Lehmann and Magidor [16], which is the main principled approach to nonmonotonic reasoning.

It is also interesting to see how the bipolar view of if-then rules can be handled by the bipolar possibilistic logic setting. Although Δ-based formulas and N-based formulas are dealt with separately in the inference machinery, their parallel processing may be of interest, as shown now. Observe that a Δ-possibilistic logic formula $[\varphi \wedge \psi, b]$ is semantically equivalent to the formula $[\psi, \min(v(\varphi), b)]$ where $v(\varphi) = 1$ if φ is true and $v(\varphi) = 0$ if φ is false. For N-possibilistic logic formulas, it is also possible to "move" a part of the formula in the weight slot [10], but in a different manner, namely $(\neg \varphi \vee \psi, a)$ is equivalent to $(\psi, \min(v(\varphi), a))$. This remark enables us to deal with the application of a set of parallel uncertain "if then" rules, say "if φ_1 then

ψ_1" and "if φ_2 then ψ_2" in presence of a disjunctive input "$\varphi_1 \lor \varphi_2$". In the bipolar view, each rule is represented by a pair of possibilistic formulas of each type, namely $([\varphi_i \land \psi_i, b_i], (\neg \varphi_i \lor \psi_i', a_i))$. Then, the rules can be rewritten as $([\psi_i, \min(v(\varphi_i), b_i)]$, $(\psi_i', \min(v(\varphi_i), a_i))$. Note that we should have here $\psi_i \mid= \psi_i'$ for ensuring consistency.

Now, observe that $[\psi_1, \min(v(\varphi_1), b_1)]$ semantically entails $[\psi_1 \land \psi_2, \min(v(\varphi_1), b_1)]$, and similarly $[\psi_2, \min(v(\varphi_2), b_2)]$ entails $[\psi_1 \land \psi_2, \min(v(\varphi_2), b_2)]$ since $[\psi, b]$ entails $[\psi \land \zeta, b]$. Finally, since $[\psi, b]$ and $[\psi, b']$ entail $[\psi, \max(b, b')]$, we obtain $[\psi_1 \land \psi_2, \max(\min(v(\varphi_1), b_1), \min(v(\varphi_2), b_2))]$.

For N-possibilistic logic formulas, we have that (φ, a) entails $(\varphi \lor \kappa, a)$ on the one hand, while (φ, b) and (φ, b') entail $(\varphi, \max(b, b'))$. So for the two rules, we obtain $(\psi_1' \lor \psi_2', \max(\min(v(\varphi_1), a_1), \min(v(\varphi_2), a_2)))$.

For $b_1 = b_2 = 1$ and $a_1 = a_2 = 1$, this expresses that if it is known that $\varphi_1 \lor \varphi_2$ is true (since $v(\varphi_1 \lor \varphi_2) = \max(v(\varphi_1), v(\varphi_2))$), then $\psi_1 \land \psi_2$ is guaranteed to be possible, and interpretations which falsifies $\psi_1' \lor \psi_2'$ are impossible. This is now illustrated.

Example 2. Let us consider the two rules R1: "if an employee is in category 1 (φ_1), his monthly salary (in euros) is necessarily in the interval [1000, 2000] (ψ_1') and typically in the interval [1500, 1800] (ψ_1)", R2: "if an employee is in category 2 (φ_2), his salary (in euros) is necessarily in the interval [1500, 2500] (ψ_2') and typically in [1700, 2000] (ψ_2)". Typical values are here the values known as being guaranteed to be possible. Let us examine the case of a person who is in category 1 or 2 ($\varphi_1 \lor \varphi_2$); then we can calculate that his salary is necessarily in the interval [1000, 2500] ($\psi_1' \lor \psi_2'$), while values in [1700, 1800] ($\psi_1 \land \psi_2$) are for sure possible for his salary. Note that one might get an empty conclusion with other numerical values, but this would not have meant a contradiction! This bipolar conclusion may, for instance, be instrumental when proposing a salary to a person whose category is ill-determinate.

Lastly, the bipolar view has been also applied to fuzzy rules "if A then B" (when A and/or B become fuzzy), and more particularly the advantages of using conjointly implicative rules (encoding negative information) and conjunctive rules (encoding positive information) in the same rule-based system, have been emphasized in [18]. Indeed, the bipolar view of "if-then" rules can be exploited for building a typology of fuzzy "if then" rules, based on multiple-valued implications or conjunctions, where each type of fuzzy rules serves a specific purpose [14].

4   Bipolarity in Flexible Querying and Databases

Flexible queries have been aroused an increasing interest for many years in the database literature and the fuzzy set-based approach, which was introduced about 25 years ago, is simple to apply and is practically useful. Flexible queries were generally thought in terms of constraints restricting possible values of attributes. By flexible queries, we mean than these constraints could be fuzzy, or prioritized. Fuzzy constraints can be viewed as preference profiles: values associated with degree 1 are fully acceptable, while values with degree 0 are completely rejected; the smaller the degree, the less acceptable the value. The violation of a (crisp) prioritized constraint leads to an upper bound of the global evaluation involving this constraint, which is all the smaller as the constraint has a higher priority. Fuzzy constraints can be viewed as collections of nested prioritized constraints. Fuzzy constraints correspond to "negative" preferences in the sense that their complements define fuzzy sets of rejected values as being non-acceptable.

However, there is another type of preferences, which has a 'positive' flavor. These preferences are not constraints, but only wishes that are more or less strong, but not compulsory [12]. If at least some of these wishes are satisfied, it should give some bonus to the corresponding solutions (provided that they also satisfy the constraints, or at least the most important ones, if it is impossible to satisfy all of them). For example, the query "find an apartment not too expensive, and if possible near the train station" involves two attributes, the price, and the distance to the station. It expresses a fuzzy constraint on the price and a wish on the distance to the station.

Bipolar queries are queries that involve two components, one pertaining to constraints and the other to simple wishes. Let $i = 1, n$ be a set of attributes. Let C_i be the subset of the domain U_i of attribute i, representing the (flexible) constraint restricting the acceptable values for i (in other words, $C_i = R_i^c$ where R_i is the (fuzzy) set of rejected values according to i). Let D_i be the subset of the domain of i, representing the values that are really wished and satisfactory for i. The consistency condition (1) is assumed for each pair (C_i, D_i), i.e.,

$$\forall u, \mu_{D_i}(u) \leq \mu_{C_i}(u) \tag{2}$$

expressing that a value cannot be more wished than it is allowed by the constraint. Note that the pair (R_i, D_i) with $\mu_{R_i} = 1 - \mu_{C_i}$ can be viewed as an Atanassov pair of membership and non-membership functions [3], since (2) then writes $\mu_{D_i} + \mu_{C_i} \leq 1$. Then a query is represented by a set of pairs $\{(C_i, D_i), i = 1, n\}$ satisfying (2).

It may happen that for some attribute i there is no constraint. In such a case $C_i = U_i$, i.e., $\forall u, \mu_{C_i}(u) = 1$. If no value is particularly wished in U_i then $D_i = \varnothing$, i.e., $\forall u, \mu_{D_i}(u) = 0$. Considering the above example of a query about an apartment we have a constraint on the price and a wish on the distance to the station. This is represented by the set of the two pairs $\{(\text{Not\_too\_expensive}, \varnothing), (U_{\text{distance}}, \text{Near})\}$, where Not_too_expensive and Near are labels of fuzzy sets.

How can a query $\{(C_i, D_i), i = 1, n\}$ be evaluated? Given a tuple $u = (u_1, \ldots, u_i, \ldots, u_n)$ of a non-fuzzy database, we can thus compute a pair of matching degrees, namely

$$(C(u), D(u)) = (\otimes_i \mu_{C_i}(u_i), \oplus_i \mu_{D_i}(u_i)),$$

where $\otimes$ (resp. $\oplus$) are a conjunctive (resp. disjunctive) combination operation. Using min and max respectively, it reflects the extent to which u satisfies all the constraints and at least one wish. The question is then to rank-order the tuples. A basic idea is to give priority to constraints, and thus to have a lexicographic ranking, by using $C(u)$ as primary criterion, and $D(u)$ as a secondary one for breaking ties. This procedure can be improved by considering that satisfying several wishes is certainly better than satisfying just one. Let σ be a permutation reordering the $\mu_{D_i}(u_i)$'s decreasingly, i.e. $\mu_{D_{\sigma(1)}}(u_{\sigma(1)}) \geq ... \geq \mu_{D_{\sigma(n)}}(u_{\sigma(n)})$. This leads to rank–order the vectors

$$(\min_i \mu_{C_i}(u_i), \mu_{D_{\sigma(1)}}(u_{\sigma(1)}), ... , \mu_{D_{\sigma(n)}}(u_{\sigma(n)}))$$

lexicographically from $(1, 1, ... , 1)$ to $(0, 0, ... , 0)$. A similar procedure can be applied first to the C- component (reordering the $\mu_{C_i}(u_i)$'s increasingly this time), which refines both the min-based ordering and the Pareto ordering.

Besides, the information stored in the database might be bipolar itself, as in Example 1. This means that for each attribute i, one has both negative information (values that are *impossible*), giving birth to a super-set NI_i of non-impossible values, and positive information under the form of a subset GP_i of values that are guaranteed to be possible (more generally one would have a pair of possibility distributions (π_i, δ_i) s. t. $\pi_i \geq \delta_i$). Let us consider a bipolar query (C_i, D_i). This gives birth to a hierarchy of situations on the basis of which answers could be ranked, namely 1) $NI_i \subseteq D_i$; 2) $NI_i \subseteq C_i$ and $GP_i \subseteq D_i$; 3) $NI_i \subseteq C_i$; 4) $GP_i \subseteq D_i$; 5) $GP_i \subseteq C_i$; 6) $GP_i \cap D_i \neq \varnothing$; 7) $NI_i \cap D_i \neq \varnothing$ and $GP_i \cap C_i \neq \varnothing$; 8) $NI_i \cap D_i \neq \varnothing$; 9) $GP_i \cap C_i \neq \varnothing$; 10) $NI_i \cap C_i \neq \varnothing$. These conditions can be further refined under the form of graded degrees of inclusion and intersection in the general case.

5 Conclusion

This paper has proposed an overview on the representation and handling of bipolar information in possibility theory framework. Bipolarity enables us to distinguish between what is possible or satisfactory from what is just not impossible or not undesirable. Bipolarity is met in many areas such as knowledge representation, learning, decision, inconsistency handling, etc.

References

1. R. Agrawal, T. Imielinski and A. Swami. Mining association rules between sets of items in large databases. *Proc. of ACM SIG-MOD*, 207-216,1993.
2. L. Amgoud, C. Cayrol and M.C. Laguasquie-Schiex. On the bipolarity in argumentation frameworks.*Proc. 10th Non Monotonic Reason. Workshop* (NMR'04), Whistler, 2004, 1-10.
3. K. T. Atanassov. Intuitionistic fuzzy sets. *Fuzzy Sets and Systems*, 20, 87-96,1986.
4. S. Benferhat, D. Dubois, S. Kaci and H. Prade. Bridging logical, comparative and graphical possibilistic representation frameworks. *Proc. 6th Europ. Conf. on Symbolic and Quantitative Approaches to Reasoning and Uncertainty* (ECSQARU'01),LNAI 2143, 422-431, 2001
5. S. Benferhat, D. Dubois, S. Kaci and H. Prade. Bipolar representation and fusion of preferences in the possibilistic logic framework. *Proc. 8th Int. Conf. on Principle of Knowledge Representation and Reasoning* (KR'02), 158-169, 2002.
6. S. Benferhat, D. Dubois, S. Kaci and H. Prade. Bipolar possibilistic representations. *Proc. 18th Int. Conf. Uncertainty in Artificial Intelligence* (UAI'02), 45-52, 2002.
7. J.T. Cacioppo and G.G. Bernston. The affect system: Architecture and operating characteristics. In *Current Directions in Psychological Science*, 8(5), 133-137, 1999.

8. B. De Finetti. La logique des probabilités. *Actes Int. Cong. de Philosophie Scientifique*, volume 5, 1-9, 1936.
9. D. Dubois, P. Hajek and H. Prade. Knowledge-Driven versus data-driven logics. *J. of Logic, Language, and Information*, 9, 65-89, 2000.
10. D. Dubois, J. Lang and H. Prade. Possibilistic logic. In *Handbook of Logic in Artificial Intelligence and Logic Programming*, 3 (D. Gabbay et al., eds), 439-513, 1994.
11. D. Dubois and H. Prade. Conditional objects as non-monotonic consequence relationships. *IEEE Trans. on Systems, Man and Cybernetics*, 24 (12), 1724-1739,1994.
12. D. Dubois and H. Prade. Bipolarity in flexible querying. In *Flexible Query Answering Systems, 5th International Conference*, (FQAS'02), Copenhagen, LNAI 2522, Springer-Verlag, 174-182, Oct. 27-29, 2002.
13. D. Dubois, H. Prade and P. Smets. "Not impossible" vs. "guaranteed possible" in fusion and revision. *Proc. 6th Europ. Conf. on Symbolic and Quantitative Approaches to Reasoning with Uncertainty* (ECSQAR U'01), LNAI 2143, 522-531, 2001.
14. D. Dubois, H. Prade and L. Ughetto. A new perspective on reasoning with fuzzy rules. *Int. J. of Intelligent Systems*, 18, 541-567, 2003.
15. W.L. Harper, R. Stalnaker and G. Pearce, eds. In *Ifs*. D. Reidel, Dordrecht, 1981.
16. S. Kraus and D. Lehmann and M. Magidor. Nonmonotonic reasoning, preferential models and cumulative logics. *Artificial Intelligence*, 44, 167-207, 1990.
17. D. Lewis. Probabilities of conditionals and conditional probabilities. *J. Phil. Logic*, 3, 1973
18. L. Ughetto, D. Dubois and H. Prade. Implicative and conjunctive fuzzy rules- a tool for reasoning from knowledge aaid examples. *Proc.16th National Conf. on Artificial Intelligence (AAAI'99), 214-219, 1999.*
19. L. Zadeh. Fuzzy sets as a basis for a theory of possibility. *Fuzzy Sets & Syst.*, 1, 3-28, 1978.

Statistical Distribution of Chemical Fingerprints

S. Joshua Swamidass[1] and Pierre Baldi[1,2]

[1] Department of Computer Science,
Institute for Genomics and Bioinformatics,
University of California, Irvine,
CA 92697-3435
Tel.: (949) 824-5809, Fax.: (949) 824-4056
sswamida@ics.uci.edu
[2] Department of Biological Chemistry, College of Medicine,
University of California, Irvine
pfbaldi@ics.uci.edu

Abstract. Binary fingerprints are binary vectors used to represent chemical molecules by recording the presence or absence of particular substructures, such as labeled paths in the 2D graph of bonds. Complete fingerprints are often reduced to a compressed format–of typical dimension $n = 512$ or $n = 1024$–by using a simple congruence operation. The statistical properties of complete or compressed fingerprints representations are important since fingerprints are used to rapidly search large databases and to develop statistical machine learning methods in chemoinformatics. Here we present an empirical and mathematical analysis of the distribution of complete and compressed fingerprints. In particular, we derive formulas that provide good approximation for the expected number of bits set to one in a compressed fingerprint, given its uncompressed version, and vice versa.

1 Introduction

As in bioinformatics, one of the most fundamental tasks of chemoinformatics is the rapid search of large repositories of molecules. In a typical task, given a query molecule or a family or query molecules and a set of additional constrains, one is interested in retrieving all the molecules contained in a large repository that may contain million of compounds, such as PubChem, ZINC [6], or ChemDB [3],aa that are similar to the query molecule(s) and satisfy the given constraints. To faciliate this process, in most chemoinformatics systems, molecules are represented by binary fingerprints ([4, 5] and references therein) and it is these fingerprints and their similarity measures (e.g Tanimoto, Tversky) [10, 8] that are used for the searches. While a sophisticated technology for deriving useful fingerprints and similarity measures has been developed, the *statistical* analyses of the properties of these fingerprints and the associated fingerprint measures and significance cutoffs have not been studied extensively. Lessons learnt in bioinformatics, for instance in the development of the BLAST [1] family of algorithms, show that these statistical properties may be important to develop better search algorithms, as well as better similarity measures and kernels for machine learning in chemoinformatics [7, 9]. Here we take a

I. Bloch, A. Petrosino, and A.G.B. Tettamanzi (Eds.): WILF 2005, LNAI 3849, pp. 11–18, 2006.

first step by studying, modeling, and aproximating the statistical distributions of both compressed and uncompressed chemical fingerprints.

2   Chemical Fingerprints

We use $\mathcal{A}, \mathcal{B}, \ldots$ to denote molecules. We assume that molecules are represented by binary fingerprints of length N, denoted by $\boldsymbol{A}$, $\boldsymbol{B}$,.... The precise interpretation of the fingerprint is irrelevant for our purpose, but to fix his ideas, the reader may think that each bit is associated with the present or absence of a labeled path of labeled atoms and bonds present in a molecule. The value of N depends on the depth of the paths, but a typical value could be something like $N = 2^{15} = 32,768$ or more. These fingerprints are compressed (or "folded") using a simple modulo operator to fingerprints $\boldsymbol{a}$, $\boldsymbol{b}$....of lenth n, with $N = nk$. Typically, in current systems, $n = 512 = 2^9$ or $n = 1024 = 2^{10}$. Ultimately, it is these short fingerprints that are used to derive similarity measures and searching databases. A bit in position j of the compressed fingerprint is set to one if and only if there is at least one bit set to one in position j modulo n in the full fingerprint of length N. For each molecule, we will use capital letters $A, B, \ldots$ to denote the number of bits set to one in the full fingerprint, and lower case letters $a, b, \ldots$ to denote the number of bits set to one in the corresponding compressed fingerprint. The corresponding count and projection operator are denoted by c and n so that: $A = c(\boldsymbol{A})$, $a = c(\boldsymbol{a})$ and $\boldsymbol{a} = n(\boldsymbol{A})$.

While in some applications it may be possible to exploit information associated with specific bits or weigh bits differently, in order to develop a probabilistic theory and consistently with common practice, here we will assume that the fingerprints are "random" in the sense that information about the labeled paths associated with each position is ignored and that the bits are being distributed randomly and uniformly along the fingerprint. In practice this requires applying a random permutation to the deterministically computed fingerprints of length N prior to compression, or, in practice, using a good hashing function to derive "randomized" fingerprints of length n. Note that for a typical users, the value of N is often unknown. Even for developers, the value of N is often not very relevant since labeled paths are indexed one molecule after the other. In this case, what is meaningful is the value of N^*, the total number of paths actually observed in a set of molecules. In some theoretical derivations we will assume first knowledge of N and then dispense from such knowledge.

3   Basic Distributions and Models

First we are interested in understanding the joint distribution of A and a (or A/N and a/n) and this is best achieved by studying the marginals and conditionsl distributions and using decompositions of the form $P(A, a) = P(A)P(a|A) = P(a)P(A|a)$. For illustration purposes, an example of empirical distribution is given in Table 1 using a random sample of 50,000 molecules extracted from the ChemDB database.

Table 1. Empirical and predicted statistics computed on a random sample of 50,000 molecules extracted from the ChemDB [3] database. Fingerprints are associated with labeled paths of length up to 8 (i.e. 8 atoms and 7 bonds). The total number of observed labeled paths is $N^* = 152,087$. The length of the compressed fingerprints is $n = 512$. Predicted values are derived using Equations 6 and 8 (see text).

	A	A/N^*	a	a/n
Mean	138.68	9.1186e-04	119.5305	0.2335
Median	136	8.9423e-04	119	0.2324
STD	53.2388	3.5006e-04	40.0672	0.0783
Variance	2.8344e+03	1.2254e-07	1.6054e+03	0.0061
Predicted Mean	136.12	8.9591e-04	121.485	o.2373
Predicted STD	52.452	3.4488e-04	40.68	0.0794

3.1 Distribution of Uncompresssed Fingerprints [A (or A/N)]

It is reasonable to assume a Gaussian or Gamma distribution approximation for $P(A)$

$$G_{\bar{A}\sigma_A}(A) = \frac{1}{\sqrt{2\pi}\sigma_A}e^{-(A-\bar{A})^2/2\sigma_A^2} \qquad (1)$$

with $\bar{A}$ being the mean (also median and mode) and σ_A the standard deviation. In the Gamma model defined for $A \geq 0$,

$$Gamma_{rs}(A) = \frac{s^r}{\Gamma(r)}A^{r-1}e^{-sA} \qquad (2)$$

with mean r/s, variance r/s^2, and mode $(r-1)/s$ (for $r \geq 1$, which is the case in our application). The two parameters that characterize these distributions can easily be fit to the data. It is worth noting, that a simple binomial model in general is not a good model for A because empirically the variance of A is larger than its mean (Table 1), whereas in a binomial distribution the variance (Npq) is necessarily less or equal to the mean (Np).

Altenatively, the distribution of $\alpha = A/N$ or $\alpha = A/N^*$ can be modeled using a Beta distribution with

$$Beta_{rs}(\alpha) = \frac{\Gamma(r+s)}{\Gamma(r)\Gamma(s)}\alpha^{r-1}(1-\alpha)^{s-1} \qquad (3)$$

The expectation of this Beta distribution is $E(\alpha) = r/(r+s)$, the variance $Var(\alpha) = rs/(r+s)^2(r+s+1)$, and the mode $(r-1)/(r+s-2)$. Figure 1.

3.2 Conditional Distribution of Compressed Fingerprints [a (or a/n) Given A (or A/N)]

As in the case of random graph theory [2], with a given value A we can consider two slightly different, but asymptotically, equivalent models: fixed density and fixed size uniform models. In the fixed density model, bits are produced by

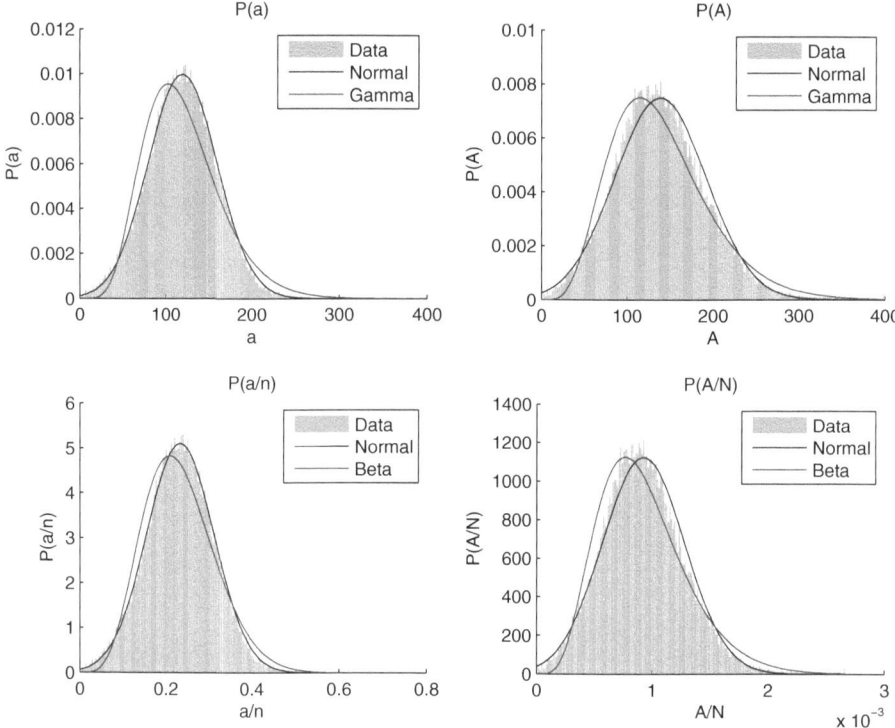

Fig. 1. Empirical distributions of a, A, a/n and A/N^* and maximum likelihood fit using Gaussian, Gamma, and Beta models

independent identically distributed coin flips with probability $\alpha = A/N$ (or α/N^*) of success. In the fixed size model, a subset of A bit is selected uniformly among all subsets of A bits of the full N bit vector. The bits in the corresponding subset are set to one, the bits in the complement are set to 0. Although both models are known to yield similar results asymptotically, the fixed size model is often less tractable due to the weak but non zero-correlations it introduces between different bits. In general, we will use the density model since it is more tractable. We use $\mathbf{A}$ and $\mathbf{a}$ to denote the random variables associated with the total number of bits set to one and write, for instance, $P(\mathbf{A} = A)$ or just $P(A)$ when the context is clear. In this case, the probability of setting a bit to 0 in $\mathbf{a}$ is $(1 - \alpha)^k$ and therefore the corresponding distribution is binomial $\mathcal{B}(n, p)$ with:

$$P(\mathbf{a} = a|\alpha) = \mathcal{B}(n, p) \quad \text{with} \quad p = 1 - (1 - \alpha)^k \qquad (4)$$

Thus although A or a do not have binomial distributions, the condition distribution of a given A can be reasonably approximated by a binomial distributions (Figure 3), although the binomial approximation here overestimates the variance.

Therefore, given α, $E(\mathbf{a}) = n[1-(1-\alpha)^k]$ and $Var(\mathbf{a}) = n[1-(1-\alpha)^k](1-\alpha)^k$. When N is large, we have:

$$\lim_{N \to \infty} [1 - (1 - \alpha)^k] = \lim_{N \to \infty} [1 - (1 - \frac{A}{N})]^{\frac{N}{n}} = e^{-A/n} \tag{5}$$

which does not depend on N (hence the irrelevance of its exact value). Thus given A, for N large a is approximately binomial with $\mathcal{B}(n, 1 - e^-A/n)$ and

$$E(a|A) \approx n(1 - e^{-A/n}) \tag{6}$$

This is a one-to-one increasing function. Using the same approximations, we get the conditional variance

$$Var(a|A) \approx n(1 - e^{-A/n})e^{-A/n} \tag{7}$$

By inverting the one to one function in Equation,6 we get also an estimate of A given a for large N

$$A \approx -n \log(1 - \frac{a}{n}) \tag{8}$$

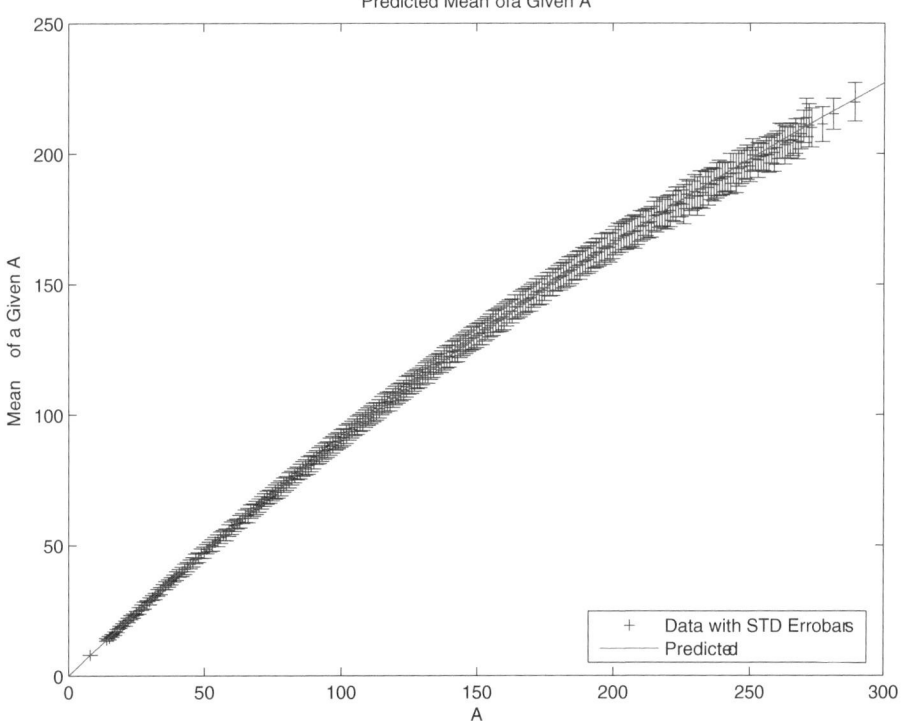

Fig. 2. Empirical mean and standard deviation of a given A. Red curve corresponds to the predicted mean derived from Equation 6.

The functions in 6 and 8 are monotone increasing, hence one-to-one, and provide good estimates of the value of a given A and vice versa in practical regimes. For instance, by applying them to the means (or modes) of the empirical distribution of a and A we get the estimates in Table 1 which are less than 5% away from the true values. Figure 2 shows, for each value of A, the predicted value of a using Equation 6, together with the empirical standard deviation of a. When applied using the mean of A, Equation 7 gives an underestimate of the variance of a. Better estimates are easily obtained by approximating A or a with Gaussian distributions and applying Equations 6 and 8 to points that are one empirical standard deviation away from the mean of A or a. This yields the estimates in the last row of Table 1.

The binomial model for the conditional distribution is thus a reasonable model that gives good first order predictions. However, as a distribution, for a given value of A it overestimates the variance of a (Figure 3) because instead of holding A fixed, it holds the $\alpha = A/N$ fixed, and therefore looks at a range of possible values for A.

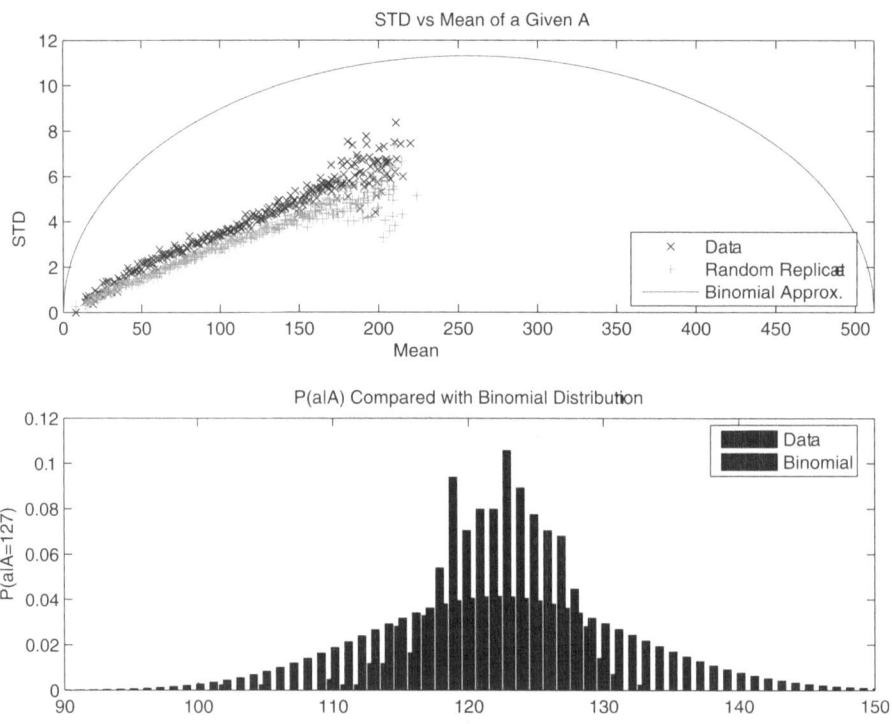

Fig. 3. Upper figure: Empirical STD versus mean of a given A. Theoretical curve, derived from Equation 7, overestimates the STD. Lower figure: Empirical distribution of a given $A = 127$, and corresponding binomial distribution associated with fixed density, rather than fixed A.

3.3   Distribution of Compressed Fingerprints [a (or a/n)]

Using continuous notation, the distribution of a is obtained by integrating over A or over α in the form

$$
\begin{aligned}
P(\mathbf{a} = a) &= \int_{-\infty}^{+\infty} P(a|A)P(A)dA = \int_0^1 P(a|\alpha)P(\alpha)d\alpha \\
&= \int_{-\infty}^{+\infty} \binom{n}{a}[1 - (1 - \frac{A}{N})^k]^a [(1 - \frac{A}{N})^k]^{n-a} P(A)dA \\
&= \int_0^1 \binom{n}{a}[1 - (1 - \alpha)^k]^a [(1 - \alpha)^k]^{n-a} P(\alpha)d\alpha
\end{aligned}
\tag{9}
$$

These integrals do not have simple closed form solutions. However, using Gaussian or Gamma distributions on A or Beta distributions on α, we can expand the polynomials, integrate term by term, and derive closed form sum expansions (given in our corresponding Technical Report).

Furthermore, for N large, by interchanging the limit and integral operators and using Equation 5 we get:

$$
\begin{aligned}
P(\mathbf{a} = a) &= \int_{-\infty}^{+\infty} \binom{n}{a}(1 - e^{-A/n})^a (e^{-A/n})^{n-a} P(A)dA \\
&= \int_0^1 \binom{n}{a}(1 - e^{-k\alpha})^a (e^{-k\alpha})^{n-a} P(\alpha)d\alpha
\end{aligned}
\tag{10}
$$

Another approach to derive an approximation to $P(a)$ is to combine the Gaussian or Gamma approximation to $P(A)$ (or Beta to A/N^*) with the relationship in Equation 8. This gives

$$
P(a) \approx \frac{n}{n-a} G_{\bar{A}\sigma_A}(-n\log(1 - \frac{a}{n})) \approx \frac{n}{n-a} Gamma_{rs}(-n\log(1 - \frac{a}{n}))
\tag{11}
$$

3.4   Conditional Distribution of Uncompressed Fingerprints [A or A/N Given a (or a/n)]

This can easily be derived indirectly by using Bayes theorem in the form

$$
P(A|a) = P(a|A)\frac{P(A)}{P(a)}
\tag{12}
$$

and using the marginal and conditional distributions derived above. A more direct model may also be possible.

4   Conclusion

Here we have taken the first steps towards studying the statistical properties of uncompressed and compressed fingerprints. Work in progress is currently aimed at studying the statistical properties of similarity measures, such as the Tanimoto or Tversky similarity measures computed on pairs of fingerprints, the properties of similarity measures computed on groups of molecules (profiles), and at the extreme value distribution of fingerprint similarities in order to define efficient significance thresholds.

Acknowledgements

Work supported by an NIH Biomedical Informatics Training grant (LM-07443-01) and an NSF MRI grant (EIA-0321390) to PB, by the UCI Medical Scientist Training Program, and by a Harvey Fellowship to S.J.S. We would like also to acknowledge the OpenBabel project and OpenEye Scientific Software for their free software academic license, and Drs. Chamberlin, Nowick, Piomelli, and Weiss for their useful feedback.

References

1. S. Altschul, T. Madden, A. Shaffer, J. Zhang, Z. Zhang, W. Miller, and D. Lipman. Gapped Blast and PSI-Blast: a new generation of protein database search programs. *Nucl. Acids Res.*, 25:3389–3402, 1997.
2. Bela Bollobas. *Random Graphs.* Academic Press, London, 1985.
3. J. Chen, S. J. Swamidass, Y. Dou, and J. Bruand P. Baldi. ChemDB: a public database of small molecules and related chemoinformatics resources. 2005. Submitted.
4. M. A. Fligner, J. S. Verducci, and P. E. Blower. A Modification of the Jaccard/Tanimoto Similarity Index for Diverse Selection of Chemical Compounds Using Binary Strings. *Technometrics*, 44(2):110–119, 2002.
5. D. R. Flower. On the properties of bit string-based measures of chemical similarity. *J. of Chemical Information and Computer Science*, 38:378–386, 1998.
6. J. J. Irwin and B. K. Shoichet. ZINC–a free database of commercially available compounds for virtual screening. *Journal of Chemical Information and Computer Sciences*, 45:177–182, 2005.
7. L. Ralaivola, S. J. Swamidass, H. Saigo, and P. Baldi. Graph kernels for chemical informatics. *Neural Networks*, 2005. Special issue on Neural Networks and Kernel Methods for Structured Domains. In press.
8. D. Rouvray. Definition and role of similarity concepts in the chemical and physical sciences. *Journal of Chemical Information and Computer Sciences*, 32(6):580–586, 1992.
9. S. J. Swamidass, J. Chen, J. Bruand, P. Phung, L. Ralaivola, and P. Baldi. Kernels for small molecules and the prediction of mutagenicity, toxicity, and anti-cancer activity. *Bioinformatics*, 21(Supplement 1):i359–368, 2005. Proceedings of the 2005 ISMB Conference.
10. A. Tversky. Features of similarity. *Psychological Review*, 84(4):327–352, 1977.

Fuzzy Transforms and Their Applications to Image Compression*

Irina Perfilieva

University of Ostrava,
Institute for Research and Applications of Fuzzy Modeling,
30.dubna 22, 701 03 Ostrava 1, Czech Republic
Irina.Perfilieva@osu.cz

Abstract. The technique of direct and inverse fuzzy (F-)transforms of
three different types is introduced and approximating properties of the
inverse F-transforms are described. A method of lossy image compression
and reconstruction on the basis of the F-transform is presented.

1 Introduction

In classical mathematics, various kinds of transforms (Fourier, Laplace, integral, wavelet) are used as powerful methods for construction of approximation
models and their further utilization. The main idea of these transforms consists
in transforming the original model into a special space where the computation
is simpler. The transform back to the original space produces an approximate
model or an approximate solution.

In this paper, we put a bridge between these well known classical methods
and methods for construction of *fuzzy* approximation models. We have developed
[4] the general method called *fuzzy transform* (or, shortly, *F-transform*) that
encompasses both classical transforms as well as approximation methods based
on elaboration of fuzzy IF-THEN rules studied in fuzzy modelling.

In this paper, we will use approximation models on the basis of three different
fuzzy transforms to data compression and decompression. A method of lossy
image compression and reconstruction on the basis of F-transforms is illustrated
on examples of pictures.

Due to the space limitations, all proofs are omitted. The interested reader can
find them in [4].

2 Fuzzy Partition of the Universe and the Direct F-Transform

The core idea of the technique proposed in this paper is a fuzzy partition of an
interval as a universe. We claim that for a sufficient (approximate) representation

---

* The paper has been partially supported by the projects 1M0572 and
MSM6198898701 of MŠMT ČR.

I. Bloch, A. Petrosino, and A.G.B. Tettamanzi (Eds.): WILF 2005, LNAI 3849, pp. 19–31, 2006.

of a function, defined on the interval, we may consider its average values within subintervals which constitute a partition of that interval. Then, an arbitrary function can be associated with a mapping from thus obtained set of subintervals to the set of average values of this function. Moreover, the set of the above mentioned average values gives an approximate (compressed) representation of the considered function. Let us give the necessary details (see [4]).

Definition 1. *Let* $x_1 < \ldots < x_n$ *be fixed nodes within* $[a, b]$, *such that* $x_1 = a$, $x_n = b$ *and* $n \geq 2$. *We say that fuzzy sets* $A_1, \ldots, A_n$, *identified with their membership functions* $A_1(x), \ldots, A_n(x)$ *defined on* $[a, b]$, *form a fuzzy partition of* $[a, b]$ *if they fulfil the following conditions for* $k = 1, \ldots, n$:

1. $A_k : [a, b] \longrightarrow [0, 1]$, $A_k(x_k) = 1$;
2. $A_k(x) = 0$ *if* $x \notin (x_{k-1}, x_{k+1})$ *where for the uniformity of denotation, we put* $x_0 = a$ *and* $x_{n+1} = b$;
3. $A_k(x)$ *is continuous*;
4. $A_k(x)$, $k = 2, \ldots, n$, *monotonically increases on* $[x_{k-1}, x_k]$ *and* $A_k(x)$, $k = 1, \ldots, n - 1$, *monotonically decreases on* $[x_k, x_{k+1}]$;
5. *for all* $x \in [a, b]$

$$\sum_{k=1}^{n} A_k(x) = 1. \tag{1}$$

The membership functions $A_1(x), \ldots, A_n(x)$ *are called* basic functions.

Let us remark that the shape of basic functions is not predetermined and therefore, it can be chosen additionally.

We say that a fuzzy partition $A_1(x), \ldots, A_n(x)$, $n > 2$, is *uniform* if the nodes $x_1, \ldots, x_n$ are equidistant, i.e. $x_k = a + h(k - 1)$, $k = 1, \ldots, n$, where $h = (b - a)/(n - 1)$, and two more properties are fulfilled for $k = 2, \ldots, n - 1$:

6. $A_k(x_k - x) = A_k(x_k + x)$, for all $x \in [0, h]$,
7. $A_k(x) = A_{k-1}(x - h)$, for all $x \in [x_k, x_{k+1}]$ and
   $A_{k+1}(x) = A_k(x - h)$, for all $x \in [x_k, x_{k+1}]$

Figure 1 shows a uniform partition by sinusoidal shaped basic functions:

$$A_1(x) = \begin{cases} 0.5(\cos \frac{\pi}{h}(x - x_1) + 1), & x \in [x_1, x_2], \\ 0, & \text{otherwise}, \end{cases}$$

$$A_k(x) = \begin{cases} 0.5(\cos \frac{\pi}{h}(x - x_k) + 1), & x \in [x_{k-1}, x_{k+1}], \\ 0, & \text{otherwise}, \end{cases}$$

where $k = 2, \ldots n - 1$, and

$$A_n(x) = \begin{cases} 0.5(\cos \frac{\pi}{h}(x - x_n) + 1), & x \in [x_{n-1}, x_n], \\ 0, & \text{otherwise}. \end{cases}$$

The following lemma [3,4] shows that, in the case of a uniform partition, the expression of F-transform components can be simplified.

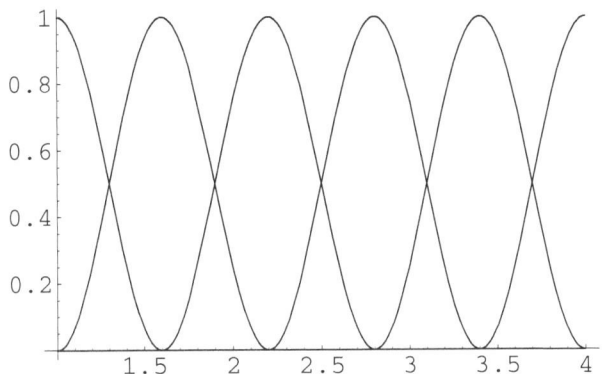

Fig. 1. An example of a uniform fuzzy partition of $[1, 4]$ by sinusoidal membership functions

Lemma 1. *Let the uniform partition of $[a, b]$ be given by basic functions A_1, $\ldots, A_n$, $n \geq 3$. Then*

$$\int_{x_1}^{x_2} A_1(x)dx = \int_{x_{n-1}}^{x_n} A_n(x)dx = \frac{h}{2}, \tag{2}$$

and for $k = 2, \ldots, n - 1$

$$\int_{x_{k-1}}^{x_{k+1}} A_k(x)dx = h. \tag{3}$$

Let $C([a, b])$ be the set of continuous functions on the interval $[a, b]$. The following definition (see also [3,4]) introduces the fuzzy transform of a function $f \in C([a, b])$.

Definition 2. *Let $A_1, \ldots, A_n$ be basic functions which form a fuzzy partition of $[a, b]$ and f be any function from $C([a, b])$. We say that the n-tuple of real numbers $[F_1, \ldots, F_n]$ given by*

$$F_k = \frac{\int_a^b f(x)A_k(x)dx}{\int_a^b A_k(x)dx}, \qquad k = 1, \ldots, n, \tag{4}$$

is the (integral) F-transform of f with respect to $A_1, \ldots, A_n$.

Denote the F-transform of a function $f \in C([a, b])$ with respect to $A_1, \ldots, A_n$ by $\mathbf{F}_n[f]$. Then according to Definitions 2, we can write

$$\mathbf{F}_n[f] = [F_1, \ldots, F_n]. \tag{5}$$

The elements $F_1, \ldots, F_n$ are called *components of the F-transform*.

It is easy to see that if the fuzzy partition of $[a, b]$ (and therefore, basic functions) is fixed then the F-transform establishes a linear mapping from $C([a, b])$ to $\mathbb{R}^n$ so that

$$\mathbf{F}_n[\alpha f + \beta g] = \alpha \mathbf{F}_n[f] + \beta \mathbf{F}_n[g]$$

At this point we will refer to [4] for some useful properties of the F-transform components. The most important one we are going to present below. We will show that the components of its F-transform are the *weighted mean values* of a given function where the weights are given by the basic functions.

Theorem 1. *Let f be a continuous function on $[a, b]$ and $A_1, \ldots, A_n$ be basic functions which form a fuzzy partition of $[a, b]$. Then the k-th component of the integral F-transform gives minimum to the function*

$$\Phi(y) = \int_a^b (f(x) - y)^2 A_k(x) dx \tag{6}$$

defined on $[f(a), f(b)]$.

Let us specially consider a discrete case, when the original function f is known (may be computed) only at some nodes $p_1, \ldots, p_l \in [a, b]$. We assume that the set P of these nodes is *sufficiently dense with respect to the fixed partition*, i.e.

$$(\forall k)(\exists j) A_k(p_j) > 0. \tag{7}$$

Then the (discrete) F-transform of f is introduced as follows.

Definition 3. *Let a function f be given at nodes $p_1, \ldots, p_l \in [a, b]$ and $A_1, \ldots, A_n$, $n < l$, be basic functions which form a fuzzy partition of $[a, b]$. We say that the n-tuple of real numbers $[F_1, \ldots, F_n]$ is the discrete F-transform of f with respect to $A_1, \ldots, A_n$ if*

$$F_k = \frac{\sum_{j=1}^l f(p_j) A_k(p_j)}{\sum_{j=1}^l A_k(p_j)}. \tag{8}$$

Similarly to the integral F-transform, we may show that the components of the discrete F-transform are the *weighted mean values* of the given function where the weights are given by the basic functions.

2.1   Inverse F-Transform

A reasonable question is the following: can we reconstruct the function from its F-transform? The answer is clear: not precisely in general because we are loosing information when passing to the F-transform. However, the function that can be reconstructed (by the inversion formula) approximates the original one in such a way that a universal convergence can be established.

Definition 4. *Let $A_1, \ldots, A_n$ be basic functions which form a fuzzy partition of $[a, b]$ and f be a function from $C([a, b])$. Let $\mathbf{F}_n[f] = [F_1, \ldots, F_n]$ be the integral F-transform of f with respect to $A_1, \ldots, A_n$. Then the function*

$$f_{F,n}(x) = \sum_{k=1}^n F_k A_k(x) \tag{9}$$

is called the inverse F-transform.

The theorem below shows that the inverse F-transform $f_{F,n}$ can approximate the original continuous function f with an arbitrary precision.

Theorem 2. *Let f be a continuous function on $[a, b]$. Then for any $\varepsilon > 0$ there exist n_ε and a fuzzy partition $A_1, \ldots, A_{n_\varepsilon}$ of $[a, b]$ such that for all $x \in [a, b]$*

$$|f(x) - f_{F,n_\varepsilon}(x)| \leq \varepsilon \tag{10}$$

where f_{F,n_ε} is the inverse F-transform of f with respect to the fuzzy partition $A_1, \ldots, A_{n_\varepsilon}$.

In the discrete case, we define the inverse F-transform only at nodes where the original function is given.

Definition 5. *Let function $f(x)$ be given at nodes $p_1, \ldots, p_l \in [a, b]$ and $\mathbf{F}_n[f] = [F_1, \ldots, F_n]$ be the discrete F-transform of f w.r.t. $A_1, \ldots, A_n$. Then the function*

$$f_{F,n}(p_j) = \sum_{k=1}^{n} F_k A_k(p_j),$$

defined at the same nodes, is the inverse discrete F-transform.

Analogously to Theorem 2, we may show that the inverse discrete F-transform $f_{F,n}$ can approximate the original function f at common nodes with an arbitrary precision.

3   F-Transforms of Functions of Two and More Variables

The direct and inverse F-transforms of a function of two and more variables can be introduced as a direct generalization of the case of one variable.

Suppose that the universe is a rectangle $[a, b] \times [c, d]$ and $x_1 < \ldots < x_n$ are fixed nodes from $[a, b]$ and $y_1 < \ldots < y_m$ are fixed nodes from $[c, d]$, such that $x_1 = a$, $x_n = b$, $y_1 = c$, $y_m = d$ and $n, m \geq 2$. Let us formally extend the set of nodes by $x_0 = a$, $y_0 = c$ and $x_{n+1} = b$, $y_{m+1} = d$. Assume that $A_1, \ldots, A_n$ are basic functions which form a fuzzy partition of $[a, b]$ and $B_1, \ldots, B_m$ are basic functions which form a fuzzy partition of $[c, d]$. Let $C([a, b] \times [c, d])$ be the set of continuous functions of two variables $f(x, y)$.

Definition 6. *Let $A_1, \ldots, A_n$ be basic functions which form a fuzzy partition of $[a, b]$ and $B_1, \ldots, B_m$ be basic functions which form a fuzzy partition of $[c, d]$. Let $f(x, y)$ be any function from $C([a, b] \times [c, d])$. We say that the $n \times m$-matrix of real numbers $\mathbf{F}_{nm}[f] = (F_{kl})$ is the (integral) F-transform of f with respect to $A_1, \ldots, A_n$ and $B_1, \ldots, B_m$ if for each $k = 1, \ldots, n$, $l = 1, \ldots, m$,*

$$F_{kl} = \frac{\int_c^d \int_a^b f(x, y) A_k(x) B_l(y) dx dy}{\int_c^d \int_a^b A_k(x) B_l(y) dx dy}. \tag{11}$$

In the discrete case, when an original function $f(x, y)$ is known only at some nodes $(p_i, q_j) \in [a, b] \times [c, d]$, $i = 1, \ldots, N$, $j = 1, \ldots, M$, the (discrete) F-transform of f can be introduced analogously to the case of a function of one variable. We assume additionally that sets $P = \{p_1, \ldots, p_N\}$ and $Q = \{q_1, \ldots, q_M\}$ of these nodes are *sufficiently dense with respect to the chosen partitions*, i.e.

$$(\forall k)(\exists j) A_k(p_j) > 0,$$
$$(\forall k)(\exists j) B_k(q_j) > 0.$$

Definition 7. *Let a function f be given at nodes $(p_i, q_j) \in [a, b] \times [c, d]$, $i = 1, \ldots, N$, $j = 1, \ldots, M$, and $A_1, \ldots, A_n$, $B_1, \ldots, B_m$ where $n < N$, $m < M$, be basic functions which form fuzzy partitions of $[a, b]$ and $[c, d]$ respectively. Suppose that sets P and Q of these nodes are sufficiently dense with respect to the chosen partitions. We say that the $n \times m$-matrix of real numbers $\mathbf{F}_{nm}[f] = (F_{kl})$ is the discrete F-transform of f with respect to $A_1, \ldots, A_n$ and $B_1, \ldots, B_m$ if*

$$F_{kl} = \frac{\sum_{j=1}^{M} \sum_{i=1}^{N} f(p_i, q_j) A_k(p_i) B_l(q_j)}{\sum_{j=1}^{M} \sum_{i=1}^{N} A_k(p_i) B_l(q_j)}. \tag{12}$$

holds for all $k = 1, \ldots, n$, $l = 1, \ldots, m$.

As in the case of functions of one variable, the elements F_{kl}, $k = 1, \ldots, n$, $l = 1, \ldots, m$, are called *components of the F-transform*.

If the partitions of $[a, b]$ and $[c, d]$ by $A_1, \ldots, A_n$ and $B_1, \ldots, B_m$ are uniform then the expression (11) for the components of the F-transform may be simplified on the basis of expressions which can be easily obtained from Lemma 1:

$$F_{11} = \frac{4}{h_1 h_2} \int_c^d \int_a^b f(x, y) A_1(x) B_1(y) dx dy,$$

$$F_{1m} = \frac{4}{h_1 h_2} \int_c^d \int_a^b f(x, y) A_1(x) B_m(y) dx dy,$$

$$F_{n1} = \frac{4}{h_1 h_2} \int_c^d \int_a^b f(x, y) A_n(x) B_1(y) dx dy,$$

$$F_{nm} = \frac{4}{h_1 h_2} \int_c^d \int_a^b f(x, y) A_n(x) B_m(y) dx dy,$$

and for $k = 2, \ldots, n - 1$ and $l = 2, \ldots, m - 1$

$$F_{k1} = \frac{2}{h_1 h_2} \int_c^d \int_a^b f(x, y) A_k(x) B_1(y) dx dy,$$

$$F_{km} = \frac{2}{h_1 h_2} \int_c^d \int_a^b f(x, y) A_k(x) B_m(y) dx dy,$$

$$F_{1l} = \frac{2}{h_1 h_2} \int_c^d \int_a^b f(x, y) A_1(x) B_l(y) dx dy,$$

$$F_{nl} = \frac{2}{h_1 h_2} \int_c^d \int_a^b f(x,y) A_n(x) B_l(y) dx dy,$$

$$F_{kl} = \frac{1}{h_1 h_2} \int_c^d \int_a^b f(x,y) A_k(x) B_l(y) dx dy.$$

Remark 1. All the properties (linearity etc.) proved for the F-transform of a function of one variable can be easily generalized and proved for the considered case too.

Definition 8. *Let $A_1, \ldots, A_n$ and $B_1, \ldots, B_m$ be basic functions which form fuzzy partitions of $[a,b]$ and $[c,d]$ respectively. Let f be a function from $C([a,b] \times [c,d])$ and $\mathbf{F}_{nm}[f]$ be the F-transform of f with respect to $A_1, \ldots, A_n$ and $B_1, \ldots, B_m$. Then the function*

$$f_{nm}^F(x,y) = \sum_{k=1}^n \sum_{l=1}^m F_{kl} A_k(x) B_l(y) \tag{13}$$

is called the the inverse F-transform.

Similarly to the case of a function of one variable we can prove that the inverse F-transform $f_{n,m}^F$ can approximate the original continuous function f with an arbitrary precision.

4 F-Transforms Expressed by Residuated Lattice Operations

Our purpose here is to introduce two new fuzzy transforms which are based on operations of a residuated lattice on $[0,1]$. These transforms lead to new approximation models which are formally represented using weaker operations than the arithmetic ones used above in the case of the (ordinary) F-transform. However, these operations are successfully used in modeling of dependencies characterized by words of natural language (e.g. fuzzy IF–THEN rules) and also, in modeling of continuous functions. Therefore, two new F-transforms that we are going to introduce in this section extend and generalize the F-transform considered above.

There is another important application of fuzzy transforms based on residuated lattice operations – an application to image processing. By this, we mean an application to image compression and reconstruction. We will discuss this later in Section 6.

Let us briefly introduce the concept of residuated lattice which will be a basic algebra of operations in the sequel.

Definition 9. *A residuated lattice is an algebra*

$$\mathcal{L} = \langle L, \vee, \wedge, *, \rightarrow, \mathbf{0}, \mathbf{1} \rangle.$$

with four binary operations and two constants such that

- $\langle L, \vee, \wedge, \mathbf{0}, \mathbf{1} \rangle$ *is a lattice where the ordering* $\leq$ *defined using operations* $\vee, \wedge$ *as usual, and* $\mathbf{0}, \mathbf{1}$ *are the least and the greatest elements, respectively;*
- $\langle L, *, \mathbf{1} \rangle$ *is a commutative monoid, that is,* $*$ *is a commutative and associative operation with the identity* $a * \mathbf{1} = a;$
- *the operation* $\rightarrow$ *is a residuation operation with respect to* $*$, *i.e.*

$$a * b \leq c \quad \text{iff} \quad a \leq b \rightarrow c.$$

A residuated lattice is complete if it is complete as a lattice.

The well known examples of residuated lattices are boolean algebras, Gödel, Lukasiewicz and product algebras. In the particular case $L = [0, 1]$, the multiplication $*$ is called t-norm. In the foregoing text we will operate with some fixed residuated lattice $\mathcal{L}$ on $[0, 1]$.

4.1   Direct $\mathbf{F}^\uparrow$ and $\mathbf{F}^\downarrow$-Transforms

Let the universe be the interval $[0, 1]$. We redefine here the notion of *fuzzy partition* of $[0, 1]$ assuming that it is given by fuzzy sets $A_1, \ldots, A_n$, $n \geq 2$, identified with their membership functions $A_1(x), \ldots, A_n(x)$ fulfilling the following (only one!) *covering property*

$$(\forall x)(\exists i) \quad A_i(x) > 0. \tag{14}$$

As above, the membership functions $A_1(x), \ldots, A_n(x)$ are called the *basic functions*. In the sequel, we fix the value of $n \geq 2$ and some fuzzy partition of $[0, 1]$ by basic functions $A_1, \ldots, A_n$.

We assume that a finite subset $P = \{p_1, \ldots, p_l\}$ of $[0, 1]$ is fixed. Moreover, we assume that P is sufficiently dense with respect to the fixed partition, i.e. (7) holds.

Definition 10. *Let a function* f *be defined at nodes* $p_1, \ldots, p_l \in [0, 1]$ *and* $A_1, \ldots, A_n$, $n < l$, *be basic functions which form a fuzzy partition of* $[a, b]$. *We say that the n-tuple of real numbers* $[F_1^\uparrow, \ldots, F_n^\uparrow]$ *is a (discrete) $F^\uparrow$-transform of* f *w.r.t.* $A_1, \ldots, A_n$ *if*

$$F_k^\uparrow = \bigvee_{j=1}^{l} (A_k(p_j) * f(p_j)) \tag{15}$$

and the n-tuple of real numbers $[F_1^\downarrow, \ldots, F_n^\downarrow]$ *is the (discrete) $F^\downarrow$-transform of* f *w.r.t.* $A_1, \ldots, A_n$ *if*

$$F_k^\downarrow = \bigwedge_{j=1}^{l} (A_k(p_j) \rightarrow f(p_j)). \tag{16}$$

Denote the $F^\uparrow$-transform of f w.r.t. $A_1, \ldots, A_n$ by $\mathbf{F}_n^\uparrow[f]$ and the $F^\downarrow$-transform of f w.r.t. $A_1, \ldots, A_n$ by $\mathbf{F}_n^\downarrow[f]$. Then we may write:

$$\mathbf{F}_n^\uparrow[f] = [F_1^\uparrow, \ldots, F_n^\uparrow], \qquad \mathbf{F}_n^\downarrow[f] = [F_1^\downarrow, \ldots, F_n^\downarrow].$$

Analogously to Theorem 1 we will show that components of the lattice based F-transforms are *lower mean values* (respectively, *upper mean values*) of an original function which give least (greatest) elements to certain sets.

Lemma 2. *Let a function f be defined at nodes $p_1, \ldots, p_l \in [0,1]$ and $A_1, \ldots, A_n$ be basic functions which form a fuzzy partition of $[0,1]$. Then the k-th component of the $F^\uparrow$-transform is the least element of the set*

$$S_k = \{a \in [0,1] \mid \quad A_k(p_j) \le (f(p_j) \to a) \text{ for all } j = 1, \ldots, l\} \qquad (17)$$

and the k-th component of the $F^\downarrow$-transform is the greatest element of the set

$$T_k = \{a \in [0,1] \mid \quad A_k(p_j) \le a \to f(p_j) \text{ for all } j = 1, \ldots, l\} \qquad (18)$$

where $k = 1, \ldots, n$.

5 Inverse $\mathbf{F}^\uparrow$ ($\mathbf{F}^\downarrow$)-Transforms

All F-transforms (the ordinary one and those based on the lattice operations) convert the respective space of functions into the space of n-dimensional real vectors. We have defined the inverse F-transform in Subsection 2.1. In this section, we will define inverse $F^\uparrow$ and inverse $F^\downarrow$-transforms and prove their approximation properties.

In the construction of the inverse $F^\uparrow$- and $F^\downarrow$-transforms we use the fact that the operations $*$ and $\to$ are mutually adjoint in a residuated lattice.

Definition 11. *Let function f be defined at nodes $p_1, \ldots, p_l \in [a,b]$ and let $\mathbf{F}_n^\uparrow[f] = [F_1^\uparrow, \ldots, F_n^\uparrow]$ be the $F^\uparrow$-transform of f and $\mathbf{F}_n^\downarrow[f] = [F_1^\downarrow, \ldots, F_n^\downarrow]$ be the $F^\downarrow$-transform of f w.r.t. basic functions $A_1, \ldots, A_n$. Then the following functions, defined at the same nodes as f, are called the* inverse $F^\uparrow$-transform

$$f_{F,n}^\uparrow(p_j) = \bigwedge_{k=1}^{n} (A_k(p_j) \to F_k^\uparrow), \qquad (19)$$

and the inverse $F^\downarrow$-transform

$$f_{F,n}^\downarrow(p_j) = \bigvee_{k=1}^{n} (A_k(p_j) * F_k^\downarrow), \qquad (20)$$

The following theorem shows that the inverse $F^\uparrow$- and $F^\downarrow$-transforms approximate the original function from above and from below.

Theorem 3. *Let function f be defined at nodes $p_1, \ldots, p_l \in [0,1]$. Then for all $j = 1, \ldots, l$*

$$f_{F,n}^\downarrow(p_j) \le f(p_j) \le f_{F,n}^\uparrow(p_j). \qquad (21)$$

Remark 2. Let us remark that similarly to Definition 7, the direct and inverse lattice based F-transforms of a function of two and more variables can be introduced as a direct generalization of the case of one variable.

6   Application of the F-Transform to Image Compression and Reconstruction

A method of lossy image compression and reconstruction on the basis of fuzzy relations has been proposed in a number of papers (see e.g. [1,2]). When analyzing these methods, we have realized that they can be expressed using F-transforms based on lattice operations. In this section, we will explain how the general technique of F-transform can be successfully applied to the compression and reconstruction of images and compare the effectiveness of all three types of the F-transform with respect to this problem. We will see that all considered examples witnessed the advantage of the ordinary F-transform (4) over the lattice based F-transforms (cf. (15) and (16)).

Let an image I of the size $N \times M$ pixels be represented by a function of two variables (a fuzzy relation) $f_I : \mathbb{N} \times \mathbb{N} \longrightarrow [0, 1]$ partially defined at nodes $(i, j) \in [1, N] \times [1, M]$. The value $f_I(i, j)$ represents an intensity range of each pixel. We propose to compress this image with the help of the one of the three discrete F-transforms of a function of two variables by the $n \times m$-matrix of real numbers

$$\mathbf{F}_{nm}[f_I] = \begin{pmatrix} F_{11} \cdots F_{1m} \\ \vdots \quad \vdots \quad \vdots \\ F_{n1} \cdots F_{nm} \end{pmatrix}.$$

The following expression reminds the components F_{kl} obtained by the ordinary F-transform of a function of two variables (11):

$$F_{kl} = \frac{\sum_{j=1}^{M} \sum_{i=1}^{N} f_I(i, j) A_k(i) B_l(j)}{\sum_{j=1}^{M} \sum_{i=1}^{N} A_k(i) B_l(j)}$$

where $A_1, \ldots, A_n, B_1, \ldots, B_m$, are basic functions which form fuzzy partitions of $[1, N]$ and $[1, M]$, respectively and $n < N$, $m < M$. We refer to Remark 2 for the case of lattice based F-transforms. The value $\rho = nm/NM$ is called a *compression ratio*.

A reconstruction of the image f_I, being compressed by $\mathbf{F}_{nm}[f_I] = (F_{kl})$ with respect to $A_1, \ldots, A_n$ and $B_1, \ldots, B_m$, is given by the inverse F-transform (13) adapted to the domain $[1, N] \times [1, M]$:

$$f_{nm}^F(i, j) = \sum_{k=1}^{n} \sum_{l=1}^{m} F_{kl} A_k(i) B_l(j).$$

On the basis of Theorem 2 we know that the reconstructed image is close to the original one and moreover, that it can be obtained with a prescribed level of accuracy. On Fig. 2 and Fig. 3 (the figures are taken from [2]), we illustrate the proposed compression method and reconstruction based on F-transforms of all three types. Let us show the advantage of the ordinary F-transform over the lattice based F-transforms in what concerns images "Bird" and "Bridge"

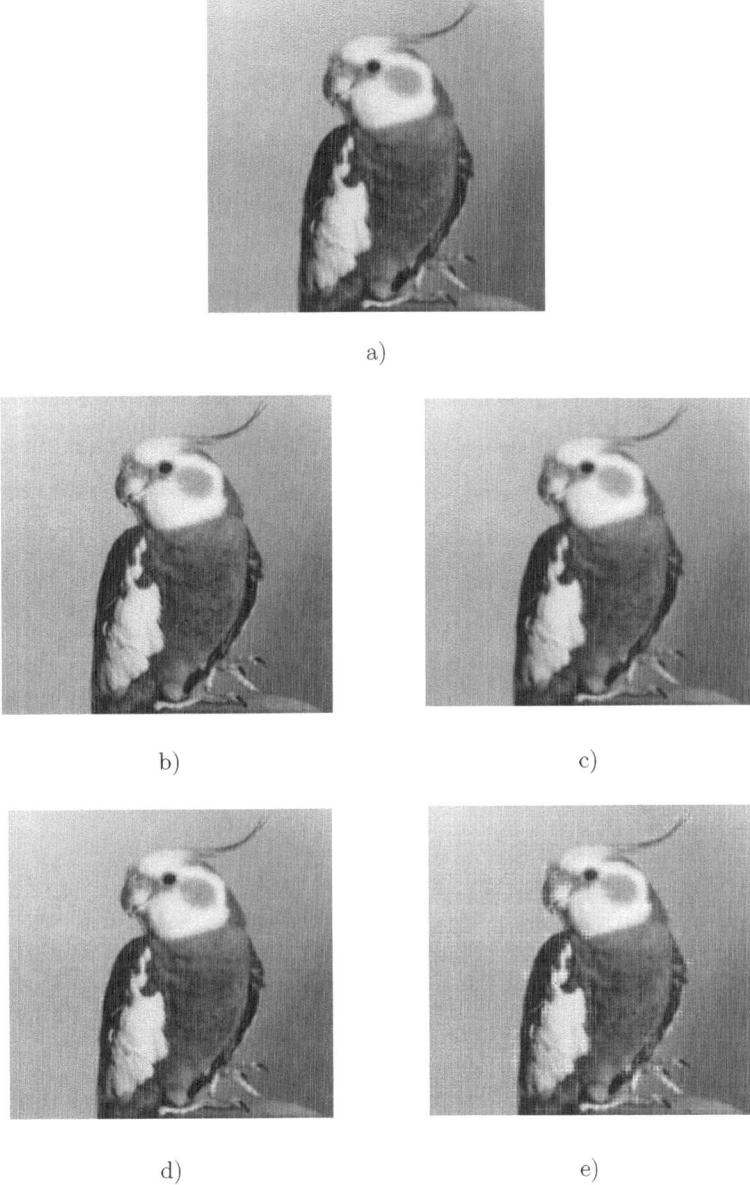

a)

b)   c)

d)   e)

Fig. 2. The original image "Bird" a) is compressed and reconstructed by the ordinary F-transform method (pictures b) and d)) and by the lattice based $F^\uparrow$-transform method (pictures c) and e)). The compression ratio $\rho = 0.56$ has been used for images on pictures b) and c) and the compression ratio $\rho = 0.39$ has been used for images on pictures d) and e).

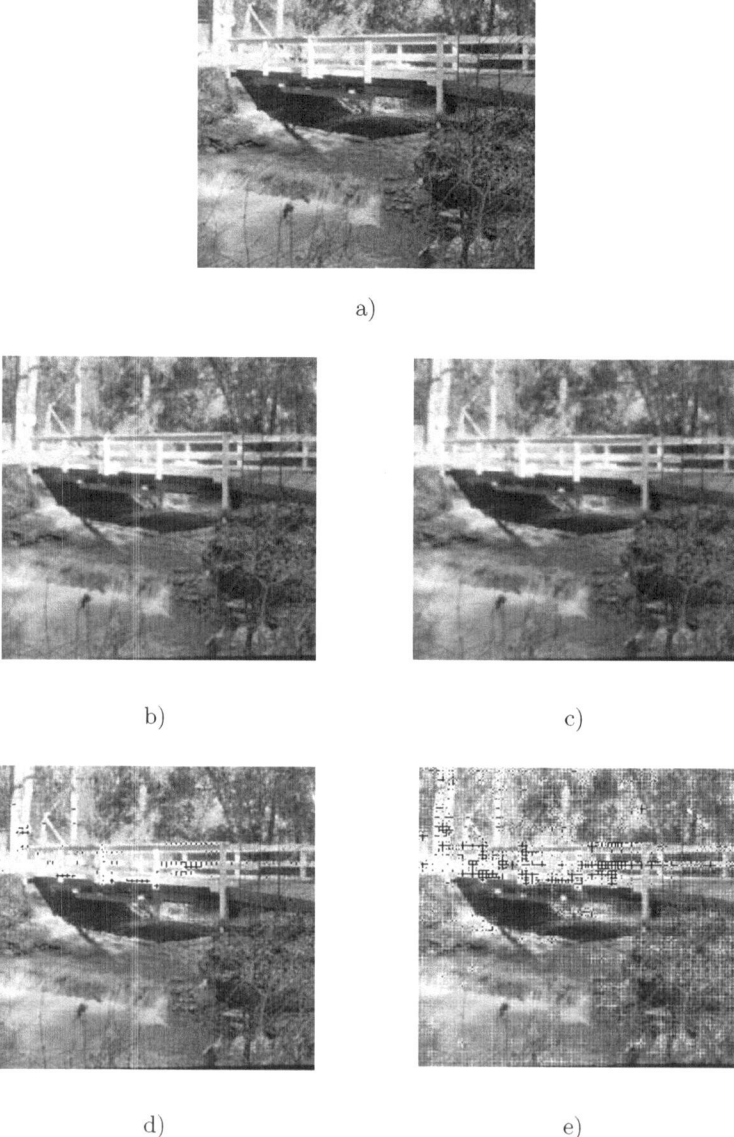

Fig. 3. The original image "Bridge" a) is compressed and reconstructed by the ordinary F-transform method (pictures b) and d)) and by the lattice based $F^\top$-transform method (pictures c) and e)). The compression ratio $\rho = 0.56$ has been used for images on pictures b) and c) and the compression ratio $\rho = 0.39$ has been used for images on pictures d) and e).

compression/reconstruction. We have evaluated the quality of the reconstructed images using PSNR value given by

$$PSNR = 20\log_{10}\frac{255}{\varepsilon}, \qquad \varepsilon = \sqrt{\frac{\sum_{i=1}^{N}\sum_{j=M}^{n}(f_I(i,j) - f_{nm}^F(i,j))}{MN}}.$$

The table below [2] contains values of PSNR for images "Bird" and "Bridge".

	"Bird"			"Bridge"	
ρ	F-transform	F†-transform	ρ	F-transform	F†-transform
0.56	38.3182	35.9521	0.56	38.3182	35.9521
0.39	36.4782	28.4277	0.39	36.4782	28.4277

On the basis of Theorem 1, we may proclaim that the PSNR-quality is the best for compression/reconstruction methods based on the ordinary F-transform in comparison with the lattice based F-transforms.

7 Conclusion

A method of lossy compression and reconstruction data on the basis of fuzzy transforms has been proposed and its advantage over the similar method based on a lattice based F-transform is discussed. The proposed method can be applied not only to pictures, but also to music and other kinds of digital data.

References

1. K. Hirota, W. Pedricz, Fuzzy relational compression, IEEE Trans. Syst., Man, Cyber. 29 (1999) 407–415.
2. Di Martino F., Sessa S., Loia V., Perfilieva I. (2005): An image coding/decoding method based on direct and inverse fuzzy transforms, IEEE Transactions of Fuzzy Systems, submitted
3. I. Perfilieva, Fuzzy Transform: Application to Reef Growth Problem, in: R. B. Demicco, G. J. Klir (Eds.), Fuzzy Logic in Geology, Academic Press, Amsterdam, 2003, pp. 275–300.
4. I. Perfilieva, Fuzzy Transforms, Fuzzy Sets and Systems, submitted, 2004.

Development of Neuro-fuzzy System for Image Mining

K. Maghooli [1] and A.M. Eftekhari Moghadam [2]

[1] Biomedical Engineering Dept., Science and Research Branch,
Islamic Azad University, Tehran, Iran
k_maghooli@yahoo.com
[2] Image Mining Research Lab., IT Applications Dept.,
Iran Telecommunication Research Center (ITRC), Tehran, Iran
eftekhari@itrc.ac.ir

Abstract. We can get much knowledge from images. This process can be done in the mind by a human, and implementation of this mind processing by a system is very difficult. This project attempt to image mining for a simple case. In this paper we develop designed neuro-fuzzy system and it is used for accident prediction in two vehicles scenario. The results show better performance respect to previous version.

Keywords: Image Mining, ANFIS, Knowledge Extraction.

1 Introduction

Recent years have witnessed the rapid increase of data all around the world. Data can be classified in two parts: Structured and Unstructured. Knowledge discovery from structured data can be done through data analysis methods. But in the case of unstructured data such as text, image and voice, special mining techniques are needed to discover the implicit knowledge of them [29]. Image mining systems are used for many applications, such as: data transmission, data saving, image studying, automatic learning of an art, event or action due an image sequential, and may be not considered by a human, automatic control for a place such as large shops, airports, security systems, and others.... Much research was done for "Image Retrieval" [4-27] but didn't consider "Knowledge Mining". This paper attempted to implement an "Image Mining" system for a simple case study. Material & methods, results and conclusion for this study have been presented in the following sections.

2 Materials and Methods

Knowledge has different levels and definitions so because of some problems we must constraint our project to some simple issues. For this purpose we attempted to Knowledge Discovery from Traffic People Behavior (KDFTPB) with simple behaviors [28] .

Any entropy, disturbance, accident, accident prediction, normal behavior, traffic rules and etc., can be considered as knowledge. As mentioned ago, at the first step we consider a simple condition and complex issue holds later. Therefore for simplicity, we

I. Bloch, A. Petrosino, and A.G.B. Tettamanzi (Eds.): WILF 2005, LNAI 3849, pp. 32–39, 2006.

considered a set of scenario with two vehicles with some normal and abnormal behaviors. Also we considered simple knowledge extraction such as accident prediction. Thus system must be got online images and at each time with the current data and the previous knowledge (extracted from the previous images), get some knowledge about the accident prediction. So the system must be having a memory and includes the previous knowledge extraction position. In the other hand it must be intelligence.

Behavior of people and vehicle are fuzzy concepts. In the other words, we can introduce some concept *'Low Velocity'* for very different velocities but at the different roads. Therefore we must have fuzzy concept for this purpose. We showed that one of the best solutions for the image mining system is to use ANFIS[1], because of its adaptive nature, capability of non-systematic rule learning, and extended input-output mapping [2, 3].

ANFIS uses a given input/output data set and constructs a fuzzy inference system (FIS) whose membership function parameters are tuned using either a back propagation algorithm or in combination with a least squares type of method. This allows fuzzy systems to learn from the data they are modeling [1].

2.1 Data Simulation

Data was simulated for ANFIS training, checking, and testing. Simulated data was divided in two categories: (1) Normal motion, and (2) Abnormal motion. Figure (1) shows normal motion for two vehicles. Two dimensions stand for road length and width that are normalized between zero and one.

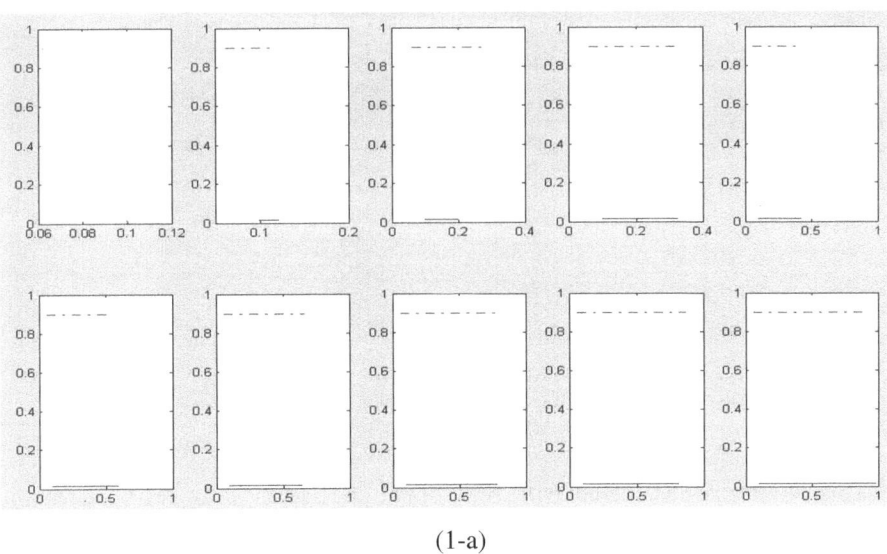

(1-a)

Fig. 1. (a)Normal motions with zero swing for vehicles with never leaves their lines (b) Small swing normal motions for vehicles with never leaves their lines, (c) Large swing normal motions for vehicles with never leaves their lines

---

[1] Adaptive Neuro Fuzzy Inference System.

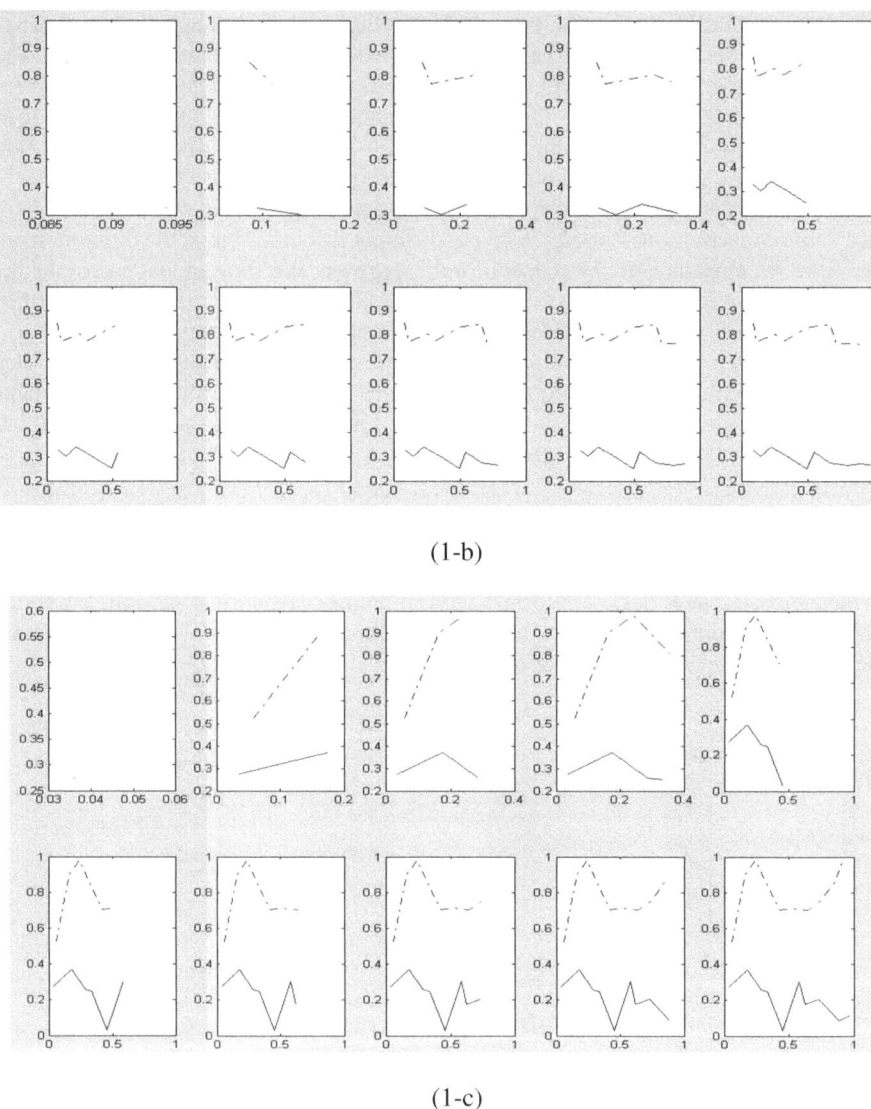

(1-b)

(1-c)

Fig. 1. (*continued*)

This motion is considered with zero, small and large swing for vehicles with never leaves their lines. Because of normalization, large swing is real for vehicle motions. Figure (2) shows abnormal motion for two vehicles. This motion is considered with large swing for vehicles with leaving their lines. Expert reviewed generated data and he/she assigned its outputs for each time. Thus we have a set of inputs/outputs for ANFIS training, checking and testing. Data can be used by ANFIS have the following format.

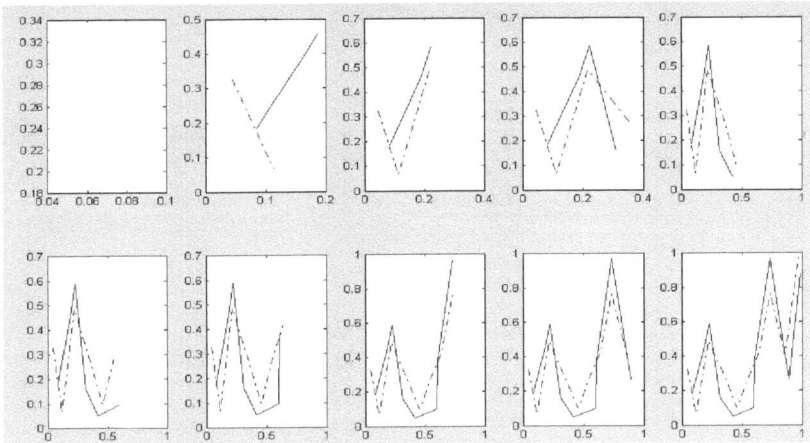

Fig. 2. Large swing abnormal motions for vehicles with leaving their lines

Image1=(x-position1, y-position1, object-velocity1, object-direction1, ..., object-direction2, desired-output, prediction -output).
INPUT VECTOR= [Image1, Image2...Image10].

2.2 ANFIS System

Figure (3) shows last block diagram for image mining by ANFIS [28]. The concept of this figure is: M images from one scenario are processed and some features extracted

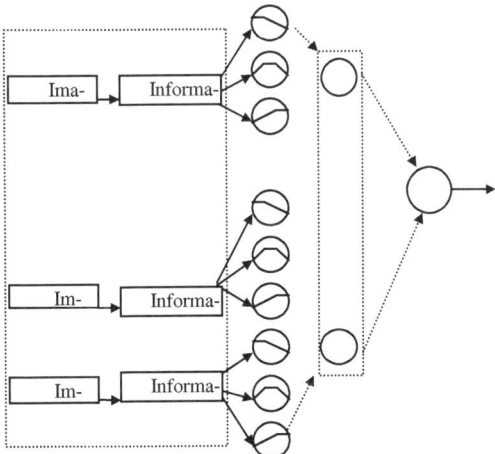

Preprocessing   Layer 1   Layer 2   Layer 3
                     Fuzzification Rule Base Defuzzification

Fig. 3. Block diagram of last version neuro-fuzzy image mining system

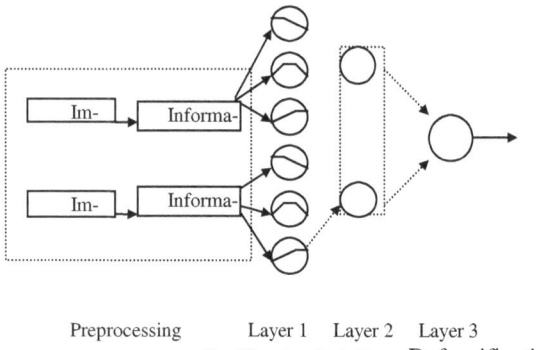

Preprocessing   Layer 1   Layer 2   Layer 3
Fuzzification Rule Base Defuzzification

Fig. 4. Block diagram of new version neuro-fuzzy image mining system

in order to feed into ANFIS. So that sequence of the images is used for knowledge extraction such as it performed in the brain for image mining by human. Because of non-deterministic human behavior and capability of ANFIS in chaotic time series prediction (Mackey-Glass) [3], it is the best choice for this purpose. At the new version (figure 4), we divided system into some steps. At each step we consider only 2 images from one scenario, and feed them to ANFIS system, therefore we break total processing into many sub- processing, that each of the input vector has one share image with previous input vector. With sharing images we can save memory or history of system, in the other words we can save capability of the system for event tracking.

3 Results and Conclusion

Designed ANFIS systems were used in three phases: (1) Training phase. (2) Testing phase (Fig. 5). (3) Checking phase. Horizontal axis stands for rules that extracted by ANFIS from input-output data pairs and vertical axis stands for outputs. Outputs in each of the figures have different symbols for system outputs (output of ANFIS system or FIS output) and target outputs (that is labeled by expert). Checking data was used for over-fitting prevention. It can be seen from results, that training was been successful, and some errors occurred for checking and testing data. It must be noticed that conceptual nature of output, and different value for each of the scenario output by different experts or one expert at different time, we must not used absolute errors for system performance validation. It must see that how much the FIS outputs and target outputs are close together, and very different values at each of the situation can be seen as an error. In the other words if the population and distribution of two outputs is similar and don't see any disagreement, it must be seen as good performance. Fig. (5-b) shows the better performance in testing data respect to Fig. (5-a). So it is clear that new design has better performance respect to last design version.

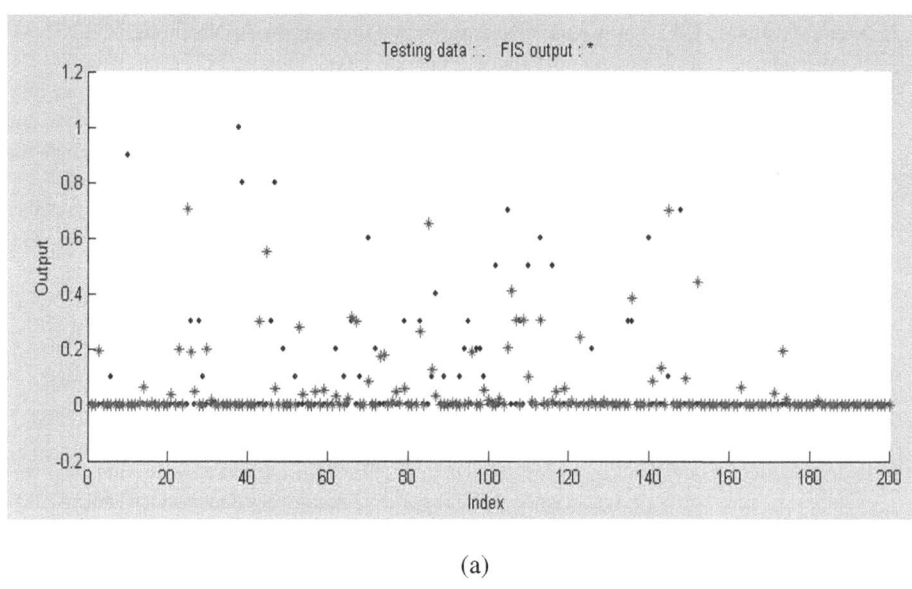

(a)

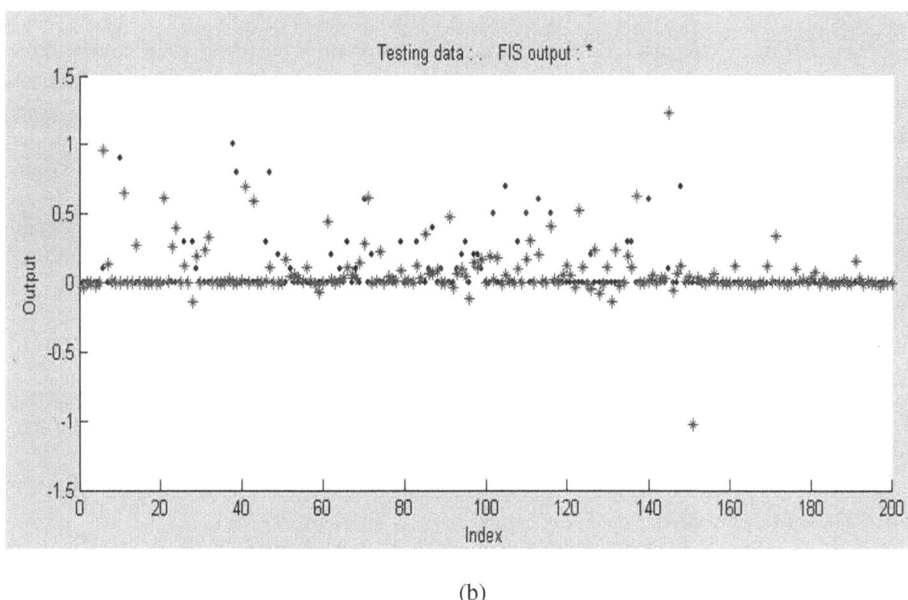

(b)

Fig. 5. Testing phase result (a) last version (b) new version

References

1. Wang, Li-Xin: A Course in Fuzzy Systems and Control, Printice Hall, 1997.
2. Jang, JYH-S.R, Sun, C.T. : Neuro-Fuzzy Modeling and control, Proceedings of the IEEE, Vol. 83, No. 3, March 1995, pp. 378-405.

3. Lin, C.T., Juang, C.F.: An Adaptive Neural Fuzzy Filter and Its Applications, IEEE Transactions on Systems, Man, and Cybernetics, Vol. 27, No. 4, August 1997, pp.635-656.
4. Marchisio, G.B., Wen-Hao, L., Sannella, M., Goldschneider, J.R.: GeoBrowse: an integrated environment for satellite image retrieval and mining, Proc. of the Geoscience and Remote Sensing Symposium (IGARSS '98), IEEE International Conference, Vol2, Page(s): 669 –673,6-10 July 1998.
5. You, J., hattacharya, P.B.: Dynamic shape retrieval by hierarchical curve matching, snakes and data mining, Proc. of the 15th International Conference on Pattern Recognition, Vol1, Page(s): 1035 –1038 3-7 Sept. 2000.
6. Olukunle, A., Ehikioya, S.: A fast algorithm for mining association rules in medical image data, Proc. Of the Canadian Conference on Electrical and Computer Engineering (IEEE CCECE 2002), May 2002.
7. Tilton, J.C., Koperski, K., Marchisio, G., Datcu, M.: Image information mining utilizing hierarchical segmentation, Proc. of International Geoscience and Remote Sensing Symposium (IGARSS '02), IEEE, June 2002.
8. Koperski, K, Marchisio, G., Aksoy, S., Tusk, C.: VisiMine: interactive mining in image databases, International Geoscience and Remote Sensing Symposium (IGARSS '02), IEEE, June 2002.
9. Roden, J., Burl, M., Fowlkes, C.: The Diamond Eye image mining system, Proc. of Eleventh International Conference on Scientific and Statistical Database Management, July 1999.
10. Ordonez, C., Omiecinski, E.: Discovering association rules based on image content, Proc. of IEEE Forum on Research and Technology Advances in Digital Libraries (ADL'99), May 1999.
11. Koperski, K., Marchisio, G., Aksoy, S., Tusk, C.: Applications of terrain and sensor data fusion in image mining, International Geoscience and Remote Sensing Symposium(IGARSS '02), IEEE, June 2002.
12. Dorado, A., Izquierdo, E.: Semi-automatic image annotation using frequent keyword mining, Proc. of Seventh International Conference on Information Visualization (IV'2003), July 2003.
13. Hongyu, W., Beng, C.O., Tung, A.K.H. : Search: mining retrieval history for content-based image retrieval, Proc. of Eighth International Conference on Database Systems for Advanced Applications (DASFAA 2003), March 2003.
14. Soh, L.K, Tsatsoulis, C.: Segmentation of Satellite Imagery of Natural Scenes Using Data Mining, IEEE Transactions on Geoscience and remote sensing, Vol37, NO2, March 1999.
15. Li, J., Narayanan, R.A.: Shape-based change detection and information mining in remote sensing, Proc. of the International Geoscience and Remote Sensing Symposium (IGARSS '02), IEEE, June 2002.
16. Sun, J.Y., Sun, Z.X., Zhou, R.H., Wang, H.F. : A semantic-based image retrieval system: VisEngine, Proc.of 2002 International Conference on Machine Learning and Cybernetics, Nov. 2002.
17. Smith, G.B., Bridges, S.M.: Fuzzy spatial data mining, Proc. of Annual Meeting of the North American Fuzzy Information Processing Society (NAFIPS'02), June 2002.
18. Qiang, X., Baozong, Y.: A new framework of CBIR based on KDD, 6th International Conference on Signal Processing, Aug. 2002.
19. Zhang, L.Q., Sun Z.X., Peng, B.B., Wang, H.F.: Image retrieval using object template, Proc. of International Conference on Machine Learning and Cybernetics, Nov. 2002.
20. Sokol, L.: Creating knowledge from heterogeneous data stove pipes, Proc. of Fifth International Conference on Information Fusion, July 2002.

21. McCaslin, S., Kulkarni, A.: Knowledge Discovery from multispectral satellite images, North American Fuzzy Information Processing Society and Special Interest Group on Fuzzy Logic and the Internet, New Orleans, LA, 2002.
22. Zhang, J., Hsu W., Lee, M.L.: Image Mining: Issues, Frameworks and Techniques, Proc. of Second International Workshop on Multimedia Data Mining (MDM/KDD'2001), San Francisco, CA, USA, August, 2001.
23. Zhang, J., Hsu W., Lee, M. L.: An Information driven Framework for Image Mining, Proc. of 12th International Conference on Database and Expert Systems Applications (DEXA), Munich, Germany, September 2001.
24. Conci, A., Castro, E.M.M.M.: Image Mining by Color Content, Proc. of 2001 ACM International Conference on Software Engineering & Knowledge Engineering (SEKE), Buenos Aires, Argentina, Jun13 -15, 2001.
25. ZHANG, Y.J.: Image Engineering and Related Publications, International Journal of Image and Graphics, Vol.2, No.3, 2002.
26. Xu, Y., ZHANG, Y.J., Image Retrieval Framework Driven by Association Feedback with Feature Element Evaluation Built in, SPIE, 2001.
27. Hsu, W., Lee M.L., Zhang, J.: Image Mining: Trends and Developments, Journal of Intelligent Information System (JISS): Special Issue on Multimedia Data Mining, Kluwer Academic, 2002.
28. Maghooli, K., E.Moghadam, A.M.: Neuro-Fuzzy Approach to Image Mining System, Third International Conference On Systems, Signals and Devices, IEEE SSD-05, 2005, Sousse, Tunisia.
29. Safara, F., E.Moghadam, A.M.: IMAGE MINING: A New Approach to Knowledge Discovery, Third International Conference On Systems, Signals and Devices, IEEE SSD-05, 2005, Sousse, Tunisia.

Reinforcement Distribution
in Continuous State Action Space
Fuzzy Q–Learning: A Novel Approach

Andrea Bonarini, Francesco Montrone, and Marcello Restelli

Politecnico di Milano Electronic and Information Department,
Piazza Leonardo da Vinci, 32 - 20133 Milan - Italy
Tel: +390223992525; Fax: +390223993411
{Bonarini, Restelli}@elet.polimi.it, fra_jcss@virgilio.it
http://www.elet.polimi.it/res/air/index.html

Abstract. Fuzzy Q–learning extends the Q–learning algorithm to work
in presence of continuous state and action spaces. A Takagi–Sugeno
Fuzzy Inference System (FIS) is used to infer the continuous executed
action and its action–value, by means of cooperation of several rules.
Different kinds of evolution of the parameters of the FIS are possible,
depending on different strategies of distribution of the reinforcement signal. In this paper, we compare two strategies: the classical one, focusing
on rewarding the rules that have proposed the actions composed to produce the actual action, and a new one we are introducing, where reward
goes to the rules proposing actions closest the ones actually executed.

Keywords: Reinforcement Learning, Fuzzy Q–learning, Fuzzy logic,
continuous state-action space, reinforcement distribution.

1 Introduction

We present a comparison between two reinforcement distribution approaches in
Fuzzy Q-Learning, a reinforcement learning algorithm for fuzzy models.

Reinforcement Learning (RL) is a paradigm that makes an agent learn the
optimal policy, namely the optimal action to take given a state of affairs, through
the direct interaction with the environment and a critic of each action taken. The
agent gets no direct information about the optimal control strategy. RL is based
on the idea that the result of an action is evaluated and the parts of the model
contributing to it are rewarded accordingly; this reward is used to compute an
accurate estimation of the model fitness to the rewarding criteria [7].

Q–Learning [9] is one of the most popular, model–free, RL algorithm. It is
based on the idea that the expected discounted sum of future reinforcements can
be estimated by a function of all the states and the actions, called Q–function,
which gives a measure of the fitness of each action in each state and can be used
to define an optimal policy.

I. Bloch, A. Petrosino, and A.G.B. Tettamanzi (Eds.): WILF 2005, LNAI 3849, pp. 40–45, 2006.

When the agent is in state $\boldsymbol{x}^t$, an action a^t is *selected* and *applied*; then the agent senses a new state, $\boldsymbol{x}^{t+1}$, and a reinforcement signal, $r^t_{\boldsymbol{x}^t,a^t}$. The action–value of such an action $\mathcal{Q}^t(\boldsymbol{x}^t, a^t)$ is updated by:

$$\mathcal{Q}^{t+1}\left(\boldsymbol{x}^t, a^t\right) = \mathcal{Q}^t\left(\boldsymbol{x}^t, a^t\right) + \alpha\left(r^t_{\boldsymbol{x}^t,a^t} + \gamma \max_a \mathcal{Q}^t(\boldsymbol{x}^{t+1}, a) - \mathcal{Q}^t\left(\boldsymbol{x}^t, a^t\right)\right) \quad (1)$$

where α is the learning rate and γ a discount factor, both between 0 and 1.

Since operating in large search spaces is time expensive (*curse of dimensionality* problem), Q–learning is often applied in domains modeled by few discrete states where a small number of actions is possible,

Several solutions have been proposed to *generalize* over the state–action space to reduce this problem. *Function approximators* are able to interpolate a function using a limited number of parameters; in addition they rely on good convergence proofs in many important cases. In particular, we use a Takagi–Sugeno *Fuzzy Inference System* (FIS) as linear function approximator, thus considering a kind of Fuzzy Q–learning.

Many algorithms that implement Fuzzy Q–learning have been presented in the last years. In [1] Seo and Youn used fuzzy variables instead of discrete values and then filled lookup tables with fuzzy values. Berenji [2] has proposed a Q–learning approach that allows to *integrate* fuzzy constraints in direct reinforcement learning. Bonarini [3] proposed a fuzzy Q-learning version with delayed reinforcement. Glorennec [5] and Jouffe [6] have given a fundamental contribution in the definition of Fuzzy Q–learning; one of the methods described in the following sections originates from their work.

2   Fuzzy Q–Learning

The goal of a FIS is to map a N–dimensional, real–valued, input state domain X to a continuous *action function* and its *action–value function*. In order to do this, a fuzzy partition is defined over X. Each *fuzzy set* of the partition is denoted by a *label*, $\mathcal{A}$, and described by a *membership function*, $\mu_{\mathcal{A}}$.

Each dimension X_k of the state space is covered by n_{X_k} fuzzy sets. Thus, we call the membership function of the j_k^{th} fuzzy set $\mathcal{X}_{j_k}$ on the kth variable X_k as $\mu_{\mathcal{X}_{j_k}}$, where $j_k = 1, \ldots, n_{X_k}$. The conjunction among the fuzzy sets $\mathcal{X}_{j_1}, \ldots, \mathcal{X}_{j_N}$ is the *premise* of the rule $\mathcal{R}_{j_1,\ldots,j_N}$. The consequent of each rule is associated to a set of $n_{j_1,\ldots,j_N}$ possible discrete actions $A_{j_1,\ldots,j_N} = \{a_{j_1,\ldots,j_N,1}, \ldots, a_{j_1,\ldots,j_N,n_{j_1,\ldots,j_N}}\}$ and the correlated action values $Q_{j_1,\ldots,j_N} = \{q_{j_1,\ldots,j_N,1}, \ldots, q_{j_1,\ldots,j_N,n_{j_1,\ldots,j_N}}\}$, so $q_{j_1,\ldots,j_N,i}$ is associated to the i_{th} discrete action available in the rule $\mathcal{R}_{j_1,\ldots,j_N}$, which can be written as follows:

$$\mathcal{R}_{j_1,\ldots,j_N} : \textbf{IF} \quad x_1 \text{ is } \mathcal{X}_{j_1} \quad \textbf{AND} \ldots \textbf{AND} \quad x_N \text{ is } \mathcal{X}_{j_N}$$

$$\textbf{THEN} \qquad y = a_{j_1,\ldots,j_N,1} \qquad \textbf{with} \quad q_{j_1,\ldots,j_N,1}$$

$$\textbf{OR} \ldots \textbf{OR} \quad y = a_{j_1,\ldots,j_N,n_{j_1,\ldots,j_N}} \quad \textbf{with} \quad q_{j_1,\ldots,j_N,n_{j_1,\ldots,j_N}}$$

where the **AND** operator is implemented as the T-norm product. At each time step, only one action per active rule will participate to the inference, while all the others remain inactive. This describes an exclusive **OR** and a *competition* among the actions of each rule. To simplify the notation, let us associate all the possible combinations of the indexes $\{j_1, \ldots, j_N\}$ of the rules to natural numbers $r = 1, \ldots, n_R$, so we can rename the indexes, and $\mathcal{R}_{j_1,\ldots,j_N}$ simply becomes $\mathcal{R}_r$.

When a crisp input vector $\boldsymbol{x}^t$ enters the system at time t, all the rules $\mathcal{R}_r$ that cover a region to which $\boldsymbol{x}^t$ partially belongs are said to be partially active by the matching degree $\phi_r(\boldsymbol{x}^t)$ of their premises. It may happen that the sum of activation degrees is larger than 1; the normalized activation degree, $\psi_r(\boldsymbol{x}^t)$ is defined as:

$$\phi_{j_1,\ldots,j_N}(\boldsymbol{x}^t) = \prod_{j_1,\ldots,j_N} \mu_{x_{j_i}}(x_i) \, . \qquad \psi_r(\boldsymbol{x}^t) = \frac{\phi_r(\boldsymbol{x}^t)}{\sum_{r=1}^{n_R} \phi_r(\boldsymbol{x}^t)} \, . \qquad (2)$$

Each active rule infers *only one* discrete action, and the action actually executed in state $\boldsymbol{x}^t$ is a composition of such discrete actions.

In each fuzzy state the agent must exploit what it already knows in order to obtain reward, but it must also explore, to discover possibly more rewarding regions in the state–action space. In the present work, exploration is implemented by a kind of ϵ-*greedy* choice of the action: for each rule the agent computes a selection parameter J by linearly combining, with parameters η_q and η_Υ, the quality of the available actions normalized in $[0,1]$, Q_r^{norm}, and a vector Υ_r of random values uniformly distributed in $[0,1]$.

$$J = \eta_q Q_r^{norm} + \eta_\Upsilon \Upsilon_r \, . \qquad \check{a}_r^t = a_{r,\, \arg\max_i\{J_{r,i}\}} \, . \qquad (3)$$

Then, it selects the action $\check{a}_r^t$ with the highest J value.

The coefficient η_Υ of Υ_r in the linear combination gets smaller and smaller with time, so the more the system learns the more greedily it chooses.

Once the degree of activation of the premises is determined and an action for each active fuzzy rule is selected, the rules *cooperate* with each other in creating a continuous action to *execute* in $\boldsymbol{x}^t$ at time t: $A^t(\boldsymbol{x}^t)$. The overall inference is computed as the sum of the actions $\check{a}_r^t$ chosen by each rule weighted by the corresponding normalized activation degrees $\psi_r(\boldsymbol{x}^t)$:

$$A^t(\boldsymbol{x}^t) = \sum_{r=1}^{n_R} \psi_r(\boldsymbol{x}^t) \, \check{a}_r^t \, . \qquad (4)$$

Notice that with $A^t(\boldsymbol{x}^t)$ we actually mean $A(\boldsymbol{x}^t, \check{a}_r^t)$. Moreover, $A^t(\boldsymbol{x}^t)$ is a *continuous action* in the sense that, given the selected discrete actions, a continuous variation of $\boldsymbol{x}^t$ implies a continuous and smooth variation of $A^t(\boldsymbol{x}^t)$, but in a continuous state $\boldsymbol{x}^t$ there is only a discrete set of possible interpolated actions.

2.1   The $\mathcal{Q}$ Function

According to Glorennec and Jouffe [5, 6], the $\mathcal{Q}$–value associated to $A^t(\boldsymbol{x}^t)$ in state $\boldsymbol{x}^t$ at time t, namely $\mathcal{Q}^t(\boldsymbol{x}^t, A^t(\boldsymbol{x}^t))$, is derived by the quality of the actions $\check{a}_r^t$ *selected* to interpolate $A^t(\boldsymbol{x}^t)$:

$$\mathcal{Q}^t \left(\boldsymbol{x}^t, A^t(\boldsymbol{x}^t) \right) = \sum_{r=1}^{n_R} \psi_r(\boldsymbol{x}^t) \, q_{r,i}^t \, . \tag{5}$$

Note that the continuous action actually executed $A^t(\boldsymbol{x}^t)$ might be very different from the discrete actions $\check{a}_r^t$ by which it obtained reward. Furthermore, the $\mathcal{Q}$–function is not computable over the whole state–action space but it is restricted only on the subdomain $A^t(\boldsymbol{x}^t)$. The function $\mathcal{Q}$ actually should be written as $\mathcal{Q}(\boldsymbol{x}^t, A^t(\boldsymbol{x}^t), \check{a}_r^t)$, so the point $(\boldsymbol{x}^t, A^t(\boldsymbol{x}^t))$ in the state–action space is not associated to any $\mathcal{Q}$–value per se, as it happens in connectionist approaches. In order to overcome the problems arising from this issue, we introduce another way of thinking the action–value function in a FIS.

Parameter update in algorithms with function approximators is often implemented as a gradient descent with respect to a function, usually the Minimum Square Error (MSE), which is quite simple to compute in the case of la inear approximator such as the fuzzy system we are considering [8]. So, we update the parameters associated to the *chosen* actions with the following equation:

$$\begin{aligned} q_{r,i}^{t+1} &= q_{r,i}^t - \alpha \varepsilon_{\mathcal{Q}}^t \boldsymbol{\nabla}_{q_{r,i}^t} \mathcal{Q}^t \left(\boldsymbol{x}^t, A^t(\boldsymbol{x}^t) \right) \\ &= q_{r,i}^t - \alpha \varepsilon_{\mathcal{Q}}^t \psi_r(\boldsymbol{x}^t) \, . \end{aligned} \tag{6}$$

where $\varepsilon_{\mathcal{Q}}^t$ is defined as:

$$\varepsilon_{\mathcal{Q}}^t = r_{\boldsymbol{x}^t, A^t(\boldsymbol{x}^t)}^t + \gamma \max_a \mathcal{Q}(\boldsymbol{x}^{t+1}, a) - \mathcal{Q}^t \left(\boldsymbol{x}^t, A^t(\boldsymbol{x}^t) \right) \, . \tag{7}$$

In order to speed the learning up we use *eligibility traces* [8], which let $\boldsymbol{\nabla}_{q_{r,i}^t} \mathcal{Q}^t \left(\boldsymbol{x}^t, A^t(\boldsymbol{x}^t) \right)$ slowly decay to zero in time and be reemployed in extra updates during the following time–steps and a meta–learning rule that allows individual components to learn more or less, depending on the stable or oscillatory nature of the component itself. We cannot give further details here due to space restrictions.

2.2   The $\mathfrak{Q}$ Function

We propose an alternative formulation of the interpolation of the action–value function, which we will indicate as $\mathfrak{Q}$–function, that is *univocally* defined over the *whole* state–action domain. We *strongly* fuzzify the action dimension in correspondence with each rule $\mathcal{R}_r$ with one fuzzy set $\mathcal{A}_{r,i}$ centered in each discrete action $a_{r,i}$, so that a generic point a^t – and in particular $A^t(\boldsymbol{x}^t)$ – is covered by the fuzzy sets of the discrete actions, in the active rules, which are most similar to it. Thus, we can compute and update the $\mathfrak{Q}$–values focusing the attention on the rules proposing actions similar to the one actually executed. We interpolate in each $\mathcal{R}_r$ the value $\tilde{\mathfrak{Q}}_r(a^t)^1$ by considering the $\mathcal{A}_{r,i}$ matching the actual action, and then we use it as in Equation 5 of Section 2.1 to interpolate $\mathfrak{Q}^t(\boldsymbol{x}^t, a^t)$:

$$\tilde{\mathfrak{Q}}_r(a^t) = \sum_{i=1}^{n_r} \mu_{\mathcal{A}_{r,i}}(a^t) \, q_{r,i}^t \, . \qquad \mathfrak{Q}^t(\boldsymbol{x}^t, a^t) = \sum_{r=1}^{n_R} \psi_r(\boldsymbol{x}^t) \, \tilde{\mathfrak{Q}}_r(a^t) \, . \tag{8}$$

---

[1] In this case it not necessary to normalize the sum since the fuzzification is strong.

The update of $\mathfrak{Q}$–values is performed by a gradient descent method as well, but in this case the updated parameters correspond to the discrete actions that are most similar to the *executed* action, and therefore responsible for the interpolation of $\tilde{\mathfrak{Q}}_r(a^t)$ and then $\mathfrak{Q}^t(\boldsymbol{x}^t, a^t)$. The update formula takes into account the fuzzy sets associated to the discrete actions in every active rule $\mathcal{R}_r$:

$$
\begin{aligned}
q_{r,i}^{t+1} &= q_{r,i}^t - \alpha \varepsilon_\mathfrak{Q}^t \boldsymbol{\nabla}_{q_{r,i}^t} \mathfrak{Q}^t(\boldsymbol{x}^t, A^t(\boldsymbol{x}^t)) \\
&= q_{r,i}^t - \alpha \varepsilon_\mathfrak{Q}^t \psi_r(\boldsymbol{x}^t) \mu_{A_{r,i}}(A^t(\boldsymbol{x}^t)) \,.
\end{aligned}
\tag{9}
$$

where $\varepsilon_\mathfrak{Q}^t$ is defined like $\varepsilon_\mathcal{Q}^t$ in Equation 7.

Eligibility traces and meta–learning rule have been adopted in this case too.

3   Experiments and Comparison Between $\mathcal{Q}$ and $\mathfrak{Q}$

We have implemented a system that can distribute reward according to both $\mathcal{Q}$ and $\mathfrak{Q}$. We have applied it on the following example: a golf player agent should decide the strength of the stroke given the distance from the hole; if it strokes too strongly, the reward of the action is very small, if it strokes the ball in the hole at once it is very large, if it takes more than one stroke it is in the middle.

As shown in Figure 1, $\mathcal{Q}$–values focus on the individual selection of the best actions available by the FIS, bound to a specific fuzzification, without dispersing

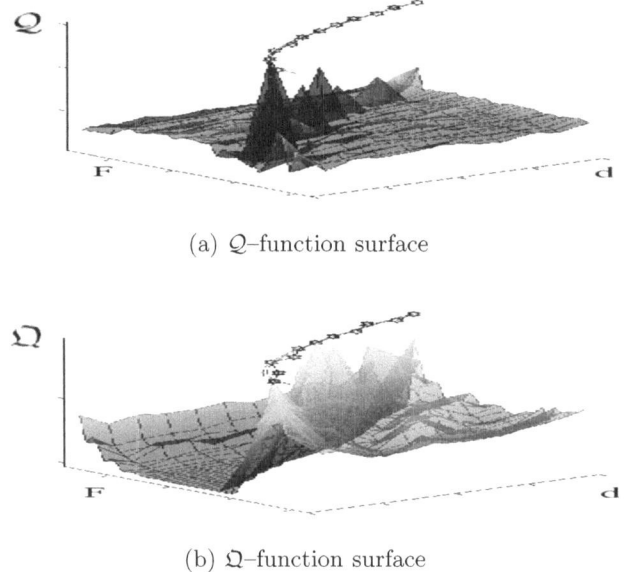

(a) $\mathcal{Q}$–function surface

(b) $\mathfrak{Q}$–function surface

Fig. 1. $\mathcal{Q}$ and $\mathfrak{Q}$ surfaces of a golf player agent: d is the distance from the hole, F is the force of the stroke. In 1(a) the $\mathcal{Q}$–values put in evidence only the best actions to select, while in 1(b) the $\mathfrak{Q}$–values characterize the whole action–value space.

the reward information. However, the parameters are usually updated by different rules. This makes them incoherent and oscillatory. Furthermore, the function Q depends on $\boldsymbol{x}^t, A^t(\boldsymbol{x}^t), \tilde{\boldsymbol{a}}_r^t$, so the point $(\boldsymbol{x}^t, A^t(\boldsymbol{x}^t))$ in the state–action space is not associated to any Q–value per se, so a real action–value function is defined only when the choice of $\tilde{a}_r^t$ is deterministic, like in the case of the optimal policy, otherwise parameters are meaningless. Thus, when the fuzzification does not suit the problem well, we cannot rely on these values.

Ω-values provide an action–value with characteristics more similar to the expected ones. These functions are *stable*, since they receive *coherent* updates from similar executed actions and, most of all, they are defined over the *whole* domain. Thus, they provide an estimation of the whole action–values distribution, so we can compare optimal and suboptimal actions, and we may be able to understand whether the fuzzification is inadequate. However, there are some drawbacks: the boundaries turn to be less frequently updated and are biased by the initial value for longer; moreover, the information about the action–value function is bounded to the shape of the membership functions, so rules covering in discontinuous regions of action–value functions may give intrinsically oscillatory results. This problem might be reduced by an accurate design of the reward function [4].

References

1. He-Sub Seo andSo Joeng Youn and Kyung-Whan Oh. A fuzzy reinforcement function for the intelligent agent to process vague goals. In *Fuzzy Information Processing Society*, pages 29–33, Atlanta, GA, USA, 2000.
2. H. R. Berenji. A reinforcement learning-based architecture for fuzzy logic control. In D. Dubois, H. Prade, and R. R. Yager, editors, *Readings in Fuzzy Sets for Intelligent Systems*, pages 368–380. Kaufmann, San Mateo, CA, 1993.
3. A. Bonarini. *Delayed Reinforcement, Fuzzy Q-Learning and Fuzzy Logic Controllers*, pages 447–466. Physica Verlag (Springer Verlag), Heidelberg, D, 1996.
4. A. Bonarini, C. Bonacina, and M. Matteucci. An approach to the design of reinforcement functions in real wolrd, agent-based applications. *IEEE Transactions on Systems. Man, and Cybernetics – Part B*.
5. P. Y. Glorennec and L. Jouffe. Fuzzy Q-learning. In *Proceedings of Fuzz-Ieee'97, Sixth International Conference on Fuzzy Systems*, pages 659–662, Barcelona, Spain, 1997.
6. L. Jouffe. *Fuzzy Inference Systems Learning by Reinforcement Methods: Application to a Pig House Atmosphere Control*. PhD thesis, Computer Science Department, Universitè de Rennes I, Rennes, France, 1997.
7. Richard S. Sutton and Andrew G. Barto. *Reinforcement Learning: An Introduction*. MIT Press, Cambridge, MA, 1998.
8. R.S. Sutton and A.G. Barto. *Reinforcement Learning: An Introduction*. Adaptive Computation and Machine Learning Series. MIT Press, Cambridge, Mass., 1998.
9. Chris Watkins and Peter Dayan. Q-learning. *Machine Learning*, 8:279–292, 1992.

A Possibilistic Approach to Combinatorial Optimization Problems on Fuzzy-Valued Matroids

Adam Kasperski[1] and Paweł Zieliński[2]

[1] Institute of Industrial Engineering and Management,
Wrocław University of Technology, Wybrzeże Wyspiańskiego 27,
50-370 Wrocław, Poland
adam.kasperski@pwr.wroc.pl
[2] Institute of Mathematics and Computer Science,
Wrocław University of Technology, Wybrzeże Wyspiańskiego 27,
50-370 Wrocław, Poland
pziel@im.pwr.wroc.pl

Abstract. In this paper several combinatorial optimization problems on fuzzy weighted matroid are considered. It is shown how to characterize the degrees of optimality of elements in the setting of the possibility theory.

1 Introduction

An important and interesting class of combinatorial optimization problems can be formulated on *matroids* (a good introduction to matroids can be found in [7]). Let $E = \{e_1, \ldots, e_n\}$ be a given, finite set and let $\mathcal{I}$ be a collection of subsets of E closed under inclusion (that is if $A \in \mathcal{I}$ and $B \subseteq A$, then $B \in \mathcal{I}$). The system $(E, \mathcal{I})$ is matroid if it satisfies the following growth property: if $A \in \mathcal{I}$, $B \in \mathcal{I}$ and $|A| < |B|$, then there exists $e \in B \setminus A$ such that $A \cup \{e\} \in \mathcal{I}$. The elements in $\mathcal{I}$ are called *independent sets*. The maximal (under inclusion) elements in $\mathcal{I}$ are called *bases* and the minimal (under inclusion) elements not in $\mathcal{I}$ are called *circuits*. It can be easily verified that all the bases of a matroid have the same cardinality. In the *matroidal combinatorial optimization problem* (OPT for short) a nonnegative weight w_e is given for every element $e \in E$ and we seek a base B, for which the cost $F(B) = \sum_{e \in B} w_e$ is maximal (or minimal).

Matroids are precisely the structures for which the very simple and efficient *greedy algorithm* works. The greedy algorithm for the maximization problem takes the following form:

1: Order elements so that $w_{e_1} \geq w_{e_2} \geq \cdots \geq w_{e_n}$.
2: $B \leftarrow \emptyset$
3: **for** $i \leftarrow 1$ **to** n **do**
4:    **if** $B \cup \{e_i\} \in \mathcal{I}$ **then** $B \leftarrow B \cup \{e_i\}$
5: **end for**
6: **return** B

I. Bloch, A. Petrosino, and A.G.B. Tettamanzi (Eds.): WILF 2005, LNAI 3849, pp. 46–52, 2006.

For the minimization problem we have to reverse the order of the elements in line 1. The running time of the greedy algorithm is $\mathcal{O}(n \log n + n f(n))$, where $f(n)$ is time required for deciding whether a given set $B \subseteq E$ contains a circuit (see line 4). Note that $f(n)$ depends on the particular structure of a matroid.

In this paper we study problem OPT in which the values of the weights are not precisely known. We use fuzzy intervals to model the ill-known weights. The evaluation by fuzzy intervals is performed in the setting of possibility theory. The possibility theory, which proposes a natural framework for handling incomplete knowledge, is fully described in [3]. If the exact values of the weights are not available, then the notion "optimal solution" also becomes imprecise. It is not possible to say *a priori* which elements of E will be a part of an optimal solution. However, using the possibility theory, we can characterize the possibility and the necessity of the event that a given element $e \in E$ will be a part of an optimal solution. Such a characterization may be useful in some practical applications. It allows for example to detect the elements which will never (will always) be a part of an optimal solution. Moreover, in the fuzzy case we can detect the elements for which the possibility (necessity) of being a part of an optimal solution is low or high. This may be used for preprocessing the problem before choosing a solution. Such a preprocessing was performed for example in [1] and [8] before calculating the optimal solution with respect to the maximal regret criterion.

2   Fuzzy Interval Weighted Matroids

Suppose that every weight w_e, $e \in E$, can take any value from a given interval $W_e = [w_e^-, w_e^+]$, where $w_e^- \geq 0$. A vector $\boldsymbol{w} = (w_e)_{e \in E}$, $w_e \in W_e$, that represents an assignment of weights w_e to elements $e \in E$ is called *configuration*. We denote by Γ the set of all the configurations, i.e. $\Gamma = \times_{e \in E} [w_e^-, w_e^+]$. We use $w_e(\boldsymbol{w})$ to denote the weight of element $e \in E$ in configuration $\boldsymbol{w} \in \Gamma$. We denote by $E(\boldsymbol{w})$ a subset of those elements of E which belong to at least one optimal base in configuration $\boldsymbol{w} \in \Gamma$. A given element $e \in E$ is said to be *possibly optimal* if and only if there exists a configuration $\boldsymbol{w} \in \Gamma$ such that $e \in E(\boldsymbol{w})$. A given element $e \in E$ is said to be *necessarily optimal* if and only if $e \in E(\boldsymbol{w})$ for all configurations $\boldsymbol{w} \in \Gamma$.

We show now how to decide whether a given element $e \in E$ is possibly (necessarily) optimal. Consider two particular configurations $\boldsymbol{w}_{\{f\}}^+$ and $\boldsymbol{w}_{\{f\}}^-$ induced by $f \in E$:

$$w_e(\boldsymbol{w}_{\{f\}}^+) = \begin{cases} w_e^+ & \text{if } e = f, \\ w_e^- & \text{otherwise} \end{cases}, \quad w_e(\boldsymbol{w}_{\{f\}}^-) = \begin{cases} w_e^- & \text{if } e = f, \\ w_e^+ & \text{otherwise} \end{cases}, e \in E.$$

If we are seek a maximum weighted base, then the following theorem can be proven [5]:

Theorem 1. *A given element $e \in E$ is possibly (resp. necessarily) optimal if and only if there exists an optimal base B in configuration $\boldsymbol{w}_{\{e\}}^+$ (resp. $\boldsymbol{w}_{\{e\}}^-$) such that $e \in B$.*

For the minimum weighted base, there is a symmetric theorem (it is enough to exchange $w^+_{\{e\}}$ and $w^-_{\{e\}}$ in Theorem 1). Using Theorem 1 one can check whether a given element e is possibly (necessarily) optimal. This can be done by executing the slightly modified greedy algorithm for configurations $w^+_{\{e\}}$ and $w^-_{\{e\}}$. The modification consists in assuring that e appears before all the other elements with the same weight (see ordering in line 1 of the greedy algorithm). The element e is then possibly (necessarily) optimal if and only if the obtained base contains e (see [5] for details). This result generalizes the results obtained in [8] for the particular minimum spanning tree problem. Thus, it is clear that we can detect all the possibly and necessarily optimal elements in $\mathcal{O}(n^2 \log n + n^2 f(n))$ time. The complexity of detecting all the possibly optimal elements can be additionally reduced. If we seek a maximum weighted base, then the following theorem holds [5]:

Theorem 2. *Let w^- be configuration where all the weights are set to their lower bounds and let B be an optimal base in w^-. An element $e \notin B$ is possibly optimal if and only if an element $f \neq e$ in the circuit $C \subseteq B \cup \{e\}$ of the minimal value of w^-_f satisfies $w^+_e \geq w^-_f$.*

The symmetric theorem holds if we seek a minimum weighted base. Using Theorem 2 we can proceed as follows. First, we compute an optimal base B in configuration w^-. All the elements $e \in B$ are then possibly optimal by definition. Then, for all $e \in E \setminus B$ we determine the element f from Theorem 2, which in a typical situation requires $\mathcal{O}(nf(n))$ time, where $f(n)$ is the time required for detecting cycle C in $B \cup \{e\}$. Thus, all the possibly optimal elements can be detected in $\mathcal{O}(n \log n + nf(n))$ time. Unfortunately there is no a counterpart of Theorem 2 for the necessarily optimal elements.

We show now how to extend the notions of possible and necessary optimality of elements to the fuzzy intervals case. Let us recall that a *fuzzy interval* $\tilde{X}$ is a fuzzy set in the space of real numbers, whose membership function $\mu_{\tilde{X}}$ is normal, quasiconcave and upper semi-continuous on $\mathbb{R}$ [3]. In this paper we work only with nonnegative fuzzy intervals ($\mu_{\tilde{X}}(x) = 0$ for $x < 0$) with bounded support. We assume that the membership function $\mu_{\tilde{X}}$ denotes the *possibility distribution* for variable X, whose value is not precisely known. The interpretation of the possibility distribution and some methods of obtaining it from the knowledge possessed about X are described in detail in [3]. It is well known that the λ-cut of $\tilde{X}$, that is the set $\tilde{X}^\lambda = \{x : \mu_{\tilde{X}}(x) \geq \lambda\}$, $\lambda \in (0,1]$, is the classical interval $[\tilde{X}^-(\lambda), \tilde{X}^+(\lambda)]$. We additionally assume that $\tilde{X}^0 = [\tilde{X}^-(0), \tilde{X}^+(0)]$ is the support of $\tilde{X}$. The intervals $\tilde{X}^\lambda$, $\lambda \in [0,1]$ can be easily calculated for a wide class of fuzzy intervals, called the fuzzy intervals of the L-R type [3].

Suppose now that the weights of the elements are modeled as fuzzy intervals $\tilde{W}_e$, $e \in E$. Then, the join possibility distribution over all the weights configurations $w = (w_e)_{e \in E}$ is $\pi(w) = \min_{e \in E}\{\mu_{\tilde{W}_e}(w_e)\}$. The degrees of possible and necessary optimality of a given element $e \in E$ are defined as follows:

$$\Pi(e \text{ is optimal}) = \sup_{w \,:\, e \in E(w)} \pi(w),$$

$$\mathrm{N}(e \text{ is optimal}) = 1 - \Pi(e \text{ is not optimal}) = \inf_{\boldsymbol{w}:\, e \notin E(\boldsymbol{w})} (1 - \pi(\boldsymbol{w})).$$

The optimality degrees of a given element $e \in E$ can be calculated in the following way. Let $\mathcal{M}^\lambda$, $\lambda \in [0,1]$ be a matroid in which the weights are given as the classical intervals $W_e = [\tilde{W}_e^-(\lambda), \tilde{W}_e^+(\lambda)]$, $e \in E$. Then, the degrees of possible and necessary optimality of e can be calculated as follows (see also [5]):

$$\Pi(e \text{ is optimal}) = \sup\{\lambda \in [0,1] \,|\, e \text{ is possibly optimal in } \mathcal{M}^\lambda\},$$
$$\mathrm{N}(e \text{ is optimal}) = 1 - \inf\{\lambda \in [0,1] \,|\, e \text{ is necessarily optimal in } \mathcal{M}^\lambda\}.$$

If e is not possibly (not necessarily) optimal in $\mathcal{M}^0$ ($\mathcal{M}^1$) then degree of possible (necessary) optimality of e is equal to 0. It can be easily verified that $\Pi(e \text{ is optimal}) < 1$ implies $\mathrm{N}(e \text{ is optimal}) = 0$, and $\mathrm{N}(e \text{ is optimal}) > 0$ implies $\Pi(e \text{ is optimal}) = 1$. The calculation of the optimality degrees of $e \in E$, with a given accuracy $\epsilon \in (0,1)$, is straightforward if there exist algorithms for the corresponding interval problems. The standard method consists in dividing the interval $[0,1]$ and solving a sequence of the classical interval cases. Thus, the difficulty of the problems lies in the classical interval case. In the next sections we will analyze in detail three particular matroidal problems. We will focus on the classical interval cases, assuming that the further generalization to the fuzzy intervals is straightforward.

3   The Minimum Spanning Tree Problem

Let $E = \{e_1, \ldots, e_n\}$ be a set of edges of a given connected and undirected graph $G = (V, E)$. The set $\mathcal{I}$ consists of all the subsets of edges $E' \subseteq E$ such that subgraph $G' = (V, E')$ is acyclic (that is $\mathcal{I}$ is the set of all forests in G). The system $(E, \mathcal{I})$ in this problem is one of the best known examples of matroids (it is so-called *graphic matroid*). A base is a *spanning tree* of G and a circuit is a subset of edges creating a simple cycle in G. In the classical minimum spanning tree problem, we seek a spanning tree for which the total weight is minimal. The greedy algorithm calculating the optimal spanning tree is known in the literature as Kruskal's algorithm. The complexity of Kruskal's algorithm is $\mathcal{O}(n \log n)$ [2].

Consider the case in which the weights of edges of G are specified as closed intervals. The problem of detecting the possibly and the necessarily optimal edges was studied in [1] and [8] (in [8] the possibly optimal edge was called *weak* and the necessarily optimal one was called *strong*). This characterization was used for preprocessing the problem before calculating the robust spanning tree. All the possibly and the necessarily optimal edges can be detected in $\mathcal{O}(n^2 \log n)$ time, using the results presented in Section 2. This method, consists in executing n times the Kruskal's algorithm, was adopted in [8]. The complexity of calculation of all the possibly optimal edges can be significantly reduced. In [1] the following proposition (which is a consequence of the more general Theorem 2) was proven:

Proposition 1 ([1]). *Let $\boldsymbol{w}^+$ be a configuration where all the weights are at their upper bounds and let T be a minimum spanning tree in $\boldsymbol{w}^+$. An edge $e = (u, v) \in E \setminus T$ is possibly optimal if and only if an edge f of the maximal weight w_f^+ on the path from u to v in T satisfies $w_e^- \le w_f^+$.*

The idea of algorithm for detecting all the possibly optimal elements, which is based on Proposition 1, is as follows. We start with computing the minimum spanning tree T in configuration $\boldsymbol{w}^+$. All the edges $e \in T$ are then possibly optimal by definition. Then, for every edge $e \in E \setminus T$ we can compute in $O(|V|)$ time the proper edge f from Proposition 1 and check whether $w_e^- \leq w_f^+$. The overall complexity becomes then $\mathcal{O}(n \log n + n|V|)$, which can be additionally reduced to $\mathcal{O}(n \log n + |V|^2)$ (see [1] for details). Observe that in dense graphs ($n \approx |V|^2$) this complexity is $\mathcal{O}(n \log n)$, which is the same as the running time of Kruskal's algorithm. Unfortunately there is no a counterpart of Proposition 1 for detecting all the necessarily optimal edges. Therefore, all these edges can be detected in $\mathcal{O}(n^2 \log n)$ time by executing n times the Kruskal's algorithm.

4   The Selecting Items Problem

Let $E = \{e_1, \ldots, e_n\}$ be a set of items. The set $\mathcal{I}$ consists of all the subsets of E with cardinality less or equal than a given number $1 \leq p < n$. It can be easily verified that system $(E, \mathcal{I})$ in this problem is a matroid (it is so called *uniform matroid*). A base is a subset A of E such that $|A| = p$ and a circuit C is a subset of A such that $|C| = p + 1$. Thus, in the selecting items problem, we seek a subset of E with cardinality exactly p, for which the total weight is maximal. The selecting items problem can be viewed as a basic resource allocation problem with linear cost function [4]. In the classical case the best solution can be easily obtained by selecting p items of the greatest weights. It can be performed in $\mathcal{O}(n)$ time in the following way: first, the $(p+1)th$ greatest weighted element $g \in E$ is calculated (this can be done in $\mathcal{O}(n)$, see e.g. [2]); then, all the items $e \in E$ such that $w_e > w_g$ are added to A; if $|A| < p$ then set A is completed to p by items $e \in E$ such that $w_e = w_g$.

Consider now the case in which the weights of elements in the problem are given as closed intervals. The following two propositions allow to detect all the possibly and necessarily optimal elements very efficiently.

Proposition 2. *Let $\boldsymbol{w}^-$ be a configuration where all the weights are at their lower bounds. Let f be the p-th greatest weighted element in $\boldsymbol{w}^-$. Then element $e \in E$ is possibly optimal if and only if $w_e^+ \geq w_f^-$.*

Proof. ($\Rightarrow$) Suppose by contradiction that e is possibly optimal and $w_e^+ < w_f^-$. Condition $w_e^+ < w_f^-$ means that there are at least p elements which weights are strictly greater than the weight of e in configuration $\boldsymbol{w}_{\{e\}}^+$. Thus e cannot be a part of the optimal base in $\boldsymbol{w}_{\{e\}}^+$. This implies that e is not possibly optimal (see Theorem 1), which is a contradiction.

($\Leftarrow$) If $w_e^+ \geq w_f^-$ the element e is one of the p elements of the greatest weights in configuration $\boldsymbol{w}_{\{e\}}^+$. Thus e is a part of the optimal base in $\boldsymbol{w}_{\{e\}}^+$ and by Theorem 1, it is possibly optimal. □

Proposition 3. *Let $\boldsymbol{w}^+$ be a configuration where all the weights are at their upper bounds. Let f be the p-th greatest weighted element and let g be the $(p+1)$-th* *greatest weighted element in $\boldsymbol{w}^+$. Then element $e \in E$ is necessarily optimal if and only if $w_e^+ \geq w_f^+$ and $w_e^- \geq w_g^+$.*

Proof. ($\Rightarrow$) Suppose by contradiction that element e is necessarily optimal and $w_e^+ < w_f^+$ or $w_e^- < w_g^+$. In the first case ($w_e^+ < w_f^+$) element e cannot be a part of the optimal base in configuration $\boldsymbol{w}^+$ and in the second case ($w_e^- < w_g^+$) it cannot be a part of the optimal base in configuration $\boldsymbol{w}_{\{e\}}^-$. This contradicts the assumption that e is necessarily optimal.

($\Leftarrow$) Conditions $w_e^+ \geq w_f^+$ and $w_e^- \geq w_g^+$ assure that element e is one of the p elements of the greatest weights in configuration $\boldsymbol{w}_{\{e\}}^-$. Thus e is a part of the optimal base in $\boldsymbol{w}_{\{e\}}^-$ and by Theorem 1, it is necessarily optimal.   □

Propositions 2 and 3 allow to detect all the possibly and necessarily optimal elements very efficiently. The k-th greatest element in a given configuration can be found in $\mathcal{O}(n)$ time (see e.g. [2]). Thus the overall complexity of the algorithm for detecting all the possibly (necessarily) optimal elements is $\mathcal{O}(n)$.

5   The Scheduling Problem $1|p_i = 1| \sum w_i U_i$

Let $J = \{1, \ldots, n\}$ be a set of jobs to be processed on a single machine. Every job $i \in J$ has unit processing time $p_i = 1$. For every job $i \in J$, there are given: a due date d_i and a weight w_i. A *schedule* is a sequence π of jobs. A job $i \in J$ is called *late* in π if its completion time in π is greater than d_i, otherwise job i is called *on-time* in π. We seek a schedule for which the sum of weights of all the late jobs is minimal. A subset of jobs $S \subseteq J$ belongs to $\mathcal{I}$ if and only if all the jobs in S are on-time in a certain schedule π. It is easy to decide whether a given set S belongs to $\mathcal{I}$. To do this, it is enough to schedule first all the jobs in S in order of nondecreasing due dates and then all the remaining jobs in an arbitrary order. Then $S \in \mathcal{I}$ if and only if all the jobs in S are on-time in the resulting schedule [2]. This schedule is said to be in the *canonical* form. It is easily seen that the problem consists now in determining the set $S \in \mathcal{I}$ with the maximal value of $F(S) = \sum_{i \in S} w_i$. The optimal solution can be obtained by constructing the corresponding canonical schedule for S. It can be proven that system $(J, \mathcal{I})$ is matroid and the optimal schedule (in the canonical form) can be found in $\mathcal{O}(n^2)$ time by the greedy algorithm [2], [6].

Before we consider the interval case, we show how to detect efficiently a circuit in the considered problem. Let $S = \{j_1, \ldots, j_k\}$, $k \leq n$, be a given base. Assume that $d_{j_1} \leq \cdots \leq d_{j_k}$. Let $i \in J$ be a job such that $i \notin S$. Then set $S \cup \{i\}$ contains a circuit C. The circuit C is the minimal subset of jobs in $S \cup \{i\}$ which cannot be all scheduled on-time. The circuit C can be detected in the following way: find the smallest number $r \in \{1, \ldots, k\}$ such that $d_{j_r} \geq d_i$ and $r + 1 > d_{j_r}$; set $C = \{j_1, \ldots, j_r, i\}$. It is not difficult to check that such number r must exist, since otherwise we could create a canonical schedule in which all the jobs in $S \cup \{i\}$ are on-time (this would contradict the fact that S is a base). To see that C is a circuit schedule jobs in C in order of nondecreasing due dates (note that we try to construct a canonical schedule) and denote the resulting schedule by σ. From the definition of r we conclude that only the last job in σ (either i or j_r) is late. Since all the processing times are equal to 1, we can remove any job from σ and we get a schedule in which all the jobs are on-time. This implies that all the subsets of C with cardinality $|C| - 1$ are independent and C is the minimal dependent subset.

Assume now that the weights of jobs are uncertain and they are given as closed intervals. A possibly optimal job is on-time job in an optimal schedule for some configuration of the weights and a necessarily optimal job is on-time job in some optimal schedule for all configurations of the weights. Such a characterization under the uncertain weights may be useful. For example, the necessarily optimal jobs should be always processed first, while non-possibly optimal ones should be processed last. In the fuzzy case, one can use the optimality indices to establish some priority rules for scheduling the jobs.

Now we can use Theorem 2 to detect efficiently all the possibly optimal jobs. We compute first in $\mathcal{O}(n^2)$ time the optimal schedule π in configuration of weights $\boldsymbol{w}^-$. The schedule π is in the canonical form, thus we can easily determine the set $S = \{\pi(1), \ldots, \pi(k)\}$ of jobs which are on-time in π. The set S is the optimal base in configuration $\boldsymbol{w}^-$, thus all the jobs $i \in S$ are possibly optimal. Suppose that $i \notin S$. Then, we can determine in $\mathcal{O}(n)$ time the smallest number r such that $d_{\pi(r)} \geq d_i$ and $r + 1 > d_{\pi(r)}$ and check whether $w_i^+ \geq \min\{w_{\pi(1)}^-, \ldots, w_{\pi(r)}^-\}$ (see Theorem 2). Therefore, all the possibly optimal jobs can be detected in $\mathcal{O}(n^2)$ time. Unfortunately, it is a difficult issue to give an $\mathcal{O}(n^2)$ algorithm for detecting all the necessarily optimal jobs. Thus, one has to run n times the greedy algorithm which gives $\mathcal{O}(n^3)$ time.

References

1. Aron, I., van Hentenryck, P.: A Constraint Satisfaction Approach to the Robust Spanning Tree with Interval Data. In: Proceedings of the 18th Conference on Uncertainty in Artificial Intelligence, Edmonton, Canada 2002.
2. Cormen, T.H., Leiserson, C.E., Rivest, R.L.: Introduction to Algorithms. MIT Press, 1994.
3. Dubois, D., Prade, H.: Possibility theory: an approach to computerized processing of uncertainty. Plenum Press, New York, 1988.
4. Ibaraki, T., Katoh, N.: Resource Allocation Problems. MIT Press, Cambridge, Massachusetts, 1988.
5. Kasperski, A., Zieliński, P.: On combinatorial optimization problems on matroids with ill-known weights. Instytut Matematyki PWr., Wroclaw 2005, raport serii PREPRINTY nr 34, submitted for publication in Eur. J. Oper. Res.
6. Lawler, E. L.: Combinatorial optimization: networks and matroids. Holt, Rinehart and Winston, 1976.
7. Oxley, J., G.: Matroid Theory. Oxford University Press, New York, 1992.
8. Yaman, H., Karaşan, O.E., Pinar, M.C.: The robust spanning tree problem with interval data. Oper. Res. Lett. 29 (2001) 31–40.

Possibilistic Planning Using Description Logics: A First Step

Célia da Costa Pereira and Andrea G.B. Tettamanzi

Università degli Studi di Milano,
Dipartimento di Tecnologie dell'Informazione,
Via Bramante 65, I-26013 Crema (CR), Italy
pereira@dti.unimi.it, andrea.tettamanzi@unimi.it

Abstract. This paper is a first step in the direction of extending possibilistic planning to take advantage of the expressive power and reasoning capabilities of fuzzy description logics. Fuzzy description logics are used to describe knowledge about the world and about actions. Fundamental definitions are given and the possibilistic planning problem is recast in this new setting.

1 Introduction

Planning is a branch of artificial intelligence which studies how to find the most suitable sequence of actions to take a system from a given initial state into a desired state, called goal.

In a classical planning problem, it is assumed that actions are deterministic, the initial state is known, and the goal is defined by a set of final states; a solution plan is then an unconditional sequence of actions that leads from the initial state to a goal state. However, most practical problems do not satisfy these assumptions of complete and deterministic information. This is why in recent years many approaches taking into account the uncertainty in the planning problem have been proposed. In particular, a possibilistic approach to planning has been proposed in [2] which is an extension of the well-known STRIPS formalism to make possible the representation of the possibilistic uncertainty. In this formalism, the representation of the states of the world and of the effect of the actions is made using sets of literals. This fact has the advantage that reasoning for solving a planning problem is decidable but it has the disavantage that the formalism thus constructed has a severely limited expressiveness.

This paper provides a first step in the direction of extending that approach to take advantage of: (i) the expressive power of description logics [1], (ii) the decidable reasoning capabilities of description logics and (iii) recent extensions of description logics to take into account uncertainty and imprecision [5].

A relevant advantage of combining fuzzy description logics and planning is that it becomes possible to describe the planning domain using formal ontologies for expressing and organizing the (possibly vague) knowledge available about the planning domain in general and the particular problem at hand. On the one

I. Bloch, A. Petrosino, and A.G.B. Tettamanzi (Eds.): WILF 2005, LNAI 3849, pp. 53–60, 2006.

hand, it may be possible to stablish a level of detail for the concepts describing the domain (the world states and the effect of actions may be represented by concepts) and, on the another hand, it may be possible to make inferences about the available knowledge. As a consequence, implicit knowledge becomes available. This is possible because a formal ontology contains a structured vocabulary (its "terminology") which defines the relations between different terms.

In this work, we consider that for each planning problem we have two ontologies. One is related to the general knowledge, valid for every planning problem; the other describes knowledge about the specific problem to solve. The ontological general formalization of a planning problem must then respect two important features:

- Analitical feature: it must be easily understandable for all planning problems, i.e, the proposed formalism must be independent of the context of the particular problem.
- Engineering feature: the formalism must be flexible in order to support and to manipulate the knowledge acquired during the planning process.

In Section 2, we provide a brief introduction to description logics. In Section 3, we present a general ontology for the planning domain and a specific ontology describing a simple planning problem in the block world. In Section 4, we define the possibilistic planning problem using fuzzy description logics and Section 5 concludes.

2   Description Logics

Description logics (DL) [1] are a family of logic-based knowledge-representation formalisms, which stem from the classical AI tradition of semantic networks and frame-based systems. Description logics are characterized by a set of constructors for building complex concepts and roles starting from the primitive ones. Concepts represent classes which are interpreted as sets of objects, and roles represent relations which are interpreted as binary relations on objects.

Semantics are expressed in terms of interpretations $\mathcal{I} = (\Delta^{\mathcal{I}}, \cdot^{\mathcal{I}})$, where:

- $\Delta^{\mathcal{I}}$ is the interpretation domain,
- $\cdot^{\mathcal{I}}$ is the interpretation function, which maps:
  - different individuals into different elements of $\Delta^{\mathcal{I}}$,
  - primitive concepts into subsets of $\Delta^{\mathcal{I}}$,
  - primitive roles into subsets of $\Delta^{\mathcal{I}} \times \Delta^{\mathcal{I}}$.

The architecture of a description-logic-based system is composed of:

- A knowledge base with: (i) terminological knowledge (TBox), containing the definitions of concepts (for example FreeBlock $\equiv$ Block $\sqcap \neg \exists$on$^-.\top$ (a free block is a block such that no other object is on it)) and (ii) the knowledge about objects (ABox) containing assertions which characterize the objects and define the relations between them (for example on(BLOCKA, TABLE),
- The Inference Engine,
- Applications

In order to represent uncertainty, incompleteness, and imprecision in knowledge, fuzzy description logics have been proposed that allow for no crisp concept description by using fuzzy sets and fuzzy relations. A fuzzy extension of the description logic $\mathcal{ALC}$ has been introduced in [5], with complete algorithms for solving the entailment problem, the subsumption problem, as well as the best truth-value bound problem.

For the purpose of our approach, we have added the inverse role construct to $\mathcal{ALC}$ yielding the $\mathcal{ALCI}$ language, which is suitable for most planning problems. If R is a role, R^- denotes its inverse, meaning $R^-(a,b) \equiv R(b,a)$, for all individuals a and b. For example, with the description logic $\mathcal{ALC}$, from the assertion on(BLOCKB, BLOCKA) it is only possible to deduce that BLOCKB is on BLOCKA. With the $\mathcal{ALCI}$ language we may also express that BLOCKA has BLOCKB on it, because on(BLOCKB, BLOCKA) = on$^-$(BLOCKA, BLOCKB)

A fuzzy description logic is identical in most respects to a classical description logic but it assigns a meaning to symbols by means of a fuzzy interpretation $\tilde{\mathcal{I}} = (\Delta^{\tilde{\mathcal{I}}}, \cdot^{\tilde{\mathcal{I}}})$ such that:

- $\Delta^{\tilde{\mathcal{I}}}$ is as in the crisp case, the interpretation domain,
- $\cdot^{\tilde{\mathcal{I}}}$ is the interpretation function mapping:
  - different individuals into different elements of $\Delta^{\tilde{\mathcal{I}}}$ as the crisp case, i.e., $a^{\tilde{\mathcal{I}}} \neq b^{\tilde{\mathcal{I}}}$ if $a \neq b$,
  - a primitive concept A into a membership function $A^{\tilde{\mathcal{I}}} : \Delta^{\tilde{\mathcal{I}}} \mapsto [0,1]$,
  - a primitive role R into a membership function $R^{\tilde{\mathcal{I}}} : \Delta^{\tilde{\mathcal{I}}} \times \Delta^{\tilde{\mathcal{I}}} \mapsto [0,1]$.

3   Domain Representation

In this section, we propose an ontological formalization for the general planning problem, with new representation for a state of the world and new representation for the actions, both using description logics. We also propose a specific ontological formalization for planning in the block world.

3.1   Representing World States

A world state is best represented as an interpretation $\mathcal{I}$ that is a model of both a TBox $\mathcal{T}$ and an ABox $\mathcal{A}$. Usually, we do not have complete information about the world, i.e., the model $\mathcal{I}$ of $\mathcal{T}$ is not known completely. All we know is some facts about the world which are represented in an ABox $\mathcal{A}$. Thus, all models of the ABox and of the TBox are considered to be possible states of the world.

Let us illustrate an example of specific ontological formalization for a planning problem in the block world. We dispose of tree blocks A, B, C and a robot arm. In the initial state (see Figure 1a), block A and block C are on the table, block B is on block A, and both block B and the robot arm are free.

To describe this problem, we define four individual names: BLOCKA, BLOCKB, BLOCKC, and TABLE, and one role on (on(a,b) means that a is on b).

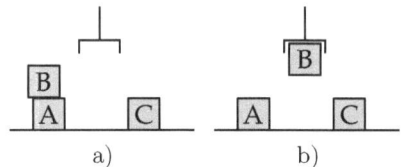

Fig. 1. Two states in the block world: a) the initial state; b) one of the effects of applying the action "grasp block B"

The TBox $\mathcal{T}$, wich represents the specific ontology, will contain definitions of a few useful concepts, namely:

$$\mathcal{T} = \{\mathsf{FreeBlock} \equiv \mathsf{Block} \sqcap \neg\exists\mathsf{on}^-.\top, \mathsf{InArm} \equiv \neg\exists\mathsf{on}.\top\}.$$

FreeBlock is a defined concept that characterizes a block which has no other object on it.

InArm is a defined concept that characterizes a block which is in the robot arm. Precisely, we suppose that if a block is not on any of the available objects, then it is in the robot arm.

In the initial state, the situation of Figure 1a will be described with an ABox $\mathcal{A}_0$ containing the follwing assertions:

$$\mathcal{A}_0 = \{\mathsf{on}(\mathsf{BLOCKB}, \mathsf{BLOCKA}), \mathsf{on}(\mathsf{BLOCKA}, \mathsf{TABLE}), \mathsf{on}(\mathsf{BLOCKC}, \mathsf{TABLE})\}.$$

Uncertainty about the world states is represented by means of possibility theory [3]. Possibility theory allows to represent the fact that, at a certain point of the planning process, a state is more possible than another. To represent incomplete and imprecise knowledge about individuals as well as about relations between individuals, we use fuzzy description logics. Consequently, an uncertain state may be represented by a possibility distribution on fuzzy interpretations $\tilde{\mathcal{I}}$.

3.2 Describing Actions

The syntax and the semantics of a deterministic action are those proposed by [4]. The effect of executing an action depends on the context in which it occurs. This kind of representation allows to group several actions into one action and thus contributes to keeping the complexity of the planning problem low. Let $\mathcal{T}$ be an acyclic TBox and $\mathcal{A}$ an ABox. A (deterministic) context-dependent action act is an n-tuple:

$$\mathsf{act} = \{\langle \mathrm{Context}_1, \mathrm{Effect}_1\rangle, \ldots, \langle \mathrm{Context}_n, \mathrm{Effect}_n\rangle\},$$

in which $\mathrm{Context}_i$ is a set of assertions describing the ith context in which action act may be executed, and Effect_i is a set of primitive assertions describing the ith conditional effect that should be obtained after the execution of the action. If all assertions in $\mathrm{Context}_i$ are satisfied before executing action act, then all assertions in Effect_i should be satisfied afterwards. For all interpretation $\mathcal{I}$, there is a unique $\mathrm{Context}_i$ such that $\mathcal{I} \models \mathrm{Context}_i$.

Definition 1 (DL Possibilistic action). *An action with possibilistic effects describes an uncertain behaviour of executing an action. Its syntax and semantic are inherited from the deterministic action described above extended to take into account uncertain effects [2]. Let $\mathcal{T}$ be an acyclic TBox. A possibilistic action* pact *for $\mathcal{T}$ is an m-tuple of possibilistic effects* pe_i:

$$\text{pact} = \big\{\, \text{pe}_i = \langle \text{Context}_i, (\pi_{i1}, \text{Effect}_{i1}), \ldots, (\pi_{in_i}, \text{Effect}_{in_i}) \rangle \,\big\}_{i=1\ldots m},$$

in which $\pi_{ij} \in (0,1]$ and:

- *for each state represented by an interpretation $\mathcal{I}$ there is a unique Context_{ij} such that $\mathcal{I} \models \text{Context}_{ij}$.*
- *for all i, $\max_{1 \leq j \leq n_i} \pi_{ij} = 1$*

The following example illustrates a possibilistic action "grasp_B" as follows:

$$\text{grasp}_\text{B} = \left\{ \begin{array}{l} \langle \{\text{FreeBlock}(\text{BLOCKB}), \\ \quad \neg\text{InArm}(\text{BLOCKA}) \sqcap \neg\text{InArm}(\text{BLOCKB}) \sqcap \neg\text{InArm}(\text{BLOCKC})\}, \\ \quad (1, \{\neg\text{on}(\text{BLOCKB}, \text{BLOCKA})\}), (0.2, \emptyset) \rangle \\[6pt] \langle \{\neg\text{FreeBlock}(\text{BLOCKB})\}, (1, \emptyset) \rangle \\[6pt] \langle \{\text{InArm}(\text{BLOCKA}) \sqcup \text{InArm}(\text{BLOCKB}) \sqcup \text{InArm}(\text{BLOCKC})\}, (1, \emptyset) \rangle \end{array} \right\}.$$

Executing the action grasp_B on a state satisfying one of the contexts of the action may results on changes on the world state. These changes are represented by updating the ABox representing the situation of the state. Precisely, applying the action grasp_B on a context in which both block B and the robot arm are free results has two outcomes:

- the robot may succeed, i.e., with possibility 1, it grasps block B. This results on the "retracting" of the assertion $\text{on}(\text{BLOCKB}, \text{BLOCKA})$ on the ABox (see Figure 1b).
- The robot may fail, with possibility 0.2, thus leaving the situation unchanged.

The other two possibilistic effects cover the remaining possibilities, and both have no effect.

3.3 Reasoning About Actions

For all primitive concept name A and role name R, the result of the execution of an action act in a state $\mathcal{I}$ considering the unique context Context_i such that $\mathcal{I} \models \text{Context}_i$, is a state $\mathcal{I}' = \text{Res}(\text{Effect}_{ik}, \mathcal{I})$ such that:

$$A^{\mathcal{I}'} = A^{\mathcal{I}} \cup \{b^{\mathcal{I}} : A(b) \in \text{Effect}_{ik}\} \backslash \{b^{\mathcal{I}} : \neg A(b) \in \text{Effect}_{ik}\}$$

$$R^{\mathcal{I}'} = R^{\mathcal{I}} \cup \{(a^{\mathcal{I}}, b^{\mathcal{I}}) : R(a,b) \in \text{Effect}_{ik}\} \backslash \{(a^{\mathcal{I}}, b^{\mathcal{I}}) : \neg R(a,b) \in \text{Effect}_{ik}\}$$

These definitions ensure that the resulting state $\mathcal{I}'$ is a model of both the ABox and TBox resulting after executing the action on the state $\mathcal{I}$.

Applying a possibilistic action pact on a deterministic state $\mathcal{I}$ results on a possibility distribution definded as follows:

$$\pi[\mathcal{I}'|\mathcal{I}, \mathsf{pact}] = \begin{cases} \max_k \pi_{ik} & \text{if } \mathcal{I} \models \text{Context}_i \text{ and } \mathcal{I}' = \text{Res}(\text{Effect}_{ik}, \mathcal{I}), \\ 0 & \text{otherwise.} \end{cases}$$

$\pi[\mathcal{I}'|\mathcal{I}, \mathsf{pact}]$ espresses the possibility of reaching a possible resulting state $\mathcal{I}'$ after executing the action pact in the state $\mathcal{I}$. If there is more than one path leading from $\mathcal{I}$ to a possible resulting state $\mathcal{I}'$, then the possibility associated to $\mathcal{I}'$ is the maximum of the possibilities of all such paths.

In the case in which there are incompleteness and imprecision on the knowledge about the current state, the reasoning made above must be adapted for allowing fuzzy states. Precisely, we must define the possibility associated to a resulting state when the concepts and roles describing the current state are fuzzy.

Definition 2 (Fuzzy state resulting from a possibilistic action). *Let $\tilde{\mathcal{I}}$ be a fuzzy interpretation for both the acyclic TBox $\mathcal{T}$ and the ABox $\mathcal{A}$. Let pact be a possibilistic action and* Context_i *the unique context such that $\mathcal{S}_{\tilde{\mathcal{I}}}(\text{Context}_i) \in (0, 1]$. For each primitive concept name A and role name R, the result of executing action pact in $\tilde{\mathcal{I}}$, $\tilde{\mathcal{I}}' = \text{Res}'(\text{Effect}_{ik}, \tilde{\mathcal{I}})$ is such that:*

$$A^{\tilde{\mathcal{I}}'}(b) = \min(\max(A^{\tilde{\mathcal{I}}}(b), \sup_{k: A(b) \in \text{Effect}_{ik}} \pi_{ik}), 1 - \sup_{k: \neg A(b) \in \text{Effect}_{ik}} \pi_{ik})$$

$$R^{\tilde{\mathcal{I}}'}(a, b) = \min(\max(R^{\tilde{\mathcal{I}}}(a, b), \sup_{k: R(a,b) \in \text{Effect}_{ik}} \pi_{ik}), 1 - \sup_{k: \neg R(a,b) \in \text{Effect}_{ik}} \pi_{ik})$$

The possibility distribution on the new fuzzy states obtained after the execution of action pact is given by:

$$\pi[\tilde{\mathcal{I}}'|\tilde{\mathcal{I}}, \mathsf{pact}] = \begin{cases} \max_k \min(\pi_{ik}, \mathcal{S}_{\tilde{\mathcal{I}}}(\text{Context}_i)) & \text{and } \tilde{\mathcal{I}}' = \text{Res}'(\text{Effect}_{ik}, \tilde{\mathcal{I}}), \\ 0 & \text{otherwise.} \end{cases}$$

where $\mathcal{S}_{\tilde{\mathcal{I}}}(\text{Context}_i)$ expresses the degree with which the state $\mathcal{I}$ satisfies the context Context_i and $\pi[\tilde{\mathcal{I}}'|\tilde{\mathcal{I}}, \mathsf{pact}]$ espresses the possibility to arrive in fuzzy state $\tilde{\mathcal{I}}'$ after executing action pact in fuzzy state $\tilde{\mathcal{I}}$.

3.4   Plan of Actions

A sequential plan is a totally ordered set of actions $\langle \mathsf{pact}_i \rangle_{i=0}^{N-1}$ such execution transforms the interpretation representing the possible initial state to one representing possible final states satisfying the desired goal.

A partially ordered plan is a pair $\mathcal{P} = (A, O)$ where A is a set of actions and O is a set of ordering constraints between these actions. A completion of $\mathcal{P}$ is a sequential plan $\mathcal{CP} = \langle \mathsf{pact}_i \rangle_{i=0}^{N-1}$ such that $A = \{\mathsf{pact}_0, \ldots, \mathsf{pact}_{N-1}\}$ and the total ordering $\mathsf{pact}_0 < \cdots < \mathsf{pact}_{N-1}$ is consistent with O.

A consistent partially ordered plan is a plan $\mathcal{P} = (A, O)$ with a consistent set O of ordering constraints.

Executing a plan $\mathcal{P}$ means executing pact_0, pact_1, ..., pact_{N-1} in sequence, where $\langle\mathsf{pact}_i\rangle_{i=0}^{N-1}$ is a completion of $\mathcal{P}$.

The possibility of reaching a given state $\mathcal{I}_N$ by executing a sequential plan of possibilistic actions $\langle\mathsf{pact}_i\rangle_{i=0}^{N-1}$ starting in $\mathcal{I}_0$, is given by:

$$\pi[\mathcal{I}_N|\mathcal{I}_0,\langle\mathsf{pact}_i\rangle_{i=0}^{N-1}] = \max_{\langle\mathcal{I}_1\dots\mathcal{I}_{N-1}\rangle}\ \min_{i=0\dots N-1}\ \pi[\mathcal{I}_{i+1}|\mathcal{I}_i,\mathsf{pact}_i],$$

where $\langle\mathcal{I}_1\dots\mathcal{I}_{N-1}\rangle$ represents a sequence of states visited from $\mathcal{I}_1$ to $\mathcal{I}_{N-1}$ and $\mathcal{I}_{i+1}$ is obtained from $\mathcal{I}_i$ by applying Definition 2.

The evaluation of a solution plan is made using the necessity measure which corresponds to the certainty of reaching a goal state after applying the plan.

Let $Goals$ be the set of the goal states, and π_{init} a possibility distribution over the possible initial states $\mathcal{I}_0$. The possibility and necessity measures (in possibility theory) to reach a goal state after the execution of the sequential plan $\langle\mathsf{pact}_i\rangle_{i=0}^{N-1}$ from $\mathcal{I}_0$ are given by:

$$\Pi[Goals|\pi_{init},\langle\mathsf{pact}_i\rangle_{i=0}^{N-1}] = \max_{\mathcal{I}_0}\min(\Pi[Goals|\mathcal{I}_0,\langle\mathsf{pact}_i\rangle_{i=0}^{N-1}],\pi_{init}(\mathcal{I}_0))$$

$$= \max_{\mathcal{I}_0,\mathcal{I}_N\in Goals}\min(\pi[\mathcal{I}_N|\mathcal{I}_0,\langle\mathsf{pact}_i\rangle_{i=0}^{N-1}],\pi_{init}(\mathcal{I}_0))$$

$$N[Goals|\pi_{init},\langle\mathsf{pact}_i\rangle_{i=0}^{N-1}] = 1 - \Pi[\overline{Goals}|\pi_{init},\langle\mathsf{pact}_i\rangle_{i=0}^{N-1}]$$

$$= \min_{\mathcal{I}_0,\mathcal{I}_N\in\overline{Goals}}\max(1-\pi_{init}(\mathcal{I}_0),1-\pi[\mathcal{I}_N|\mathcal{I}_0,\langle\mathsf{pact}_i\rangle_{i=0}^{N-1}])$$

4   Possibilistic Planning Problem

Given a possibilistic planning described by means of a specific ontology, our objective is to construct a plan whose execution leads the world from an initial possible state to a state satisfying the goals with a given certainty. It corresponds to finding an optimal sequence of transition relation on interpretations that transforms an initial ABox representing the initial state to another ABox representing the final possible states which satisfy the goal conditions.

Definition 3 (Possibilistic Planning Problem). *A possibilistic planning problem Δ is a triple $\langle\pi_{init},Goals,A\rangle$ where π_{init} is the possibility distribution associated to the initial state, Goals is a set of possible goal states and A is the set of available possibilistic actions.*

Given a possibilistic problem, two criteria may be considered to define a *solution plan* $\mathcal{P}$:

- $\mathcal{P}$ is a γ-*acceptable plan* if $N[Goals|\pi_{init},\mathcal{P}] \geq \gamma$;
- $\mathcal{P}$ is an *optimally safe plan*, or simply, *optimal plan* if $N[Goals|\pi_{init},\mathcal{P}]$ is maximal among all possible sequential plans.

This definition can be extended to partially ordered sets of actions. Let $\mathcal{P}$ be a consistent partially ordered plan :

- $\mathcal{P}$ is a γ-**acceptable plan** if $N[Goals|\mathcal{I}_0, \mathcal{CP}] \geq \gamma$ for all totally ordered completion $\mathcal{CP}$ of $\mathcal{P}$;
- $\mathcal{P}$ is an **optimal plan** if $N[Goals|\mathcal{I}_0, \mathcal{CP}]$ is maximal among all possible sequential plans for all totally ordered completion $\mathcal{CP}$ of $\mathcal{P}$.

5 Conclusions

In this paper, we have proposed an initial framework for integrating description logics and possibilistic planning. The fundamental definitions for approaching possibilistic planning when knowledge about the world and actions is represented by means of fuzzy DLs have been provided, and the possibilistic planning problem has been recast in this setting. In particular, the framework we propose in this work allows the use of conceptual knowledge for describing both the states of the world and the possibilistic action. A state is represented as an interpretation in description logics; the contexts in which a possibilistic action may be executed and the effects of the action are both represented using concepts and assertions.

We have also defined the reasoning problem for the possibilistic actions thus represented. We have first considered the crisp case with a deterministic representation for a state. In a second time, we have proposed an extension of this representation for taking into account the case in which the knowledge about the state of the world is uncertain.

The main advantage of using fuzzy DLs to represent knowledge about the world and actions is that represetations may be more concise and efficient reasoning algorithms can be exploited to infer implicit knowledge.

References

1. Franz Baader, Diego Calvanese, Deborah McGuinness, Daniele Nardi, and Peter Patel-Schneider, editors. *The Description Logic Handbook: Theory, implementation and applications*. Cambridge, 2003.
2. C. da Costa Pereira, F. Garcia, J. Lang, and R. Martin-Clouaire. Planning with graded nondeterministic actions: a possibilistic approach. *Int. Journal of Intelligent Systems*, 12:935–962, 1997.
3. D. Dubois and H. Prade. *Théorie des possibilités*. Masson, 1988.
4. J. S. Penberthy and D. S. Weld. Ucpop: A sound, complete, partial order planner for adl. In B. Nebel, C. Rich, and W. Swartout, editors, *Principles of Knowledge Representation and Reasoning: Proc. of the Third International Conference (KR'92)*, pages 103–114. Kaufmann, San Mateo, CA, 1992.
5. Umberto Straccia. Reasoning within fuzzy description logics. *Journal of Artificial Intelligence Research*, 14:137–166, 2001.

Multi-lattices as a Basis for Generalized Fuzzy Logic Programming*

Jesús Medina, Manuel Ojeda-Aciego, and Jorge Ruiz-Calviño

Dept. Matemática Aplicada, Universidad de Málaga
{jmedina, aciego, jorgerucal}@ctima.uma.es

Abstract. A prospective study of the use of ordered multi-lattices as underlying sets of truth-values for a generalised framework of logic programming is presented. Specifically, we investigate the possibility of using multi-lattice-valued interpretations of logic programs and the theoretical problems that this generates with regard to its fixed point semantics.

1 Introduction

Weakening the structure of the underlying set of truth-values for logic programming has been studied extensively in the recent years. There are approaches which are based either on the structure of lattice (residuated lattice [4, 13] or multi-adjoint lattice [9]), or more restrictive structures, such as bilattices or tri-lattices [7], or even more general structures such as algebraic domains [11]. One can also find some attempts aiming at weakening the restrictions imposed on a (complete) lattice, namely, the "existence of least upper bounds and greatest lower bounds" is relaxed to the "existence of *minimal* upper bounds and *maximal* lower bounds". In this direction, Benado [1] and Hansen [5] proposed definitions of a structure so-called multi-lattice.

Recently an alternative notion of multi-lattice was introduced [2, 8] as a theoretical tool to deal with some problems in the theory of mechanised deduction in temporal logics. This kind of structure also arises in the research area concerning fuzzy extensions of logic programming: for instance, one of the hypotheses of the main termination result for sorted multi-adjoint logic programs [3] can be weakened only when the underlying set of truth-values is a multi-lattice (the question of providing a counter-example on a lattice remains open).

As far as we know, there have been no attempts to use multi-lattices in the context of extended fuzzy logic programming; our aim in this work is precisely to study the computational capabilities of this new structure in that framework and, specifically, in relation to its fixed point semantics.

The structure of the paper is as follows: In Section 2 the definition and preliminary theoretical results about multi-lattices are introduced; later, the syntax and semantics of our extended logic programs are presented in Section 3; then, an initial proposal for fixed point semantics for these extended logic programs is given in Section 4. Finally, in the last section we present some conclusions and prospects for future work.

* Partially supported by Spanish DGI project TIC2003-09001-C02-01.

I. Bloch, A. Petrosino, and A.G.B. Tettamanzi (Eds.): WILF 2005, LNAI 3849, pp. 61–70, 2006.

2   Preliminary Results

Recall that a lattice is a poset such that the set of upper (lower) bounds has a unique minimal (maximal) element, that is, a *minimum* (*maximum*). In a multi-lattice, this property is relaxed in the sense that minimal elements for the set of upper bounds should exist, but the uniqueness condition is dropped.

Definition 1. *A* complete multi-lattice *is a partially ordered set, $\langle M, \leq \rangle$, such that for every subset $X \subseteq M$, the set of upper (lower) bounds of X has minimal (maximal) elements, which are called* multi-suprema *(*multi-infima*).*

Note that, by definition, it follows that the sets multinf(X) and multisup(X) are *antichains* (non-empty sets consisting of pair-wise incomparable elements).

It is remarkable that, under suitable conditions, the set of fixed points of a mapping from M to M does have a minimum and a maximum.

Definition 2. *A mapping $f: P \longrightarrow Q$ between two posets is said to be* isotone *if $x \leq y$ implies $f(x) \leq f(y)$; a mapping $g: P \longrightarrow P$ is* inflationary *if $x \leq g(x)$ for all $x \in P$.*

Theorem 1. *Let $f: M \longrightarrow M$ be an isotone and inflationary mapping on a multi-lattice, then its set of fixed points is non-empty and has a minimum element.*

Proof. Let us write $X = \{x \mid f(x) = x\}$, this set is nonempty since inflation forces $\top$ to be a fixed point; now, consider $a \in$ multinf(X) a maximal lower bound of X, and let us prove that a is a fixed point of f.

As a is a lower bound for all $x \in X$, we have $a \leq x$ and, by isotonicity, $f(a) \leq f(x) = x$ for all $x \in X$ (the equality follows by definition of X); thus, $f(a)$ is also a lower bound of X. Moreover, a is maximal and, by inflation, we have $a \leq f(a)$; thus, we also have $f(a) \leq a$ and a should be a fixed point, that is $a \in X$.

Consider $a, b \in$ multinf(X), and recall that we have just proved that $a, b \in X$. As both are lower bounds of X, then $a \leq b$ and $b \leq a$. Thus, multinf(X) is a singleton consisting of the minimum element of X, that is, the minimum fixed point. □

As by assumption, our sets will not necessarily have a supremum but a *set* of multi-suprema, we will need to work with some ordering between subsets of posets. Three different (pre-)orderings are usually considered in the literature, the Hoare ordering, the Smyth ordering and the Egli-Milner ordering:

Definition 3. *Consider $X, Y \subseteq 2^M$:*

- *$X \sqsubseteq_H Y$ iff for all $x \in X$ exists $y \in Y$ such that $x \leq y$.*
- *$X \sqsubseteq_S Y$ iff for all $y \in Y$ exists $x \in X$ such that $x \leq y$.*
- *$X \sqsubseteq_{EM} Y$ iff $X \sqsubseteq_H Y$ and $X \sqsubseteq_S Y$*

Regarding computational properties of multi-lattices, it is interesting to impose certain conditions on the sets of upper (lower) bounds of a given set X. Specifically, we would like to ensure that any upper (lower) bound is greater (less) than a minimal (maximal); this condition enables to work on the set of multi-suprema (multi-infima) as a set of "generators" of the bounds of X. The formalisation of these concepts is given as follows, where $UB(X)$ (resp. $LB(X)$) denotes the set of upper (lower) bounds of X:

Definition 4. *A multi-lattice is said to be* consistent *if the following set of inequalities hold for all $X \subseteq M$:*

$$LB(X) \sqsubseteq_{EM} \text{multinf}(X) \qquad\qquad \text{multisup}(X) \sqsubseteq_{EM} UB(X)$$

Note that in the two items above, one part of the Egli-Milner ordering is trivial, since any multi-infimum is a lower bound and any multi-supremum is an upper bound. It is not difficult to provide examples of non-consistent multi-lattices:

Example 1. A non-consistent multi-lattice is showed on the left of Fig. 1, where

$$UB(\{a,b\}) = \{\top, d\} \cup \{c_n \mid n \in \mathbb{N}\}$$

in which element d is minimal in $UB(\{a,b\})$; however, the elements c_n fail to be greater than one minimal upper bound.

Another reasonable condition to require on a multi-lattice is that it should not contain infinite sets of mutually incomparable elements (antichains) since, semantically, it makes little sense to consider infinitely many incomparable truth-values. Consistent multi-lattices without infinite antichains have interesting computational properties: to begin with, recall that the sets of multi-suprema or multi-infima for totally ordered subsets (also called *chains*) always have a supremum and an infimum.

Lemma 1. *Let M be a consistent multi-lattice without infinite antichains, then any chain in M has a supremum and an infimum.*

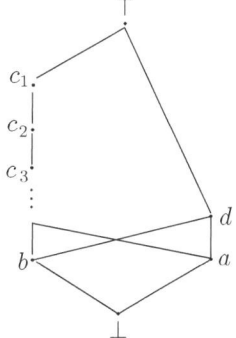 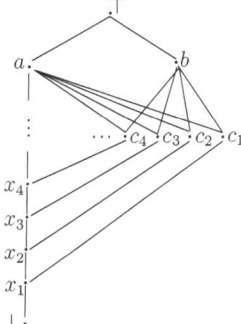

Fig. 1.

Proof. Let $\{x_i\}_{i \in I} \subset M$ be a chain and, assume that $a, b \in \mathrm{multisup}(\{x_i\})$. We will show that there is an element $c \in \mathrm{multinf}(\{a, b\})$ which is an upper bound of the chain.

As there are no infinite antichains in M, the set $\mathrm{multinf}(\{a, b\})$ is finite, and we can write

$$\mathrm{multinf}(\{a, b\}) = \{c_1, \ldots, c_n\}$$

If $n = 1$, as any x_i is a lower bound of $\{a, b\}$, by the hypothesis of consistency we would have $x_i \leq c_1$ for all $i \in I$.

If $n > 1$, by contradiction, assume that no c_j, with $j = 1, \ldots, n$, is an upper bound of the chain; then, for all j we choose an element x_j which is not upper bounded by c_j. Now, as $\{x_i\}$ is a chain, let us consider the greatest of $x_1, \ldots, x_n$, say x_{j_0}. By consistency, there is c_k which is greater than x_{j_0}, but then

$$x_k \leq x_{j_0} \leq c_k$$

which would contradict the choice of x_k.

Summarising, we have proved the existence of $c \in \mathrm{multinf}(\{a, b\})$ which, moreover, is an upper bound of the chain. Now, $c \in \mathrm{multinf}(\{a, b\})$ implies the inequalities $c \leq a$ and $c \leq b$; on the other hand, as c is also an upper bound of $\{x_i\}$ and a and b are multi-suprema of $\{x_i\}$, then $a \leq c$ and $b \leq c$, resulting that $a = b = c$, which proves that $\mathrm{multisup}(\{x_i\})$ is a singleton, hence the supremum of the chain.

The proof for the infimum is similar.   □

All the hypotheses are necessary for the existence of supremum and infimum of chains; in particular, the condition on infinite antichains cannot be dropped.

Example 2. The poset on the right of Fig. 1 is a consistent multi-lattice; however, the set of upper bounds of the increasing sequence $\{x_n\}$ does not have a minimum, but two minimals, namely, a and b.

We will assume in the rest of the paper that our underlying multi-lattices are complete, consistent and without infinite antichains.

3   Extended Logic Programs

In this section we provide a first approximation of the definition of an extended logic programming paradigm in which the underlying set of truth-values is assumed to have structure of multi-lattice. The proposed schema is an extension of the monotonic logic programs of [4]. The definition of logic program is given, as usual, as a set of rules and facts.

Definition 5. *An* extended logic program *is a set $\mathbb{P}$ of rules of the form $A \leftarrow \mathcal{B}$ such that:*

1. *A is a propositional symbol of Π, and*
2. *$\mathcal{B}$ is a formula of $\mathfrak{F}$ built from propositional symbols and elements of M by using isotone operators.*

An interpretation is an assignment of truth-values to every propositional symbol in the language.

Definition 6. *An* interpretation *is a mapping* $I: \Pi \to M$. *The set of all interpretations is denoted* $\mathcal{I}$.

Note that by the unique homomorphic extension theorem, any interpretation I can be uniquely extended to the whole set of formulas (the extension will be denoted as $\hat{I}$). The ordering $\leq$ of the truth-values M can be extended point-wise to the set of interpretations as usual.

A rule of an extended logic program is satisfied whenever the truth-value of the head of the rule is greater or equal than the truth-value of its body. Formally:

Definition 7. *Given an interpretation* I, *a rule* $A \leftarrow B$ *is* satisfied *by* I *iff* $\hat{I}(B) \leq I(A)$. *An interpretation* I *is said to be a* model *of an extended logic program* $\mathbb{P}$ *iff all rules in* $\mathbb{P}$ *are satisfied by* I, *then we write* $I \models \mathbb{P}$.

Example 3. Let us consider the following program on the multi-lattice in Fig. 2:

$$E \leftarrow A$$
$$E \leftarrow B$$
$$A \leftarrow a$$
$$B \leftarrow b$$

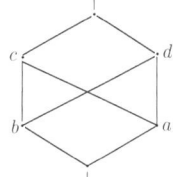

Fig. 2.

It is easy to check that the interpretation defined as $I(E) = \top$, $I(A) = a$, $I(B) = b$ is a model of the program.

Every extended program $\mathbb{P}$ has the top interpretation $\triangledown$ as a model; regarding minimal models, it is possible to prove the following technical lemma.

Lemma 2. *A chain of models* $\{I_k\}_{k \in K}$ *of* $\mathbb{P}$ *has an infimum in* $\mathcal{I}$ *which is a model of* $\mathbb{P}$.

Proof. Given a propositional symbol A, the existence of $\inf_k\{I_k(A)\}$ is guaranteed by Lemma 1, thus we can safely define an interpretation I_ω as follows:

$$I_\omega(A) = \inf_{k \in K}\{I_k(A)\}$$

Now, let us show that I_ω is a model of $\mathbb{P}$:

Given a rule $A \leftarrow @[B_1, \ldots, B_n]$ in $\mathbb{P}$, where @ denotes the composition of the operators occurring in the body of the rule, and the B_i's are the variables occurring in it; by isotonicity of @ we obtain the following chain of inequalities for all $i \in K$:

$$\hat{I}_i(B) = @[I_i(B_1), \ldots, I_i(B_n)] \geq @\left[\inf_{k \in K}\{I_k(B_1)\}, \ldots, \inf_{k \in K}\{I_k(B_n)\}\right] = \hat{I}_\omega(B)$$

As I_i is a model for all i we obtain:

$$I_i(A) \geq \hat{I}_i(\mathcal{B}) \geq \hat{I}_\omega(\mathcal{B})$$

thus, by definition of infimum, we have

$$I_\omega(A) = \inf_{k \in K} \{I_k(A)\} \geq \hat{I}_\omega(\mathcal{B})$$

so I_ω is a model of $\mathbb{P}$.   □

Theorem 2. *There exist minimal models for any extended logic program $\mathbb{P}$.*

Proof. Let $\mathcal{M}$ be the set of models of $\mathbb{P}$. By Zorn's lemma, we only have to prove that any chain in $\mathcal{M}$ is lower bounded, but this follows from the previous lemma since the infimum of a chain of models is also a model.   □

Example 4. Continuing with the program in the previous example, it is easy to check that the program does not have a minimum model but two minimal ones:

$$\begin{array}{ll} I_1(E) = c & I_2(E) = d \\ I_1(A) = a & I_2(A) = a \\ I_1(B) = b & I_2(B) = b \end{array}$$

4   Fix-Point Semantics

An interesting technical problem arises when trying to extend the definition of the immediate consequences operators to the framework of multi-lattice-based logic programs. One of the several possible approaches to provide a fixed point semantics for the extended logic programs is presented and analysed.

The main theoretical tool for the study of the fixed point semantics of programming languages is Knaster-Tarski theorem in some of its constructive versions, although some other fixed point theorems are also of use, see [6].

Given a logic program $\mathbb{P}$ valued on a *lattice*, the operator $T_\mathbb{P} \colon \mathcal{I} \to \mathcal{I}$, maps interpretations to interpretations, and can be defined by considering

$$T_\mathbb{P}(I)(A) = \sup\{\hat{I}(\mathcal{B}) \mid A \leftarrow \mathcal{B} \in \mathbb{P}\}$$

Note that all the suprema involved in the definition do exist provided that we are assuming a complete lattice structure on the underlying set of truth-values; however, this needs not hold for a multi-lattice.

In order to work this problem out, we consider the following definition.

Definition 8. *Given an extended logic program $\mathbb{P}$, an interpretation I and a propositional symbol A; we can define $T_\mathbb{P}(I)(A)$ as*

$$\mathrm{multisup}\left(\{I(A)\} \cup \{\hat{I}(\mathcal{B}) \mid A \leftarrow \mathcal{B} \in \mathbb{P}\}\right)$$

Some properties of this definition of the $T_\mathbb{P}$ operator are stated below, where $\sqsubseteq_S$ denotes the Smyth-ordering between subsets of a poset:

Lemma 3. *If $I \sqsubseteq J$, then $T_{\mathbb{P}}(I)(A) \sqsubseteq_S T_{\mathbb{P}}(J)(A)$ for all propositional symbol A.*

Proof. Let us write X_I to denote the set $\{I(A)\} \cup \{\hat{I}(\mathcal{B}) \mid A \leftarrow \mathcal{B} \in \mathbb{P}\}$, then the hypothesis states that $X_J^{\uparrow} \subseteq X_I^{\uparrow}$, where the $\uparrow$ denotes the upwards-closure of a set. Now, consider $b \in T_{\mathbb{P}}(J)(A)$, then b is an element of $X_J^{\uparrow} \subseteq X_I^{\uparrow}$; thus, by consistency, considering any minimal a of $X_I^{\uparrow}$ below b leads to the existence of an element $a \in T_{\mathbb{P}}(I)(A)$.   □

The definition of $T_{\mathbb{P}}$ proposed above generates some coherence problems, in that the resulting 'value' is not an element, but a subset of the multi-lattice. A possible solution to this problem would be to consider a *choice function* $()^*$ which, given an interpretation, for any propositional symbol A selects an element in $T_{\mathbb{P}}(I)(A)$; this way, $T_{\mathbb{P}}(I)^*$ represents actually an interpretation.

Regarding particular properties of the composition of the $T_{\mathbb{P}}$ operator with suitable choice functions, the first property one can obtain, directly from the definition, is that the composition leads to an inflationary operator.

Lemma 4. *Given an interpretation I and a choice function $()^*$, then $I(A) \le T_{\mathbb{P}}(I)^*(A)$ for all propositional symbol A.*

Note that, for some choice functions, the resulting operator $T_{\mathbb{P}}^*$ might not be monotone in the set of interpretations, since it can lead to incomparable interpretations; the multi-lattice of Fig. 2 can be used to construct a counter-example.

Example 5. Consider the following program with just two facts $\{A \leftarrow a, A \leftarrow b\}$ and interpretations $I(A) = \bot$ and $J(A) = c$; obviously $I \sqsubseteq J$. Now, we have that $T_{\mathbb{P}}(I)(A) = \{c, d\}$ and $T_{\mathbb{P}}(J)(A) = \{c\}$. Thus, the choice function $()^*$ which selects d in $T_{\mathbb{P}}(I)(A)$ generates incomparable interpretations $T_{\mathbb{P}}(I)^*$ and $T_{\mathbb{P}}(J)^*$.

We are interested in computing models of our extended programs by successive iteration of $T_{\mathbb{P}}^*$. Therefore, we should characterise the models of $\mathbb{P}$ in terms $T_{\mathbb{P}}$. The following result, which characterises the models of our extended programs in terms of properties of $T_{\mathbb{P}}$, can be proved:

Proposition 1. *The four statements below are equivalent:*

1. *I is a model of $\mathbb{P}$.*
2. *$T_{\mathbb{P}}(I)(A) = \{I(A)\}$ for all $A \in \Pi$.*
3. *$T_{\mathbb{P}}(I)^* = I$ for all choice function.*
4. *$I \in T_{\mathbb{P}}(I)$,[1] (i.e. I is a fixed point of $T_{\mathbb{P}}$ as a non-deterministic operator).*

Proof.
$(1 \Rightarrow 2)$. Let us assume that I is a model of $\mathbb{P}$; then, we have that $I(A) \ge \hat{I}(\mathcal{B})$ for all rule $A \leftarrow \mathcal{B} \in \mathbb{P}$. This implies that $I(A)$ is the maximum of the set

$$\{I(A)\} \cup \{\hat{I}(\mathcal{B}) \mid A \leftarrow \mathcal{B} \in \mathbb{P}\}$$

hence, the only multi-supremum.

---

[1] Abusing notation this means that $I(A) \in T_{\mathbb{P}}(I)(A)$ for all $A \in \Pi$.

$(2 \Rightarrow 1)$. The hypothesis implies that $I(A)$ is an upper bound of

$$\{\hat{I}(\mathcal{B}) \mid A \leftarrow \mathcal{B} \in \mathbb{P}\}$$

as a result, $I(A) \geq \hat{I}(\mathcal{B})$ for all rule $A \leftarrow \mathcal{B} \in \mathbb{P}$ and $I \models \mathbb{P}$.
$(2 \Leftrightarrow 3 \Leftrightarrow 4)$. Trivial. $\qquad\qquad\square$

Regarding the iterated application of the $T_\mathbb{P}$ operator, the use of choice functions is essential. Let us consider a model I, that is, a fixed point of $T_\mathbb{P}$, then for all propositional variable A, we have that $T_\mathbb{P}(I)(A) = \{I(A)\}$. Lemma 3 guides us in the choice after each application of $T_\mathbb{P}$ as follows:

- For the base case, we have[2] $\triangle \sqsubseteq I$, then $T_\mathbb{P}(\triangle)(A) \sqsubseteq_S T_\mathbb{P}(I)(A) = \{I(A)\}$. This means that there exists an element $m_1(A) \in T_\mathbb{P}(\triangle)(A)$ such that

$$m_1(A) \leq I(A)$$

  This way we obtain an interpretation m_1 satisfying $m_1 \sqsubseteq I$ such that for any propositional variable A, $m_1(A)$ is an element of $T_\mathbb{P}(\triangle)(A)$.
- This argument applies also to any successor ordinal: given $m_k \sqsubseteq I$, there exists an element $m_{k+1}(A) \in T_\mathbb{P}(m_k)(A)$ such that

$$m_k(A) \leq m_{k+1}(A) \leq I(A)$$

  where the first inequality holds by the definition of $T_\mathbb{P}$ and the second inequality follows from Lemma 3.
- For a limit ordinal α, Lemma 1 states that for all A the increasing sequence $\{m_n(A)\}$ has a supremum, which is considered, by definition, to be $m_\alpha(A)$.

As a result of the discussion above we obtain that we can choose suitable elements in the sets generated by the application of $T_\mathbb{P}$ in such a way that we can construct a transfinite sequence of interpretations m_k satisfying

$$m_1 \sqsubseteq m_2 \sqsubseteq \cdots \sqsubseteq m_k \sqsubseteq \cdots \sqsubseteq I$$

Note that the sequence of interpretations above, can be interpreted as the Kleene sequence which allows to reach the least fixed point of $T_\mathbb{P}$ in the classical case.

Interestingly enough, if I is a minimal model of $\mathbb{P}$, the previous sequence of interpretations can be proved to converge to I.

Theorem 3. *Let I be a minimal model of $\mathbb{P}$, then the previous construction leads to a Kleene sequence $\{m_\lambda\}$ which converges to I.*

Proof. A cardinality-based argument suffices to show that $\{m_\lambda\}$ is eventually constant and equal to I:

Let β be the least ordinal greater than the cardinal of the set of interpretations, for all $\lambda < \beta$ we can consider the interpretation m_λ and, thus, define the following map

$$h \colon \beta \longrightarrow \mathcal{I}$$
$$\lambda \mapsto m_\lambda$$

---

[2] Here, as usual, $\triangle$ denotes the minimum interpretation.

If the transfinite sequence were strictly increasing, then h would be injective, obtaining a contradiction with the choice of β. As a result, we have proved the existence of an ordinal α such that $m_\alpha = m_{\alpha+1}$.

Recall that, by definition, we have $m_\alpha \sqsubseteq I$ and $m_{\alpha+1} \in T_{\mathbb{P}}(m_\alpha)$, therefore $m_\alpha \in T_{\mathbb{P}}(m_\alpha)$ and, by Proposition 1, m_α is a model of $\mathbb{P}$. By minimality of I we have that $m_\alpha = I$.   □

Example 6. Continuing with the previous example, let us consider the minimal model I_1, and let us construct a sequence of approximating interpretations as stated in the theorem above.

	$\triangle$	$T_{\mathbb{P}}(\triangle)$	m_1	$T_{\mathbb{P}}(m_1)$	m_2
A	$\perp$	$\{a\}$	a	$\{a\}$	a
B	$\perp$	$\{b\}$	b	$\{b\}$	b
E	$\perp$	$\{\perp\}$	$\perp$	$\{c,d\}$	c

5   Conclusions and Future Work

A fixed point semantics has been presented for multi-lattice-based logic programming, together with some initial and encouraging results: in particular, we have proved the existence of minimal models for any extended program and that any minimal model can be attained by some Kleene-like sequence.

However, a number of theoretical problems have to be investigated in the future: such as the constructive nature of minimal models (is it possible to construct suitable choice functions which generate convergent sequence of interpretations with limit a minimal model?). Possible answers should on a general theory of fixed points, relying on some of the ideas related to fixed points in partially ordered sets [10] or, perhaps, in fuzzy extensions of Tarski's theorem [12].

References

1. M. Benado. Les ensembles partiellement ordonnés et le théorème de raffinement de Schreier, II. Théorie des multistructures. *Czech. Math. J.*, 5(80):308–344, 1955.
2. P. Cordero, G. Gutiérrez, J. Martínez, and I. P. de Guzmán. A new algebraic tool for automatic theorem provers. *Ann. Math. and Artif. Intelligence*, 42(4):369–398, 2004.
3. C. Damásio, J. Medina, and M. Ojeda-Aciego. Sorted multi-adjoint logic programs: termination results and applications. *Lect. Notes in Artificial Intelligence* 3229:260–273, 2004.
4. C. Damásio and L. Pereira. Monotonic and residuated logic programs. *Lecture Notes in Computer Science*, 2143:748–759, 2001.
5. D. Hansen. An axiomatic characterization of multi-lattices. *Discrete Mathematics*, 1:99–101, 1981.
6. M.A. Khamsi and D. Misane. Fixed point theorems in logic programming. *Ann. Math. and Artif. Intelligence*, 21:231–243, 1997.

7. L. Lakhsmanan and F. Sadri. On a theory of probabilistic deductive databases. *Theory and Practice of Logic Programming*, 1(1):5–42, 2001.
8. J. Martínez, G. Gutiérrez, I.P. de Guzmán and P. Cordero. Generalizations of lattices looking at computation. *Discrete Mathematics*, 295:107–141, 2005.
9. J. Medina, M. Ojeda-Aciego, and P. Vojtáš. Multi-adjoint logic programming with continuous semantics. *Lect. Notes in Artificial Intelligence*, 2173:351–364, 2001.
10. A. Ran and M. Reurings. A fixed point theorem in partially ordered sets and some applications to matrix equations. *Proc. of the AMS*, 132(5):1435–1443, 2003.
11. W. Rounds and G.-Q. Zhang. Clausal logic and logic programming in algebraic domains. *Inform. and Computation*, 171:183–200, 2001.
12. A. Stouti. A fuzzy version of Tarski's fixpoint theorem. *Archivum mathematicum*, 40:273–279, 2004.
13. P. Vojtáš. Fuzzy logic programming. *Fuzzy sets and systems*, 124(3):361–370, 2001.

A Method for Characterizing Tractable Subsets of Qualitative Fuzzy Temporal Algebrae

Marco Falda

Dept. of Information Engineering,
via Gradenigo, 6 - 35131 Padova, Italy
Tel./Fax: +39 049 827 7856 / 7826
marco.falda@unipd.it

Abstract. Allen's interval algebra allows one to formulate problems that are, in the general case, intractable; for this reason several tractable sub-algebrae have been proposed. In this paper the attention is focused on the fuzzy counterparts of those sub-algebrae and a different method to identify their relations is shown: rules for identifying fuzzy tractable relations starting from the knowledge of the classic tractable relations. Enumeration is used to verify the rules and quantify expressiveness, and algebraic considerations adopted to bind the enumeration itself.

Keywords: Fuzzy Sets, Possibility Theory, Representation of Vague and Imprecise Knowledge.

1 Introduction

Temporal reasoning problems are usually formulated as CSPs (Constraint Satisfaction Problems) where the variables represent temporal entities (points or intervals) and the constraints are disjunctions of atomic temporal relations. The general formulation of Allen's Interval Algebra is $\mathcal{NP}$−complete, therefore many studies have been devoted to the identification of subsets that are computationally affordable and enough expressive. The first tractable sub-algebra discovered is the pointizable algebra SA [12], that is the subset of Allen's relations that can be expressed by means of relations in the Point Algebra PA [11].

In [12] another tractable sub-algebra, which is based on convex point algebra PA_c, is identified; it is called SA_c.

If a network is built with relations in PA, then the path-consistency is sufficient to assure the global consistency and the 4-consistency to find the minimal network.

The maximal tractable sub-algebra that includes all the Allen's atomic relations is unique and it has been identified in [10] as the set of $ORD - Horn$ formulae $\mathcal{H}$. Applying the path-consistency algorithm to a network whose constraints belong to $\mathcal{H}$ is enough to decide its global consistency.

To deal with uncertain temporal knowledge, the Possibility Theory [4] can be introduced in qualitative temporal relations by associating a degree of preference to each atomic relation. In [5] it is stated that a set of fuzzy relations is tractable if

I. Bloch, A. Petrosino, and A.G.B. Tettamanzi (Eds.): WILF 2005, LNAI 3849, pp. 71–80, 2006.

all its α-cuts [14] are classic tractable relations; besides, from [5], if all the classic sets coming from the α-cuts are algebrae then also the original fuzzy set is an algebra. This principle will be used for rules discovery.

The fuzzy tractable sub-algebrae considered in this paper are the counterparts of the following classic sub-algebrae:

1. PA_c which is the subset of convex relations of the point algebra PA^1
2. $\mathcal{V}_{23}$, a subset of point-interval set of relations described in [6]: this is not an algebra[2], because the composition relation it is not closed, but it is useful to build the qualitative algebra QA [9]; however operations are not needed in the following, so this "algebra" will be studied as well;
3. SA, the pointizable algebra cited before;
4. SA_c, the sub-algebra of IA analogous to the SA but based on PA_c relations;
5. $ORD - Horn$, the maximal tractable sub-algebra of IA [10].

The first two items of the previous list can be easily considered by hand, because the number of possible combinations are 8 and 32 respectively (and indeed the $PA_c{}^{fuz}$ relations have already been explicitly listed). The last three sub-algebrae however are all based on IA, whose classic relations amount to 8 192, a number of combinations that only a computer can manage. In this paper a method for finding sets of rules to identify tractable relations is presented in Section 2 and an analytical formulation to calculate the cardinality of an algebra in Section 3. The correctness of the rules is verified using Prolog scripts that enumerate the relations taking into account the bounds from Section 3; the results are reported in Section 4.

2 The Discovery of Rules

Qualitative aspect of temporal knowledge can be represented using qualitative algebrae and CSPs [1, 11, 9]; a temporal constraint is a binary relation between a pair of intervals or points, represented by a disjunction of *atomic relations*:

$$X_1 \ (b_1, \ldots, b_m) \ X_2$$

where each b_i can be one of the thirteen mutually exclusive atomic relations that may exist between two intervals (such as *equal, before, meets* et c.), or one of the three possible relations between two points ($<$, $=$, $>$) or one of the five possible relations between an interval and a point [9].

In [3] Allen's Interval Algebra has been extended to the framework of Possibility Theory by defining a new algebra IA^{fuz} where a degree α_i has been attached to every atomic relation b_i. It indicates the *preference degree* of the corresponding assignment among the others

$$X_1 \ R \ X_2 \ \text{with} \ R = (b_1[\alpha_1], \ldots, b_{13}[\alpha_{13}])$$

---

[1] PA is already tractable, it is used just for testing.
[2] Recall that an algebra is a set of relations closed under inversion, conjunction and composition.

where α_i is the preference degree of b_i $(i = 1, \dots, 13)$. Preferences degrees can be defined in the interval $[0, 1]$; if we restrict them to the set $\{0, 1\}$, then the classic approach is re-obtained.

Intervals are interpreted as ordered pairs $(x, y) : x \leq y$ of $\Re^2$, and soft constraints between them as fuzzy subsets of $\Re^2 \times \Re^2$ in such a way that the pairs of intervals that are in relation rel_k have membership degree α_k.

In this framework, different types of temporal constraints can be represented such as:

- soft constraints, that enables us to express preferences among solutions;
- prioritized constraints, the priority indicating how essential it is that a constraint be satisfied;
- uncertain constraints.

We have extended in a similar way Meiri's Qualitative Algebra [2].

The function "$deg\{\}$" over a set of relations will indicate the preference degrees of the elements of that set, so $deg\{b_1, \dots, b_k\} = \{deg(b_1), \dots, deg(b_k)\}$.

In the crisp case there are several studies that have characterized tractable subsets of relations (see [6, 7, 8, 10, 13]); the fuzzy extensions of qualitative temporal algebrae add another "dimension" to the search space, because also the correlations among the preference degrees must be taken into account. In this paper an algorithm to evaluate those correlations and write rules that characterize tractable subclasses is developed exploiting the theorems in [5], as stated before. It starts from the knowledge of the atomic relations, of the base relations (that is the relations of the super-algebra) and of the classic tractable relations; it can be briefly described as follows. For each base relation $\{rel_i, i = 1 \dots n\}$:

1. if it is in the list of tractable relations right, otherwise
2. add to the relation each one of the atomic relations $rel_{k+1}, k \geq n$ that neither belongs to the current set nor it is already considered;
   (a) check if the new set of relations $\{rel_i, i = 1 \dots n\} \cup \{rel_{k+1}\}$ is now in the list of tractable relations and, if so, associate it with the conditions

$$\{rel_i, i = 1 \dots n\} \leftarrow deg(rel_{k+1}) \geq min_{i=1\dots n}\{deg\{rel_i\}\} \qquad (1)$$

3. if the iterations in step 2 do not produce any additional condition then the original relation $\{rel_i, i = 1 \dots n\}$ cannot be included in the tractable relations, therefore add the condition

$$\{rel_i, i = 1 \dots n\} \leftarrow min_{i=1\dots n}\{deg\{rel_i\}\} \leq max_{k:k \geq n}\{deg\{rel_k\}\} \qquad (2)$$

The heads of the rules (on the left of the arrow) represent the relations that have been considered, and can be thought just as labels that simplify the following reduction steps (indeed all the relevant information is in the right hand side). Two reduction steps can be applied on the set of relations that have an associated condition to be satisfied and on the excluded relations. First, if an associated condition is a superset of another one then the whole condition can be skipped; this means that it will be satisfied whenever the other is satisfied because the relation with fewer terms is stricter. This check can be done reasoning on the labels.

For example, let us consider these two relations with conditions:

$$\{a, c, d\} \leftarrow deg(b) \geq min\{deg\{a, c, d\}\}$$
$$\{a, d\} \leftarrow deg(b) \geq min\{deg\{a, d\}\}$$

The first condition is a superset of the second one, so the whole first condition can be omitted:

$$\{a, d\} \leftarrow deg(b) \geq min\{deg\{a, d\}\}$$

To see this it is possible to rewrite the above conditions as

$$(deg(b) \geq deg(a) \wedge deg(b) \geq deg(c) \wedge deg(b) \geq deg(d))$$
$$\wedge$$
$$(deg(b) \geq deg(a) \wedge deg(b) \geq deg(d))$$

and easily see that if both these conditions are true for a given set then $deg(b) \geq deg(c)$ and they can be reduced to

$$(deg(b) \geq deg(a) \wedge deg(b) \geq deg(d))$$

The second possible reduction is analogous to the previous one, but is applied to the conditions of excluded relations. Here is an example: given the following conditions

$$\{a, c, e\} \leftarrow min\{deg\{a, c, e\}\} \leq max\{deg\{b, d\}\}$$
$$\{a, c\} \leftarrow min\{deg\{a, c\}\} \leq max\{deg\{b, d, e\}\}$$

the left hand side of the first condition ($min\{$ deg\{a, c, e\} $\}$) is a superset of the second one and the whole first condition can be omitted:

$$\{a, c\} \leftarrow min\{deg\{a, c\}\} \leq max\{deg\{b, d, e\}\}$$

To prove this it is possible to rewrite the above conditions as

$$((deg(a) \leq deg(b) \vee deg(a) \leq deg(d))$$
$$\vee (deg(c) \leq deg(b) \vee deg(c) \leq deg(d))$$
$$\vee (deg(e) \leq deg(b) \vee deg(e) \leq deg(d)))$$
$$\wedge$$
$$((deg(a) \leq deg(b) \vee deg(a) \leq deg(d) \vee deg(a) \leq deg(e))$$
$$\vee (deg(c) \leq deg(b) \vee deg(c) \leq deg(d) \vee deg(c) \leq deg(e)))$$

and notice that if both these conditions are true for a given set then it cannot be the case that the only true clause is

$$(deg(e) \leq deg(b) \vee deg(e) \leq deg(d)) \wedge (deg(a) \leq deg(e) \vee deg(c) \leq deg(e))$$

because, by distributing $\vee$ over $\wedge$ we obtain

$$a \leq b \vee c \leq b \vee a \leq d \vee c \leq d$$

therefore $(deg(e) \leq deg(b) \wedge deg(e) \leq deg(d))$ can be ignored and the second condition is sufficient.

2.1 Results

The algorithm has been implemented in Prolog. It is able to build the $PA_c{}^{fuz}$ and also the $\mathcal{V}_{23}{}^{fuz}$ deduced by hand starting from the list of the classic relations $(\mathcal{V}_{23})$ in [6].

For example let consider an algebra that is based on four atomic relations $a, b, c,$ d, therefore having $2^4 = 16$ base relations, and let exist 9 tractable relations

$$\{(a), (b), (c), (d), (a, d), (b, d), (a, b, d), (b, c, d), (a, b, c, d)\}$$

From step 2 we obtain the following five conditions:

$$\{a, b\} \leftarrow deg(d) \geq min\{deg(a), deg(b)\}$$
$$\{b, c\} \leftarrow deg(d) \geq min\{deg(b), deg(c)\}$$
$$\{c, d\} \leftarrow deg(b) \geq min\{deg(c), deg(d)\}$$
$$\{a, b, c\} \leftarrow deg(d) \geq min\{deg(a), deg(b), deg(c)\}$$
$$\{a, c, d\} \leftarrow deg(b) \geq min\{deg(a), deg(c), deg(d)\}$$

From step 3 this one:

$$\{a, c\} \leftarrow min\{deg(a), deg(c)\} \leq max\{deg(b), deg(d)\}$$

the first reduction gives three conditions:

$$\{a, b\} \leftarrow deg(d) \geq min\{deg(a), deg(b)\}$$
$$\{b, c\} \leftarrow deg(d) \geq min\{deg(b), deg(c)\}$$
$$\{c, d\} \leftarrow deg(b) \geq min\{deg(c), deg(d)\}$$

the second reduction in this case is not useful (indeed there is only a relation).

The results are summarized in Table 1. Here for every set the cardinality of the classic base set, the cardinality of the classic subset, the number of conditions found in step 2 (1) and their amount after the first reduction, the number of additional conditions found in step 3 (2) and their amount after the second reduction

Table 1. Statistics on the example algebra

base	set	cnd.	r. cnd.	excl.	r. excl.	s
16	9	5	3 (67%)	1	1 (0%)	0.07

are shown. The computations have been performed on a Pentium IV PC at 2 GHz, and the algorithms implemented in SWI Prolog, SICStus Prolog and YAP Prolog.

A comparison between the third and the sum of the fifth and seventh columns gives an idea of the effectiveness of the rules in identifying the relations and also, in a sense, of the "regularity" of the sub-algebra:

$$\frac{9 - (1 + 3)}{9} = 0.556 \cong 56\%$$

By applying the algorithm to the considered algebrae results in Table 2 are obtained.

Table 2. Statistics on the computations

	base	set	excl.	r. excl.	cnd.	r. cnd.	s
$PA_c{}^{fuz}$	8	7	0	0 (0%)	1	1 (0%)	0.00
$\mathcal{V}_{23}{}^{fuz}$	32	23	0	0 (0%)	9	4 (56%)	0.01
$SA_c{}^{fuz}$	8 192	83	7 872	50 (99%)	237	28 (88%)	3.96
SA^{fuz}	8 192	188	7 368	38 (99%)	636	52 (92%)	7.31
H^{fuz}	8 192	868	4 874	30 (99%)	2 450	36 (98%)	13.75

The ratios are in Table 3.

Table 3. Reductions introduced by the rules

$PA_c{}^{fuz}$	$\mathcal{V}_{23}{}^{fuz}$	$SA_c{}^{fuz}$	SA^{fuz}	H^{fuz}
86%	83%	6%	52%	92%

The fuzzy tractable sub-algebra Alg^{fuz} of a generic fuzzy qualitative algebra Alg is

$$Alg^{fuz} = (Alg, Cnd1, Cnd2)$$

where $Cnd1$ and $Cnd2$ represent the sets of conditions to be checked. $Cnd1$ is built from the conditions found in step 2 and the first reduction step, $Cnd2$ from step 3 and the second reduction step.

3   Equivalent Relations

In order to obtain an upper-bound of the relations to be checked in this section a general formula to obtain the exact number of full-relations is proposed. It is based on equivalence classes because the relations themselves are infinite, being infinite the set of real numbers in the range $[0, 1]$; there is however a finite number of ways to order the relations, to distinguish n relations indeed only n distinct degrees of preference are needed. Besides, as far as tractability is concerned, they can safely abstracted further by considering only "full relations", that is relations with n elements in an algebra with n elements (in this context the term "element" means "atomic relation"). In fact the "full" relations are always tractable, so a relation with $k < n$ elements is equivalent to the following:

$$\{r_1[d_1], \ldots, r_k[d_k], r_{k+1}[d_{k+1}], \ldots, r_{k+2}[d_{k+1}], \ldots, r_n[d_{k+1}]\}$$

with $d_{k+1} < min_{i=1\ldots k}\{d_i\}$.

In qualitative relations the ordering among relations is the only aspect relevant for tractability; therefore, the relations can be ordered using their preference degrees in a decreasing manner and study only the ">" and "=" orderings. To take into account the symmetry of "=", the partitions $P(n)$ of a number n in addenda will be used; the partitions are written as a set of sets, for example $P(2) = \{\{1,1\}, \{2\}\}$ and $P_{1,2} = 1$.

For a given algebra with n elements there are

$$nr = n! \sum_{j=0}^{n-1} \sum_{i=1}^{|P(j)|} \frac{1}{\varphi(P_i(j))} \chi(P_i(j))\mu(P_i(j)) \tag{3}$$

unique full relations to be checked for tractability, where

$$\chi(P_i(j)) = C_{|P_i(j)|}^{n-j} = \frac{(n-j)!}{|P_i(j)|\,(n-j-|P_i(j)|)!}$$

represents the number of combinations of $n - j$ elements in groups of $|P_i(j)|$ elements,

$$\mu(P_i(j)) = \frac{|P_i(j)|!}{\prod_{k=0}^{j-1} \zeta_{P_i(j)}(k)!}$$

is the multinomial of $|P_i(j)|$ elements in $j - 1$ groups of $\zeta_{P_i(j)}(k)$ elements

$$\zeta_{P_i(j)}(k) = |\{x_h : x_h = P_{ih}(j) \wedge P_{ih}(j) = k, h = 1 \ldots |P_i(j)|\}|$$

and

$$\varphi(P_i(j)) = \prod_{k=1}^{|P_i(j)|} (P_{ik}(j) + 1)!$$

counts the equivalent relations.

Applying the formula (3) to the super-algebrae in exam the numbers in Table 4 are obtained.

Table 4. Cardinality of fuzzy full algebrae

alg.	classic rel.	fuzzy rel.
PA^{fuz}	3	13
PI^{fuz}	5	541
IA^{fuz}	13	526 858 348 381

4 Results

In order to verify the rules the actual relations are needed; the enumeration has been performed using Prolog scripts. A first script uses a "brute force" approach that generates exhaustively all the relations and then tests if they are unique; the second one identifies "a priori" the duplicates and prunes them, exploiting the fact that, working with relations ordered by preference degrees, once a prefix has been proven to be intractable all the relations that share the same prefix will be intractable as well, by definition of α-cut [14], so:

$$\{r_1[d_1],\dots,r_k[d_k]\} \text{ intractable} \Leftrightarrow \{r_1[d_1],\dots,r_k[d_k]\}\cup \bigcup_{j:j>k} \{r_j[d_j]\} \text{ intractable}$$

Unfortunately, neither of the two scripts is able to enumerate all the unique full relations for the Allen's algebra, which has thirteen atomic relations, for the number of possible combinations is too high (the formula gives 526 858 348 381, as reported in Table 4). The partial results are in Table 5 and they stop at $n = 8$.

The alternative is to generate just the "right" relations by checking in the pruning step if the prefix belongs to a tractable classic relation; in this way it has been possible to enumerate relations in Table 6 (H^{fuz} is still unreachable).

This last approach is the most effective, but it does not explicitly enumerate all the relations.

Table 5. Pruning speed-ups

n.	permut.	s (spd-up)
3	13	0.00
4	75	0.00
5	541	0.01 (50%)
8	545 835	28.96 (22%)

Table 6. Full tractable fuzzy relations

s-alg.	class. rel.	fuzzy rel.	s
$PA_c{}^{fuz}$	7	10	0.00
$V_{23}{}^{fuz}$	23	290	0.01
$SA_c{}^{fuz}$	83	68 216	34.87
SA^{fuz}	188	915 564	406.5

These relations are relevant because they synthesize the essential expressiveness of the algebrae; in fact even if the fuzzy relations are infinite, due to the infinite degrees of preference, there are only a finite number of distinguishable relations (reported in Table 5), and all the remaining relations could be "grouped" by assigning a degree of priority and capture in this way the infinity. To return to the original equivalence classes it is sufficient to multiply the cardinality by 2, because each full relation includes exactly a partial relation:

$$\{r_1[d_1], \ldots, r_n[d_n]\} : d_n = min_{i=1\ldots n}\{d_i\}$$
$$= \{r_1[d_1], \ldots, r_{n-1}[d_{n-1}]\} \cup I[d_n]$$

where $I[d_n] = \{r_1[d_n], \ldots, r_n[d_n]\}$.

The relations found by applying the rules, by enumeration and by manual verification (when possible, that is for $n < 6$) are identical; therefore the algorithms can reasonably deduce and enumerate the right sets of fuzzy relations from the classic ones, and discover consistent rules.

5   Conclusions

Tractable sub-algebrae are important in order to build solvers able to scale well with the problem size. We are working on a temporal constraint solver that is able to manage qualitative and metric constraints also affected by uncertainty using Possibility Theory [2]. We use convex metric relations, namely trapezoidal distributions, in order to assure the existence of a fix point in the resolution process and the user can choice singleton constraints to obtain a result in a polynomial amount of time. However for the qualitative constraints there is not an easy way to make a problem tractable (obviously a user could constrain the relations to belong to PA, but it would be rather restrictive); the rules proposed in this article are an easy tool for the tractability assessment. Tractable qualitative sub-algebrae can be exploited in two ways: a first idea is to guide the input of the user in order to filter out the constraints not belonging to a particular tractable sub-algebra; a more sophisticated method is to build a Branch & Bound algorithm that partitions the intractable relations in tractable subsets.

A further development could be the application of the method to the QA extended as QA^{fuz} in [2] whose tractable subsets have been identified in [7], and possibly to other tractable sub-algebrae extended by means of Possibility Theory.

References

1. Allen, J.F.: Maintaining knowledge about temporal intervals. Communications of the ACM **26** (1983) 832–843
2. Badaloni, S., Falda, M., Giacomin, M.: Integrating quantitative and qualitative constraints in fuzzy temporal networks. AI Communications **17** (2004) 183–272
3. Badaloni, S., Giacomin, M.: Flexible temporal constraints. In: Proc. of IPMU 2000, Madrid, Spain (2000) 1262–1269
4. Dubois, D., Fargier, H., Prade, H.: Possibility theory in constraint satisfaction problems: Handling priority, preference and uncertainty. Applied Intelligence **6** (1996) 287–309
5. Giacomin, M.: From crisp to fuzzy constraint networks. In: Proc. of CP '01 "Workshop Modelling and Solving Problems with Soft Constraints", Paphos, Cyprus (2001)
6. Jonsson, P., Drakengren, T., Backström, C.: Computational complexity of relating time points with intervals. Artificial Intelligence **109** (1999) 273–295

7. Krokhin, A., Jonsson, P.: Extending the point algebra into the qualitative algebra. In: 9th International Symposium on Temporal Representation and Reasoning (TIME-2002), Manchester, UK, Springer Verlag (2002) 28–35
8. Ligozat, G.: A new proof of tractability for ORD-Horn relations. In: Actes AAAI96: workshop on Spatial and Temporal Reasoning, Portland, Oregon (1996) 395–401
9. Meiri, I.: Combining qualitative and quantitative constraints in temporal reasoning. Artificial Intelligence **87** (1996) 343–385
10. Nebel, B., Bürckert, H.J.: Reasoning about temporal relations: a maximal tractable subclass of Allen's interval algebra. Journal of the ACM **42** (1995) 43–66
11. Vilain, M., Kautz, H., van Beek, P.: Constraint propagation algorithms for temporal reasoning: a revised report. In D. S. Weld, J.d.K., ed.: Readings in Qualitative Reasoning about Physical Systems, San Mateo, CA, Morgan Kaufmann (1989) 373–381
12. van Beek, P., Cohen, R.: Exact and approximate reasoning about temporal relations. Computational Intelligence **6** (1990) 132–144
13. van Beek, P., Manchak, D.W.: The design and experimental analysis of algorithms for temporal reasoning. Journal of Artificial Intelligence Research **4** (1996) 1–18
14. Zadeh, L.A.: Fuzzy sets. Information and Control **8** (1965) 338–353

Reasoning and Quantification in Fuzzy Description Logics

Daniel Sánchez[1] and Andrea G.B. Tettamanzi[2]

[1] University of Granada, Department of Computer Science and Artificial Intelligence,
Periodista Daniel Saucedo Aranda s/n, 18071 Granada, Spain
daniel@decsai.ugr.es
[2] University of Milan, Department of Information Technologies,
Via Bramante 65, I-26013 Crema (CR), Italy
andrea.tettamanzi@unimi.it

Abstract. In this paper we introduce reasoning procedures for $\mathcal{ALCQ}_F^+$, a fuzzy description logic with extended qualified quantification. The language allows for the definition of fuzzy quantifiers of the absolute and relative kind by means of piecewise linear functions on $\mathbb{N}$ and $\mathbb{Q} \cap [0,1]$ respectively. In order to reason about instances, the semantics of quantified expressions is defined based on recently developed measures of the cardinality of fuzzy sets. A procedure is described to calculate the fuzzy satisfiability of a fuzzy assertion, which is a very important reasoning task. The procedure considers several different cases and provides direct solutions for the most frequent types of fuzzy assertions.

1 Introduction

Description logics (DL) [1] are a family of logic-based knowledge-representation formalisms, which stem from the classical AI tradition of semantic networks and frame-based systems. DLs are well-suited for the representation of and reasoning about terminological knowledge, configurations, ontologies, database schemata, etc.

The need of expressing and reasoning with imprecise knowledge and the difficulties arising in classifying individuals with respect to an existing terminology is motivating research on nonclassical DL semantics, suited to these purposes. To cope with this problem, fuzzy description logics have been proposed that allow for imprecise concept description by using fuzzy sets and fuzzy relations. For instance, a fuzzy extension of the description logic $\mathcal{ALC}$ has been introduced in [6], with complete algorithms for solving the entailment problem, the subsumption problem, as well as the best truth-value bound problem.

In [5] we introduced $\mathcal{ALCQ}_F^+$, a fuzzy description logic with extended qualified quantification[1] that allows for the definition of fuzzy quantifiers of the absolute

---

[1] In keeping with DL naming conventions, the superscript plus is to suggest that, in addition to qualified number restrictions available in the description logic $\mathcal{ALCQ}$ introduced by De Giacomo and Lenzerini [2], we provide also more general fuzzy linguistic quantifiers. The subscript F means that the language deals with infinitely many truth-values, as in the language $\mathcal{ALC}_{F_M}$ of Tresp and Molitor [7].

I. Bloch, A. Petrosino, and A.G.B. Tettamanzi (Eds.): WILF 2005, LNAI 3849, pp. 81–88, 2006.

and relative kind by means of piecewise linear functions on $\mathbb{N}$ and $\mathbb{Q} \cap [0,1]$ respectively. These quantifiers extends the usual (qualified) $\exists$, $\forall$ and number restriction.

Incorporating fuzzy quantification into fuzzy description logics is important by several reasons. On the one hand, number restriction is a kind of quantification that arises very frequently in concept description, so it is necessary to extend it to the fuzzy case. But another important reason is that not only concepts, but also quantifiers are imprecise in many cases (e.g. "around two", "most").

For example, suppose you are the marketing director of a supermarket chain. You are about to launch a new line of low-calorie products. In order to set up your budget, you need to project the sales of this new line of products. This can be done either by means of an expensive market research, or by means of some kind of inference based on your knowledge of customer habits. For instance, you could expect prospective buyers of this new line of products to be essentially faithful customers who mostly buy foods with low energy value. We have here all the ingredients of imprecise knowledge: a "faithful customer" is a fuzzy concept; "low" energy value is a linguistic value, which might be modelled as a fuzzy number; to "mostly" buy a given kind of product is equivalent to a quantified statement of the form "most of the bought products are of this kind", where "most" is an imprecise quantifier.

Zadeh [8] showed that imprecise quantifiers can be defined by using fuzzy sets, and by incorporating them into the language and providing the tools to define their semantics we can provide a very powerful knowledge representation tool, with greater expressive power, and closer to the humans' way of thinking.

2   The Language $\mathcal{ALCQ}_F^+$

The language $\mathcal{ALCQ}_F^+$ has the following syntax:

⟨concept_description⟩ ::= ⟨atomic_concept⟩ |
$\top$ | $\bot$ | ¬⟨concept_description⟩ |
⟨concept_description⟩⊓⟨concept_description⟩ |
⟨concept_description⟩⊔⟨concept_description⟩ |
⟨quantification⟩
⟨quantification⟩ ::= ⟨quantifier⟩⟨atomic_role⟩.⟨concept_description⟩
⟨quantifier⟩ ::= "(" ⟨absolute_quantifier⟩ ")" | "(" ⟨relative_quantifier⟩ ")" |
$\exists$ | $\forall$
⟨absolute_quantifier⟩ ::= ⟨abs_point⟩ | ⟨abs_point⟩ + ⟨absolute_quantifier⟩
⟨relative_quantifier⟩ ::= ⟨fuzzy_degree⟩/u | ⟨fuzzy_degree⟩/u + ⟨piecewise_fn⟩
⟨piecewise_fn⟩ ::= ⟨rel_point⟩ | ⟨rel_point⟩ + ⟨piecewise_fn⟩
⟨abs_point⟩ ::= ⟨val⟩/⟨natural_number⟩
⟨rel_point⟩ ::= ⟨val⟩/⟨[0,1]-value⟩
⟨val⟩ ::= [⟨fuzzy_degree⟩ ◁]⟨fuzzy_degree⟩[▷ ⟨fuzzy_degree⟩]

In this extension, the semantics of quantifiers is defined by means of piecewise-linear membership functions. In the case of absolute quantifiers, the quantifier is obtained by restricting the membership function to the naturals. The semantics of fuzzy assertions in $\mathcal{ALCQ}_F^+$ is given by the standard fuzzy conjunction, disjunction and negation, as well as method GD [3] for quantified expressions.

The piecewise-linear functions are defined by means of a sequence of points. These points are expressed as $\alpha \lhd \beta \rhd \gamma / x$, where x is the cardinality value, β is the membership degree of x, and α and γ are the limit when the membership function goes to x from the left and from the right, respectively. When the function is continuous, this can be summarized as β/x (since $\alpha = \beta = \gamma$), whereas discontinuities on the left ($\alpha \neq \beta = \gamma$) or right ($\alpha = \beta \neq \gamma$) can be summarized as $\alpha \lhd \beta/x$ and $\beta \rhd \gamma/x$, respectively.

2.1 An Example

Let us go back to the example of the marketing director of a supermarket chain about to launch a line of low-calorie products.

The knowledge base describing the business of running a supermarket chain could contain, among others, the following terminological axioms:

$$\mathsf{FaithfulCustomer} \sqsubseteq \mathsf{Customer} \sqsubseteq \top$$
$$\mathsf{FoodProduct} \sqsubseteq \mathsf{Product} \sqsubseteq \top$$
$$\mathsf{LowCalorie} \sqsubseteq \mathsf{EnergyMeasure} \sqsubseteq \top$$
$$\mathsf{LowCalorieFood} \equiv \mathsf{FoodProduct} \sqcap \forall \mathsf{energyValue}.\mathsf{LowCalorie}$$

The ABox describing facts about your supermarket chain might contain TVBs which we might summarize as follows:

– given an individual customer c and a product p, $\mathsf{buys}(c, p)$ might be interpreted as
$$\mathsf{buys}(c, p) = f(\mathrm{weeklyrevenue}(c, p)),$$

where $f : \mathbb{R} \to [0, 1]$ is nondecreasing, and $\mathrm{weeklyrevenue}(c, p) : \mathsf{Customer}^{\mathcal{I}} \times \mathsf{Product}^{\mathcal{I}} \to \mathbb{R}$ returns the result of a database query which calculates the average revenue generated by product p on customer c in all the stores operated by the chain;

– given an individual customer c, $\mathsf{FaithfulCustomer}(c)$ might be interpreted as

$$\mathsf{FaithfulCustomer}(c) = g(\mathrm{weeklyrevenue}(c)),$$

where $g : \mathbb{R} \to [0, 1]$ is nondecreasing, and $\mathrm{weeklyrevenue}(c) : \mathsf{Customer}^{\mathcal{I}} \to \mathbb{R}$ returns the result of a database query which calculates the average revenue generated by customer c in all the stores operated by the chain;

– finally, $\mathsf{LowCalorie}(x)$, where x is an average energy value per 100 g of product measured in kJ, could be interpreted as

$$\mathsf{LowCalorie}(x) = \begin{cases} 1 & x < 1000, \\ \frac{2000 - x}{1000} & 1000 \le x \le 2000, \\ 0 & x > 2000. \end{cases}$$

Table 1. The energy value, membership in the LowCalorieFood, and the degree to which customer CARD0400009324198 buys them for a small sample of products

Product	Energy [kJ/hg]	LowCalorieFood($\cdot$)	buys(CARD$\ldots$,$\cdot$)
GTIN8001350010239	1680	0.320	0.510
GTIN8007290330987	1475	0.525	0.050
GTIN8076809518581	1975	0.025	0.572
GTIN8000113004003	1523	0.477	0.210
GTIN8002330006969	498	1.000	1.000
GTIN8005410002110	199	1.000	1.000
GTIN017600081636	1967	0.033	0.184

By using the $\mathcal{ALCQ}_F^+$ language, it is now possible to express the notion of a faithful customer who mostly buys food with low energy value as

$$C \equiv \mathsf{FaithfulCustomer} \sqcap (\mathsf{Most})\mathsf{buys}.\mathsf{LowCalorieFood},$$

where $(\mathsf{Most}) \equiv (0/u + 0/0.5 + 1/0.75)$.

A useful deduction this new axiom allows you to make is, for instance, calculating the extent to which a given individual customer or, more precisely, a fidelity card, say CARD0400009324198, is a C. For instance, you could know that

$$\mathsf{FaithfulCustomer}(\mathsf{CARD0400009324198}) = 0.8,$$

and, by querying the sales database, you might get all the degrees to which that customer buys each product. For sake of example, we give a small subset of those degrees of truth in Table 1, along with the energy values of the relevant products.

According to the semantics of $\mathcal{ALCQ}_F^+$,

$$C(\mathsf{CARD0400009324198}) \approx 0.742$$

i.e., the degree to which most of the items purchased by this customer are low-calorie is around 0.742. This seems to be in accordance with the data in Table 1,

Table 2. Percentage of purchased items that are low-calorie at significant levels

Level	Percentage
1.000	$1.000 = 2/2$
0.572	$0.667 = 2/3$
0.510	$0.500 = 2/4$
0.320	$0.750 = 3/4$
0.210	$0.800 = 4/5$
0.184	$0.667 = 4/6$
0.050	$0.714 = 5/7$
0.033	$0.857 = 6/7$

where we can see that four products (those products p in rows 2, 4, 5, and 6) verify

$$\mathsf{buys}(\mathsf{CARD0400009324198}, p) \leq \mathsf{LowCalorieFood}(p)$$

while for the products in rows 1 and 7 the difference between being purchased and being low-calorie food is not so high. Only the item in row 3 seems to be a clear case of item purchased but not low-calorie.

As another justification of why this result appears in agreement with the data, in Table 2 we show the percentage of purchased items that are low-calorie at α-cuts of the same level. At any other level, the percentage obtained is one of those shown in Table 2.

At many levels the percentage is above 0.75, therefore fitting the concept of Most as we have defined it. At level 0.050 the percentage is almost 0.75. The only level that clearly doesn't fit Most is 0.510, but at the next level (0.320) we have again 0.75 and $\mathsf{Most}(0.75) = 1$.

3   Reasoning with $\mathcal{ALCQ}_F^+$

Of course, the purpose of a knowledge representation system goes beyond storing concept definitions and assertions. A knowledge representation system based on fuzzy DLs should be able to perform specific kinds of reasoning. One particularly important reasoning task is to calculate the fuzzy satisfiability of a fuzzy assertion Ψ, i.e., the interval of values $\mathcal{S}(\Psi) = [\beta_\Psi, \tau_\Psi]$ such that for any interpretation $\mathcal{I}$, the degree of truth of Ψ under $\mathcal{I}$, noted $\mathrm{truth}_\mathcal{I}(\Psi)$, verifies $\mathrm{truth}_\mathcal{I}(\Psi) \in [\beta_\Psi, \tau_\Psi]$ (i.e., the maximum interval $[\beta, \tau]$ such that Ψ is $[\beta, \tau]$–satisfiable in the sense of Navara's definition [4]).

In this work we introduce a PSPACE-complete algorithm to calculate the fuzzy satisfiability of a fuzzy assertion. Though infinite interpretations are taken into account in the definition of the semantics of the language $\mathcal{ALCQ}_F^+$, in practice and due to the physical limitations of computers, we are going to deal with a finite number of individuals. The same limitations put a bound on the number of different membership degrees we can deal with. Therefore we shall calculate the fuzzy satisfiability of a fuzzy assertion up to a certain precision degree, given as a number of decimals p.

In addition to the general algorithm, we have obtained some results showing that we can calculate the fuzzy satisfiability of a fuzzy assertion directly in some (the most common) cases. Some of these results are based on a previous result about *independence* of fuzzy assertions. We introduce the following definition:

Definition 1. *Two concepts A and B are* independent *of each other if, for all $\alpha \in \mathcal{S}(A)$ and $\beta \in \mathcal{S}(B)$, there exists an interpretation $\mathcal{I}$ containing an individual $d \in \Delta_\mathcal{I}$ such that*

$$A^\mathcal{I}(d) = \alpha \quad and \quad B^\mathcal{I}(d) = \beta.$$

In other words, A and B are independent if the degree of truth of A does not affect the degree of truth of B and *vice versa*.

In order to determine whether two concepts are independent, we provide the following results:

Proposition 1. *Two concepts A and B, not containing quantifiers, which are neither tautologies nor contradictions, are independent if and only if the following four concepts are satisfiable in the crisp sense: $A \sqcap B$; $A \sqcap \neg B$; $\neg A \sqcap B$; $\neg A \sqcap \neg B$.*

Proposition 2. *Let D_1 and D_2 be two independent concepts. Then, no atomic concept and no role appears in the expansion of both.*

Proposition 3. *Let D_1 be a concept that doesn't contain quantifiers in its expansion, and let $D_2 \equiv QR.C$. Then, D_1 and D_2 are independent, regardless of whether D_1 and C are independent or not.*

Proposition 4. *Let $D_1 \equiv Q_1 R_1.C_1$ and let $D_2 \equiv Q_2 R_2.C_2$. If $R_1 \neq R_2$ or C_1 and C_2 are independent, then D_1 and D_2 are independent.*

A general procedure to determine whether two concepts are independent can be obtained from Proposition 1 using the crisp procedure to check unsatisfiability. However, Propositions 2, 3, and 4 can make things easier in some cases.

On this basis, we introduce the following results on the calculation of satisfiability of fuzzy assertions:

- If A is an atomic concept, then $\mathcal{S}(A) = [0, 1]$.
- (Negation) If $\mathcal{S}(D) = [\beta_D, \tau_D]$, then $\mathcal{S}(\neg D) = [1 - \tau_D, 1 - \beta_D]$. This verifies $\mathcal{S}(\neg\neg D) = \mathcal{S}(D)$ and:
  - if A is atomic, $\mathcal{S}(\neg A) = \mathcal{S}(A) = [0, 1]$;
  - if $C = C_1 \sqcap \cdots \sqcap C_r$, then $\mathcal{S}(\neg C) = \mathcal{S}(\neg C_1 \sqcup \cdots \sqcup \neg C_r)$;
  - if $C = C_1 \sqcup \cdots \sqcup C_r$, then $\mathcal{S}(\neg C) = \mathcal{S}(\neg C_1 \sqcap \cdots \sqcap \neg C_r)$;
  - if $D = QR.C$,

$$\mathcal{S}(\neg D) = \mathcal{S}(\neg(QR.C)) = \mathcal{S}((\neg Q)R.C). \tag{1}$$

- Let Q be an absolute quantifier such that $\text{core}(Q) \neq \emptyset$ and $\mathbb{N}\backslash\text{supp}(Q) \neq \emptyset$. If $D \equiv QR.C$ and $\mathcal{S}(C) = [\beta_C, \tau_C]$, $\mathcal{S}(D) = \mathcal{S}(QR.C) = [\beta_D, \tau_D]$, with

$$\beta_D = (1 - \tau_C)Q(0) \tag{2}$$
$$\tau_D = \max\{(\tau_C + (1 - \tau_C)Q(0)), Q(0)\} \tag{3}$$

- Let Q be a relative quantifier such that $\text{core}(Q) \neq \emptyset$ and $[0, 1] \cap \mathbb{Q}\backslash\text{supp}(Q) \neq \emptyset$ with $u^Q = Q(x/0)$ (i.e., u^Q is the value returned by the quantifier when the relative cardinality is undefined). If $D \equiv QR.C$ and $\mathcal{S}(C) = [\beta_C, \tau_C]$,
  - If $Q(0) = 0$ and $Q(1) = 1$,

$$\mathcal{S}(D) = [\min\{u^Q, \beta_C\}, \tau_C + (1 - \tau_C)u^Q]$$

In particular, for quantifiers $\exists$ and $\forall$ we have $u^\exists = 0$, $u^\forall = 1$, and

$$\mathcal{S}(\exists R.C) = [0, \tau_C],$$
$$\mathcal{S}(\forall R.C) = [\beta_C, 1].$$

- If $Q(0) = 1$ and $Q(1) = 0$,

$$\mathcal{S}(QR.C) = [1 - (\tau_C + (1 - \tau_C)u^Q), 1 - \min\{u^Q, \beta_C\}].$$

- If $Q(0) = 0$ and $Q(1) = 0$,

$$\mathcal{S}(D) = [0, \max\{u^Q, (1 - \tau_C)u^Q + (\tau_C - \beta_C)\}].$$

- If $Q(0) = 1$ and $Q(1) = 1$,

$$\mathcal{S}(QR.C) = [1 - \max\{u^Q, (1 - \tau_C)u^Q + (\tau_C - \beta_C)\}, 1].$$

The remaining cases are solved by means of an $O(1)$ algorithm with a fixed precision of p decimals.

- If $D \equiv D_1 \sqcup D_2 \sqcup \cdots \sqcup D_s$ with $\mathcal{S}(D_i) = [\beta_{D_i}, \tau_{D_i}]$, then $\mathcal{S}(D) = [\beta_D, \tau_D]$, with

$$\beta_D \geq \max_{i \in \{1,\ldots,s\}} \{\beta_{D_i}\},$$

$$\tau_D = \max_{i \in \{1,\ldots,s\}} \{\tau_{D_i}\}.$$

In particular, if the fuzzy assertions D_i are pairwise independent,

$$\beta_D = \max_{i \in \{1,\ldots,s\}} \{\beta_{D_i}\},$$

$$\tau_D = \max_{i \in \{1,\ldots,s\}} \{\tau_{D_i}\}.$$

The value of β_D with a precision degree of p decimals, when the D_i are not independent, is obtained by means of an algorithm that performs a dichotomic search after guessing values of the atomic concepts and roles that appear in the D_i's.

- The satisfiability of conjunctions can be obtained by using De Morgan's laws and the latter result.

4   Conclusions

$\mathcal{ALCQ}_F^+$ allows for concept description involving fuzzy linguistic quantifiers of the absolute and relative kind, and using qualifiers. We have introduced algorithms to perform two important reasoning tasks with this logic: reasoning about instances, and calculating the fuzzy satisfiability of a fuzzy assertion. In addition, we have defined *independence* of fuzzy assertions and obtained some results that speed up the calculation of fuzzy satisfiability in some (the most common) cases.

References

1. Franz Baader, Diego Calvanese, Deborah McGuinness, Daniele Nardi, and Peter Patel-Schneider, editors. *The Description Logic Handbook: Theory, implementation and applications*. Cambridge, 2003.

2. Giuseppe De Giacomo and Maurizio Lenzerini. A uniform framework for concept definitions in description logics. *Journal of Artificial Intelligence Research*, 6:87–110, 1997.
3. M. Delgado, D. Sánchez, and M.A. Vila. Fuzzy cardinality based evaluation of quantified sentences. *International Journal of Approximate Reasoning*, 23:23–66, 2000.
4. Mirko Navara. Satisfiability in fuzzy logics. *Neural Networks World*, 10(5):845–858, 2000.
5. Daniel Sánchez and Andrea G. B. Tettamanzi. Generalizing quantification in fuzzy description logics. In *Proceedings 8th Dortmund Fuzzy Days*, Dortmund, Germany, 2004.
6. Umberto Straccia. Reasoning within fuzzy description logics. *Journal of Artificial Intelligence Research*, 14:137–166, 2001.
7. Christopher B. Tresp and Ralf Molitor. A description logic for vague knowledge. In *Proceedings of the 13th biennial European Conference on Artificial Intelligence (ECAI'98)*, pages 361–365, Brighton, UK, 1998. J. Wiley and Sons.
8. L. A. Zadeh. A computational approach to fuzzy quantifiers in natural languages. *Computing and Mathematics with Applications*, 9(1):149–184, 1983.

Programming with Fuzzy Logic and Mathematical Functions*

Ginés Moreno and Vicente Pascual

Department of Computer Science,
University of Castilla-La Mancha,
02071 Albacete, Spain
{gmoreno, vpascual}@info-ab.uclm.es

Abstract. This paper focuses on the integration of the (also integrated) declarative paradigms of functional logic and fuzzy logic programming, in order to obtain a richer and much more expressive framework where mathematical functions cohabit with fuzzy logic features. In this sense, this paper must be seen as a first stage in the development of this new research line. Starting with two representative languages from both settings, namely Curry and Likelog, we propose an hybrid dialect where a set of rewriting rules associated to the functional logic dimension of the language, are accompanied with a set of similarity equations between symbols of the same nature and arity, which represents the fuzzy counterpart of the new environment. We directly act inside the kernel of the operational mechanism of the language, thus obtaining a fuzzy variant of *needed narrowing* which fully exploits the similarities collected in a given program. A key point in the design of this last operational method is that, apart from computing at least the same elements of the crisp case, all similar terms of a given goal are granted to be completely treated too while avoiding the risk of infinite loops associated to the intrinsic (reflexive, symmetric and transitive) properties of similarity relations.

Keywords: Fuzzy Logic, Similarity, Functional Logic Programming.

1 Introduction

Logic Programming [12] has been widely used for problem solving and knowledge representation in the past. Nevertheless, traditional logic programming languages do not incorporate techniques or constructs in order to treat explicitly uncertainty and approximated reasoning.

Fuzzy Logic provides a mathematical background for modeling uncertainty and/ or vagueness. Fuzzy logic relays on the concept of fuzzy set. Given a set U, a *fuzzy subset A of U* is a function $A : U \rightarrow [0, 1]$. The function A is called the *membership function*, and the value $A(x)$ represents the *degree of membership* of x in the fuzzy set A. Different functions A can be considered for a fuzzy concept and, in general,

---

* This work has been partially supported by the EU, under FEDER, and the Spanish Science and Education Ministry (MEC) under grant TIN 2004-07943-C04-03.

I. Bloch, A. Petrosino, and A.G.B. Tettamanzi (Eds.): WILF 2005, LNAI 3849, pp. 89–98, 2006.

they will present a soft shape instead of the characteristic function's crisp slope of an ordinary set. A recent introduction on fuzzy logic is [17].

Fuzzy Logic Programming is an interesting and still growing research area that agglutinates the efforts for introducing Fuzzy Logic into Logic Programming. During the last decades, several fuzzy logic programming systems have been developed, where the classical inference mechanism of SLD–Resolution is replaced with a fuzzy variant which is able to handle partial truth and to reason with uncertainty. Most of these systems implement the fuzzy resolution principle introduced by Lee in [10], such as the Prolog–Elf system [9], the Fril Prolog system [4] and the F–Prolog language [11]. Other fuzzy variants of Prolog can be found in [19], [6], and [13].

In general, there is no common method for introducing fuzzy concepts into logic programming. However, we have found two major, and rather different, approaches:

- The first approach, represented by languages as Likelog [3, 18], replaces the syntactic unification mechanism of classical SLD–resolution by a fuzzy unification algorithm, based on similarity relations (over constants and predicates). The fuzzy unification algorithm provides an extended most general unifier as well as a numerical value, called *unification degree*. Intuitively, the unification degree represents the truth degree associated with the (query) computed instance. Programs written in this kind of languages consist, in essence, in a set of ordinary (Prolog) clauses jointly with a set of "similarity equations" which play an important role during the unification process.
- For the second approach, programs are fuzzy subsets of (clausal) formulas, where the *truth degree* of each clause is explicitly annotated. The work of computing and propagating truth degrees relies on an extension of the resolution principle, whereas the (syntactic) unification mechanism remains untouched. Examples of this kind of languages are the ones described in [19], [6] and [13].

In this paper we are specially interested in the first class of fuzzy logic languages explained before, as well as in modern functional logic languages, such as Curry. Functional logic programming languages combine the operational principles of the most important declarative programming paradigms, namely functional and logic programming (see [7] for a survey). The operational semantics of such integrated languages is usually based on narrowing, a combination of variable instantiation and reduction, where efficient demand-driven functional computations are amalgamated with the flexible use of logical variables providing for function inversion and search for solutions.

The structure of this paper is as follows. The following section is devoted to detail the main features of both fuzzy logic programming and functional logic programming. In Section 3 we explain in detail the original *needed narrowing* strategy used by Curry, in order to furthermore extend it with similarity relations in Section 4. Finally, we give our conclusions in Section 5.

2 Fuzzy Logic and Functional Logic Programming

Similarity relation is a mathematical notion strictly related with equivalence relations and, then, to closure operators, that provides a way to manage alternative instances of an entity that can be considered "equal" with a given degree [20]. Let us recall that a T-norm $\wedge$ in $[0, 1]$ is a binary operation $\wedge : [0, 1] \times [0, 1] \rightarrow [0, 1]$ associative, commutative, nondecreasing in both the variables, and such that $x \wedge 1 = 1 \wedge x = x$ for any $x \in [0, 1]$. In order to simplify our developments, and similarly to other approaches in fuzzy logic programming [18], in the sequel, we assume that $x \wedge y$ is the minimum between the two elements $x, y \in [0, 1]$.

Definition 1. *A similarity relation $\Re$ on a domain $\mathcal{U}$ is a fuzzy subset $\Re :$ $\mathcal{U} \times \mathcal{U} \rightarrow [0, 1]$ of $\mathcal{U} \times \mathcal{U}$ such that the following properties hold:*

1. *Reflexivity: $\Re(x, x) = 1$, $\forall x \in \mathcal{U}$*
2. *Symmetry: $\Re(x, y) = \Re(y, x)$, $\forall x, y \in \mathcal{U}$*
3. *Transitivity: $\Re(x, z) \geq \Re(x, y) \wedge \Re(y, z)$, $\forall x, y, z \in \mathcal{U}$.*

A very simple, but effective way, to introduce similarity relations into pure logic programming, generating one of the most promising ways for the integrated paradigm of fuzzy logic programming, consists in modeling them by a set of the so-called *similarity equations* of the form $eq(s1, s2) = \alpha$, whith the intended meaning that $s1$ and $s2$ are predicate/function symbols of the same arity with a similarity degree α. This approach is followed, for instance, in the fuzzy logic language Likelog [3], where a set of usual Prolog clauses are accompanied by a set of similarity equations which play an important role at (fuzzy) unification time. Of course, the set of similarity equations is assumed to be safe in the sense that each equation connects two symbols of the same arity and nature (both predicates or both functions) and the properties of Definition 1 are not violated, as occurs, for instance, with the wrong set $\{eq(a, b) = 0.5,\ eq(b, a) = 0.9\}$ which, apart for introducing risks of infinite loops when treated computationally, in particular, it does not verify the symmetric property.

Example 1. Given the following Prolog program, composed by a single clause with empty body (fact) $p(a)$, then the evaluation of goal $p(X)$ gives the computed answer (substitution) $X \mapsto a$, with the intended meaning that, since constant a satisfies predicate p in the program, then when variable X is linked to a in the proposed goal, then it is also satisfied.

However, if we add now the similarity equation $eq(a, b) = 0.5$ to the previous fact, then we obtain a Likelog program for which the evaluation of goal $p(X)$ gives two computed answers (incorporating now to the corresponding Prolog-like substitution, the associated truth degree): $X \mapsto a$ with truth degree 1, and $X \mapsto b$ with truth degree 0.5.

It is important to note, that since Likelog is oriented to manipulate inductive databases, where no function symbols of arity greater than 0 are allowed, then, similarity equations only consider similarities between two predicates or two constants (that is, function symbols with no parameters) of the same arity. In this paper, we drop out this last limitation by also allowing similarity equations between any pair of (defined or constructor) function symbols which do not necessarily be constants. Moreover, since our base language does not treat with proper predicate symbols, we allow similarity equations between boolean (among any other kind of defined or constructor function) symbols, which is a quite natural way to model predicates in functional logic languages.

In the following, we assume that the intended similarity relation $\Re$ associated to a given program $\mathcal{R}$, is induced from the (safe) set of similarity equations of $\mathcal{R}$, verifying that the similarity degree of two symbols s_1 and s_2 is 1 if $s_1 \equiv s_2$ or, otherwise, it is defined recursively as the transitive closure of the equational set defined as: $T_r^t(R) = \bigcup_{f=1\ldots\infty} R^f$ *where* $R^{f+1} = R^f \circ^t R$, for a given T-norm t. Moreover, it can be demonstrated that, if the domain is a finite set with n elements, then only n-1 powers must be calculated. Finally, by simply assuming that the set of similarity equations in $\mathcal{R}$ is trivially extended by reflexivity, then $\Re = T_r(R) = R^{(n-1)}$ [5].

Example 2. In the following pair of matrixes, we are considering similarities between four arbitrary constant symbols. The second matrix has been obtained after applying the algorithm described in [5] to the first one.

$$\begin{pmatrix} 1 & .7 & .6 & .4 \\ .7 & 1 & .8 & .9 \\ .6 & .8 & 1 & .7 \\ .4 & .9 & .7 & 1 \end{pmatrix} \xrightarrow[\text{Similarity}]{R} \begin{pmatrix} 1 & .7 & .7 & .7 \\ .7 & 1 & .8 & .9 \\ .7 & .8 & 1 & .8 \\ .7 & .9 & .8 & 1 \end{pmatrix}$$

In what follows, we propose the combined use of similarity equations together with rewriting rules (instead of Horn clauses) typically used in pure functional (Haskell) and integrated functional–logic (Curry) languages.

We consider a *signature* Σ partitioned into a set $\mathcal{C}$ of *constructors* and a set $\mathcal{F}$ of *defined* functions. The set of *constructor terms* (with *variables*) is obtained by using symbols from $\mathcal{C}$ (and a set of variables $\mathcal{X}$). The set of variables occurring in a term t is denoted by $Var(t)$. We write $\overline{o_n}$ for the *list* of objects $o_1, \ldots, o_n$. A *pattern* is a term of the form $f(\overline{d_n})$ where $f/n \in \mathcal{F}$ and $d_1, \ldots, d_n$ are constructor terms (with variables). A term is *linear* if it does not contain multiple occurrences of one variable. A *position* p in a term t is represented by a sequence of natural numbers (Λ denotes the empty sequence, i.e., the root position). Positions are ordered by the *prefix* ordering: $p \le q$, if $\exists w$ such that $p.w = q$. $t|_p$ denotes the *subterm* of t at a given position p, and $t[s]_p$ denotes the result of *replacing the subterm* $t|_p$ by the term s. We denote by $\{x_1 \mapsto t_1, \ldots, x_n \mapsto t_n\}$ the *substitution* σ with $\sigma(x_i) = t_i$ for $i = 1, \ldots, n$ (with $x_i \ne x_j$ if $i \ne j$), and $\sigma(x) = x$ for all other variables x. The application of a substitution σ to an expression e is denoted by $\sigma(e)$. The empty substitution is denoted by id.

A set of rewriting rules $l \to r$ such that $l \notin \mathcal{X}$, and $Var(r) \subseteq Var(l)$ is called a *term rewriting system* (TRS). The terms l and r are called the *left-hand side* (lhs) and the *right-hand side* (rhs) of the rule, respectively. A TRS $\mathcal{R}$ is left-linear if l is linear for all $l \to r \in \mathcal{R}$. A TRS is constructor–based (CB) if each left-hand side is a pattern. In the remainder of this paper, a functional logic *program* is a left-linear CB-TRS without overlapping rules (i.e. the lhs's of two different program rules do not unify).

A *rewrite step* is an application of a rewriting rule to a term, i.e., $t \rightarrow_{p,R} s$ if there exists a position p in t, a rewriting rule $R = (l \rightarrow r)$ and a substitution σ with $t|_p = \sigma(l)$ and $s = t[\sigma(r)]_p$. The operational semantics of modern integrated languages is usually based on *(needed) Narrowing*, which can be seen as a combination of variable instantiation and reduction. Formally, $s \rightsquigarrow_{p,R,\sigma} t$ is a *narrowing step* if p is a non-variable position in s and $\sigma(s) \rightarrow_{p,R} t$. We denote by $t_0 \rightsquigarrow^*_\sigma t_n$ a sequence of narrowing steps $t_0 \rightsquigarrow_{\sigma_1} \ldots \rightsquigarrow_{\sigma_n} t_n$ with $\sigma = \sigma_n \circ \cdots \circ \sigma_1$. Needed narrowing is currently an optimal, correct/complete strategy for modern, first-order, lazy functional logic languages and it its the operational basis of the language Curry [2].

3   Needed Narrowing

A challenge in the design of functional logic languages is the definition of a "good" narrowing strategy, i.e., a restriction on the narrowing steps issuing from a term without losing completeness. *Needed narrowing* [2] is currently the best known narrowing strategy due to its optimality properties w.r.t. the length of the derivations and the number of computed solutions. It extends Huet and Lévy's notion of a needed reduction [8].

The definition of needed narrowing [2] uses the notion of a *definitional tree* [1], which refines the standard matching trees of functional programming. However, differently from left-to-right matching trees used in either Hope, Miranda, or Haskell, definitional trees deal with *dependencies* between arguments of functional patterns. Roughly speaking, a definitional tree for a function symbol f is a tree whose leaves contain all (and only) the rules used to define f and whose inner nodes contain information to guide the (optimal) pattern matching during the evaluation of expressions. Each inner node contains a pattern and a variable position in this pattern (the *inductive position*) which is further refined in the patterns of its immediate children by using different constructor symbols. The pattern of the root node is simply $f(\overline{x_n})$, where $\overline{x_n}$ are different variables.

A defined function is called *inductively sequential* if it has a definitional tree. A rewrite system $\mathcal{R}$ is called *inductively sequential* if all its defined functions are inductively sequential. An inductively sequential TRS can be viewed as a set of definitional trees, each one defining a function symbol. There can be more than one definitional tree for an inductively sequential function. In the following, we assume that there is a fixed definitional tree for each defined function.

Example 3. It is often convenient and simplifies the understanding to provide a graphic representation of definitional trees, where each node is marked with a pattern, the inductive position in branches is surrounded by a box, and the leaves contain the corresponding rules. For instance, given the following program (right column) defining functions "f" and "g", the definitional trees for both function symbols can be depicted as follows:

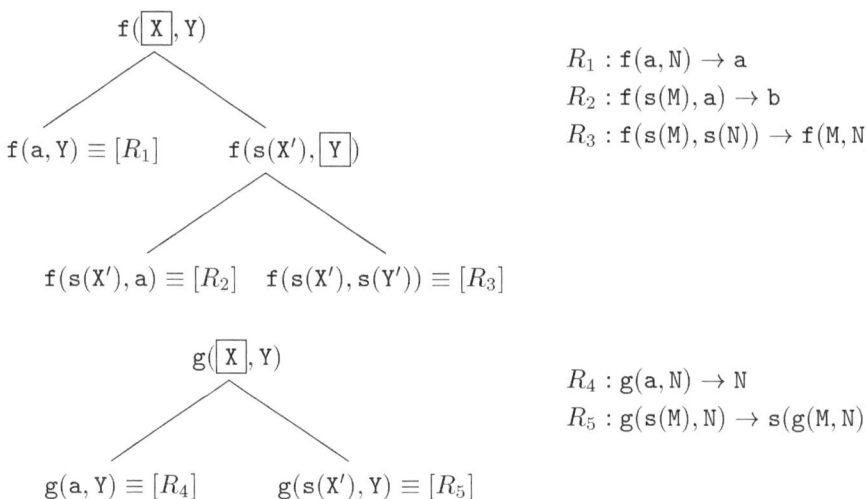

$$R_1 : f(a, N) \to a$$
$$R_2 : f(s(M), a) \to b$$
$$R_3 : f(s(M), s(N)) \to f(M, N)$$

$$R_4 : g(a, N) \to N$$
$$R_5 : g(s(M), N) \to s(g(M, N))$$

For the definition of needed narrowing, we assume that $t \equiv f(\overline{s_n})$ is an operation-rooted term and $\mathcal{P}_f$ is a definitional tree with root π such that $\pi \leq t$. Hence, when π is a leaf, i.e., $\mathcal{P}_f = \{\pi\}$, we have that $R : \pi \to r$ is a variant of a rewriting rule. On the other hand, if π is a branch, we consider the inductive position o of π and we say that the pattern $\pi_i \equiv \pi[c_i(\overline{x_n})]_o \in \mathcal{P}_f$, is a child of π in $\mathcal{P}_f$. Moreover, the definitional (sub-)tree of $\mathcal{P}_f$ rooted with π_i, (i.e., where all patterns are instances of π_i) is denoted by $\mathcal{P}_f^{\pi_i} = \{\pi' \in \mathcal{P}_f \mid \pi_i \leq \pi'\}$, whereas $children(\mathcal{P})$ refers to the set of children of the root of a definitional (sub)-tree $\mathcal{P}$. For instance, in the previous definitional tree for f, the children of $\mathcal{P}_f$ (or, equivalently $\mathcal{P}_f^{f(X,Y)}$) is the set $children(\mathcal{P}_f) = \{f(a, Y), f(s(X'), Y)\}$, whereas $children(\mathcal{P}_f^{f(s(X'),Y)}) = \{f(s(X'), a), f(s(X'), s(Y'))\}$.

We define now a function λ_{crisp} from terms to sets of tuples (position, rule, substitution) which uses an auxiliary function λ for explicitly referring to the appropriate definitional tree in each case. Then, $\lambda_{crisp}(t) = \lambda(t, \mathcal{P}_f)$ returns the least set satisfying:

LR (LEAF-RULE) CASE: $\lambda(t, \mathcal{P}_f) = \{(\Lambda, R, id)\}$

BV (BRANCH-VAR) CASE: If $t|_o = x \in \mathcal{X}$, then $\lambda(t, \mathcal{P}_f) = \{(p, R, \sigma \circ \tau) \mid \pi_i \equiv \pi[c_i(\overline{x_n})]_o \in children(\mathcal{P}_f)$ and $\tau = \{x \mapsto c_i(\overline{x_n})\}$ and $(p, R, \sigma) \in \lambda(\tau(t), \mathcal{P}_f^{\pi_i})\}$

BC (BRANCH-CONS) CASE: If $t|_o = c_i(\overline{t_n})$, where $c_i \in \mathcal{C}$, then $\lambda(t, \mathcal{P}_f) = \{(p, R, \sigma \circ id) \mid \pi_i \equiv \pi[c_i(\overline{x_n})]_o \in children(\mathcal{P}_f)$ and $(p, R, \sigma) \in \lambda(t, \mathcal{P}_f^{\pi_i})\}$

BF (BRANCH-FUNC) CASE: If $t|_o = g(\overline{t_n})$, where $g \in \mathcal{F}$, then $\lambda(t, \mathcal{P}_f) = \{(o.p, R, \sigma \circ id) \mid (p, R, \sigma) \in \lambda(t|_o, \mathcal{P}_g)\}$

When none of the previous cases is satisfied, we assume that function λ returns $\emptyset$. Informally speaking, needed narrowing directly applies a rule if the term is an instance of some left-hand side (LR case), or checks the subterm corresponding to the inductive position of the branch: if it is a variable (BV case), it is instantiated to the constructor of each one of the children; if it is already a constructor (BC case), we proceed with the corresponding child; if it is a function (BF case), we evaluate it by recursively applying needed narrowing.

In contrast to more traditional narrowing strategies, needed narrowing does not compute *most general* unifiers. In each recursive step during the computation of λ, we compose the current substitution with the local substitution of this step (which can be the identity *id*). As in proof procedures for logic programming, we assume that definitional trees always contain new variables if they are used in a narrowing step. Then, $t \leadsto_{\sigma}^{p,R} t'$ is a *needed narrowing step* for all $(p, R, \sigma) \in \lambda_{crisp}(t)$.

Example 4. If we consider again the rules for f and g in Example 3 then we have $\lambda_{crisp}(\mathtt{f(X,g(X,X))}) = \{(\Lambda, R_1, \{\mathtt{X} \mapsto \mathtt{a}\}),\ (2, R_5, \{\mathtt{X} \mapsto \mathtt{s(X')}\})\}$ which enables the following pair of needed narrowing steps (for readability reasons, in our examples we underline the subterm exploited in each narrowing step and, when needed, we also accompany it with a superindex/subindex representing respectively its associated rule/binding):

$$\underline{\mathtt{f(X,g(X,X))}} \qquad \leadsto_{\{\mathtt{X} \mapsto \mathtt{a}\}}^{\Lambda, R_1} \qquad \mathtt{a}$$

$$\mathtt{f(X,\underline{g(X,X)})} \qquad \leadsto_{\{\mathtt{X} \mapsto \mathtt{s(X')}\}}^{2, R_5} \qquad \mathtt{f(s(X'), s(g(X', s(X'))))}$$

4   Extending Needed Narrowing with Similarity Relations

In this section we define a fuzzy variant of needed narrowing with similarity relations. It is important to note that the approach presented here largely improves the older one presented in [16], where similarities were exploited in a much more limited form by simply relaxing the notion of "strict equality" (typical of lazy functional languages) to the richer one of "similar equality".

We extend the notion of computed answer for also reporting now (apart for the classical components of substitution and value), a real number in the interval $[0, 1]$ indicating the similarity degree computed along the corresponding derivation. Hence, we re-define function λ from terms and definitional trees to sets of tuples (position, rule, substitution, similarity_degree). If $t \equiv f(\overline{s_n})$ is the operation-rooted term we consider for being processed, in the initial call to λ_{fuzzy}, we must guarantee that any term (including t itself), rooted with a symbol similar to f be will be treated. So,

$$\lambda_{fuzzy}(t) = \{(p, R, \sigma, min(\alpha, \beta)) \mid \Re(f,g) = \alpha \text{ and } (p, R, \sigma, \beta) \in \lambda(g(\overline{s_n}), \mathcal{P}_g\}$$

and now, the extended definition for λ is:

LR (LEAF-RULE) CASE: $\lambda(t, \mathcal{P}_f) = \{(\Lambda, \pi \rightarrow r, id, 1)\}$

BV (BRANCH-VAR) CASE: if $t|_o = x$, where $x \in \mathcal{X}$, then $\lambda(t, \mathcal{P}_f) =$
   $\{(p, R, \sigma \circ \tau, \alpha) \mid \pi_i \equiv \pi[c_i(\overline{x_n})]_o \in children(\mathcal{P}_f)$ and
   $\tau = \{x \mapsto c_i(\overline{x_n})\}$ and $(p, R, \sigma, \alpha) \in \lambda(\tau(t), \mathcal{P}_f^{\pi_i})\}$

BC (BRANCH-CONS) CASE: if $t|_o = d(\overline{t_n})$, where $d \in \mathcal{C}$, then $\lambda(t, \mathcal{P}_f) =$
   $\{(p, R, \sigma, min(\alpha, \beta)) \mid \Re(d, c_i) = \alpha$ and $\pi_i \equiv \pi[c_i(\overline{x_n})]_o \in children(\mathcal{P}_f)$
   and $(p, R, \sigma, \beta) \in \lambda((t[c_i(\overline{x_n})]_o, \mathcal{P}_f^{\pi_i}))\}$

BF (BRANCH-FUNC) CASE: if $t|_o = g(\overline{t_n})$, where $g \in \mathcal{F}$, then $\lambda(t, \mathcal{P}_f) =$
   $\{(o.p, R, \sigma, min(\alpha, \beta) \mid \Re(g, h) = \alpha$ and $(p, R, \sigma, \beta) \in \lambda(h(\overline{t_n}), \mathcal{P}_h\}$

As we can see, LR and BV cases are very similar to the corresponding ones presented in Section 3, but propagating now the corresponding similarity degrees. Moreover, closely related to the initial call to λ_{fuzzy} seen before, the last case (BF) performs recursive calls to λ for evaluating the operation–rooted subterm at the considered inductive position, as well as all other (almost identical) subterms rooted with defined function symbols similars to g. Something almost identical occurs with the BC case, but the intention now is to treat all subterms whose constructor symbols at the inductive position are similars to d.

Example 5. Consider again the same program of Example 4 augmented with the new rule $R_6 : \texttt{h(r(X),Y)} \rightarrow \texttt{r(Y)}$ together with the similarity equations $S_1 : eq(g, h) = 0.7$ and $S_2 : eq(s, r) = 0.5$. Then, $\lambda_{fuzzy}(\texttt{f(X,g(X,X))}) = \lambda(\texttt{f(X,g(X,X))}, \mathcal{P}_f) = $ [see BV1] $\cup$ [see BV2] $= \{(\Lambda, R_1, \{\texttt{X} \mapsto \texttt{a}\}, 1), (2, R_5, \{\texttt{X} \mapsto \texttt{s(X')}\}, 1), (2, R_6, \{\texttt{X} \mapsto \texttt{s(X')}\}, min(0.7, 0.5))\}$.

BV1 The first alternative in this BV case, consists in generating the binding $\tau_1 = \{\texttt{X} \mapsto \texttt{a}\}$ and then computing $\lambda(\tau_1(\texttt{f(X,g(X,X))}), \mathcal{P}_f^{\texttt{f(a,Y)}}) = \lambda(\texttt{f(a,g(a,a))}, \mathcal{P}_f^{\texttt{f(a,Y)}})$. Since this last call represents a LR case, it returns $\{(\Lambda, R_1, id, 1)\}$. Then, after applying the binding τ_1 to the third element of this last tuple, the returned set for this case is $\{(\Lambda, R_1, \{\texttt{X} \mapsto \texttt{a}\}, 1)\}$.

BV2 After generating the second binding $\tau_2 = \{\texttt{X} \mapsto \texttt{s(X')}\}$, we must compute $\lambda(\tau_2(\texttt{f(X,g(X,X))}), \mathcal{P}_f^{\texttt{f(s(X'),Y)}}) = \lambda(\texttt{f(s(X'),g(s(X'),s(X')))}, \mathcal{P}_f^{\texttt{f(s(X'),Y)}}) = $ [see BF1]$= \{(2, R_5, id, 1), (2, R_6, id, min(0.7, 0.5))\}$. Now, we simply need to apply τ_2 to the last component of the tuples obtained in BF1, hence returning $\{(2, R_5, \{\texttt{X} \mapsto \texttt{s(X')}\}, 1), (2, R_6, \{\texttt{X} \mapsto \texttt{s(X')}\}, min(0.7, 0.5))\}$.

BF1 In this BF case, where the considered inductive position is 2, we perform the following two recursive calls (observe that the second one exploit the similarity equation S_1): $\lambda(\texttt{g(s(X'),s(X'))}, \mathcal{P}_g) \cup \lambda(\texttt{h(s(X'),s(X'))}, \mathcal{P}_h) = $ [see BC1] $\cup$ [see BC2] $= \{(\Lambda, R_5, id, 1), (\Lambda, R_6, id, 0.5)\}$. And then, since obviously position $\Lambda.2$ coincides directly with position 2, and the similarity between $\texttt{g}$ and $\texttt{h}$ is 0.7, the set returned in this case is $\{(2, R_5, id, 1), (2, R_6, id, min(0.7, 0.5))\}$.

BC1: This BC case, immediately evolves to the following LR case $\lambda(\mathtt{g}(\mathtt{s}(\mathtt{X}'),\mathtt{s}(\mathtt{X}'))$, $\mathcal{P}_{\mathtt{g}}^{\mathtt{g}(\mathtt{s}(\mathtt{M}),\mathtt{N})}) = \{(\Lambda, R_5, id, 1)\}$. Now, since $\Re(\mathtt{s},\mathtt{s}) = 1$, and $min(1,1) = 1$, the returned tuple in this case is $(\Lambda, R_5, id, 1)$ itself.

BC2: By exploiting the second similarity equation $S_2 : \mathtt{eq}(\mathtt{s},\mathtt{r}) = 0.5$, this BC case also computes the LR case $\lambda(\mathtt{h}(\mathtt{r}(\mathtt{X}'),\mathtt{s}(\mathtt{X}'))$, $\mathcal{P}_{\mathtt{h}}^{\mathtt{h}(\mathtt{r}(\mathtt{X}),\mathtt{Y})}) = \{(\Lambda, R_6, id, 1)\}$. Finally, since $min(0.5,1) = 0.5$, then $\lambda(\mathtt{h}(\mathtt{s}(\mathtt{X}'),\mathtt{s}(\mathtt{X}'))$, $\mathcal{P}_{\mathtt{h}}) = \{(\Lambda, R_6, id, 0.5)\}$.

Now, since $\lambda_{fuzzy}(\mathtt{f}(\mathtt{X},\mathtt{g}(\mathtt{X},\mathtt{X}))) = \{(\Lambda, R_1, \{\mathtt{X} \mapsto \mathtt{a}\}, 1), (2, R_5, \{\mathtt{X} \mapsto \mathtt{s}(\mathtt{X}')\}, 1),$ $(2, R_6, \{\mathtt{X} \mapsto \mathtt{s}(\mathtt{X}')\}, min(0.7, 0.5))\}$, then we can apply the three following derivation steps (note that similarity degrees corresponds to the second sub-index of each narrowing step):

$$\underline{\mathtt{f}(\mathtt{X},\mathtt{g}(\mathtt{X},\mathtt{X}))} \qquad \leadsto^{\Lambda, R_1}_{\{\mathtt{X} \mapsto \mathtt{a}\}, 1} \qquad \mathtt{a}$$

$$\mathtt{f}(\mathtt{X},\underline{\mathtt{g}(\mathtt{X},\mathtt{X})}) \qquad \leadsto^{2, R_5}_{\{\mathtt{X} \mapsto \mathtt{s}(\mathtt{X}')\}, 1} \qquad \mathtt{f}(\mathtt{s}(\mathtt{X}'),\mathtt{s}(\mathtt{g}(\mathtt{X}',\mathtt{s}(\mathtt{X}'))))$$

$$\mathtt{f}(\mathtt{X},\underline{\mathtt{g}(\mathtt{X},\mathtt{X})}) \qquad \leadsto^{2, R_6}_{\{\mathtt{X} \mapsto \mathtt{s}(\mathtt{X}')\}, 0.5} \qquad \mathtt{f}(\mathtt{s}(\mathtt{X}'),\mathtt{r}(\mathtt{s}(\mathtt{X}')))$$

As our example reveals, there are two important properties enjoyed by our extended definition of needed narrowing:

- λ_{fuzzy} is conservative w.r.t. λ_{crisp} since, the first two tuples computed before are the same to those ones obtained in the crisp case (see example 4), but accompanied now with the maximum truth degree 1, and
- moreover, similarity equations between defined/constructor function symbols are exploited as much as possible (in the initial call and BF/BC cases), which is the key point to obtain the third tuple in our example, by exploiting the two similarity equations collected in our program.

5   Conclusions and Future Work

In this paper we have highlighted the main problems appearing when introducing fuzziness into functional logic programming. Focusing on practical aspects, we have been mainly concerned with syntactic and operational subjects, also providing representative examples and discussions about our all contributions.

The next step in our ongoing research is centered in the development of a theoretical framework defining a declarative semantics for the new paradigm and the corresponding correctness/completenes proofs for our approach. Moreover, we are also planning to investigate other integration lines, such as those based on SLDE-resolution, programs with weighted clauses, and so on.

Finally, since in our research group we have experience in the development of sophisticated (fold/unfold based) optimizations tools for functional logic and fuzzy logic programs (see [14, 15], respectively), for the future we are also planning to adapt these techniques to the new setting proposed in this paper.

References

1. S. Antoy. Definitional trees. In *Proc. of the 3rd Int'l Conference on Algebraic and Logic Programming, ALP'92*, pages 143–157. Springer LNCS 632, 1992.
2. S. Antoy, R. Echahed, and M. Hanus. A needed narrowing strategy. In *Journal of the ACM*, volume 47(4), pages 776–822, 2000.
3. F. Arcelli and F. Formato. Likelog: A logic programming language for flexible data retrieval. In *Proc. of the ACM Symposium on Applied Computing (SAC'99)*, pages 260–267. ACM, Artificial Intelligence and Computational Logic, 1999.
4. J. F. Baldwin, T. P. Martin, and B. W. Pilsworth. *Fril- Fuzzy and Evidential Reasoning in Artificial Intelligence*. John Wiley &; Sons, Inc., 1995.
5. L. Garmendia and A. Salvador. Comparing transitive closure with other new T-transivization methods. In *Proc. Modeling Decisions for Artificial Intelligence*, pages 306–315. Springer LNAI 3131, 2004.
6. S. Guadarrama, S. Muñoz, and C. Vaucheret. Fuzzy Prolog: A new approach using soft constraints propagation. *Fuzzy Sets and Systems, Elsevier*, 144(1):127–150, 2004.
7. M. Hanus. The Integration of Functions into Logic Programming: From Theory to Practice. *Journal of Logic Programming*, 19&20:583–628, 1994.
8. G. Huet and J.J. Lévy. Computations in orthogonal rewriting systems, Part I + II. In J.L. Lassez and G.D. Plotkin, editors, *Computational Logic – Essays in Honor of Alan Robinson*, pages 395–443. The MIT Press, Cambridge, MA, 1992.
9. M. Ishizuka and N. Kanai. Prolog-ELF Incorporating Fuzzy Logic. In Aravind K. Joshi, editor, *Proceedings of the 9th International Joint Conference on Artificial Intelligence (IJCAI'85). Los Angeles, CA, August 1985.*, pages 701–703. Morgan Kaufmann, 1985.
10. R.C.T. Lee. Fuzzy Logic and the Resolution Principle. *Journal of the ACM*, 19(1):119–129, 1972.
11. Deyi Li and Dongbo Liu. *A fuzzy Prolog database system*. John Wiley & Sons, Inc., 1990.
12. J.W. Lloyd. *Foundations of Logic Programming*. Springer-Verlag, Berlin, 1987. Second edition.
13. J. Medina, M. Ojeda-Aciego, and P. Vojtáš. Similarity-based unification: a multi-adjoint approach. *Fuzzy Sets and Systems*, 146(1):43–62, 2004.
14. G. Moreno. Transformation Rules and Strategies for Functional-Logic Programs. *AI Communications, IO Press (Amsterdam)*, 15(2):3, 2002.
15. G. Moreno. Building a Fuzzy Transformation System. *Proc. of the 32nd Conference on Current Trends in Theory and Practice of Computer Science, SOFSEM'06*, 10 pages. Springer LNCS 3831 (to appear), 2006.
16. G. Moreno and V. Pascual. Functional Logic Programming with Similarity. In F. López-Fraguas, editor, *Proc. of the V Jornadas sobre Programación y Lenguajes, PROLE'05*, pages 121–126. University of Granada, 2005.
17. H.T. Nguyen and E.A. Walker. *A First Course in Fuzzy Logic*. Chapman & Hall/CRC, Boca Ratón, Florida, 2000.
18. M.I. Sessa. Approximate reasoning by similarity-based SLD resolution. *Fuzzy Sets and Systems*, 275:389–426, 2002.
19. P. Vojtáš and L. Paulík. Soundness and completeness of non-classical extended SLD-resolution. In R. Dyckhoff et al, editor, *Proc. ELP'96 Leipzig*, pages 289–301. LNCS 1050, Springer Verlag, 1996.
20. L. A. Zadeh. Similarity relations and fuzzy orderings. *Informa. Sci.*, 3:177–200, 1971.

Efficient Methods for Computing Optimality Degrees of Elements in Fuzzy Weighted Matroids

Jérôme Fortin[1], Adam Kasperski[2], and Paweł Zieliński[3]

[1] IRIT/UPS 118 route de Narbonne, 31062, Toulouse, cedex 4, France
fortin@irit.fr
[2] Institute of Industrial Engineering and Management, Wrocław University of
Technology, Wybrzeże Wyspiańskiego 27, 50-370 Wrocław, Poland
adam.kasperski@pwr.wroc.pl
[3] Institute of Mathematics and Computer Science, Wrocław University of Technology,
Wybrzeże Wyspiańskiego 27, 50-370 Wrocław, Poland
pziel@im.pwr.wroc.pl

Abstract. In this paper some effective methods for calculating the exact degrees of possible and necessary optimality of an element in matroids with ill-known weights modeled by fuzzy intervals are presented.

1 Introduction

In combinatorial optimization problems, we are given a set of elements E and a weight w_e is associated with each element $e \in E$. We seek an object composed of the elements of E for which the total weight is maximal (minimal). In the deterministic case the elements of E can be divided into two groups: those which belong to an optimal solution (*optimal elements*) and those which do not belong to an optimal one. In this paper, we consider the case in which the weights are imprecise and they are modeled by the classical intervals and fuzzy intervals. In the interval-valued case the elements form three groups: those that are optimal for sure (*necessarily optimal* elements), those that are not optimal for sure and the elements whose optimality is unknown (*possibly optimal* elements). In the fuzzy-valued case the weights of elements are modeled by *possibility distributions*[1]. In this case the notions of possible and necessary optimality can be extended and every element can be characterized by degrees of possible and necessary optimality.

In this paper, we wish to investigate the combinatorial optimization problem, which can be formulated on a *matroid* (a good introduction to matroids can be found in [4]). The case of interval-valued weights is first addressed and it is then extended to fuzzy intervals. The main results of this paper are two effective algorithms, based on *profile approach* [2], for calculating the exact values of degrees of possible and necessary optimality of a given element.

2 Preliminaries

Consider a system $(E, \mathcal{I})$, where $E = \{e_1, \dots, e_n\}$ is a nonempty ground set and $\mathcal{I}$ is a collection of subsets of E closed under inclusion, i.e. if $B \in \mathcal{I}$ and

I. Bloch, A. Petrosino, and A.G.B. Tettamanzi (Eds.): WILF 2005, LNAI 3849, pp. 99–107, 2006.

$A \subseteq B$ then $A \in \mathcal{I}$. The system $(E, \mathcal{I})$ is a *matroid* (see e.g. [4]) if it satisfies the following *growth property*: if $A \in \mathcal{I}$, $B \in \mathcal{I}$ and $|A| < |B|$, then there exists $e \in B \setminus A$ such that $A \cup \{e\} \in \mathcal{I}$. The maximal (under inclusion) independent sets in $\mathcal{I}$ are called *bases*. The minimal (under inclusion) sets not in $\mathcal{I}$ are called *circuits*. The construction of any base is not a difficult issue. If σ specifies an order of elements of E, then the corresponding base B_σ can be constructed by simple Algorithm 1. We call B_σ, the *base induced by* σ.

Algorithm 1. Constructing a base of a matroid

Input: A matroid $\mathcal{M} = (E, \mathcal{I})$, a sequence $\sigma = (e_1, \ldots, e_n)$ of E.
Output: A base B_σ of $\mathcal{M}$.
$B_\sigma \leftarrow \emptyset$
for $i \leftarrow 1$ **to** n **do**
    if $B_\sigma \cup \{e_i\} \in \mathcal{I}$ **then** $B_\sigma \leftarrow B_\sigma \cup \{e_i\}$
return B_σ

The running time of Algorithm 1 is $\mathcal{O}(nf(n))$, where $f(n)$ is time required for deciding whether set $B \cup \{e_i\}$ contains a circuit, which depends on the particular structure of a matroid.

Let us denote by $pred(e, \sigma)$ the elements which precede element e in sequence σ. The following property of matroids can be proven [3].

Proposition 1. *Let σ and ρ be two sequences of the elements of E. Let $e \in E$ be an element such that $pred(e, \sigma) \subseteq pred(e, \rho)$. If $e \notin B_\sigma$ then $e \notin B_\rho$.*

In the *combinatorial optimization problem* on matroid, a nonnegative weight w_e is given for every element $e \in E$ and we seek a base B, for which the cost $\sum_{e \in B} w_e$ is maximal. This problem can be solved by means of a *greedy algorithm*, that is Algorithm 1 with sequence σ, in which the elements are sorted in the nonincreasing order of their weights. The greedy algorithm constructs the *optimal base* in $\mathcal{O}(n \log n + nf(n))$ time.

Evaluating whether a given *element $f \in E$ is optimal*, i.e. whether f is a part of an optimal base, is not a difficult issue. Let $\sigma^*(\boldsymbol{w}, f)$, $f \in E$, denote a special sequence of elements of E, in which the elements are sorted in the nonincreasing order of their weights w_e, $e \in E$, $\boldsymbol{w}$ is the vector of weights. Moreover, if $w_f = w_e$, $e \neq f$, then element f precedes element e in this sequence. The following proposition gives the necessary and sufficient condition for establishing whether a given element is optimal [3].

Proposition 2. *A given element f is optimal if and only if f is a part of the optimal base $B_{\sigma^*(\boldsymbol{w}, f)}$ induced by $\sigma^*(\boldsymbol{w}, f)$.*

Proposition 2 suggests an $\mathcal{O}(n \log n + nf(n))$ method for evaluating the optimality of f, where $\mathcal{O}(n \log n)$ is the time required for forming $\sigma^*(\boldsymbol{w}, f)$. This complexity can be improved. Let $\sigma(\boldsymbol{w}, f)$, $f \in E$, denote a sequence, such that $pred(f, \sigma(\boldsymbol{w}, f)) = \{e \in E : w_e > w_f\}$. Clearly, $\sigma(\boldsymbol{w}, f)$ can be obtained in $\mathcal{O}(n)$

time since it is not necessary to order the elements (we only require elements $e \in E$, such that $w_e > w_f$ to appear before f).

Proposition 3. *A given element f is optimal if and only if f is a part of a base $B_{\sigma(\boldsymbol{w}, f)}$ induced by $\sigma(\boldsymbol{w}, f)$.*

Proof. It is clear that $pred(f, \sigma(\boldsymbol{w}, f)) = pred(f, \sigma^*(\boldsymbol{w}, f))$. Thus, by Propositions 1, $f \in B_{\sigma^*(\boldsymbol{w}, f)}$ if and only if $f \in B_{\sigma(\boldsymbol{w}, f)}$. Hence, by Propositions 2, f is optimal if and only if $f \in B_{\sigma(\boldsymbol{w}, f)}$.   □

From Proposition 3, we immediately obtain a method for evaluating the optimality of an element. It requires $\mathcal{O}(n + nf(n)) = \mathcal{O}(nf(n))$, where $\mathcal{O}(n)$ is time for forming sequence $\sigma(\boldsymbol{w}, f)$ and $\mathcal{O}(nf(n))$ is time for constructing base $B_{\sigma(\boldsymbol{w}, f)}$ by Algorithm 1.

In the case when element $f \in E$ is not optimal (i.e. it is not a part of an optimal base), a natural question arises: how far is f from optimality. In other words, what is the minimal nonnegative real number δ_f that added to the weight of f makes it optimal. Clearly, δ_f can be calculated as follows: $\delta_f = \max_{B \in \mathcal{B}} \sum_{e \in B} w_e - \max_{B \in \mathcal{B}_f} \sum_{e \in B} w_e$, where $\mathcal{B}$ is the set of all bases and $\mathcal{B}_f$ is the set of all the bases containing f.

3   Evaluating the Optimality of Elements in Interval-Valued Matroids

Consider now the case in which the values of weights are only known to belong to intervals $W_e = [w_e^-, w_e^+]$, $e \in E$. We define a *configuration* as a precise instantiation of the weights of each element $e \in E$, i.e. $\boldsymbol{w} = (w_e)_{e \in E}$, $w_e \in W_e$. We denote by Γ the set of all the configurations, i.e. $\Gamma = \times_{e \in E}[w_e^-, w_e^+]$. We use $w_e(\boldsymbol{w})$ to denote the weight of element $e \in E$ in configuration $\boldsymbol{w} \in \Gamma$. Among the configurations of Γ, we distinguish two *extreme* ones. Namely, the configurations $\boldsymbol{w}_{\{f\}}^+$ and $\boldsymbol{w}_{\{f\}}^-$ such that:

$$w_e(\boldsymbol{w}_{\{f\}}^+) = \begin{cases} w_e^+ & \text{if } e = f, \\ w_e^- & \text{otherwise} \end{cases}, \quad w_e(\boldsymbol{w}_{\{f\}}^-) = \begin{cases} w_e^- & \text{if } e = f, \\ w_e^+ & \text{otherwise} \end{cases}, e \in E. \quad (1)$$

A given element $f \in E$ is *possibly optimal* if and only if it is optimal in some configuration $\boldsymbol{w} \in \Gamma$. A given element $f \in E$ is *necessarily optimal* if and only if it is optimal in all configurations $\boldsymbol{w} \in \Gamma$. Instead of being optimal or not, like in the deterministic case, elements now form three groups: those that are for sure optimal despite uncertainty (necessarily optimal elements), those that are for sure not optimal, and elements whose optimality is unknown (possibly optimal elements). Note that, if an element $f \in E$ is necessarily optimal, then it is also possibly optimal but the converse statement is not true.

We can obtain more information about optimality of $f \in E$. Let $\delta_f(\boldsymbol{w})$, $\boldsymbol{w} \in \Gamma$, denote the minimal nonnegative real number such that f with weight $w_f(\boldsymbol{w}) + \delta_f(\boldsymbol{w})$ becomes optimal in configuration $\boldsymbol{w}$. Let us define $\delta_f^- = \min_{\boldsymbol{w} \in \Gamma} \delta_f(\boldsymbol{w})$

and $\delta_f^+ = \max_{\boldsymbol{w} \in \Gamma} \delta_f(\boldsymbol{w})$. Now the interval $\varDelta_f = [\delta_f^-, \delta_f^+]$ indicates how far f is from being possibly (resp. necessarily) optimal. There are some obvious connections between the notions of optimality and the bounds δ_f^- and δ_f^+ of element $f \in E$.

Proposition 4. *An element f is possibly (resp. necessarily) optimal if and only if $\delta_f^- = 0$ (resp. $\delta_f^+ = 0$).*

The following theorems characterize the possibly and the necessarily optimal elements.

Theorem 1. *Element $f \in E$ is possibly optimal if and only if f is a part of a base $B_{\sigma(\boldsymbol{w}_{\{f\}}^+, f)}$ induced by $\sigma(\boldsymbol{w}_{\{f\}}^+, f)$.*

Proof. ($\Rightarrow$) From the possible optimality of f, it follows that there exits configuration $\boldsymbol{w} \in \Gamma$ such that f is optimal. Proposition 3 implies f is a part of a base $B_{\sigma(\boldsymbol{w}, f)}$ induced by $\sigma(\boldsymbol{w}, f)$. It is easy to observe that $pred(f, \sigma(\boldsymbol{w}_{\{f\}}^+, f)) \subseteq pred(f, \sigma(\boldsymbol{w}, f))$. Proposition 1 now yields $f \in B_{\sigma(\boldsymbol{w}_{\{f\}}^+, f)}$.

($\Leftarrow$) If $f \in B_{\sigma(\boldsymbol{w}_{\{f\}}^+, f)}$ then f is optimal under configuration $\boldsymbol{w}_{\{f\}}^+$, by Proposition 3, and thus f is possibly optimal. $\square$

Theorem 2. *Element $f \in E$ is necessarily optimal if and only if f is a part of a base $B_{\sigma(\boldsymbol{w}_{\{f\}}^-, f)}$ induced by $\sigma(\boldsymbol{w}_{\{f\}}^-, f)$.*

Proof. ($\Rightarrow$) If f is necessarily optimal then it is optimal for all configurations, in particular for $\boldsymbol{w}_{\{f\}}^-$. By Proposition 3, $f \in B_{\sigma(\boldsymbol{w}_{\{f\}}^-, f)}$.

($\Leftarrow$) Suppose $f \in B_{\sigma(\boldsymbol{w}_{\{f\}}^-, f)}$. Consider any configuration $\boldsymbol{w} \in \Gamma$. It is easy to see that $pred(e, \sigma(\boldsymbol{w}, f)) \subseteq pred(e, \sigma(\boldsymbol{w}_{\{f\}}^-, f))$. We conclude from Proposition 3 that $f \in B_{\sigma(\boldsymbol{w}, f)}$ and by Proposition 1 f is optimal in $\boldsymbol{w}$. Accordingly, f is optimal for all configurations $\boldsymbol{w} \in \Gamma$ and thus it is necessarily optimal. $\square$

Making use of Theorems 1 and 2, one can easily evaluate the possible and necessary optimality of an element f. In order to assert whether f is possibly optimal, we apply Algorithm 1 in which the order of elements is specified by $\sigma(\boldsymbol{w}_{\{f\}}^+, f)$. Element f is then possibly optimal if the obtained base contains f. Otherwise, it is not possibly optimal. In the same way, we assert whether f is necessarily optimal. The running time of both methods is $\mathcal{O}(nf(n))$.

Theorems 1 and 2 allow also to determine interval $\varDelta_f = [\delta_f^-, \delta_f^+]$. If $f \notin B_{\sigma(\boldsymbol{w}_{\{f\}}^+, f)}$ (which indicates that $\delta_f^- > 0$) then set $B_{\sigma(\boldsymbol{w}_{\{f\}}^+, f)} \cup \{f\}$ contains an unique circuit C. We can find an element $g \in C \setminus \{f\}$ of the minimal value of w_g^-. Then, from Theorem 1, it follows that $\delta_f^- = w_g^- - w_f^+$. Similarly, if $f \notin B_{\sigma(\boldsymbol{w}_{\{f\}}^-, f)}$ (which indicates that $\delta_f^+ > 0$) then set $B_{\sigma(\boldsymbol{w}_{\{f\}}^-, f)} \cup \{f\}$ contains an unique circuit C. We can find an element $g \in C \setminus \{f\}$ of the minimal value of w_g^+ and, by Theorem 2, $\delta_f^+ = w_g^+ - w_f^-$. It is easily seen that both values δ_f^- and δ_f^+ for a given element $f \in E$ can be computed in $\mathcal{O}(nf(n))$ time.

4   Some Methods of Computing the Optimality Degrees of Elements in Fuzzy-Valued Matroids

We now generalize the concepts of interval-valued matroids to the fuzzy-valued ones and provide a *possibilistic* formulation of the problem (see [1]). The weights of elements of E are ill-known and they are modeled by fuzzy intervals $\tilde{W}_e$, $e \in E$. Let us recall that a fuzzy interval is a fuzzy set in the space of real numbers $\mathbb{R}$, whose membership function $\mu_{\tilde{W}_e}$ is normal, quasiconcave and upper semi-continuous on $\mathbb{R}$ (see for instance [1]), $\mu_{\tilde{W}_e} : \mathbb{R} \to [0,1]$. The membership function $\mu_{\tilde{W}_e}$, $e \in E$, expresses the *possibility distribution* of the weight of element $e \in E$ (see [1]). Let $\boldsymbol{w} = (w_e)_{e \in E}$, $w_e \in \mathbb{R}$, be a configuration of the weights. The configuration $\boldsymbol{w}$ represents a certain state of the world. Assuming that the weights are unrelated, the joint possibility distribution over configurations, induced by the $\tilde{W}_e$, $e \in E$, is as follows: $\pi(\boldsymbol{w}) = \min_{e \in E} \mu_{\tilde{W}_e}(w_e)$. Hence, the degrees of possibility and necessity that an element $f \in E$ is optimal are defined as follows:

$$\Pi(f \text{ is optimal}) = \sup_{\boldsymbol{w}:\, f \text{ is optimal in } \boldsymbol{w}} \pi(\boldsymbol{w}), \qquad (2)$$

$$\mathrm{N}(f \text{ is optimal}) = \inf_{\boldsymbol{w}:\, f \text{ is not optimal in } \boldsymbol{w}} (1 - \pi(\boldsymbol{w})). \qquad (3)$$

The degrees of optimality can be generalized by fuzzyfying the quantity δ_f, $f \in E$ in the following way:

$$\mu_{\tilde{\Delta}_f}(x) = \Pi(\delta_f = x) = \sup_{\boldsymbol{w}:\, x = \delta_f(\boldsymbol{w})} \pi(\boldsymbol{w}).$$

The following relations hold:

$$\Pi(f \text{ is optimal}) = \Pi(\delta_f = 0) = \mu_{\tilde{\Delta}_f}(0),$$

$$\mathrm{N}(f \text{ is optimal}) = \mathrm{N}(\delta_f = 0) = 1 - \sup_{x > 0} \mu_{\tilde{\Delta}_f}(x).$$

Every fuzzy weight $\tilde{W}_e$, $e \in E$, can be decomposed into its λ-cuts, that is the sets $\tilde{W}_e(\lambda) = \{x \,|\, \mu_{\tilde{W}_e}(x) \geq \lambda\}$, $\lambda \in (0,1]$. It is well known that $\tilde{W}_e(\lambda)$, $\lambda \in (0,1]$, is the classical interval $[\tilde{W}_e^-(\lambda), \tilde{W}_e^+(\lambda)]$. We assume that $[\tilde{W}_e^-(0), \tilde{W}_e^+(0)]$ is the support of $\tilde{W}_e$. Functions $\tilde{W}_e^- : [0,1] \to \mathbb{R}$ and $\tilde{W}_e^+ : [0,1] \to \mathbb{R}$, are called *left* and *right profiles* of $\tilde{W}_e$ [2], respectively (see Fig. 1a). Thus, the profiles can be seen as a parametric representations of the left and right hand sides of a fuzzy interval. We assume additionally that both profiles are strictly monotone. This assumption holds for the fuzzy intervals of L-R type [1]. Therefore, it is not restrictive. Let $\mathcal{M}(\lambda) = (E, \mathcal{I})$, $\lambda \in [0,1]$, be the interval-valued matroid with weights $\tilde{W}_e(\lambda)$, $e \in E$, being the λ-cuts of the fuzzy weights. A link between the interval case and the fuzzy one resulting from formulae (2) and (3) and the fact that if $\alpha < \beta$ then $\tilde{W}_e(\beta) \subseteq \tilde{W}_e(\alpha)$, $e \in E$, is as follows:

$$\Pi(f \text{ is optimal}) = \sup\{\lambda \,|\, f \text{ is possibly optimal in } \mathcal{M}(\lambda)\}, \qquad (4)$$

$$\mathrm{N}(f \text{ is optimal}) = 1 - \inf\{\lambda \,|\, f \text{ is necessarily optimal in } \mathcal{M}(\lambda)\}. \qquad (5)$$

Equations (4) and (5) form the theoretical basis for calculating the values of the optimality indices. They suggest a standard bisection method for determining the optimality degrees (2) and (3) of a fixed element with a given accuracy ϵ via the use of λ-cuts. At each iteration the possible (necessary) optimality of the element is evaluated in the interval-valued matroid $\mathcal{M}(\lambda)$ according to Theorem 1 (Theorem 2). The calculations take $\mathcal{O}(|\log \epsilon| nf(n))$ time. Unfortunately, this method gives only an approximate values of the optimality degrees. Further in this section, we propose some polynomial algorithms which give the exact values of the degrees of optimality.

First, we need to extend two extreme configurations (1) to the fuzzy case. The fuzzy counterparts $\tilde{\boldsymbol{w}}_{\{f\}}^{+}$ and $\tilde{\boldsymbol{w}}_{\{f\}}^{-}$ are vectors of the left and right profiles defined as follows:

$$\tilde{W}_e(\tilde{\boldsymbol{w}}_{\{f\}}^{+}) = \begin{cases} \tilde{W}_e^{+} & \text{if } e = f, \\ \tilde{W}_e^{-} & \text{otherwise} \end{cases}, \quad \tilde{W}_e(\tilde{\boldsymbol{w}}_{\{f\}}^{-}) = \begin{cases} \tilde{W}_e^{-} & \text{if } e = f, \\ \tilde{W}_e^{+} & \text{otherwise} \end{cases}, e \in E. \quad (6)$$

Assume that we intend to calculate the value of $\Pi(f \text{ is optimal})$, $f \in E$. The key observation is that in order to do this, it is enough to analyze only the fuzzy configuration $\tilde{\boldsymbol{w}}_{\{f\}}^{+}$. Moreover, it is sufficient to take into account only the intersection points of profile $\tilde{W}_f^{+}$ with profiles $\tilde{W}_e^{-}$, $e \neq f$, in configuration $\tilde{\boldsymbol{w}}_{\{f\}}^{+}$. Let $e_{\lambda_1}, \ldots, e_{\lambda_m}$ be elements in E, whose left profiles intersect with the right profile of element f. The numbers $\lambda_1, \ldots, \lambda_m \in [0, 1]$ denote the cuts such that $\tilde{W}_{e_{\lambda_i}}^{-}(\lambda_i) = \tilde{W}_f^{+}(\lambda_i)$, $i = 1, \ldots, m$. We assume that $\lambda_1 \leq \cdots \leq \lambda_m$, $\tilde{W}_{e_{\lambda_1}}^{-}(\lambda_1) \leq \cdots \leq \tilde{W}_{e_{\lambda_m}}^{-}(\lambda_m)$. Let us also distinguish elements $v_1, \ldots, v_r$ in E whose left profiles are entirely on the left hand side of $\tilde{W}_f^{+}$ and elements $u_1, \ldots, u_q$ whose left profiles are entirely on the right hand side of $\tilde{W}_f^{+}$. The 3-partition of elements is of the form $E = \{u_1, \ldots, u_q\} \cup \{e_{\lambda_1}, \ldots, e_{\lambda_m}\} \cup \{v_1, \ldots, v_r\}$ (see Fig. 1b).

Let us now define sequences $\sigma_0, \sigma_1, \ldots \sigma_m$ in the following way:

$$\sigma_0 = (u_1, \ldots, u_q, \boldsymbol{f}, e_{\lambda_1}, \ldots, e_{\lambda_m}, v_1, \ldots, v_r),$$
$$\sigma_i = (u_1, \ldots, u_q, e_{\lambda_1}, \ldots, e_{\lambda_i}, \boldsymbol{f}, e_{\lambda_{i+1}} \ldots, e_{\lambda_m}, v_1, \ldots, v_r), \quad i = 1, \ldots, m-1,$$
$$\sigma_m = (u_1, \ldots, u_q, e_{\lambda_1}, \ldots, e_{\lambda_m}, \boldsymbol{f}, v_1, \cdots, v_r).$$

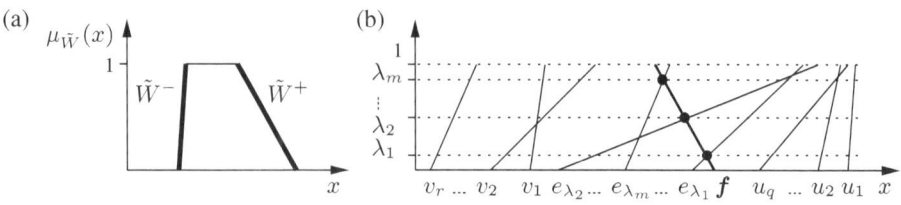

Fig. 1. (a) The left and right profiles of fuzzy interval $\tilde{W}$ (in bold). (b) The partition of E with respect to the intersection points of profile $\tilde{W}_f^{+}$ with profiles $\tilde{W}_e^{-}$, $e \neq f$, in configuration $\tilde{\boldsymbol{w}}_{\{f\}}^{+}$.

Note that the sequences differ from each other only in the position of the element f, which depends on the cut λ_i. Let us define $\lambda_0 = 0$.

Observation 1. *If $f \in B_{\sigma_{i-1}}$, then f is possibly optimal in matroid $\mathcal{M}(\lambda)$, $\lambda \in [0, \lambda_i]$, $i = 1, \ldots, m$.*

Proof. Observe, that it is sufficient to show that f is possibly optimal in interval weighted matroid $\mathcal{M}(\lambda_i)$. It is easy to see that the extreme configuration $\boldsymbol{w}_{\{f\}}^+$ in $\mathcal{M}(\lambda_i)$ is as follows: $w_f(\boldsymbol{w}_{\{f\}}^+) = \tilde{W}_f^+(\lambda_i)$ and $w_e(\boldsymbol{w}_{\{f\}}^+) = \tilde{W}_e^-(\lambda_i)$ if $e \neq f$. From the construction of the sequence σ_i, it follows that $pred(\sigma(\boldsymbol{w}_{\{f\}}^+, f), f) \subseteq pred(\sigma_{i-1}, f)$. Thus, from Proposition 1 and the assumption $f \in B_{\sigma_{i-1}}$, we see that f is a part of the base $B_{\sigma(\boldsymbol{w}_{\{f\}}^+, f)}$ in $\mathcal{M}(\lambda_i)$ and, by Theorem 1, it is possibly optimal in $\mathcal{M}(\lambda_i)$. $\square$

Observation 2. *If $f \notin B_{\sigma_i}$ then f is not possibly optimal in matroid $\mathcal{M}(\lambda)$, $\lambda \in (\lambda_i, 1]$, $i = 0, \ldots, m$.*

Proof. From the definition of sequence σ_i it follows that $\tilde{W}_e^-(\lambda_i) \geq \tilde{W}_f^+(\lambda_i)$ for all $e \in pred(f, \sigma_i)$ (see also Fig. 1b). Let $\lambda > \lambda_i$. From the strict monotonicity of the right and left profiles we obtain $\tilde{W}_e^-(\lambda) > \tilde{W}_f^+(\lambda)$ for all $e \in pred(f, \sigma_i)$. Thus in the interval weighted matroid $\mathcal{M}(\lambda)$ all the elements $e \in pred(f, \sigma_i)$ must also precede f in the corresponding sequence $\sigma(\boldsymbol{w}_{\{f\}}^+, f)$ in matroid $\mathcal{M}(\lambda)$. Therefore, according to Propositions 1 and Theorem 1, element f is not possibly optimal in $\mathcal{M}(\lambda)$. $\square$

Observations 1 and 2, together with formula (4) yield:

Proposition 5. *If $f \in B_{\sigma_m}$ then $\Pi(f \text{ is optimal}) = 1$. Otherwise, let k be the smallest index in $\{0, 1, \ldots, m\}$ such that $f \notin B_{\sigma_k}$. Then $\Pi(f \text{ is optimal}) = \lambda_k$.*

Proposition 5 allows to construct an effective algorithm (Algorithm 2) for computing the value of $\Pi(f \text{ is optimal})$ of a given element $f \in E$. The key of Algorithm 2 is that there is no need to apply Algorithm 1 for each sequence $\sigma_0, \ldots, \sigma_m$ for checking whether f is a part of base B_{σ_i} induced by σ_i, $i = 0, \ldots, m$. Using the fact that the sequences differ from each other only in the position of element f, which depends on the cut λ_i, we only need to test if f can be added to a base constructed in Algorithm 2 after each choosing of an element e_{λ_i}. It is easily seen that Algorithm 2 implicitly check whether $f \in B_{\sigma_i}$, for $i = 0, \ldots, m$ deciding this way if f is possibly optimal. Hence, Algorithm 2 is equivalent to one course of Algorithm 1. Since finding all the intersection points requires $\mathcal{O}(n)$ time if all the fuzzy intervals are of the L-L type [1], it is easily seen that Algorithm 2 takes $\mathcal{O}(n \log n + nf(n))$ time, where $\mathcal{O}(n \log n)$ is time required for sorting $(\lambda_i, e_{\lambda_i})$ with respect to the values of λ_i.

An approach to computing $N(f \text{ is optimal})$ of a given element $f \in E$ is symmetric. In this case one need to consider fuzzy configuration $\tilde{\boldsymbol{w}}_{\{f\}}^-$ and take into account intersection points of profile $\tilde{W}_f^-$ with profiles $\tilde{W}_e^+$, $e \neq f$, in configuration $\tilde{\boldsymbol{w}}_{\{f\}}^-$ (see Fig. 2). The numbers $\lambda_1, \ldots, \lambda_m$ denote the cuts such

---

Algorithm 2. Computing $\Pi(f$ is optimal$)$.

Input: A fuzzy weighted matroid $\mathcal{M} = (E, \mathcal{I})$, a distinguished element $f \in E$.
Output: $\Pi(f$ is optimal$)$.
Find all pairs $(\lambda_i, e_{\lambda_i})$
Sort $(\lambda_i, e_{\lambda_i})$ in nondecreasing order with respect to the values of λ_i
Form $\sigma_0 = (u_1, \ldots, u_q, f, e_{\lambda_1}, \ldots, e_{\lambda_m}, v_1, \ldots, v_r)$
$B \leftarrow \emptyset$
for $i \leftarrow 1$ **to** q **do**
   ⌊ **if** $B \cup \{u_i\} \in \mathcal{I}$ **then** $B \leftarrow B \cup \{u_i\}$
if $B \cup \{f\} \notin \mathcal{I}$ **then return** 0 /*$f \notin B_{\sigma_0}$ */
for $i \leftarrow 1$ **to** m **do**
   ⌊ **if** $B \cup \{e_{\lambda_i}\} \in \mathcal{I}$ **then** $B \leftarrow B \cup \{e_{\lambda_i}\}$
      if $B \cup \{f\} \notin \mathcal{I}$ **then return** λ_i /*$f \notin B_{\sigma_i}$ */
return 1 /*$f \in B_{\sigma_m}$ */

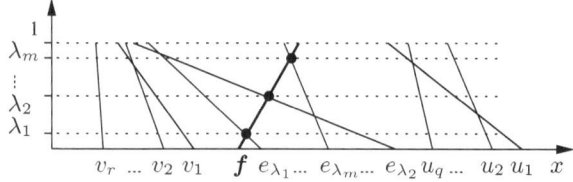

Fig. 2. The partition of E with respect to the intersection points of profile $\tilde{W}_f^-$ with profiles $\tilde{W}_e^+$, $e \neq f$, in configuration $\tilde{w}_{\{f\}}^-$

that $\tilde{W}_{e_{\lambda_i}}^+(\lambda_i) = \tilde{W}_f^-(\lambda_i)$, $i = 1, \ldots, m$, under the assumption that $\lambda_1 \leq \cdots \leq \lambda_m$, $\tilde{W}_{e_{\lambda_1}}^+(\lambda_1) \leq \cdots \leq \tilde{W}_{e_{\lambda_m}}^+(\lambda_m)$.

Similarly, we define $\sigma_1, \ldots, \sigma_{m+1}$ with respect to elements $e_{\lambda_1}, \ldots, e_{\lambda_m}$ whose right profiles intersect with the left profile of element f.

$$\sigma_1 \ = (u_1, \ldots, u_q, e_{\lambda_m}, \ldots, e_{\lambda_1}, f, v_1, \ldots, v_r),$$
$$\sigma_i \ = (u_1, \ldots, u_q, e_{\lambda_m}, \ldots, e_{\lambda_i}, f, e_{\lambda_{i-1}} \ldots, e_{\lambda_1}, v_1, \ldots, v_r), \ i = 2, \ldots, m,$$
$$\sigma_{m+1} = (u_1, \ldots, u_q, f, e_{\lambda_m}, \ldots, e_{\lambda_1}, v_1, \cdots, v_r).$$

Set $\lambda_{m+1} = 1$. The following proposition is symmetric to Proposition 5 (the proof goes in the similar manner).

Proposition 6. *If $f \in B_{\sigma_1}$ then* $\mathrm{N}(f$ *is optimal*$) = 1$. *Otherwise, let k be the largest index in $\{1, \ldots, m+1\}$ such that $f \notin B_{\sigma_k}$. Then* $\mathrm{N}(f$ *is optimal*$) = 1 - \lambda_k$.

Algorithm 3 is similar in a spirit to Algorithm 2. Here, there is also no need to apply Algorithm 1 for each sequence $\sigma_{m+1}, \ldots, \sigma_1$ for checking whether f is a part of base B_{σ_i} induced by σ_i, $i = m + 1, \ldots, 1$, according to Theorem 2. Similarly, Algorithm 3 implicitly check whether $f \in B_{\sigma_i}$, for $i = m + 1, \ldots, 1$ evaluating this way the necessary optimality of f, which is due to the fact that

---

Algorithm 3. Computing N(f is optimal).

Input: A fuzzy weighted matroid $\mathcal{M} = (E, \mathcal{I})$, a distinguished element $f \in E$.
Output: N(f is optimal).
Find all pairs $(\lambda_i, e_{\lambda_i})$
Sort $(\lambda_i, e_{\lambda_i})$ in nondecreasing order with respect to the values of λ_i
Form $\sigma_{m+1} = (u_1, \ldots, u_q, f, e_{\lambda_m}, \ldots, e_{\lambda_1}, v_1, \ldots, v_r)$
$B \leftarrow \emptyset$
for $i \leftarrow 1$ **to** q **do**
  ⌊ **if** $B \cup \{u_i\} \in \mathcal{I}$ **then** $B \leftarrow B \cup \{u_i\}$
if $B \cup \{f\} \notin \mathcal{I}$ **then return** 0   /*$f \notin B_{\sigma_{m+1}}$ */
for $i \leftarrow m$ **downto** 1 **do**
  ⌈ **if** $B \cup \{e_{\lambda_i}\} \in \mathcal{I}$ **then** $B \leftarrow B \cup \{e_{\lambda_i}\}$
  ⌊ **if** $B \cup \{f\} \notin \mathcal{I}$ **then return** $1 - \lambda_i$ /*$f \notin B_{\sigma_i}$ */
return 1 /*$f \in B_{\sigma_1}$ */

---

sequences $\sigma_{m+1}, \ldots, \sigma_1$ differ from each other only in the position of element f. Obviously, computing N(f is optimal) also requires $\mathcal{O}(n \log n + nf(n))$ time.

References

1. Dubois, D., Prade, H.: Possibility theory: an approach to computerized processing of uncertainty. Plenum Press, New York 1988.
2. Dubois, D., Fargier, H., Fortin, J.: A generalized vertex method for computing with fuzzy intervals. In: Fuzz'IEEE (2004) 541-546.
3. Kasperski, A., Zieliński, P.: On combinatorial optimization problems on matroids with ill-known weights. Instytut Matematyki PWr., Wroclaw 2005, raport serii PREPRINTY nr 34, submitted for publication in Eur. J. Oper. Res.
4. Oxley, J., G.: Matroid Theory. Oxford University Press, New York, 1992.

Imprecise Temporal Interval Relations

Steven Schockaert, Martine De Cock, and Etienne E. Kerre

Department of Applied Mathematics and Computer Science,
Fuzziness and Uncertainty Modelling Research Unit,
Ghent University, Krijgslaan 281 (S9), B-9000 Gent, Belgium
Tel: +32 9 2644772; Fax: +32 9 2644995
Steven.Schockaert@UGent.be
http://www.fuzzy.ugent.be

Abstract. When the time span of an event is imprecise, it can be represented by a fuzzy set, called a fuzzy time interval. In this paper we propose a representation for 13 relations that can hold between intervals. Since our model is based on fuzzy orderings of time points, it is not only suitable to express precise relationships between imprecise events ("the mid 1930's came *before* the late 1930's") but also imprecise relationships ("the late 1930's came *long before* the early 1990's"). Furthermore we show that our model preserves many of the properties of the 13 relations Allen introduced for crisp time intervals.

Keywords: Temporal Reasoning, Fuzzy Relation, Fuzzy Ordering.

1 Introduction

A significant part of the work on temporal representation and reasoning is concerned with time intervals. Allen [1] defined 13 qualitative relations that may hold between two intervals $A = [a^-, a^+]$ and $B = [b^-, b^+]$. Table 1 shows how these relations are expressed by means of constraints on the boundaries of the intervals. The relations are mutually exclusive and exhaustive, i.e. for any two intervals, exactly one of the relations holds. Temporal information is however often ill–defined, e.g. because the definition of some historical events is inherently subjective (e.g. the Renaissance), or because historical documents are usually written in a vague style (e.g. "in the late 1930s"). Ill–defined time intervals can either be intervals with uncertain boundaries [2], or imprecise intervals [4, 5]. In this paper we will focus on the latter, i.e. we will assume that we have complete knowledge about the time span of an event, but that it has a gradual beginning and/or ending. This kind of time intervals can be represented as fuzzy sets.

To our knowledge, Nagypál and Motik [4] were the first to extend Allen's work to fuzzy time intervals, generalizing the relations of Table 1 to fuzzy relations. However, their approach suffers from a number of important disadvantages, in particular concerning the relations e, m, s and f. For example, e is not reflexive in general; if A is a continuous fuzzy set in $\mathbb{R}$, it holds that $e(A, A) = s(A, A) = f(A, A) = 0.5$ while one would expect $e(A, A) = 1$ and $s(A, A) = f(A, A) = 0$. Furthermore they only consider precise relationships. The approach proposed

I. Bloch, A. Petrosino, and A.G.B. Tettamanzi (Eds.): WILF 2005, LNAI 3849, pp. 108–113, 2006.

Table 1. Allen's temporal interval relations

Name		Definition
1. before	$b(A, B)$	$a^+ < b^-$
2. overlaps	$o(A, B)$	$a^- < b^-$ and $b^- < a^+$ and $a^+ < b^+$
3. during	$d(A, B)$	$b^- < a^-$ and $a^+ < b^+$
4. meets	$m(A, B)$	$a^+ = b^-$
5. starts	$s(A, B)$	$a^- = b^-$ and $a^+ < b^+$
6. finishes	$f(A, B)$	$a^+ = b^+$ and $b^- < a^-$
7. equals	$e(A, B)$	$a^- = b^-$ and $a^+ = b^+$

Inverse relations	
8. $bi(A, B) = b(B, A)$	11. $mi(A, B) = m(B, A)$
9. $oi(A, B) = o(B, A)$	12. $si(A, B) = s(B, A)$
10. $di(A, B) = d(B, A)$	13. $fi(A, B) = f(B, A)$

by Ohlbach [5] allows to express some imprecise temporal relations (e.g. *A more or less* finishes *B*), but it does not deal with imprecise constraints such as "*A was long before B*". Moreover, many desirable properties that hold for Allen's relationships are not preserved in this fuzzification.

In Section 2 of this paper we introduce a generalization of Allen's 13 interval relations that can not only be used when the time intervals are fuzzy ("the mid 1930's came *before* the late 1930's), but is even powerful enough to express imprecise relationships ("the late 1930's came *long before* the early 1990's"). The magical ingredients are fuzzy orderings of time points; they are lifted into relationships between fuzzy time intervals through the use of relatedness measures for fuzzy sets. In Section 3, we show that our model preserves important properties regarding (ir)reflexivity, (a)symmetry and transitivity. To our knowledge, we are the first to introduce a generalization of Allen's relations that can be used for precise as well as imprecise temporal relationships between fuzzy time intervals, and at the same time preserves so many desirable properties.

2 Fuzzy Temporal Interval Relations

Throughout this paper, we represent time points as real numbers. A real number can, for example, be interpreted as the number of milliseconds since January 1, 1970. Because we want to model imprecise temporal relations, we need a way to express that a certain time point a is long before a time point b, and a way to express that a is before or approximately at the same time as b. Fuzzy relations are particularly well suited for this purpose, due to the vague nature of these concepts.

Definition 1 (Fuzzy Ordering of Time Points). *For $\beta \in]0, +\infty[$, the fuzzy relation $L_\beta^{\lll}$ in $\mathbb{R}$ is defined as*

$$L_\beta^{\lll}(a, b) = \begin{cases} 1 & \text{if } b - a > \beta \\ 0 & \text{if } b - a \leq 0 \\ \frac{b-a}{\beta} & \text{otherwise} \end{cases} \tag{1}$$

for all a and b in $\mathbb{R}$. $L_0^{\ll}$ *is defined by* $L_0^{\ll}(a, b) = 1$ *if* $a < b$ *and* $L_0^{\ll}(a, b) = 0$ *otherwise. The fuzzy relation* $L_\beta^{\preceq}$ *in* $\mathbb{R}$ *is defined as*

$$L_\beta^{\preceq}(a, b) = 1 - L_\beta^{\ll}(b, a) \tag{2}$$

$L_\beta^{\ll}(a, b)$ represents the extent to which a is much smaller than b. Note that the parameter β defines how the concept "much smaller than" should be interpreted. Likewise, $L_\beta^{\preceq}(a, b)$ represents the extent to which b is not "much smaller than a", in other words, the extent to which a is smaller than or approximately equal to b. Moreover, $L_0^{\preceq}(a, b) = 1$ if $a \leq b$ and $L_0^{\preceq}(a, b) = 0$ otherwise, i.e $L_\beta^{\preceq}$ is a generalization of the crisp ordering $\leq$. We use these ordering relations between time points as a stepping stone for the representation of imprecise relations that may hold between fuzzy time intervals.

Proposition 1. *Let* $\beta \geq 0$; *it holds that for every* a, b *and* c *in* $\mathbb{R}$

$$T_W(L_\beta^{\ll}(a, b), L_\beta^{\ll}(b, c)) \leq L_\beta^{\ll}(a, c) \tag{3}$$

$$T_W(L_\beta^{\preceq}(a, b), L_\beta^{\preceq}(b, c)) \leq L_\beta^{\preceq}(a, c) \tag{4}$$

$$T_W(L_\beta^{\preceq}(a, b), L_\beta^{\ll}(b, c)) \leq L_\beta^{\ll}(a, c) \tag{5}$$

$$T_W(L_\beta^{\ll}(a, b), L_\beta^{\preceq}(b, c)) \leq L_\beta^{\ll}(a, c) \tag{6}$$

$$T_W(L_\beta^{\ll}(a, b), L_\beta^{\preceq}(b, a)) = 0 \tag{7}$$

where T_W *denotes the Lukasiewicz t–norm* $T_W(x, y) = \max(0, x + y - 1)$.

Recall that a fuzzy set A in $\mathbb{R}$ is convex and upper semicontinuous iff for each α in $]0, 1]$ the set $\{x | A(x) \geq \alpha\}$ is a closed interval.

Definition 2 (Fuzzy Time Period). *A fuzzy time period is a normalised fuzzy set in* $\mathbb{R}$ *which is interpreted as the time span of some event. A fuzzy (time) interval is a convex and upper semicontinuous normalised fuzzy set in* $\mathbb{R}$. *A fuzzy time period* A *is called nondegenerate w.r.t.* β *iff* $A \circ L_\beta^{\ll} \circ A = 1$, *i.e. if the beginning of* A *is long before the end of* A.

As recalled in Section 1, Allen's definitions are based on constraints on the boundaries of the intervals. If A and B are fuzzy time intervals, the boundaries of A and B can be gradual. Hence, we cannot refer to these boundaries in the same way we refer to the boundaries of crisp intervals. Therefore, as shown in Table 2, we propose using relatedness measures to express the relations between the boundaries of fuzzy intervals without actually referring to these boundaries. For an arbitrary fuzzy relation $\mathbb{R}$, these relatedness measures are defined as [3]:

$$A \circ_T R \circ_T B = \sup_{v \in \mathbb{R}} T(B(v), \sup_{u \in \mathbb{R}} T(A(u), R(u, v))) \tag{8}$$

$$A \triangleleft_I R \triangleright_I B = \inf_{v \in \mathbb{R}} I(B(v), \inf_{u \in \mathbb{R}} I(A(u), R(u, v))) \tag{9}$$

$$(A \triangleleft_I R) \circ_T B = \sup_{v \in \mathbb{R}} T(B(v), \inf_{u \in \mathbb{R}} I(A(u), R(u, v))) \tag{10}$$

Table 2. Relation between the boundaries of the crisp intervals $[a^-, a^+]$ and $[b^-, b^+]$, and the fuzzy intervals A and B

Crisp	Fuzzy	Crisp	Fuzzy
$a^- < b^-$	$A \circ (L_\beta^{\lll} \rhd B)$	$a^- \leq b^-$	$(A \circ L_\beta^{\prec}) \rhd B$
$a^+ < b^+$	$(A \lhd L_\beta^{\lll}) \circ B$	$a^+ \leq b^+$	$A \lhd (L_\beta^{\prec} \circ B)$
$a^+ < b^-$	$A \lhd L_\beta^{\lll} \rhd B$	$a^+ \leq b^-$	$A \lhd L_\beta^{\prec} \rhd B$
$a^- < b^+$	$A \circ L_\beta^{\lll} \circ B$	$a^- \leq b^+$	$A \circ L_\beta^{\prec} \circ B$

$$A \lhd_I (R \circ_T B) = \inf_{u \in \mathbb{R}} I(A(u), \sup_{v \in \mathbb{R}} T(B(v), R(u,v))) \tag{11}$$

$$(A \circ_T R) \rhd_I B = \inf_{v \in \mathbb{R}} I(B(v), \sup_{u \in \mathbb{R}} T(A(u), R(u,v))) \tag{12}$$

$$A \circ_T (R \rhd_I B) = \sup_{u \in \mathbb{R}} T(A(u), \inf_{v \in \mathbb{R}} I(B(v), R(u,v))) \tag{13}$$

where T is a left–continuous t–norm and I its residual implicator. For example $A \circ (L_\beta^{\lll} \rhd B)$ expresses the degree to which there is an element in A that is much smaller than all elements in B. In the remainder of this paper we assume that T is the Lukasiewicz t–norm and I its residual implicator $I_W(x,y) = \min(1, 1-x+y)$.

Note how the appearance of $<$ (resp. $\leq$) in Table 2 corresponds to the use of $L_\beta^{\lll}$ (resp. $L_\beta^{\prec}$). If $\beta > 0$, the relations from Table 2 become imprecise relations (e.g. the beginning of A is *long* before the beginning of B). Using the expressions from Table 2, we define the temporal relations for fuzzy intervals as shown in Table 3. For convenience, we use the same notation for the temporal relations when fuzzy intervals are used instead of crisp intervals.

3   Properties

When A and B are crisp intervals and $\beta = 0$, our definitions are equivalent to Allen's original definitions. Note that in Table 3 we have used the minimum to generalize the conjunctions that appear in the crisp definitions. The use of the minimum as t–norm makes it possible to prove the following proposition.

Proposition 2 (Exhaustivity). *Let A and B be fuzzy time periods. It holds that*

$$S_W(b(A,B), bi(A,B), o(A,B), oi(A,B), d(A,B), di(A,B), m(A,B), mi(A,B),$$
$$s(A,B), si(A,B), f(A,B), fi(A,B), e(A,B)) = 1 \tag{14}$$

where S_W is the Lukasiewicz t–conorm defined by $S_W(x,y) = \min(1, x+y)$ for all x and y in $[0,1]$.

Proposition 3 (Mutual Exclusiveness). *Let A and B be nondegenerate fuzzy time periods w.r.t. β. Moreover, let R and S both be one of the 13 fuzzy temporal relations. If $R \neq S$, then it holds that*

$$T_W(R(A,B), S(A,B)) = 0 \tag{15}$$

Table 3. Fuzzy temporal interval relations

Name	Definition
$b(A, B)$	$A \triangleleft L_\beta^{\lll} \triangleright B$
$o(A, B)$	$\min(A \circ (L_\beta^{\lll} \triangleright B), B \circ L_\beta^{\lll} \circ A, (A \triangleleft L_\beta^{\lll}) \circ B)$
$d(A, B)$	$\min(B \circ (L_\beta^{\lll} \triangleright A), (A \triangleleft L_\beta^{\lll}) \circ B)$
$m(A, B)$	$\min(A \triangleleft L_\beta^{\prec\!\!\prec} \triangleright B, B \circ L_\beta^{\prec\!\!\prec} \circ A)$
$s(A, B)$	$\min((A \circ L_\beta^{\prec\!\!\prec}) \triangleright B, (B \circ L_\beta^{\prec\!\!\prec}) \triangleright A, (A \triangleleft L_\beta^{\lll}) \circ B)$
$f(A, B)$	$\min(A \triangleleft (L_\beta^{\prec\!\!\prec} \circ B), B \triangleleft (L_\beta^{\prec\!\!\prec} \circ A), B \circ (L_\beta^{\lll} \triangleright A))$
$e(A, B)$	$\min((A \circ L_\beta^{\prec\!\!\prec}) \triangleright B, (B \circ L_\beta^{\prec\!\!\prec}) \triangleright A, A \triangleleft (L_\beta^{\prec\!\!\prec} \circ B), B \triangleleft (L_\beta^{\prec\!\!\prec} \circ A))$

Proposition 4 (Reflexivity and Symmetry). *The relations b, bi, o, oi, d, di, s, si, f and fi are irreflexive and asymmetric w.r.t. T_W, i.e. let R be one of the aforementioned fuzzy relations and let A and B be fuzzy time periods. It holds that*

$$R(A, A) = 0 \tag{16}$$
$$T_W(R(A, B), R(B, A)) = 0 \tag{17}$$

Furthermore, it holds that

$$e(A, A) = 1 \tag{18}$$
$$e(A, B) = e(B, A) \tag{19}$$
$$m(A, A) = A \triangleleft L_\beta^{\prec\!\!\prec} \triangleright A \tag{20}$$
$$T_W(m(A, B), m(B, A)) \leq \min(A \triangleleft L_\beta^{\prec\!\!\prec} \triangleright A, B \triangleleft L_\beta^{\prec\!\!\prec} \triangleright B) \tag{21}$$

The crisp meets relation m (between crisp intervals) is irreflexive, provided that the beginning of each interval is strictly before the end of the interval, i.e. provided singletons (time points) are not allowed as time intervals; (20)–(21) is a generalization of this observation in the sense that our meets relation (and therefore also mi) is irreflexive and asymmetric if the beginning of A (resp. B) is not approximately equal to the end of A (resp. B). From (14)–(21) it can be concluded that the fuzzy temporal interval relations are mutually exclusive and exhaustive w.r.t. the Łukasiewicz t–norm and t–conorm. Moreover, the reflexivity and symmetry properties of our definitions are in accordance with the corresponding properties of the temporal relations between crisp intervals.

Proposition 5 (Transitivity). *The relations b, bi, d, di, s, si, f, fi and e are T_W–transitive, i.e. let R be one of the aforementioned fuzzy relations and let A, B and C be fuzzy time periods. It holds that*

$$T_W(R(A, B), R(B, C)) \leq R(A, C)$$

No kind of transitivity holds for o, oi, m and mi in general. Thus the transitivity properties of our definitions are in accordance with the transitivity properties of the (crisp) temporal relations between crisp intervals.

The properties in this section are valid for arbitrary fuzzy time periods. In practice however, it seems often more natural to consider only fuzzy time intervals in this context.

4   Concluding Remarks

In this paper we have introduced a new approach to define possibly imprecise, temporal interval relations between fuzzy time intervals. It can be shown that, unlike in previous approaches, generalizations of all the important properties of the crisp interval relations are valid. Further work will focus on the use of our approach for temporal reasoning. The reader can verify that, for example

$$T_W(d(A, B), b(B, C)) \leq b(A, C)$$

which expresses that from "A takes place during B", and "B happens before C", we deduce that "A takes place before C".

Acknowledgments

Steven Schockaert and Martine De Cock would like to thank the Fund for Scientific Research – Flanders for funding their research.

References

1. J. F. Allen. Maintaining Knowledge about Temporal Intervals. *Communications of the ACM*, Vol. 26 No. 11, pp. 832–843, 1983.
2. D. Dubois, H. Prade. Processing Fuzzy Temporal Knowledge. *IEEE Transactions on Systems, Man, and Cybernetics*, Vol. 19 No. 4, pp. 729–744, 1989.
3. S. Schockaert, M. De Cock, C. Cornelis. Relatedness of Fuzzy Sets. *To appear in: Journal of Intelligent and Fuzzy Systems*.
4. G. Nagypál, B. Motik. A Fuzzy Model for Representing Uncertain, Subjective and Vague Temporal Knowledge in Ontologies. *Proc. of the Int. Conf. on Ontologies, Databases and Applications of Semantics*, LNCS 2888, Springer–Verlag, pp. 906–923, 2003.
5. H. J. Ohlbach. Relations Between Fuzzy Time Intervals. *Proc. of the 11th Int. Symp. on Temporal Representation and Reasoning*, pp. 44–51, 2004.

A Many Valued Representation and Propagation of Trust and Distrust

Martine De Cock[1] and Paulo Pinheiro da Silva[2]

[1] Ghent University, Dept. of Applied Mathematics and Computer Science,
Krijgslaan 281 (S9), 9000 Gent, Belgium
Martine.DeCock@UGent.be
http://www.fuzzy.UGent.be
[2] Stanford University, Knowledge Systems,
AI Laboratory, Stanford CA 94305, USA
pp@ksl.stanford.edu
http://iw.stanford.edu

Abstract. As the amount of information on the web grows, users may find increasing challenges in trusting and sometimes distrusting sources. One possible aid is to maintain a network of trust between sources. In this paper, we propose to model such a network as an intuitionistic fuzzy relation. This allows to elegantly handle together the problem of ignorance, i.e. not knowing whether to trust or not, and vagueness, i.e. trust as a matter of degree. We pay special attention to deriving trust information through a trusted third party, which becomes especially challenging when distrust is involved.

Keywords: network of trust, propagation, semantic web, intuitionistic fuzzy relation, interval valued fuzzy relation.

1 Introduction

There is an increasing amount of information sources available to applications and users on the web. As information source breadth increases, users may find increasing challenges in trusting and sometimes distrusting sources. We expect a systematic support for trusting information sources to be one of the keys to a functional semantic web [2]. Trust in general has become an important interdisciplinary research area. We refer to [7] for a recently published collection of contributions which also shows an emerging interest in the notion of distrust. Existing computational models usually deal with trust in a binary way: they assume that a source is to be trusted or not, and they compute the probability or the belief that the source can be trusted (see e.g. [8], [10]). Besides full trust or no trust at all, in reality we also encounter partial trust. This is reflected in our everyday language when we say for example "this source is rather trustworthy" or "I trust this source very much". In this paper we focus on (1) representing trust as a matter of degree, including the case that an agent may fully trust (or have blind faith) or distrust a source, and (2) on deriving trust information obtained through a trusted third party (TTP).

I. Bloch, A. Petrosino, and A.G.B. Tettamanzi (Eds.): WILF 2005, LNAI 3849, pp. 114–120, 2006.
© Springer-Verlag Berlin Heidelberg 2006

The first issue pertains to situations where sources can not be divided in the trustworthy ones and the malicious ones in a clear cut way, but they can be trusted to a certain extent. Think of trust as a matter of degree, i.e. instead of computing the probability that a source can be trusted, we are interested in the degree to which a source can be trusted. Whereas the existing probabilistic approach is suitable for problems where security is at stake and malicious sources need to be discerned from trustworthy ones, our approach leans itself better for the computation of trust when the outcome of an action can be positive to some extent, e.g., when provided information can be right to some degree, as opposed to being either right or wrong. In [6], it is argued that trust and distrust are distinct, opposite concepts. Trust and distrust can clearly coexist, e.g. among politicians who trust each other enough to cooperate, but at the same time maintain a "healthy level of distrust". In Section 2, we introduce a model that takes into account partial trust, distrust and ignorance simultaneously, as different but related concepts.

The second problem can informally be described as: if the trust value of source a in source b is p, and the trust value of b in source c is q, what information can be derived about the trust value of a in c? This problem of atomic trust propagation has been well researched in a probabilistic setting, where multiplication is used as the main operation to combine trust values. However, when distrust is involved as well, the need for a new, not necessarily commutative propagation operator arises. We discuss this in Section 3.

2   Trust Network Between Sources

Trust is a multi-faceted concept, it can be full or partial, it depends on the context, it depends on the purpose, etc. Developing a computational model forces us to make some initial simplifying assumptions. One aspect is the domain dependency of trust: e.g. we may trust the website of a store on information about location and opening hours, but that does not imply that we also take for granted everything they say in their advertisements. In this paper, we assume that we are dealing with trust in a single domain, expressed between a set of sources $\mathcal{A}$. Another aspect is the purpose of trusting a source: in this paper, we are not dealing with trust to support a decision. For instance, we do not provide or discuss the use of trust-related thresholds that along with trust values may be used for decision making.

Since trust may be a matter of degree, we use a number t between 0 and 1 to express the degree of trust of a in b. This value is not a probability nor a belief. In a probabilistic setting, a higher trust level corresponds to a higher probability that a source can be trusted, while in our interpretation it corresponds to a higher trust. Both approaches are complementary.

In our approach, 0 corresponds to total absence of trust. Roughly speaking, this can occur in either one of the following situations: (1) a has reason to distrust b fully, or (2) a has no information about b and hence no reason to trust b, but also no reason to distrust b. Taking into account the fundamental difference between the two situations, and the fact that distrust is no less important than trust in relying on a source, we propose to represent distrust d simultaneously with trust as a couple (t, d), in which both t and d are numbers between 0 and 1. Trust and distrust do not have to sum up to 1, but we assume that they satisfy the restriction $t + d \leq 1$. Omitting this restriction would result in allowing inconsistency — this is an interesting option for future development that is however not further considered in this paper. As a result, the network of trust between sources is represented by an intuitionistic fuzzy relation (IFR for short).

Intuitionistic fuzzy set theory [1] is an extension of fuzzy set theory that defies the claim that from the fact that an element x "belongs" to a given degree $\mu_A(x)$ to a fuzzy set A, naturally follows that x should "not belong" to A to the extent $1 - \mu_A(x)$, an assertion implicit in the concept of a fuzzy set. On the contrary, an intuitionistic fuzzy set (IFS for short) assigns to each element x of the universe both a degree of membership $\mu_A(x)$ and one of non–membership $\nu_A(x)$ such that

$$\mu_A(x) + \nu_A(x) \leq 1 \tag{1}$$

thus relaxing the enforced duality $\nu_A(x) = 1 - \mu_A(x)$ from fuzzy set theory. Obviously, when $\mu_A(x) + \nu_A(x) = 1$ for all elements of the universe, the traditional fuzzy set concept is recovered. Formally an IFS A in a universe X is a mapping from X to the lattice L^* defined by [3]:

$$L^* = \{(t, d) \in [0, 1]^2 \mid t + d \leq 1\}$$
$$(t_1, d_1) \leq_{L^*} (t_2, d_2) \Leftrightarrow t_1 \leq t_2 \text{ and } d_1 \geq d_2$$

An IFR in $\mathcal{A}$ is an IFS in $\mathcal{A} \times \mathcal{A}$.

Definition 1. A trust network is a couple $(\mathcal{A}, R)$ such that $\mathcal{A}$ is a set of sources and R is an IFR in $\mathcal{A}$. For all a and b in $\mathcal{A}$:

- $R(a, b)$ is called the trust value of a in b
- $\mu_R(a, b)$ is called the trust degree of a in b
- $\nu_R(a, b)$ is called the distrust degree of a and b
- $1 - \mu_R(a, b) - \nu_R(a, b)$ is the hesitation of a towards b

IFS theory has been shown to be formally equivalent to interval valued fuzzy set (IVFS) theory [4]. This is another extension of fuzzy set theory in which the membership degrees are subintervals instead of numbers from $[0, 1]$ (see [9]). A couple (t, d) of trust t and distrust d corresponds to the interval $[t, 1 - d]$, indicating that the trust degree ranges from t to $1 - d$. The hesitation degree from IFS theory corresponds to the length of the interval. The longer the interval, the more doubt about the actual trust value.

Table 1 illustrates this by means of some examples. $(0, 1)$ and $(1, 0)$ are respectively the smallest and the biggest element of L^*, corresponding to full distrust and full trust; obviously in these situations there is no hesitation. In the case of no knowledge, namely $(0, 0)$, the hesitation is 1. The most wide spread approach (see column 2) only takes into account the degree of trust, and can not make

Table 1. Examples of trust values

	trust	trust and distrust		
		IFS	IVFS	Guha[5]
	t	(t, d)	$[t, 1 - d]$	$t - d$
full trust	1.0	(1.0,0.0)	[1.0,1.0]	1.0
full distrust	0.0	(0.0,1.0)	[0.0,0.0]	-1.0
no knowledge	0.0	(0.0,0.0)	[0.0,1.0]	0.0
partial trust	0.2	(0.2,0.0)	[0.2,0.8]	0.2
partial trust and distrust	0.6	(0.6,0.4)	[0.6,0.6]	0.2
inconsistency		(1.0,1.0)		0.0

the distinction between a case of full distrust and a case of no knowledge. In [5] the distrust degree d is subtracted from the trust degree t, giving rise to a trust value on a scale from -1 to 1. The examples $(0.2, 0)$ and $(0.6, 0.4)$ illustrate that valuable information is lost in this mapping process. Indeed $(0.6, 0.4)$ expresses a strong opinion to trust a source to degree 0.6 but not more, while $(0.2, 0)$ suggests to trust to degree 0.2 but possibly more because there is a lot of doubt in this case (the hesitation degree is 0.8). In [5], both cases are mapped to the same value, namely 0.2.

3   Trust and Distrust Propagation

As recalled in the introduction, in a probabilistic framework, trust is propagated by means of the multiplication operation. This can be straightforwardly adapted to a fuzzy setting by using a t–norm, i.e. an increasing, commutative and associative $[0, 1]^2 - [0, 1]$ mapping that satisfies $\mathcal{T}(1, x) = x$ for all x in $[0, 1]$. Hence

$$\mathcal{T}(\mu_R(a, b), \mu_R(b, c)) \tag{2}$$

is the trust degree of a in c, derived from the trust degree of a in b and the trust degree of b in c. Possible choices for $\mathcal{T}$ are $\mathcal{T}_M(x, y) = \min(x, y)$, $\mathcal{T}_P(x, y) = x \cdot y$ and $\mathcal{T}_L(x, y) = \max(0, x + y - 1)$.

However if, instead of only the trust degree, we consider the complete trust value, i.e. both the trust and the distrust degree, propagation is not straightforward at all anymore. In this case the propagation operator is an $(L^*)^2 - L^*$ mapping Prop. An example shows that Prop is not necessarily commutative. Suppose that a has full trust in b and b has full distrust in c, than intuitively we infer that a has full distrust in c, i.e.

$$\mathsf{Prop}((1, 0), (0, 1)) = (0, 1) \tag{3}$$

However, if a has full distrust in b and b has full trust in c, more than one approach is possible. The full distrust of a in b might lead a to ignoring b, i.e. no knowledge is inferred

$$\mathsf{Prop}((0, 1), (1, 0)) = (0, 0) \tag{4}$$

(3) and (4) together illustrate the non commutative behavior. However, the distrust of a in b might encourage a to take on the contrary of what b is saying, in other words to trust c fully, i.e.

$$\mathsf{Prop}((0,1),(1,0)) = (1,0) \tag{5}$$

For both approaches (4) and (5) a reasonable motivation can be given. This example only lifts part of the veil of the complex problem which propagation scheme to choose. Our aim in this paper is not to provide a clear cut answer to that question, but rather to provide some propagation operators that can be used in different schemes. Recall that a t–conorm $\mathcal{S}$ is an increasing, commutative and associative $[0,1]^2 - [0,1]$ mapping that satisfies $\mathcal{S}(0,x) = x$ for all x in $[0,1]$. Possible choices are $\mathcal{S}_\mathrm{M}(x,y) = \max(x,y)$, $\mathcal{S}_\mathrm{P}(x,y) = x + y - x \cdot y$, and $\mathcal{S}_\mathrm{L}(x,y) = \min(1, x+y)$. A negator $\mathcal{N}$ is a decreasing $[0,1] - [0,1]$ mapping satisfying $\mathcal{N}(0) = 1$ and $\mathcal{N}(1) = 0$. The most commonly used one is $\mathcal{N}_s(x) = 1 - x$.

Definition 2. The propagation operators Prop_1, Prop_2, and Prop_3 are defined by

$$\mathsf{Prop}_1((t_1,d_1),(t_2,d_2)) = (\mathcal{T}(t_1,t_2), \mathcal{T}(t_1,d_2))$$
$$\mathsf{Prop}_2((t_1,d_1),(t_2,d_2)) = (\mathcal{S}(\mathcal{T}(t_1,t_2),\mathcal{T}(d_1,d_2)), \mathcal{S}(\mathcal{T}(t_1,d_2),\mathcal{T}(d_1,t_2)))$$
$$\mathsf{Prop}_3((t_1,d_1),(t_2,d_2)) = (\mathcal{T}(t_1,t_2), \mathcal{T}(\mathcal{N}(d_1),d_2))$$

for all (t_1, d_1) and (t_2, d_2) in L^*.

The following proposition shows that all three propagation operators copy the information given by a fully trusted third party. It also proves that Prop_1 and Prop_3 are in accordance with (4) because they derive no knowledge through a third party that they distrust, while Prop_2 takes on exactly the opposite information given by a distrusted source and hence is in accordance with (5). Prop_1 and Prop_2 derive no knowledge through an unknown third party, while Prop_3 displays a paranoid behavior in taking on some distrust information even from an unknown third party.

Proposition 1. For all (t, d) in L^* it holds that

$$\mathsf{Prop}_1((1,0),(t,d)) = (t,d) \qquad \mathsf{Prop}_1((0,1),(t,d)) = (0,0)$$
$$\mathsf{Prop}_2((1,0),(t,d)) = (t,d) \qquad \mathsf{Prop}_2((0,1),(t,d)) = (d,t)$$
$$\mathsf{Prop}_3((1,0),(t,d)) = (t,d) \qquad \mathsf{Prop}_3((0,1),(t,d)) = (0,0)$$

$$\mathsf{Prop}_1((0,0),(t,d)) = (0,0)$$
$$\mathsf{Prop}_2((0,0),(t,d)) = (0,0)$$
$$\mathsf{Prop}_3((0,0),(t,d)) = (0,d)$$

Using $\mathcal{T}_\mathrm{P}$ and $\mathcal{S}_\mathrm{P}$, Prop_1 and Prop_2 take on the following form

$$\mathsf{Prop}_1((t_1,d_1),(t_2,d_2)) = (t_1 \cdot t_2, t_1 \cdot d_2)$$
$$\mathsf{Prop}_2((t_1,d_1),(t_2,d_2)) = (t_1 \cdot t_2 + d_1 \cdot d_2 - t_1 \cdot t_2 \cdot d_1 \cdot d_2,$$
$$t_1 \cdot d_2 + d_1 \cdot t_2 - t_1 \cdot d_2 \cdot d_1 \cdot t_2)$$

This particular form of Prop_1 has previously been proposed in [8] to combine pairs of beliefs and disbeliefs. Subtracting the distrust degree from the trust degree, the operations above reduce respectively to $t_1 \cdot (t_2 - d_2)$ and $(t_1 - d_1) \cdot (t_2 - d_2)$, which are the two distrust propagation schemes put forward in [5].

4 Conclusion

In this paper we have introduced a many valued approach for a network of trust between sources. We represent trust values as couples (t, d) in which t corresponds to a trust degree, d to a distrust degree, and $1 - t - d$ to an ignorance degree. As such, to our knowledge, we are the first to introduce a model that takes into account partial trust, distrust and ignorance simultaneously. We have also presented a collection of three operators used for atomic propagation of trust, distrust and ignorance. These operators are generic enough to be used in several "trust" schemes, including those where trust, distrust and ignorance are either full or partial, and those where propagation is commutative or not. The ability to take into account ignorance and to propagate trust become extremely meaningful in a large web where the trustworthiness of many sources is initially unknown to a user, which does not imply that the user distrusts all of them, but that the user may eventually gather evidences to trust or distrust some sources and still ignore others.

The representation and propagation solutions presented in this paper are pre-liminary since there is a lot more to computing trust on the web, such as further propagation (longer chains) and aggregation (combining the trust information received from several TTP's). Yet one step further is to update the trust network. Another aspect not yet mentioned in the paper is that it is important to be able to calculate trust in a distributed manner taking into consideration both efficiency and privacy.

Acknowledgments

Martine De Cock would like to thank the Fund for Scientific Research–Flanders for funding her research, and the members of the Knowledge Systems Lab at Stanford University for their hospitality and the inspiring cooperation, leading to the current paper.

References

1. Atanassov, K. T.: Intuitionistic Fuzzy Sets. Fuzzy Sets and Systems **20** (1986) 87–96
2. Berners-Lee, T., Hendler, J., Lassila O.: The Semantic Web, Scientific American, May 2001.
3. Deschrijver, G., Cornelis, C., Kerre E. E.: Intuitionistic Fuzzy Connectives Revisited. Proceedings of IPMU2002 (9th International Conference on Information Processing and Management of Uncertainty in Knowledge-Based Systems) (2002) 1839–1844

4. Deschrijver, G., Kerre, E. E.: On the relationship between some extensions of fuzzy set theory. Fuzzy Sets and Systems **133** (2003) 227–235

5. Guha, R., Kumar, R., Raghavan, P., Tomkins, A.: Propagation of Trust and Distrust. Proceedings of WWW2004 (2004) 403–412

6. Harrison McKnight, D., Chervany, N. L.: Trust and Distrust Definitions: One Bite at a Time. Lecture Notes in Artificial Intelligence **2246** (2001) 27–54

7. Herrmann, P., Issarny, V., Shiu, S.: Trust Management, Third International Conference, iTrust 2005. Lecture Notes in Computer Science 3477 (2005)

8. Jøsang, A., Knapskog, S. J.: A metric for trusted systems. Proceedings of the 21st National Security Conference, NSA (1998)

9. Türksen, I. B.: Interval Valued Sets Based on Normal Forms. Fuzzy Sets and Systems **20** (1986) 191–210

10. Zaihrayeu, I., Pinheiro da Silva, P., and McGuinness, D. L.: IWTrust: Improving User Trust in Answers from the Web. Lecture Notes in Computer Science 3477 (2005) 384–392

SVM Classification of Neonatal Facial Images of Pain

Sheryl Brahnam[1], Chao-Fa Chuang[2], Frank Y. Shih[2], and Melinda R. Slack[3]

[1] Missouri State University, Computer Information Systems, 901 South National,
Springfield MO 65804, USA
Shb757f@smsu.edu
[2] New Jersey Institue of Technology, University Heights, Newark, NJ 07102
{cxc1235, shih}@njit.edu
[3] Medical Director of Neonatology, St. John's Hospital, 1235 E. Cherokee,
Springfield, MO 65894, USA
Melinda_slack@pediatrix.com

Abstract. This paper reports experiments that explore performance differences in two previous studies that investigated SVM classification of neonatal pain expressions using the Infant COPE database. This database contains 204 photographs of 26 neonates (age 18-36 hours) experiencing the pain of heel lancing and three nonpain stressors. In our first study, we reported experiments where representative expressions of all subjects were included in the training and testing sets, an experimental protocol suitable for intensive care situations. A second study used an experimental protocol more suitable for short-term stays: the SVMs were trained on one sample and then evaluated on an unknown sample. Whereas SVM with polynomial kernel of degree 3 obtained the best classification score (88.00%) using the first evaluation protocol, SVM with a linear kernel obtained the best classification score (82.35%) using the second protocol. However, experiments reported here indicate no significant difference in performance between linear and nonlinear kernels.

1 Introduction

Accurate assessment of pain in neonates is a difficult yet crucial task. The clinical definition of pain assumes the person experiencing pain has the ability to articulate the location, duration, quality, and intensity of their pain experience. Although nonverbal self reporting methods have been devised that allow preverbal children to indicate their pain levels by pointing to abstract renditions of facial expressions expressive of increasing levels of discomfort, neonates must rely exclusively on the proxy judgments of others [3].

Several pain assessment measures have been developed to assist clinicians in diagnosing neonatal pain. Most of these instruments rely on the neonate's facial displays. Facial displays are considered the gold standard of pain assessment [4] because they are the most specific and consistent indicators of pain. The facial characteristics of neonatal pain displays include prominent forehead, eye squeeze, naso-labial furrow, taut tongue, and an angular opening of the mouth [5]. Despite the fact that neonatal facial displays of pain are the most reliable source of pain assessment, instruments

I. Bloch, A. Petrosino, and A.G.B. Tettamanzi (Eds.): WILF 2005, LNAI 3849, pp. 121 – 128, 2006.
© Springer-Verlag Berlin Heidelberg 2006

based on facial displays are unsatisfactory because clinicians tend to underrate pain intensity [6] and often fail to utilize all the information available to them in the infants facial signals [7].

In an attempt to bypass the unreliable observer, our research group is investigating the potential benefits face recognition technology would offer pediatric clinicians in diagnosing neonatal pain. Applying face recognition techniques to medical problems is a novel application area. Gunaratne and Sato [17] have used a mesh-based approach to estimate asymmetries in facial actions to determine the presence of facial motion dysfunction for patients with Bell's palsy, and Dai et al. [12] have proposed a method for observing the facial expressions of patients in hospital beds. The facial images used in the Dai et al. study, however, were not of actual patients but rather of subjects responding to verbal cues suggestive of medical procedures and conditions. Our work with neonatal pain expressions is the only other research we are aware of that uses face recognition techniques to diagnose medical problems.

We began work on this problem by developing the Infant COPE database. The facial displays of 26 neonates between the ages of 18 hours and 3 days old were photographed experiencing the pain of a heel lance and a variety of stressors, including transport from one crib to another, an air stimulus on the nose, and friction on the external lateral surface of the heel.

In our initial study [1], three face classification techniques, Principal Component Analysis (PCA), Linear Discriminant Analysis (LDA), and Support Vector Machines (SVMs), were used to classify the faces into two categories: pain and nonpain. The training and testing sets contained multiple samples of each subject in each expression category. No two samples were identical as each varied slightly in angle and facial configuration. While, ideally, as is the case with speech recognition software, samples of individual subjects would be available to personalize the classifier, in a clinical setting this is not practical as the typical newborn's stay is short-term. The evaluation protocol used in our first study would probably only be applicable in intensive care situations where neonates have longer stays that present opportunities for collecting facial samples. It is more realistic to assume that the classifier will need to be trained on one set of subjects and then applied out of the box to future newborns. In [2], an evaluation protocol was developed that evaluated trained classifiers using unknown subjects.

Results of the two studies were contradictory in terms of the best kernel to use with SVM. An SVM with polynomial kernel of degree 3 obtained the best classification score (88.00%) in the first study, and an SVM with a linear kernel obtained the best classification score (82.35%) in the second study. Sampling error caused by the small number of images in the sample pool is one possible explanation for this discrepancy. A set of new experiments using the first protocol was designed to explore sample error. The results of these experiments, reported in section 4, suggest that there is no significant difference in the performance of an SVM with a linear kernel and an SVM with a polynomial kernel of degree 3.

In section 2, we describe of the facial displays in the infant COPE database more completely. In section 3, we outline the two experimental protocols, designated A and B, used in the earlier studies. In section 4, we compare SVM classification rates reported in the two studies, along with baseline PCA and LDA rates. We then present the results of a new study that varies the size of the sample pool. We conclude the paper, in section 5, by pointing out some limitations in our current work and by offering suggestions for future research.

2   The Infant COPE Database

The Infant COPE Database, described more completely in [1] and [2], contains 204 facial images of 26 neonates experiencing the pain of a heel lance and three nonpain stressors: transport from one crib to another (a stressor that triggers crying that is not in response to pain), an air stimulus on the nose (a stressor that provokes eye squeeze), and friction on the surface of the heel (a stressor that produces facial expressions of distress that are similar to the expressions of pain). In addition to these four facial displays, the database includes images of the neonates in the neutral state of rest.

Fig. 1 provides two example sets of the five neonatal expressions of rest, cry, air stimulus, friction, and pain included in the Infant COPE database. Of the 204 images in the database, 67 are rest, 18 are cry, 23 are air stimulus, 36 are friction, and 60 are pain.

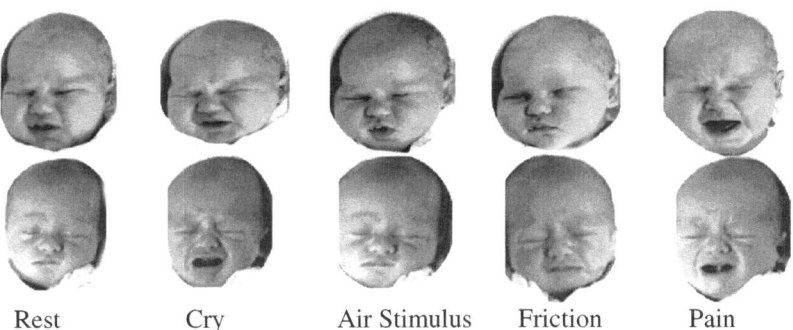

| Rest | Cry | Air Stimulus | Friction | Pain |

Fig. 1. Examples of the five facial expressions in the Infant COPE database

The data collection process complied with the protocols and ethical directives for research involving human subjects at Missouri State University and St. John's Health System, Inc. Informed consent was obtained from a parent, usually the mother in consultation with the father. Most parents were recruited in the neonatal unit of a St. John's Hospital sometime after delivery. Only mothers who had experienced uncomplicated deliveries were approached. The subjects were born in a large Midwestern hospital in the United States of America. All neonates used in the study were Caucasian, evenly split between genders (13 boys and 12 girls), and in good health. The interested reader is referred to [1] and [2] for more information on the data collection design.

3 Evaluation Protocols

In [1] and [2], images of the five facial expressions in the Infant COPE database were grouped into two categories: pain and nonpain. The set of nonpain images combined the rest, cry, air stimulus, and friction images and contained a total of 144 images. The set of pain images consisted of the remaining 60 images.

The evaluation protocol used in the first study, designated here as protocol A, focused on facial expression representation. The two classes of pain and nonpain facial expressions included representative images of all 26 subjects. Using a cross-validation technique, classification was a four step process. In step 1, the images were randomly divided into ten segments. In step 2, nine out of the ten segments were used in the training session. The remaining segment was used in testing, and an average classification score was obtained from the testing set of images. In step 3, steps 1 and 2 were repeated ten times. Finally, in step 4, the ten classification scores were averaged to obtain a final performance score for each classifier.

In the second study, we trained the classifiers on one set of subjects and tested them on another. Using protocol B, twenty-six experiments were performed, one for each subject. The facial images of 25 subjects formed the testing set, and the images of the remaining subject formed the testing set. The 26 classification scores were averaged to obtain a final performance score for each classifier.

4 Experimental Results

In this section, we compare the SVM performance results reported in the first two studies. We also introduce a new set of experiments designed to determine whether the performance differences in the earlier studies are due to sampling error.

SVMs with five kernels (linear, RBF, polynomial degree 2, polynomial degree 3, and polynomial degree 4) were assessed using protocols A and B. The regularization parameter, C, used in the SVM experiments was determined using a grid search. Since the recognition rates in our experiments were not significantly different in terms of different values for C, we adopted the regularization parameter C=1. The bandwidth parameter in SVM using RBF kernels was also optimized using a grid search. For comparison purposes, baseline PCA and LDA using the sum of absolute differences, or L1 distance metric, were also evaluated.

The SVM, PCA, and LDA experiments were processed in the MATLAB environment under the Windows XP operating system using a Pentium 4 – 2.80 GHz processor. SVM was implemented using the OSU SVM Classifier MATLAB Toolbox developed by Ohio State University.

The general experimental procedures used in all our experiments can be divided into the following stages: preprocessing, feature extraction, and classification. In the preprocessing stage, the original images were cropped, rotated, and scaled. Eyes were aligned roughly along the same axis. The original 204 images, size 3008 x 2000 pixels, were also reduced to 100 x 120 pixels. In the feature extraction stage, facial features were centered within an ellipse and color information was discarded. The rows within the ellipse were concatenated to form a feature vector of dimension 8383 with entries ranging in value between 0 and 255. PCA was then used to reduce the dimensionality of the feature vectors further. The first 70 principle components resulted in the best classification scores. Finally, in the classification stage, the feature vectors were used as inputs to the classifiers.

Table 1 compares the average classification scores obtained using the two protocols. Referring to Table 1, the average classification score for PCA was 80.36% and for LDA 80.32%. SVM, as expected, outperformed both PCA and LDA, except in the case of RBF kernel. Given previous reports in facial expression classification using SVM (see, for instance, [8]), we did not expect the RBF kernel performance to be as low as it was. An SVM with polynomial degree 3 provided the best recognition rate of 88.00% in the experiments using protocol A. An SVM with linear kernel provided the best recognition rate of 82.35% using protocol B.

Table 1. Comparison of SVM classification rates using protocol A and B

Type of svm	Protocol A	Protocol B	Average (A & B)
Linear	83.67%	**82.35%**	83.01%
Polynomial degree = 2	86.50%	79.90%	83.20%
Polynomial degree = 3	**88.00%**	80.39%	84.20%
Polynomial degree = 4	82.17%	72.06%	77.12%
RBF	70.00%	70.10%	70.05%
PCA with L1 distance	80.33%	80.39%	80.36%
LDA with L1 distance	83.67%	76.96%	80.32%

There are several possible explanations for the kernel performance differences in the two studies. The most likely cause for the discrepancy is sampling error due to the small number of images in the sample pool. The average performance of the SVMs using the two kernels, for instance, is very close, the difference being only 1.18%. However, since the data in the training sets used in the two sets of experiments differ only in a few inputs (approximately 15%), we questioned this assumption.

To determine if the difference in kernel performance is the result of sampling error, we performed new experiments that varied the size of the sample pool. We did this by comparing SVM classification of pain expressions to each of the other four facial displays. This resulted in pool sizes of 83 images for pain versus air stimulus, 78 images for pain versus cry, 96 images for pain versus friction, and 127 images for pain versus rest. Only protocol A was used in these experiments, as splitting expressions for each subject (protocol B) resulted in pool sizes that were too small for training.

Tables 2-4 present the results of the new set of experiments. The average performance of the four experiments using SVM with a linear kernel is 85.51%, and the average performance of SVM with a polynomial kernel of degree 3 is 87.74%. The difference in kernel performance (2.23%) is half that in [1] (4.33%), which also used protocol A. This leads us to believe that sample error is most likely the cause of kernel performance differences. As far as neonatal facial expressions are concerned, the results of the new set of experiments suggest that there is no significant classification difference in SVMs using a linear kernel versus a polynomial kernel of degree 3

Table 2. Pain vs. Air stimulus

Method	Classification score
Linear	90.00%
Polynomial degree = 2	77.78%
Polynomial degree = 3	83.33%
Polynomial degree = 4	78.89%
RBF	66.67%

Table 3. Pain vs. Cry

Method	Classification score
Linear	71.25%
Polynomial degree = 2	78.75%
Polynomial degree = 3	80.00%
Polynomial degree = 4	76.25%
RBF	75.00%

Table 4. Pain vs. Friction

Method	Classification score
Linear	90.00%
Polynomial degree = 2	96.00%
Polynomial degree = 3	93.00%
Polynomial degree = 4	92.00%
RBF	60.00%

Table 5. Pain vs. Rest

Method	Classification score
Linear	90.77%
Polynomial degree = 2	84.62%
Polynomial degree = 3	94.62%
Polynomial degree = 4	86.15%
RBF	53.85%

Kernel. This conclusion is consistent with [8], which examined SVM expressing classification performance using a number of adult facial databases.

5 Conclusion

This paper reports new experiments intended to explore performance differences in two pervious studies that investigated SVM classification of neonatal pain expressions using the Infant COPE database. This database contains 204 photographs of 26 neonates (age 18-36 hours) experiencing the acute pain of a heel lance and three non-pain stressors.

The SVM classifiers were trained and tested using images divided into two sets: pain and nonpain. Two separate evaluation protocols, designated in this paper as A and B, were also used. Protocol A, described in [1], assumes that samples of neonates are available for customizing the classifier. Representative expression samples of all 26 subjects were thus included in both the training and the testing sets. Protocol B, described in [2], assumes that the classifiers will be trained on one sample and tested on another. The facial images of 25 subjects formed the training set, and the images of the remaining subject formed the testing set. A total of 26 experiments were thus performed using protocol B, one for each subject. An SVM of polynomial kernel degree 3 obtained the best classification score of 88.00% using protocol A, and an SVM with a linear kernel obtained the best classification score of 82.35% using protocol B.

We assumed that the difference in kernel performance was due to sample error. A set of new experiments that varied the size of the sample pool was performed to test our assumption. In these experiments, which used protocol A, the average performance of SVM with a linear kernel was 85.51%. With polynomial kernel of degree 3 it was 87.74%. The difference in kernel performance is half that reported in [1], which also used protocol A. This leads us to believe that there is no significant performance difference using SVM with a linear kernel and polynomial kernel of degree 3.

We would like to conclude this paper with some general remarks concerning the limitations, future directions, and significance of our research in neonatal pain classification.

There are a number of limitations in our current work. First, these studies use two-dimensional still photographs and do not consider the dynamic and multidimensional nature of facial expressions. The classification rates reported in these two studies, however, are consistent with facial expression classification rates reported using video displays of adult facial expressions. For example, [9] reports classification rates between 88%-89%. Second, we have yet to explore facial shape information in the facial displays. Third, the focus thus far has been on acute pain. We have not examined facial expressions in reaction to repeated pain experiences.

In terms of future directions, we are working on addressing the limitations noted above. We are currently collecting video data of neonates experiencing additional stressors and two types of pain: acute and repeated pain. We are also working on experiments that incorporate shape information. In addition, we are examining the classification performance of a number of neural network architectures. For instance, the performance rate of NNSOA, a neural network simultaneous optimization algorithm, using protocol B is reported in [2].

Finally, in terms of significance, we expected that the performance of SVMs in the first study that used protocol A would be better than SVMs in the second study that used protocol B. What we did not know is how well SVM performance would hold up using protocol B. SVM results compare well, and the classification rates in both studies indicate a high potential for applying standard face recognition technology to this problem domain. We believe the results of the SVM experiments encourage further explorations using more sophisticated face recognition technologies.

References

1. Brahnam, S., Chuang, C., Shih, F.Y., and Slack, M.R. Machine Recognition and Representation of Neonate Facial Displays of Acute Pain. International Journal of Artificial Intelligence in Medicine (AIIM), (in press and available on publisher website)
2. Brahnam, S., Chuang, C., Sexton, R., Shih, F.Y., and Slack, M.R. Machine Assessment of Neonatal Facial Expressions of Acute Pain. Decision Support Systems, (in revision)
3. Wong, D. and Baker, C. Pain in Children: Comparison of Assessment Scales. Pediatric Nursing, Vol. 14: 1 (1988) 9017
4. Craig, K.D. The Facial Display of Pain in Infants and Children. In: Craig, K.D. (ed.). The Facial Display of Pain in Infants and Children, IASP Press, Seattle, (1998) 103-121
5. Grunau, R.E., Grunau, R.V.E., and Craig, K.D. Pain Expression in Neonates: Facial Action and Cry. Pain, Vol. 28: 3 (1987) 395-410
6. Prkachin, K.M., Solomon, P., Hwang, T., and Mercer, S.R. Does Experience Influence Judgments of Pain Behaviour? Evidence from Relatives of Pain Patients and Therapists. Pain Research and Management, Vol. 6: 2 (2001) 105-112
7. Prkachin, K.M., Berzins, S., and Mercer, S.R. Encoding and Decoding of Pain Expressions: A Judgement Study. Pain, Vol. 58: 2 (1994) 253-59
8. Littlewort, G., Bartlett, M.S., Fasel, I., Susskind, J., and Movellan, J. Dynamics of Facial Expression Extracted Automatically from Video. The First IEEE Workshop on Face Processing in Video. Washington, DC. (2004)
9. Cohen, I., Sebe, N., Garg, A., Chen, L.S., and Huang, T.S. Facial Expression Recognition from Video Sequences: Temporal and Static Modeling. Computer Vision and Image Understanding, Vol. 91: 1-2 (2003) 160-187

Performance Evaluation of a Hand Gesture Recognition System Using Fuzzy Algorithm and Neural Network for Post PC Platform

Jung-Hyun Kim, Yong-Wan Roh, Jeong-Hoon Shin, and Kwang-Seok Hong

School of Information and Communication Engineering, Sungkyunkwan University, 300,
Chunchun-dong, Jangan-gu, Suwon, KyungKi-do, 440-746, Korea
{kjh0328, elec1004}@skku.edu, only4you@chol.com,
kshong@skku.ac.kr
http://hci.skku.ac.kr

Abstract. In this paper, we implement hand gesture recognition system using fuzzy algorithm and neural network for Post PC (the embedded-ubiquitous environment using blue-tooth module, embedded i.MX21 board and smart gate-notebook computer). Also, we propose most efficient and reasonable hand gesture recognition interface for Post PC through evaluation and analysis of performance about each gesture recognition system. The proposed gesture recognition system consists of three modules: 1) gesture input module that processes motion of dynamic hand to input data, 2) Relational Database Management System (hereafter, RDBMS) module to segment significant gestures from input data and 3) 2 each different recognition module: fuzzy max-min and neural network function recognition module to recognize significant gesture of continuous / dynamic gestures. Experimental result shows the average recognition rate of 98.8% in fuzzy max-min module and 96.7% in neural network recognition module about significantly dynamic gestures.

1 Introduction

The Post PC refers to a wearable PC that has the ability to process information and networking power. The Post PC that integrates sensors and human interface technologies will provide human-centered services and excellent portability and convenience [1]. The Post PC service natural user interface, so user can get more convenient and realistic information service and also includes sensor technology such as haptic and gesture device. The haptic technology can analyze user's intention distinctly, such as finding out whether user gives instruction by intention, judging and taking suitable response for the user's intention. There are many specific technologies under this category, such as gesture realization, action pattern grasping, brow realization, living body signal realization, and hand movement [2], [3].

In this paper, we implemented systems that recognizes significant dynamic gesture of user in real-time and anytime-anywhere manner using fuzzy algorithm and neural network. Also, we propose most efficient and reasonable hand gesture recognition interface for Post PC through evaluation and analysis of performance (whole size of system, the processing speed-recognition time and recognition rate etc.) about each

I. Bloch, A. Petrosino, and A.G.B. Tettamanzi (Eds.): WILF 2005, LNAI 3849, pp. 129 – 138, 2006.
© Springer-Verlag Berlin Heidelberg 2006

gesture recognition system. The proposed gesture recognition system consists of three modules: 1) gesture input module that processes motion of dynamic hand to input data, 2) RDBMS module to segment significant gestures from input data, and 3) 2 each different hand gesture recognition module: fuzzy max-min and neural network recognition module to recognize significant gesture of continuous / dynamic gestures and to enhance the extensibility of recognition.

2   Hand Gesture Input Module and RDBMS Module

In this paper, we develop an input method by which human can convey idea rapidly, and can express various intentions via dynamic hand gestures. The focus of this study is to suggest a way to capture user's intention easily and to take suitable response for the user's intension. Especially, we will apply this system for the Post PC and its platform. Also, we use 5 DT Company's 5th Data Glove System (wireless) which is one of the most popular input devices in haptic application field. It can sense data of hand gesture. 5th Data Glove System is basic gesture recognition equipment that can capture the degree of finger stooping using fiber-optic flex sensor and acquires data through this. Also, because it has pitch and roll sensor inside, the pitch and roll of wrist can be also recognized without other equipment possible. In order to implement gesture recognition interface, this study choose 15 basis hand gestures through priority "Korean Standard Sign Language Dictionary [4]" analysis. And 9 overlapped gestures are classified as pitch and roll degree. Thus, we developed 24 significant gesture recognition models. Fig. 1 shows a classification of basis hand gesture used in this study.

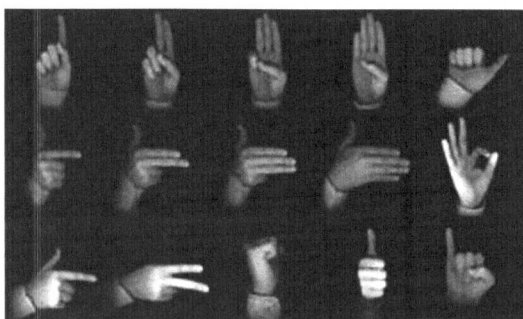

Fig. 1. The classification of basis hand gesture

The captured dynamic gesture data of various users is transmitted to embedded i.MX21 board and server (Oracle 10g RDBMS) through smart gate (notebook computer). The gesture data transmitted to server is used to as a training data for gesture recognition model by Oracle 10g RDBMS's SQL Analytic Function. And, gesture data that is transmitted to embedded i.MX21 is used as the input to fuzzy and neural network recognition module for significant gesture recognition. The architecture of gesture input module is shown in Fig. 2.

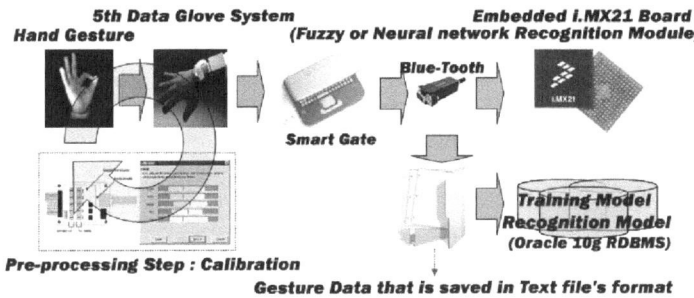

Fig. 2. The architecture of gesture input module

We can use a file-system or spreadsheet programs such as Ms-Office Excel to train and construct the recognition model database. But this approach has several disadvantages. First, many tedious manual steps for data processing, including data inserting, updating and deleting, may cause a lot of trial and error. And, as the database size increases, it takes a lot of time to process the data. Moreover, the administrative works for the source code, data management and manageability are burdens for the analysis. In order to overcome the above disadvantages, we adopt an RDBMS technology for database management. The RDBMS is used to classify stored gesture document data from gesture input module into valid gesture record set and invalid record set (that is, status transition record set) and to efficiently analyze valid record set. In particular, we found that the SQL language, which is the standard language for managing and analyzing data saved in RDBMS, can be greatly helpful in the analysis and segmentation of input gesture data. The analytic function features, which are recently introduced in the SQL language, perfectly provide the analysis power for gesture validity analysis. Fig. 3 shows segmentation result of the input data using the analytic function in SQL.

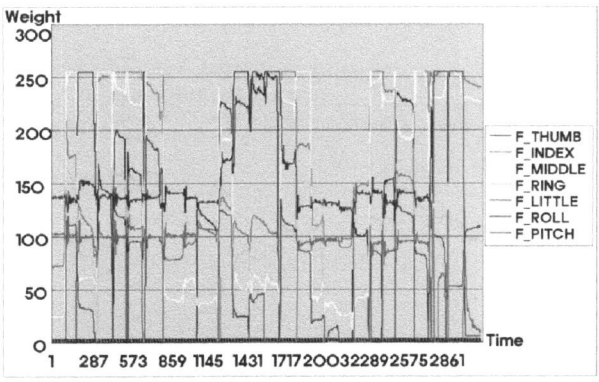

Fig. 3. The segmentation of input data by Analytic Function of SQL

3 Fuzzy Max-Min Module for Hand Gesture Recognition

The fuzzy logic is a powerful problem-solving methodology with a myriad of applications in embedded control and information processing. Fuzzy provides a remarkably simple way to draw definite conclusions from vague, ambiguous or imprecise information. Fuzzy Logic is a paradigm for an alternative design methodology which can be applied in developing both linear and non-linear systems for embedded control and has been found to be very suitable for embedded control applications [5]. Also, fuzzy logic is used in system control and analysis design, because it shortens the time for engineering development and sometimes, in the case of highly complex systems, is the only way to solve the problem and is a structured, model-free estimator that approximates a function through linguistic input/output associations and Fuzzy Logic provides a simple way to arrive at a definite conclusion based upon vague, ambiguous, imprecise, noisy, or missing input information.

In a fuzzy set, all the elements have their grades of membership to that set. This concept would make it possible to give strong flexibility to the conventional set concept. Also, The standard set of truth degrees is the real interval [0,1] with its natural ordering ≤ (1 standing for absolute truth , 0 for absolute falsity) ; but one can work with different domains , finite or infinite , linearly or partially ordered [6]. Namely, fuzzy set defines membership of its elements from U: → [0, 1] and in fuzzy set theory the membership grade can be taken as a value intermediate between 0 and 1 although in the normal case of set theory membership the grade can be taken only as 0 or 1.

The Fuzzy Logic System(hereafter, FLS) consists of the fuzzy set, fuzzy rule base, fuzzy reasoning engine, fuzzifier, and defuzzifier and the design of a FLS includes the design of a rule base, the design of input scale factors, the design of output scale factors, and the design of the membership functions. Also, because each fuzzy relation is a generalization of relation enemy ordinarily, the fuzzy relations can be composed. The fuzzy max-min CRI for fuzzy relation that is proposed in this paper is defined in Fig. 4.

The training and recognition model by the RDBMS is used with input variable of fuzzy algorithm (fuzzy max-min composition), and recognizes user's dynamic gesture through efficient and rational fuzzy reasoning process. Therefore, we decide to take

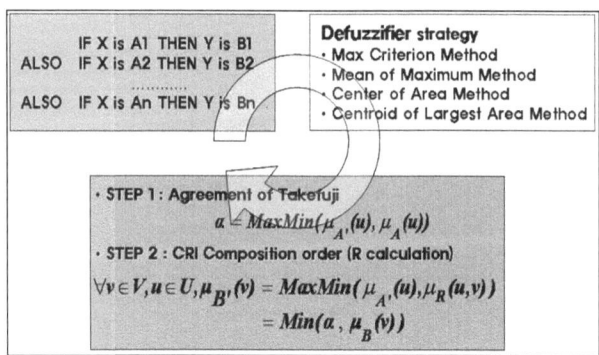

Fig. 4. Fuzzy Max-Min CRI (Direct Method)

the characteristics of gesture input data as the average value over repeated experiment result values, where the repetition number is controlled by several parameters. Input scale factors transform the real inputs into normalized values, and output scale factors transform the normalized outputs into real values. Also, we decide to give a weight to each parameter and do fuzzy max-min reasoning through comparison with recognition model constructed using RDBMS module. The proposed membership function of the fuzzy set is defined as in the following formula (1) and Fig. 5 describes the formula (1) pictorially (if express membership functions partially in case fuzzy input variable is various: "S" in Korean Standard Sign Language Dictionary).

$$
\mu_{tz} = \begin{bmatrix} \dfrac{1}{(s-p)}(x-s)+1 & p<x\leq s \\ 1 & s<x\leq t \\ -\dfrac{1}{(q-t)}(x-t)+1 & t<x\leq q \end{bmatrix} \tag{1}
$$

If compare easily, "S" in Korean Standard Sign Language Dictionary is hand gesture that corresponds to rock in game of "scissors-rock kerchief". The hand gesture of all three types about "S" is currently shown in the Fig. 5: 1) first case of "S" is that holds up thumb finger on the index and middle finger in state of the other all fingers stoop naturally (where, This gesture is defined as "THUMB_UP" for hand motion), 2) second case of "S" is that holds up the thumb finger on index finger (where, This gesture is defined as "THUMB_OUT"), and 3) finally, "S" is that puts the thumb in the remainder stooped 4 fingers (where, This gesture is defined as "THUMB_IN"). And according to stooping degree (spread-extend, stoop and normal) of the thumb, we prescribed 9 kinds of hand actions about each hand motions.

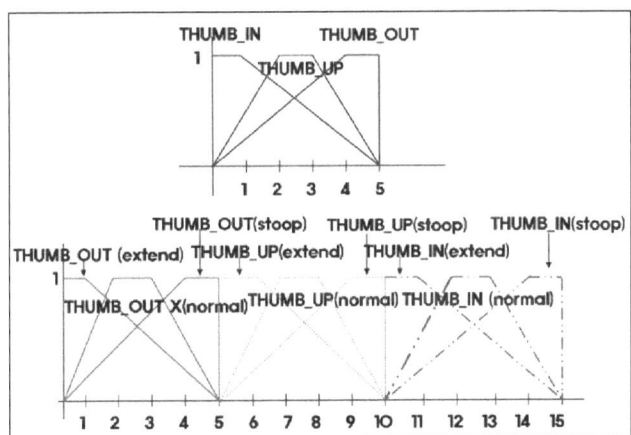

Fig. 5. The fuzzy membership functions

4 Neural-Network Module for Hand Gesture Recognition

An Artificial Neural Network (hereafter, ANN) is an information processing paradigm that is inspired by the way biological nervous systems, such as the brain, process information and is configured for a specific application, such as pattern recognition or data classification, through a learning process. Neural networks, with their remarkable ability to derive meaning from complicated or imprecise data, can be used to extract patterns and detect trends that are too complex to be noticed by either humans or other computer techniques [7]. Also, neural network is a powerful data modeling tool that is able to capture and represent complex input/output relationships. The true power and advantage of neural networks lies in their ability to represent both linear and non-linear relationships and in their ability to learn these relationships directly from the data being modeled. ANN is composed of a large number of highly interconnected processing elements working in unison to solve specific problems and the most common neural network model is the multilayer perceptron (MLP). The proposed algorithm for perceptron training and Generalized Delta Rule (hereafter, GDR) for weight conversion process are defined as in the following formula (2).

$$y(t) = F(W_i \cdot A_i - \theta) \qquad W_i(t+1) = W_i(t) + @\cdot(d_i - y_i)\cdot A_i(t)$$

$$
\begin{bmatrix}
where, \\
y(t): output\ of\ R\ layer's\ PE \\
F: fucntion\ that\ using\ in\ R\ layer's\ PE \\
W_i: weights\ of\ A\ layer's\ \& R\ layer's\ PE \\
A_i: input\ of\ A\ layer's\ PE \\
\theta: critical\ value\ of\ R\ layer's\ PE
\end{bmatrix}
\begin{bmatrix}
where, \\
W_i(t+1): weight\ after\ tarining \\
W_i(t): weight\ from\ input\ i\ (time\ is\ t) \\
@: training\ rate \\
(d_i - y_i): error \\
W_i(t): input\ pattern
\end{bmatrix}
\quad (2)
$$

The gesture data is repeatedly presented to the neural network. With each presentation, the error between the network output and the desired output is computed and fed back to the neural network. The neural network uses this error to adjust its weights such that the error will be decreased. This sequence of events is usually repeated until an acceptable error has been reached or until the network no longer appears to be learning.

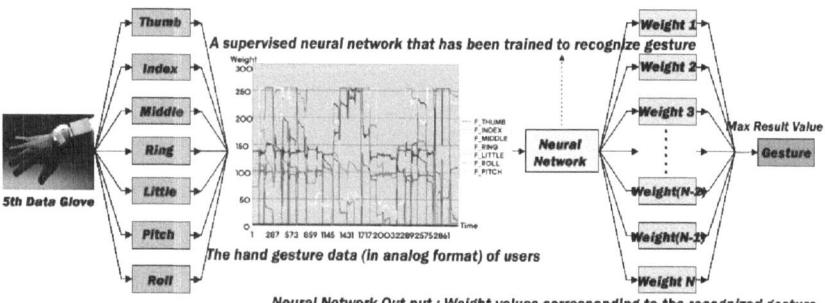

Fig. 6. The basic concept of gesture recognition system using neural network

The captured dynamic gesture data of various users is transmitted to embedded i.MX21 board and saved as a document (text type). The significant hand gesture data (in analog format) of users is then fed into a neural network that has been trained to make the association between the gesture data and a weight value that corresponds to the gesture recognition model by RDBMS. The output from the neural network is then calculated by weight value's summation and this is result of user's dynamic gesture recognition. The basic concept of gesture recognition system using neural network is shown in Fig. 6.

5 Experiments and Results - Performance Evaluation

Experimental environment consists of blue-tooth module with i.MX21 board under embedded LINUX and implemented wireless-ubiquitous environment. This can be analyze by one method to solve conditionality in space, complexity between transmission mediums (cable elements), limitation of motion and incommodiousness on use soon in wire transmission method. In the proposed 2 recognition system, gesture input module uses blue-tooth module for wireless communication. And for RDBMS-based segmentation module, we use Oracle 10g RDBMS. The operating system platform is LINUX. 5th Data Glove System transmits 7 kinds of data (5 gesture data, 2 Pitch & Roll data) to embedded i.MX21 board via smart-gate notebook computer with transmission speed of maximum 19,200 bps in case of blue-tooth module.

The proposed fuzzy gesture recognition system's file size is 141 Kbytes and it can calculate 200 samples per seconds on Wearable Personal Station (hereafter, WPS)-embedded i.MX21 board. The overall process of recognition system consists of three major steps. In the first step, the user prescribes user's hand gesture using 5th Data Glove System (wireless), and in the second step, the gesture input module captures user's data and change characteristics of data by parameters, using RDBMS module to segment recognition model. In the last step, it recognizes user's dynamic gesture through a fuzzy reasoning process and a union process in fuzzy min-max recognition module, and can react, based on the recognition result, to user via speech and graphics offer. And also the appropriately segmented data becomes the input data for fuzzy max-min recognition module. Finally, the calculated fuzzy value for gesture recognition is used as a judgment for user action. The process of automatic human gesture recognition is shown in Fig. 7.

The proposed neural network gesture recognition system's file size is 215 Kbytes and it can calculate 175 samples per seconds on WPS. The overall system process of training module using neural network are as following: 1) after initialization of critical value and Connection weight, input training pattern and target pattern by RDBMS, 2) calculate output of R layer's PE, and compare target output and actuality output and 3) if target output and actuality output are same, it end training. Otherwise, train last pattern after error control. Also, The overall system process of recognition module using neural network are as following: 1) After it loading connection weight, and input recognition pattern using 5th Data Glove System, 2) it calculate output results, and after compare with the recognition rate, remember maximum value, 3) If number of trained pattern conforms, it display recognition result. The process of gesture training and recognition in neural network recognition system are shown in Fig. 8 and Fig. 9.

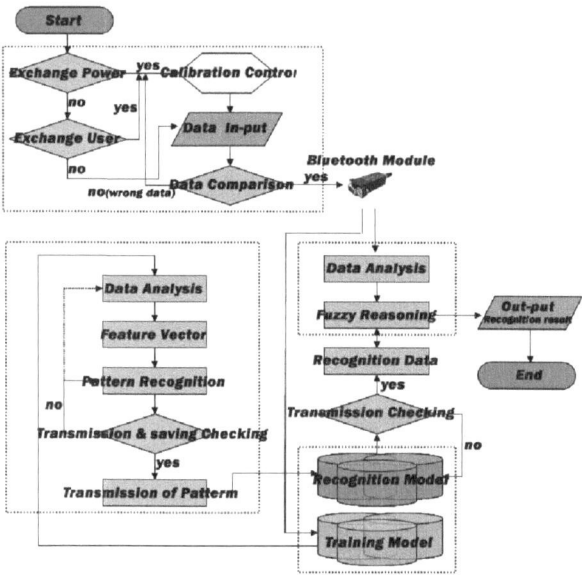

Fig. 7. The flow-chart of fuzzy gesture recognition system

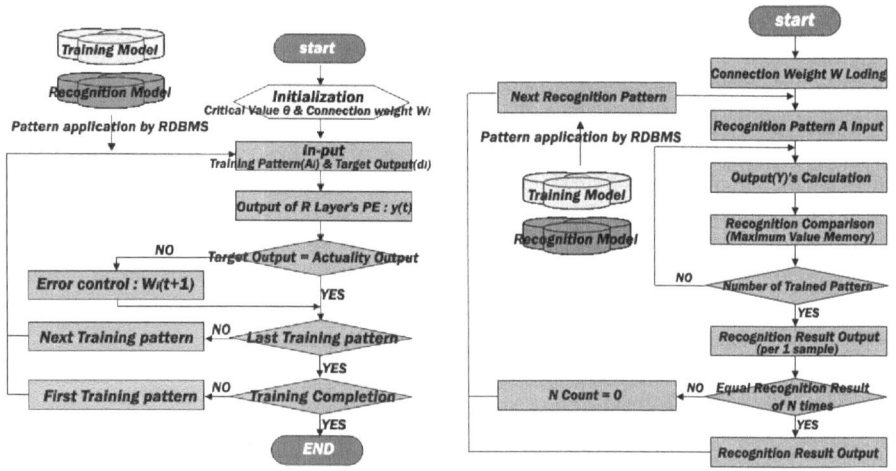

Fig. 8. The flow-chart of training module   **Fig. 9.** The flow-chart of recognition module

Experimental set-up is as follows. The distance between server (Oracle 10g RDBMS) and smart-gate is about radius 10M's ellipse form. As the gesture, we move 5th Data Glove System to prescribed position. For every 20 reagents, we repeat this action 10 times. Experimental result, Fig. 10 shows the average recognition rate of 98.8% in fuzzy max-min module and 96.7% in neural network recognition module about significantly dynamic gestures. Also, Fig. 11 shows the average recognition

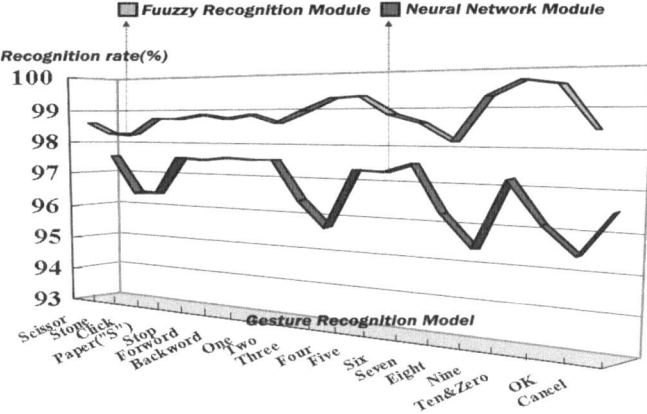

Fig. 10. The average recognition rate

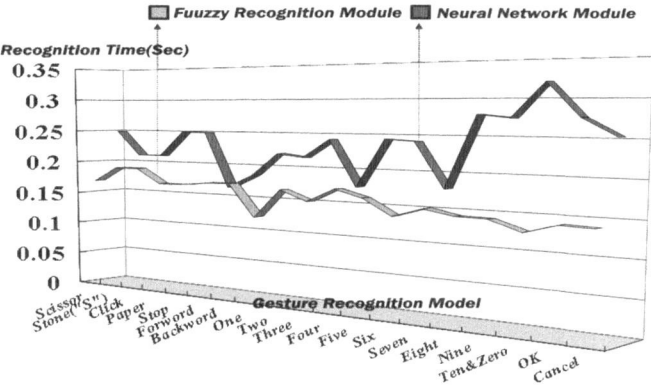

Fig. 11. The average recognition time

time of 0.16 seconds in fuzzy max-min module and 0.23 seconds in neural network recognition module about each recognition models. The root causes of errors between recognition of gestures are various: that is, imprecision of prescribed actions, user's action inexperience, and the changes in experiment environment physical transformation at 5th Data Glove System fiber-optic flex sensor.

6 Conclusions

The Post-wearable PC is subset of ubiquitous computing that is the embedding of computers in the world around us. In this paper, we implemented the hand gesture recognition interface that can recognize the dynamic gesture of the user in real-time manner under wireless integrated environment using blue-tooth module. In this paper, through analysis and evaluation of performances such as recognition rate, recognition time(processing speed of system) and recognition system's whole size about 2 each gesture recognition systems, we could certify that gesture recognition system using fuzzy algorithm is reasonable and effective gesture recognition module than using neural network in the Post PC. In the future, by integrating hand gesture recognition system with other haptics such as smell, taste, hearing and sight, we would like to develop advanced multi-modal HCI technology.

Acknowledgement

This work was supported by the MIC (The Ministry of Information & Communication) and NGV (Next Generation Vehicle technology).

References

1. http://www.iita.re.kr
2. Jong-Sung Kim, Won Jang, Zeungnam Bien.: A Dynamic Gesture Recognition System for the Korean Sign Language (KSL). Ph.D. Thesis (1996)
3. M. Ernst and M. Banks.: Humans integrate visual and haptic information in a statistically optimal fashion. Letters to Nature, 415 (2002)
4. Seung-Guk Kim.: Korean Standard Sign Language Dictionary, Osung publishing company (1995) 107–112
5. http://www.aptronix.com/fide/whyfuzzy.htm
6. http://plato.stanford.edu/entries/logic-fuzzy/#1
7. http://www.doc.ic.ac.uk/~nd/surprise_96/journal/vol4/cs11/report.html

Implementation and Performance Evaluation of Glove-Based HCI Methods: Gesture Recognition Systems Using Fuzzy Algorithm and Neural Network for the Wearable PC

Jeong-Hoon Shin, Jung-Hyun Kim, and Kwang-Seok Hong

School of Information and Communication Engineering, Sungkyunkwan University, 300, Chunchun-dong, Jangan-gu, Suwon, KyungKi-do, 440-746, Korea
Only4you@chol.com, kjh0328@skku.edu, kshong@skku.ac.kr
http://only4you.or.kr, http://hci.skku.ac.kr

Abstract. Ubiquitous computing is a new era in the evolution of computers. After the mainframe and PC (personal computers) phases, the phase of ubiquitous computing device begins. In this paper, we implement and evaluate glove-based HCI (Human Computer Interaction) methods using fuzzy algorithm and neural network for post PC in the ubiquitous computing. Using glove, we implement hand gesture recognition systems for the wearable PC. One system uses combination of fuzzy algorithm and RDBMS (Relational Database Management System) module, the other system uses neural network. Both systems are implemented on the platform of minimized wearable computer (based on i.MX21). After implementation, we conduct some performance evaluation in the mobile condition. And then we discuss strength and weakness of each method. Finally, we suggest possible improvements methods for HCI based on the wearable computers in the mobile condition.

1 Introduction

The idea of ubiquitous computing first arose from contemplating the place of today's computer in actual activities of everyday life. Ubiquitous computing is about networked microprocessors embedded in everyday objects: not just cellular phones and home appliances but also books, bookshelves, bus stops and bathtubs - all talking to each other over wireless links. Hundreds of internet enabled computers per human being, none of them resemble a conventional keyboard and monitor [1][2].

A common focus shared by researchers in mobile, ubiquitous and wearable computing is the attempt to break away from the traditional desktop computing paradigm. Computational services need to become as mobile as their users [3][4][5].

In this paper, we implement gesture recognition systems using fuzzy algorithm and neural network for wearable PC. And then, we conduct efficiency test comparing two systems. Most of the applications using wearable computers need user's mobility. For this reason, we evaluate the performance of gesture recognition systems at the posture of repose and at the posture of movement, respectively. After that, we analyze the problems of the implemented systems and suggest improving HCI method for wearable computers.

I. Bloch, A. Petrosino, and A.G.B. Tettamanzi (Eds.): WILF 2005, LNAI 3849, pp. 139–145, 2006.
© Springer-Verlag Berlin Heidelberg 2006

2   Implementation of Gesture Recognition Systems

In this paper, we use 5DT's Data Glove for hand gesture input device. This device is one of the most popular input devices in gesture recognition application field. And, we also use Motorola's i.MX21 ADS board for wearable PC platform, using this platform we minimized the size of the platform for the purpose of handiness. Fig. 1 and Fig.2 shows Motorola's i.MX21 ADS board and minimized wearable PC platform, respectively.

Actual size of the platform which we used in this paper is as small as name card. Using these device and platform, we implemented 24 significant gestures recognition systems.

Fig. 1. i.MX21 ADS Board

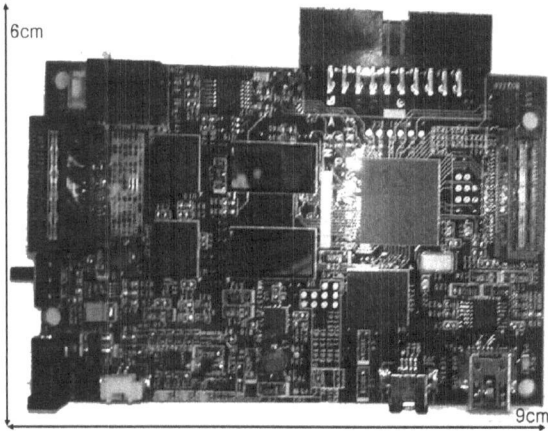

Fig. 2. Minimized wearable PC platform (based on i.MX21 ADS board)

Gesture signals from the glove are consists of 9 bytes packet and has the following structure: header, finger1, finger2, finger3, finger4, finger5, pitch, roll and checksum. The header value is always 0x80 (128 in decimal). For the purpose of indicating user's gesture, each finger value is ranged from 0 to 255. Low value indicates inflexed finger, and a high value indicates flexed finger. According to the different user's hand size, gesture signals to the same gesture may different. To meet this problem effectively, following formula (1) software calibration could be used.

$$Dynamic\_range = value\_max - value\_min$$

$$Value\_scaled$$

$$= (value\_measured - value\_min) * (\frac{255}{dynamic\_range})$$

(1)

2.1 Gesture Recognition System Using Fuzzy Algorithm

Fig. 3 shows whole configuration of the proposed gesture recognition system using Fuzzy algorithm. The captured dynamic gesture signals of various users using glove are transmitted to the minimized wearable PC platform and server (Oracle 10g RDBMS) at the same time. The transmitted signals to the server are used for training data for gesture recognition model by Oracle 10g RDBMS's SQL Analytic Function. And, transmitted signals to the minimized wearable PC platform are inputted to the fuzzy recognition module for significant gesture recognition.

The fuzzy max-min CRI for fuzzy relations which is proposed in this paper is shown in Fig. 4.

Our gesture recognition system endows weight value to the gesture signals and handles fuzzy max-min reasoning through the process of comparing recognition

Fig. 3. Configuration of proposed gesture recognition system using Fuzzy algorithm

model constructed in RDBMS module. The proposed position function of fuzzy set is defined as in the following formula (2).

$$\mu_{tz} = \left[\begin{array}{cc} \dfrac{1}{(s-p)}(x-s)+1 \quad p<x\leq s, \ -\dfrac{1}{(q-t)}(x-t)+1 \quad t<x\leq q \\ 1 \qquad\qquad s<x\leq t \end{array} \right] \tag{2}$$

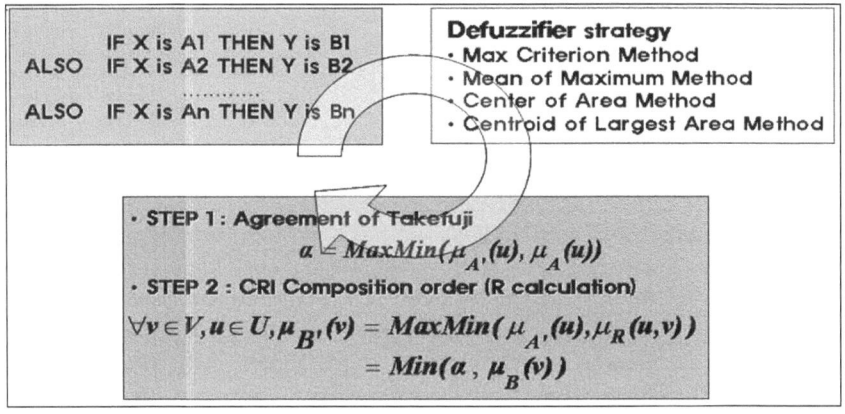

Fig. 4. Fuzzy Max-Min CRI (Direct Method)

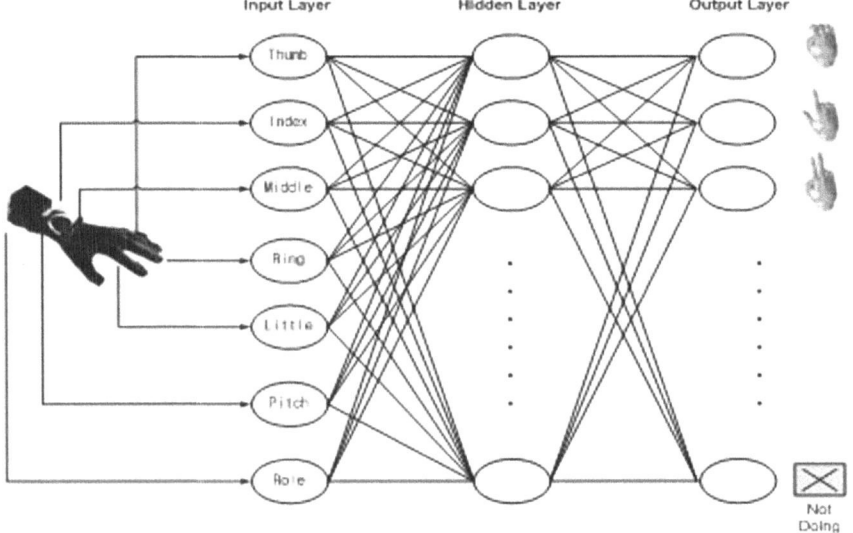

Fig. 5. Conceptual diagram of gesture recognition system using Multi Layer Perceptron

2.2 Gesture Recognition System Using Neural Network

In this paper, we implement gesture recognition system using multilayer perceptron and generalized delta rule on the basis of minimized wearable PC platform. Conceptual diagram of whole system is shown in Fig. 5.

We use Levenberg-Marquardt Back-Propagation (LM-BP) learning algorithm. Learning algorithm used in this paper is shown in formula (3).

$$W_{m+1} = W_m - [J^T(W_m)J(W_m) + \lambda_m I]^{-1} J^T(W_m)e(W_m)$$

$$where\ J_{ki} = \frac{\partial e_k}{\partial w_i} \qquad Jacobian matrix \tag{3}$$

3 Experiments and Results

The most important thing to HCI method for wearable computer is accurate recognition of user's intention, because most of users have a tendency to move when they use wearable computers. For this reason, we analyzed the recognition rate of gesture recognition systems according to user's walking velocity. 20 subjects and 15 gestures were selected for the test. As the user's walking velocity increased, wearing condition of the glove grows from bad to worse. Fig. 6 shows the configuration of our experiments.

Fig. 7 shows recognition rate according to walking velocity. As shown in Fig. 7, as the user's walking velocity increased, average recognition rate of implemented gesture recognition system falls off. These results come from the change of glove wearing condition.

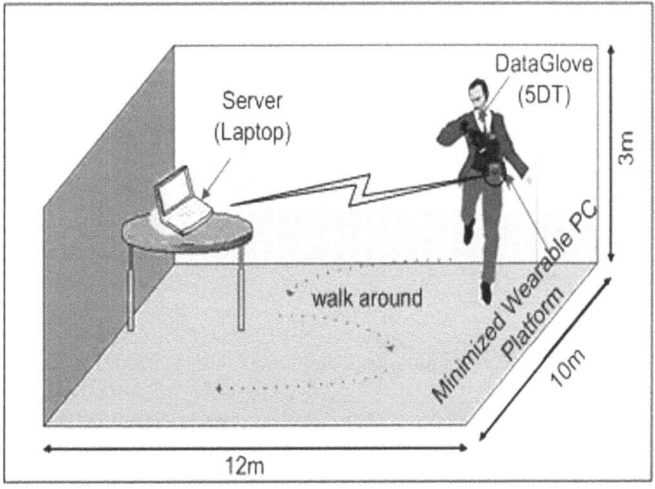

Fig. 6. Configuration of conducted experiments

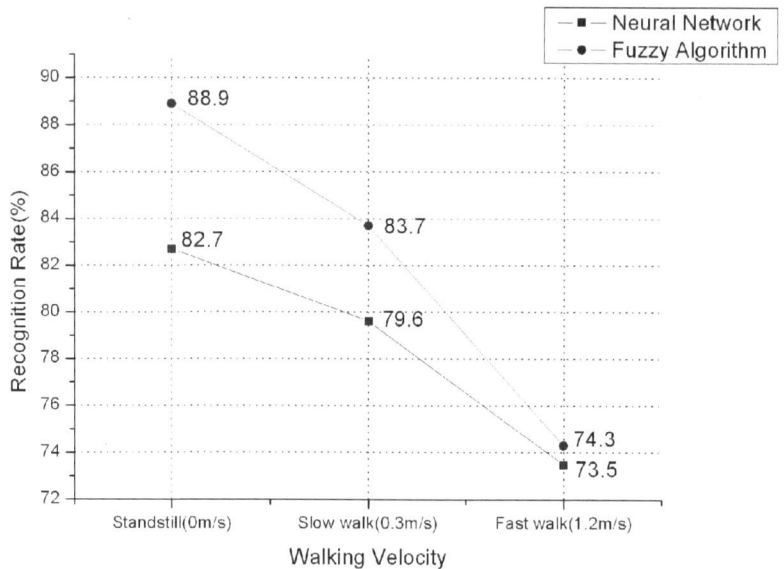

Fig. 7. Recognition rate according to walking velocity

4 Conclusions

Future computing environments promise to break the paradigm of desktop computing. To do this, new trends of wearable computer need convenient and easy user interface in mobile condition [5][6][7].

In this paper, we implemented gesture recognition systems using glove as a novel HCI method for wearable computer. And then, we conducted efficiency test in mobile condition. Though, our experiment results show somewhat lower recognition rate in mobile condition, these kinds of methods could be used effectively in mobile condition, ubiquitous computing and wearable computing.

To improve the accuracy of user interface using glove in mobile condition, the problems result from the different hand size should be solved. We hope that our efforts in this field encourage other researchers to focus more on HCI methods for wearable computers.

References

1. "Wearable Computing Meets Ubiquitous Computing", http://alumni.media.mit.edu/~rhodes/Papers/wearhive.html
2. "Ubiquitous and wearable computing", http://project.cyberpunk.ru/idb/ubicomp.html
3. Pratt, V. R.: Thumbcode: A Device-Independent Digital Sign Language. http:// boole. stanford.edu/ thumbcode (1998)

4. Balaniuk R, Laugier C : Haptic interfaces in generic virtual reality systems.In: Proceedings of the IEEE/RSJ International Conference on Intelligent Robots and Systems (2000) 1310–1315
5. M. Weiser: The computer for the twenty-first century, Scientific American, (1991) 94-104
6. M. Weiser and J. S. Brown: The coming age of calm technology", http://www.ubiq.com/hypertext/weiser/acmfuture2endnote.htm
7. D. Roy, et al.: Wearable audio computing: A survey of interaction techniques, http://www.media.mit.edu/~nitin/NormadicRadio/AudioWearables.html

A Hybrid Warping Method Approach to Speaker Warping Adaptation

Yong-Wan Roh, Jung-Hyun Kim, Dong-Joo Kim, and Kwang-Seok Hong

School of Information and Communication Engineering, Sungkyunkwan University, 300,
Chunchun Dong, Jangan-gu, Suwon, Kyungki-do, 440-746 Korea
elec1004@hotmail.com, kjh0328@skku.edu,
radioguy@korea.com, kshong@skku.ac.kr
http://hci.skku.ac.kr

Abstract. The method of speaker normalization has been known as the successful method for improving the speech recognition at speaker independent speech recognition system. This paper propose a new power spectrum warping approach to making improvement of speaker normalization better than a frequency warping. The power spectrum warping uses Mel-frequency cepstral of Mel filter bank in MFCC. Also, this paper proposes the hybrid VTN combined the power spectrum warping and a frequency warping. Experiment of this paper did a comparative analysis about the recognition performance of the SKKU PBW DB applied each the power spectrum is 3.06%, and hybrid VTN is 4.07% word error rate reduction as word recognition performance of baseline system.

1 Introduction

Generally speech recognition system intends spoken independent system to depend on restrictive training model. In this case speech recognition system is made worse performance because of all speakers not considering vocal form variation [1]. Speaker normalization method designed complement for this problem and various papers are introduced several method. Especially speech recognition system based HMM is shown performance improvement. Normalizing variation of vocal shape is VTN (VTN; Vocal Tract Normalization) that is closely related to vocal length variation of each speaker. Vocal length has difference according to sex distinction. Since the vocal tract length can vary from approximately 13cm for adult females to over 18cm for adult males, formant center frequencies can vary by as much as 25% between speakers. A speech spectrum frequency warping method use for VTN and this method have introduced a lot of papers. The method easily realized MFCC feature analysis through linear warping of frequency axis in MFB [2][3].

This paper proposed power spectrum warping method that has good recognition performance better than existing intra-speaker normalization. The proposed power spectrum warping is similar to existing frequency warping, that MFCC feature analysis adjusts MBF. Existing frequency warping method warp frequency axis of MFB while the proposed power spectrum warping method warp power spectrum axis of MFB.

Also, this paper proposed hybrid VTN that combines frequency warping method and power spectrum warping method. Existing frequency warping method deals with

I. Bloch, A. Petrosino, and A.G.B. Tettamanzi (Eds.): WILF 2005, LNAI 3849, pp. 146–155, 2006.

two important points. The first point estimates warping factor. The second, recognition procedure deals with efficiently estimated warping factor. The proposing power spectrum warping and hybrid VTN consider two important factors. The power spectrum warping factor for estimation is used the maximum likelihood method in general used frequency warping. The multiple-pass procedure method is used recognition procedure for training in generally used frequency warping.

We proposed to recognition of performance evaluation that we used speech DB of Korean word unit. Each recognition performance comparison and analysis based on recognition performance of base line system and frequency warping system that we evaluate performance of intra-speaker normalization.

2   Frequency Warping

In this chapter, broadly known frequency warping method for speaker normalization deals with introduced content from Li Lee and L.Weilling [2].

Frequency warping method is representative method of vocal track length normalization that decreases speech signal variation of each speaker.

2.1   MFB for Frequency Warping

Generally, vocal track normalization is performed through feature vector variation in front end. Especially frequency warping method based spectrum analysis in speech feature vector analysis method that it implements easily using MFCC [4]. Frequency warping is normalization method of vocal track length variation with frequency axis warping. This method executes frequency axis warping to adjust MFB in MFCC procedure process [5][6]. According to frequency warping factor, warped form of Mel frequency is shown Fig. 1. If the warping factor value is α less than 1, this case is the same to extend frequency area of spectrum. If the warping factor value is α more

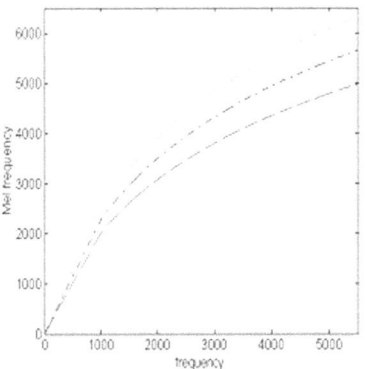
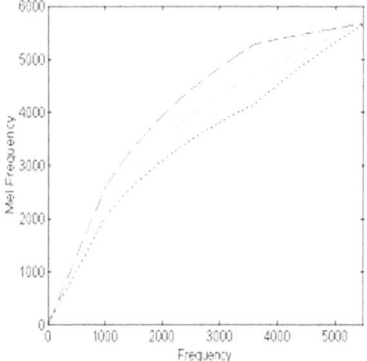

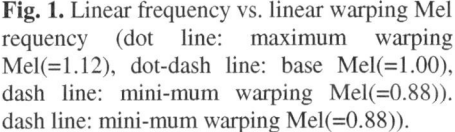

Fig. 1. Linear frequency vs. linear warping Mel requency (dot line: maximum warping Mel(=1.12), dot-dash line: base Mel(=1.00), dash line: mini-mum warping Mel(=0.88)). dash line: mini-mum warping Mel(=0.88)).

Fig. 2. Inear frequency vs. piecewise linear warping Mel frequency (dot line: maxi mum warping Mel(=1.12), dot-dash line: base Mel(=1.00), dash line: mini-mum

than 1, this case is the same to compress extend frequency area of spectrum. If frequency area extends above line in Fig.1, MFB area has effect out of part in original spectrum area. If frequency area compress below line in Fig.1, MFB area have problem not including all kind of information in original spectrum area. A piecewise linear warping method is devised to solve the problem that generally linear warping is known to show more robust speech recognition performance. Fig. 2 is shown that Mel frequency is applied to piecewise linear warping method. In this paper, we used piecewise linear warping for frequency warping procedure. It is shown MFB procedure part in general MFCC procedure that is follow as formula (1).

$$\widetilde{S}[l] = \sum_{k=0}^{k=K/2} S[k] \cdot M_l[k] \qquad l = 0,1,\ldots, L\text{-}1 \tag{1}$$

Where $S[k]$ is power spectrum, $M_l[k]$ is mel triangle filter, L is mel triangle band-pass filter number, K is resolution of FFT. Piecewise linear warping is followed as formula (2).

$$\widetilde{S}^{\alpha}[l] = \sum_{k=0}^{k=K/2} S[k] \cdot M^{\alpha}{}_l[k] \qquad l = 0,1,\ldots, L\text{-}1 \tag{2}$$

2.2 Estimation of Frequency Warping Factor

One of important things in frequency warping method look for frequency factor that are to decide normalization of vocal track for specification speaker. Estimation of frequency factor is to decide vocal track length normalization factor. Frequency warping factor can be presented by rate of a standard vocal track length and specification speaker vocal track length. In generally, vocal track length is in inverse proportion to frequency warping factor. We present estimation of frequency warping factor by mathematical model. First of all, feature vector about speaking of baseline is not applied to frequency warping factor that is followed as formula (3).

$$X_i = \{x_i(0),\ldots, x_i(T)\} \tag{3}$$

$$X^{\alpha}{}_i = \{ x^{\alpha}{}_i(0),\ldots, x^{\alpha}{}_i(T)\} \tag{4}$$

Where i is utterance speaker, T is total number of feature vector. Feature vector from applied frequency warping is executed formula (2) in 2.1 section. Feature vector from applied frequency factor is fallowed as formula (4). In addition a mark of all kinds of recognition candidate words (transcription set) is fallowed as (6).

The optimal frequency factor of speaker i gets to adjust maximum likelihood method in decoding step as formula (6).

$$W = \{\omega_1, \omega_2,\ldots,\omega_N\} \tag{5}$$

$$\hat{\alpha} = \arg\max P(X^{\alpha}{}_i | \lambda, W) , \quad for \ \alpha \tag{6}$$

$$0.88 \leq \alpha \leq 1.12 \ , spaced \ 0.02 \tag{7}$$

Optimum frequency warping factor a about HMM acoustic model and all kind of recognition candidate words w can be define maximum likelihood of warped utterance feature vector. A of searching area for optimal frequency warping factor defines using 25% difference of adults vocal track length variation that is followed as formula (7).

2.3 Recognition Processing of Frequency Warping

In front of section, we introduced how to estimation of optimum frequency warping factor. This section applies to optimum frequency warping factor form utterance speech of speaker I that deals with speech recognition processing parts.

This part is demanded effective processing because it is affected by recognition performance and processing time. This paper considers multiple-pass processing method for effective recognition processing. Estimation method of warping factor based mixture for processing time improvement of recognition processing introduce a lot of paper while this method shows low recognition performance less then multiple-pass processing method. Multiple-pass processing applied frequency warping consists of three steps that is followed as next.

We train HMM decoding about feature vector of not warped utterance speech that gets candidate word ω of the highest score. We estimate optimum frequency warping factor a using formula (6) about each frequency warping factor $\hat{\alpha}$ applied feature vector X_i^{α}. We decide the final recognition word w executing again step1 about optimum frequency warping factor a applied feature vector $X_i^{\hat{\alpha}}$.

$$\hat{\alpha} = \arg\max P(X^{\alpha}{}_i \mid \lambda, \omega) \ , for \ \alpha \tag{8}$$

$$\hat{w} = \arg\max P(X^{\hat{\alpha}}{}_i \mid \lambda, w) \ , for \ w \tag{9}$$

3 Power Spectrum Warping

In this paper, we deal with power spectrum warping method that proposed paper freshly for performance improvement of speaker normalization. Power spectrum warping have similar mechanism to existing frequency warping method that can realize how to adjusting of MBF in MFCC. The proposed power spectrum warping is similar to existing frequency warping, that MFCC feature analysis adjusts MBF. In general, we can confirm to show up different spectral envelope and formant because vocal track shape variation according to speaker in same word utterance. Especially formant position and spectral envelope of man and woman confirm easily difference that is shown Fig. 3. Existing frequency warping method performs spectrum information warping of frequency axis emphasis on formant position information of speech spectrum for each speaker of vocal track normalization. The other side, spectrum warping method performs normalization of each speaker spectral envelope and formant position because power spectrum warping is applied to Mel scale power spectrum axis warping of speech spectrum.

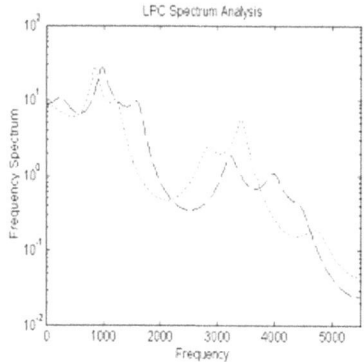

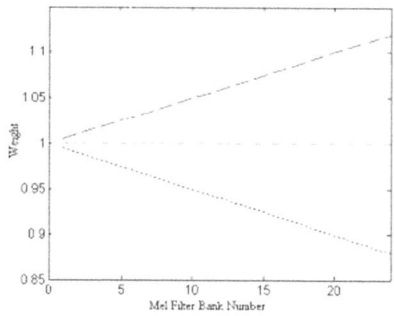

Fig. 3. Male and female LPC spectrum analysis of utterance /a/ (dash line: female, dot line: male)

Fig. 4. Mel filter bank number vs. weight value of power spectrum (dash line: maximum warping weight value(=1.12), dot-dash line: base weight value(=1.00), dot line: minimum warping weight value(=0.88), the number of filter bank = 24)

3.1 Power Spectrum Warping for MFB

Power spectrum warping can normalize vocal track variation by power spectrum warping of each mel filter bank (MFB). Power spectrum warping factor β in MFB is applied to linear warping function. That result graph of function is followed as Fig. 4. Compared frequency warping, first case β less than 1, this case affect to compress frequency axis and the other case effect to extend frequency axis.

Linear warping function of power spectrum warping $w_\beta(l)$ expresses formula that is followed as formula (10).

We substitute linear warping function of formula (10) for doing power spectrum warping in DCT processing of MFCC.

For effective possessing, warping function $w_\beta(l)$ subtract out of log that is written as formula (11).

$$\omega_\beta(l) = \frac{\beta-1}{L} \cdot (l+1) + 1 \qquad l = 0,1,\ldots, L\text{-}1 \tag{10}$$

$$c[j] = \sum_{i=0}^{L-1} \ln(\tilde{S}[l]^{\omega_\beta(l)} \cdot \cos[\frac{\pi}{L}(j+0.5)] \qquad j = 0, 1, \ldots, C\text{-}1 \tag{11}$$

$$c[j] = \sum_{i=0}^{L-1} \omega_\beta(l) \cdot \ln(\tilde{S}[l]) \cdot \cos[\frac{\pi}{L}(j+0.5)] \qquad j = 0, 1, \ldots, C\text{-}1 \tag{12}$$

3.2 Estimation of Power Spectrum Warping Factor

Frequency warping is in inverse proportion ratio of warping factor according to vocal track length and power spectrum warping is in proportion ratio of warping factor according to vocal track length. Processing of power spectrum warping factor estimation is the same processing of frequency warping factor estimation. Feature vector X^{β}_{i} from applied power spectrum factor is fallowed as formula (13).

$$X^{\beta}_{i} = \{ x^{\beta}_{i}(0), ..., x^{\beta}_{i}(T) \} \tag{13}$$

$$\hat{\beta} = \arg \max P(X^{\beta}_{i} | \lambda, W_{i}) \, , for \ \beta \tag{14}$$

$$0.88 \le \beta \le 1.12 \quad , spaced \ 0.02 \tag{15}$$

Optimum power spectrum warping factor B of frequency warping can get using maximum likelihood in HMM decoding that is followed as formula (14). Searching area to decide optimum power spectrum warping factor is same searching area of frequency warping that is followed as formula (15).

3.3 Recognition Processing of Power Spectrum Warping Factor

This section deals with recognition processing part applying optimum power spectrum warping factor $\hat{\beta}$ to deal with in front of section from utterance speaker of speaker i. This part of 2.3 section apply to multiple-pass same speech recognition processing part of frequency warping. Multiple-pass processing applied power spectrum warping is consisted of three steps that are followed as next.

(1) We train HMM decoding about feature vector of not warped utterance speech that gets candidate word $\hat{\omega}$ of the highest score.

(2) We estimate optimum power spectrum warping factor α using formula (14) about each power spectrum warping factor a applied feature vector X^{β}_{i}.

(3) We decide the final recognition word $\hat{\omega}$ executing again step1 about optimum power spectrum warping factor $\hat{\beta}$ applied feature vector $X^{\hat{\beta}}_{i}$.

4 Hybrid VTN

This chapter deals with hybrid VTN that combine frequency warping method and power spectrum warping of chapter 3. We propose second time hybrid VTN for performance improvement of speaker normalization.

4.1 Estimation of Hybrid VTN Warping Factor

Warping processing of MFB in MFCC to train hybrid VTN equals to definite frequency warping of section 2.1 and power spectrum warping of section 3.1.

Hybrid VTN can estimate optimum hybrid VTN warping factor $\hat{\gamma}$ considering two methods. Two methods of optimum warping factor estimation deal with content in section 2.2 and 2.3. But this paper estimate in serial two optimum warping factors ($\hat{\alpha}, \hat{\beta}$) because two warp factors coincident to estimate is required a lot of calculation. First, estimation of hybrid VTN warping factor γ estimates optimum warping factor of frequency warping. And next estimation train to estimate optimum warping factor of power spectrum warping. In this way, reason in order of decision is localized only warping of vocal track length (formant center frequency) for vocal track normalization to frequency warping while power spectrum warping consider not only but also spectrum envelop normalization. Optimum hybrid VTN warping factor $\hat{\gamma}$ of speaker i is followed as formula (16).

$$\hat{\alpha} = \arg\max P(X^{\alpha}_i \mid \lambda, W_i) \text{ , for } \alpha$$

$$\hat{\beta} = \arg\max P(X^{\beta}_i \mid \lambda, W_i) \text{ , for } \beta \qquad (16)$$

$$\hat{\gamma} = (\hat{\alpha}, \hat{\beta})$$

4.2 Recognition Processing of Hybrid VTN

In front of section, we deal with how to estimate optimum hybrid VTN warping factor $\hat{\gamma}$. In this section, we explain recognition processing part applying to optimum hybrid VTN warping factor $\hat{\gamma}$ from speaker i. Recognition processing of hybrid VTN apply to multiple-pass same speech recognition processing part of 2.3 section and 3.3 section.

(1) We train HMM decoding about feature vector X_i of not warped utterance speech that gets candidate word $\widetilde{w}$ of the highest score.

(2) First, we estimate optimum frequency warping factor α a using formula (17a) about each frequency factor a applied feature vector X^{α}_i. And then we apply to optimum frequency warping factor a using about each frequency factor $\hat{\alpha}$ applied power spectrum $X^{\hat{\alpha}}_i$. We estimate optimum power spectrum warping factor a using formula (17b) about each power spectrum factor a applied feature vector $X^{\hat{\alpha}.\beta}_i$. At the last, we apply to formula (17a) and (17b) and then we define optimum hybrid VTN warping factor r.

$$\hat{\alpha} = \arg\max P(X^{\alpha}_i \mid \lambda, \widetilde{\omega}) \text{ , for } \alpha \qquad (17a)$$

$$\hat{\beta} = \arg\max P(X^{\hat{\alpha}.\beta}_i \mid \lambda, \widetilde{\omega}) \text{ , for } \beta \qquad (17b)$$

$$\hat{\gamma} = (\hat{\alpha}, \hat{\beta}) \qquad (17c)$$

(3) We decide the final recognition word $\hat{\omega}$ executing again step1 about optimum hybrid VTN warping factor $\hat{\gamma}$ applied feature vector $X^{\hat{\gamma}}{}_i$.

$$\hat{\omega} = \arg \max P(X^{\hat{\gamma}}{}_i \mid \lambda, \omega_i) \ , \ for \ \omega \tag{18}$$

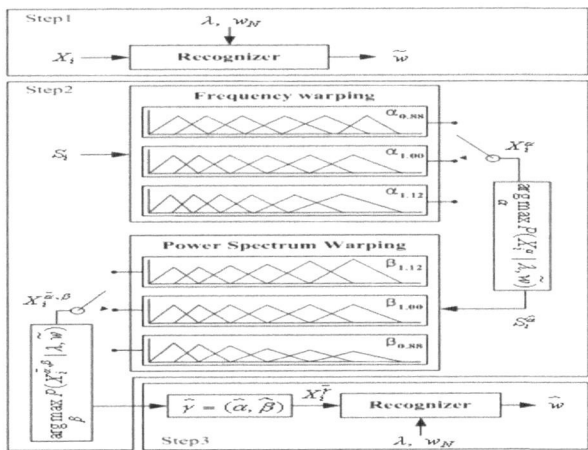

Fig. 5. HMM recognition procedures for hybrid VTN

5 Experimental Results

This paper using baseline system is consisted of as after. First of all, front end com-
poses of basic feature vector including 12 order MFCC feature vector and energy feature vector.

Also we use whole 39 order feature vector applied first order and second order differential coefficient for considering dynamic property of speech signal. Next time
codebook generation of VQ(vector Quantization) use general LBG algorithm. The
Number of codeword for considering codebook size of VQ and implication of speaker
normalization consider 256 and 512. Component density of output probability in
HMM model consists of 256 and 512 having acoustic model. We use Korean word
speech DB, SKKU PBW for HMM acoustic model and training. SKKU PBW (Pho-
netical Balanced Word) DB consist of 1001 of word spoken each sixty male and female. Table 1 is content of each DB to use training. Whole SKKU PBW DB is

Table 1. Korean speech DB for training

	Speakers	Utterances	Training
SKKU PBW	60/60	1001	30/15

Table 2. Korean speech DB for recognition testing - No contains training speech DB

	Speakers	Utterances	Testing
SKKU PBW	60/60	1001	30/30

Table 3. Performance of speaker normalization procedures(word error rates(%), SKKU PRW)

Codebook Size	Baseline	α	β	γ
256	15.40	12.00	10.00	8.70
512	10.07	8.01	7.02	6.00

sampled 11.025k. Half of whole speech DB use to training, the reminder DB use to-recognition experiment. SKKU PBW DB using recognition performance experiment is shown Table 2. Recognition experiment result about speaker normalization method of introduced three methods in this paper is shown Table 3.

It seems that recognition experiment results of Table 3, Difference of Codebook size in VQ is shown different word error rate. We analyzed recognition performance of speaker normalization with 512 codebook size that baseline system not applied speaker normalization is shown the heist error rate 10.07% and existing frequency warping well-known vocal track normalization is shown high error rate 8.01%. And suggesting power spectrum warping is approximately 1.00% higher than existing frequency warping that is error rate 7.02%. Therefore proposing power spectrum warping gets more excellent existing frequency warping. Second proposing hybrid VTN in this instance is 4.07% of word error rate less than baseline system and 2.01% of word error rate less than existing frequency. And that is 1.02% of word error rate less than power spectrum warping. In this results, hybrid VTN get excellent speaker normalization property because hybrid VTN using frequency warping and power spectrum warping at once. Therefore power spectrum warping proposed this paper get performance of speaker normalization more excellent existing frequency warping and connected two methods VTN get more robust speaker normalization than others speaker normalization.

6   Conclusions

This paper introduced well-known frequency warping for present speaker normalization, new proposing power spectrum warping, and connected two methods hybrid VTN. We compared recognition performance of each speaker normalization method using Korea speech DB (SKKU PBW) and we analyzed each speaker normalization property. According to recognition experiment results, proposing power spectrum warping has 1.00% word error rate reduction as word recognition performance of existing frequency warping. Power spectrum warping method gets more excellent frequency warping. Also, hybrid VTN has most robust speaker normalization property and hybrid VTN has 4.07% word error rate reduction as word recognition performance of baseline system. Hybrid VTN has 2.01% word error rate reduction as word recognition performance of existing frequency warping. However processing time of hybrid VTN is one and half more than before because hybrid VTN is connected two speaker normalization methods. Processing time of hybrid VTN will improve to need study on fast hybrid VTN.

Acknowledgement

This work was supported by the MIC (The Ministry of Information & Communication) and NGV (Next Generation Vehicle technology).

References

1. Li Lee, Richard Rose:A Frequency Warping Approach to Speaker Normalization, IEEE Transactions on Speech and Audio Processing, Vol 6. No. 1, January 1998
2. L. Welling, H. Ney, S. Kanthak,: Speaker Adaptive Modeling by Vocal Tract Normalization, IEEE Transaction on Speech and Audio Processing, Vol. 10, No. 6, September 2002
3. A. Andreou, T. Kam, and J. Cohen: Experiments in Vocal Tract Normalization, in Proc. CAIP Workshop: Frontiers in Speech Recognition II, 1994
4. Michael Seltzer: SPHINX III Signal Processing Front End Specification, CMU Speech Group, August 1999
5. Y. Linde, A. Duzo, R. M. Gray: An Algorithm for Vector Quantizer Design, IEEE Transaction on COM., Vol. 28, January 1980
6. J.S. Youn, K.W. Chung and K.S. Hong: A Continuous Digit Speech Recognition Applied Vowel Sequence and VCCV Unit HMM, Proceeding of the Acoustical Society of Korea, Vol. 20, No. 2, 2001
7. T.D. Rossing, P. Wheeler and F.R. Moore: The Science of Sound", Addition Wesley, 2002
8. R. Roth et al: Dragon systems'1994 Large Vocabulary Continuous Speech Recognizer, in Proc. Spoken Language Systems Technology Workshop, 1995

Genetic Programming for Inductive Inference of Chaotic Series

I. De Falco[1], A. Della Cioppa[2], A. Passaro[3], and E. Tarantino[1]

[1] Institute of High Performance Computing and Networking,
National Research Council of Italy (ICAR–CNR),
Via P. Castellino 111, 80131 Naples, Italy
{ivanoe.defalco, ernesto.tarantino}@na.icar.cnr.it
[2] Natural Computation Lab - DIIIE,
University of Salerno, Via Ponte don Melillo 1,
84084 Fisciano (SA), Italy
adellacioppa@unisa.it
[3] Department of Computer Science, University of Pisa,
Largo B. Pontecorvo 3, 56127 Pisa, Italy
a.passaro@unipi.it

Abstract. In the context of inductive inference Solomonoff complexity plays a key role in correctly predicting the behavior of a given phenomenon. Unfortunately, Solomonoff complexity is not algorithmically computable. This paper deals with a Genetic Programming approach to inductive inference of chaotic series, with reference to Solomonoff complexity, that consists in evolving a population of mathematical expressions looking for the 'optimal' one that generates a given series of chaotic data. Validation is performed on the Logistic, the Henon and the Mackey–Glass series. The results show that the method is effective in obtaining the analytical expression of the first two series, and in achieving a very good approximation and forecasting of the Mackey–Glass series.

Keywords: Genetic programming, Solomonoff complexity, chaotic series.

1 Introduction

Inductive Inference is a fundamental problem both in science and engineering. Its aim is to find a functional model of a system, in symbolic form, by determining its fundamental properties from its observed behavior. This model is a mathematical idealization that is used as a paradigm of the system, and it is chosen to well fit the experimental data according to a chosen evaluation criterion.

In [1], Solomonoff supposed that the observed data of a given phenomenon can be encoded by means of a string x of symbols on a given alphabet. Then, the Inductive Inference problem can be faced either by searching the shortest computer program u that provides the string x as output or by estimating the *a priori* probability of all the programs that generate x as output. Moreover, differently from Kolmogorov [2], rather than emphasizing the shortest program

I. Bloch, A. Petrosino, and A.G.B. Tettamanzi (Eds.): WILF 2005, LNAI 3849, pp. 156–163, 2006.

u which will produce exactly x, Solomonoff considers all programs u which will cause a universal Turing machine to produce output having x as a prefix and such that no proper prefix of u will produce x. Hence, the Inductive Inference system described above has to look for a program (functional expression) that has the highest *a priori* probability, i.e., that with the shortest length. Unfortunately, this task is not algorithmically computable, i.e., for any string (sequence of data) x defined on a given alphabet it is not possible to compute the program u with the highest *a priori* probability that yields x as output. In other words, the Solomonoff complexity is not a recursive function, i.e., it is not in the class of algorithmically computable functions. It should be noted that all the programs generating strings with x as prefix have to be considered, in that the aim is to consider x as an encoding of observed data related to a given phenomenon. Besides, the string x can be simply a substring (prefix) of a string with either finite or infinite length. This is why it is impossible to consider only programs providing as output the string x, since other descriptions of a phenomenon could be neglected, e.g. programs that generate strings whose length is larger and that have x as prefix. The above discussion holds true independently of the kind of dynamics exhibited by the phenomenon. However, while for simple dynamics, such as deterministic linear ones, the Inductive Inference problem seems to be somehow approachable, the problem gets an additional twist if we consider chaotic series. In fact, we can wonder whether the series generated by chaotic dynamics are complex or not. In this respect there exists a relation between Shannon entropy and algorithmic complexity [3]. Given that chaotic dynamics strongly depend on both the initial conditions and the system parameters, *chaotic systems produce algorithmically complex sequences*. As a consequence, the problem of Inductive Inference cannot be solved at all both in theory and in practice. Nonetheless, for real applications we can obtain approximations by making use of heuristic procedures.

This paper deals with a Genetic Programming (GP) approach [4] to Inductive Inference of chaotic series with reference to Solomonoff complexity. This consists in evolving a population of functional expressions that fit given series of data while looking for the 'optimal' one, i.e. the one with the 'shortest length'. The validation is effected on the Logistic, the Henon and the Mackey–Glass series.

2   Genetic Programming Approach

Starting from the above considerations we, very naturally, turn to evolutionary methods. In fact, evolutionary algorithms work on populations of individuals rather than on single solutions, thus searching the problem space in a parallel manner. In such a way, the above difficulties can be effectively reduced. It should be noted that since evolutionary algorithms hopefully give a good approximation, but do not guarantee the convergence to the global optimum in finite time, the drawback is to obtain an approximation of the solution. However, this drawback can be drastically reduced by setting a sufficiently high number of generations.

GP [4] is well suited for our aim. It should perform a search for functional expressions that fit all the data, while minimizing their length. However, particular attention must be devoted to the genetic generation and preservation of valid programs in order to overcome the possible lack of closure property GP is subject to, i.e, possible lack of closure property. To this aim, many researchers have presented GPs making use of grammars. Whigham, as an example, created a GP based on Context–Free Grammars (CFGs) [5]. He used derivation trees of CFGs as genotypes, the phenotypes being the programs generated by those trees. He also showed that CFGs are an efficient approach to introduce bias into evolutionary process.

Following [5], our GP is based on expression generator that provides the starting population with a set of programs different in terms of size, shape and functionality. The expression generator is implemented by means of a CFG which ensures the syntactic correctness of the programs. Evaluation procedure divides series into four consecutive intervals: *seed* ($\mathcal{I}_S$), *training* ($\mathcal{I}_T$), *validation* ($\mathcal{I}_V$) and *prediction* ($\mathcal{I}_P$). $\mathcal{I}_S$ is the set of the initial series values which are given as seed to the expressions to be evaluated. $\mathcal{I}_T$ is the set of values on which the fitness is actually evaluated: the sequence of values obtained as output for this interval is compared to real series values so as to compute fitness of the individual. $\mathcal{I}_V$ is the set used to select the overall best expression at the end of the inference process, while $\mathcal{I}_P$ contains the values forecasted by the expressions.

2.1 Encoding and Genetic Operators

The genotypes are functional expressions which state dependence of a value of the series on the previous ones and are encoded as derivation trees of the adopted CFG. This kind of encoding is very appealing in that the actions performed by the genetic operators can be easily implemented as simple operations on the derivation trees. In fact, crossover operates by randomly choosing a nonterminal node in the first individual to be crossed and then by randomly selecting the same nonterminal node in the second individual. Finally, it swaps the derivation subtrees. If a corresponding nonterminal node cannot be found in the second parent, the crossover has no effect. Mutation works by randomly choosing a nonterminal node in the offspring and then the corresponding production rule is activated in order to generate a new subtree, thus resulting either in a complete substitution (macro–mutation) or in variation of a leaf node (micro–mutation).

2.2 Fitness Function

The fitness is the sum of two weighted terms. The former accounts for the difference between computed and actual series values on $\mathcal{I}_T$, while the latter is proportional to the number of nodes of the derivation tree:

$$F(p) = \frac{1}{\sigma^2} \frac{1}{l_{\mathcal{I}_T}} \sum_{i \in \mathcal{I}_T} |p(i) - s_i|^2 + w N_p \qquad (1)$$

where p is the program to evaluate, σ is the standard deviation for the series, $l_{\mathcal{I}_T}$ is the length of $\mathcal{I}_T$, $p(i)$ is the value computed on the i–th point and s_i its actual value, w is the weight of the term related to program complexity. N_p is the number of nodes making up the derivation tree for p.

It is to note that the number of nodes N_p of a particular tree is inversely tied to the *a–priori* probability of generating p according to the chosen CFG. The higher the number of nodes which constitute a tree the lower the *a–priori* probability of generating it. The scale factor w takes on very small values and its aim is to allow that, during the first phase of expression discovery, the error term is predominant. So the algorithm is free to increase the tree length in such a way to ease the search towards the exact match of the series. Then, the system will exploit the obtained solutions to achieve shorter and shorter expressions. Even if the usage of a penalty term can seem trivial, it should be noted that no smarter techniques exist to face this problem. In fact, the only actual alternative would require the computation of the SC function which is not computable.

As mentioned before, $\mathcal{I}_V$ set allows us to select the overall best program discovered during the entire inference process. Therefore, at each generation for all the individuals in the population, the difference V of output values from $\mathcal{I}_V$ values is then evaluated and this is added to the term relating to complexity, so as to obtain a performance index P:

$$ P(p) = V(p) + wN_p, \quad V(p) = \frac{1}{\sigma^2} \frac{1}{l_{\mathcal{I}_V}} \sum_{i \in \mathcal{I}_V} |p(i) - s_i|^2 \tag{2} $$

Hence, the result of a run is the individual with the best performance index achieved in all the generations making up the run.

3   Experimental Findings

The system has been tested on three chaotic series, namely the Logistic, the Henon and the Mackey–Glass ones. The series have been recursively generated starting from a set of seeds by means of their generating equations.

Any program execution is determined by a set of parameters, among which those related to evolutionary process and those specifying widths of intervals the series is divided into. Another important parameter is the weight w present in (1) and related to program complexity: it should be chosen so as to favor evolution of simpler individuals, while also allowing creation of programs complex enough to adhere to the original series.

For any series 10 runs with different random seeds have been carried out with the same parameters. We have used a population size of 500, a tournament selection mechanism with size of 5, a crossover operator taking place with a 100% probability and a mutation operator (with 30% probability) which distinguishes between macro– and micro–mutations, applying them with probabilities equal to 30% and 70% respectively. The maximum number of generations allowed for all the runs is set to 2000. After a preliminary tuning, the parameter w has been set to 10^{-4}. For the Logistic and the Henon series a maximum tree depth of 10 has been considered and we have set widths for $\mathcal{I}_S$, $\mathcal{I}_T$, $\mathcal{I}_V$ and $\mathcal{I}_P$ to 10, 70, 10 and 10 respectively. Table 1 shows the grammar used for all the series considered.

Table 1. The grammar used for the runs

Rule no.	Rule	
1	S	$\rightarrow f(t) = E$
2	E	$\rightarrow f(t - E) \mid f(t - N) \mid (EOE) \mid R \mid t$
3	O	$\rightarrow + \mid - \mid * \mid /$
4	R	$\rightarrow 0 \mid 0.10 \mid \ldots \mid 3.90 \mid 4$
5	N	$\rightarrow 1 \mid 2 \mid \ldots \mid l_S$

3.1 Logistic Series

The standard form of the so called logistic function is given by: $f(t) = \mu \cdot t \cdot (1-t)$, where μ is the growth rate when the equation is being used to model population growth in a biological species.

R. May in 1976 introduced the Logistic series as an example of a very simple nonlinear equation being able to produce very complex dynamics [6]:

$$f(t) = \mu \cdot f(t - 1) \cdot (1 - f(t - 1)) \tag{3}$$

μ is the growth rate.

While the initial conditions don't matter, this series behaves in a way that depends on the value of μ. In particular:

- *Extinction.* If $\mu < 1$ then $\lim_{t \to \infty} f(t) = 0$.
- *Fixed point.* If $1 < \mu < 3$ the series tends to a single value. It is not important how it reaches this value, but generally it oscillates around the fixed point. Unlike a mass spring system, the series generally tends to rapidly approach fixed points.
- *Periodic.* The system alternates between 2 states for $\mu = 3$. For values greater than 3.44948, the system alternates among 4 states. The system jumps between these states, while it does not pass through intermediary values. The number of states steadily increases in a process called period doubling as μ grows.
- *Chaotic.* In this state, the system can evaluate to any position at all with no apparent order. The system undergoes increasingly frequent period doubling until it enters the chaotic regime at about 3.56994. Below $\mu = 4$ the states are bound between $[0, 1]$, above 4 the system can evaluate to $[0, \infty]$. A period of 3 surprisingly appears for $3.8284 < \mu < 3.8415$.

For the experiments we have used $\mu = 3.5$.

In all runs the same solution (see (3)) with a derivation tree of 17 nodes has been obtained, apart from a possible swap between the terms. In the best case the solution has been achieved in 27 generations, and on average in 176.

As in the most time–consuming run (Fig. 1 (left)), the evolution consists of two distinct phases. The former is characterized by search for an expression

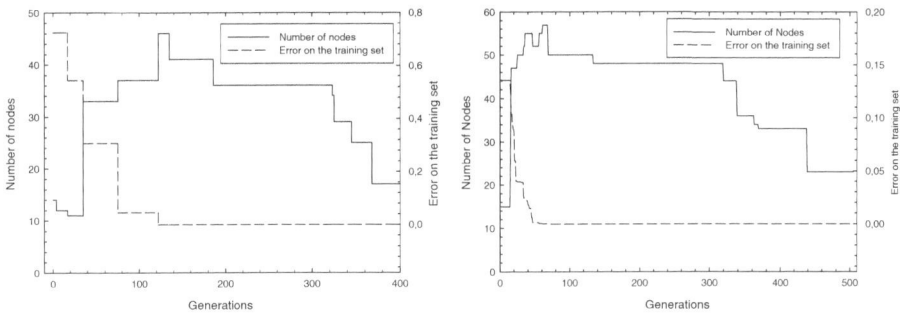

Fig. 1. Most time–consuming run for Logistic (left) and Henon (right) series

which better and better approximates the series. In this phase, larger and larger expressions are found which provide lower and lower errors on $\mathcal{I}_T$. This takes place until about generation 120. The latter phase, instead, begins when an exact yet complex solution emerges, and consists in achieving other shorter solutions. At the end of this phase, at about generation 360, the optimal solution is obtained. Even though this "simplification" is an effect of the evolution process, which tends to favor simpler solutions, it has a behavior very similar to that which could be obtained by a human. In fact, in all the runs GP has been able to discover intermediate solutions equivalent to (3). Once such solutions have emerged, then, GP has evolved them towards the optimal one.

3.2   Henon Series

Henon series is a 2–D iterated map with chaotic solutions proposed by M. Henon (1976) [7] as a simplified model of the Poincare map for the Lorenz model:

$$\begin{cases} f(t) = 1 + g(t-1) - a \cdot f^2(t-1) \\ g(t) = b \cdot f(t-1) \end{cases} \tag{4}$$

where a and b are (positive) bifurcation parameters. The parameter b is a measure of the rate of area contraction (dissipation) and the Henon series is the most general 2–D quadratic map with the property that the contraction is independent of f and g. For $b = 0$, Henon series reduces to the quadratic map, which is linked to the Logistic series. Bounded solutions exist for the Henon series over a range of a and b values, and some yield chaotic solutions. Numerical evidence of chaotic behavior can be found for $a = 1.4$ and $b = 0.3$ and such values have been used for the experiments.

Only in the 40% of the 10 runs, effected with the same parameters as the previous series, the canonical solution (see (4)) has been obtained, apart from a possible swap between the terms, thus confirming a greater difficulty in discovering the

Henon law with respect to the Logistic one. In fact the Henon series strongly depends on the boundary conditions (seeds). In the best case, the solution has been achieved in 153 generations, and on average in 289. In the remaining runs, however, a good approximation has been achieved. The evolution evidences two phases as described for the Logistic series (see Fig. 1 (right)).

3.3 Mackey–Glass Series

Mackey–Glass series was proposed by Mackey and Glass in [8] and aims to describe blood cells generation in a patient with leukemia:

$$\frac{d}{dt} z(t) = a \frac{z(t - \Delta)}{1 + [z(t - \Delta)]^c} - bz(t) \tag{5}$$

where a, b, c and Δ are constant values. Value of Δ plays a key role in series behavior, since attractor dimension depends on it. For $\Delta < 4.53$ the system reaches a stable equilibrium (fixed–point attractor), for $4.53 < \Delta < 13.3$ a cyclic behavior is obtained, for $13.3 < \Delta < 16.8$ the series shows a double cycle. Since chaotic behavior is achieved for $\Delta > 16.8$, we shall make reference to $\Delta = 17$, $a = 0.2$, $b = 0.1$ and $c = 10$, thus in a chaotic regime. The data series have been generated by using the Runge–Kutta Method of the 4–th order with a sampling rate of 6.

After a tuning phase, the maximum tree depth chosen is 15. Length for $\mathcal{I}_S$ has been set equal to 50, that for $\mathcal{I}_T + \mathcal{I}_V$ to 420, 30% of which to be used as $\mathcal{I}_V$, and that for $\mathcal{I}_P$ to 30. With respect to the previous series, some complications take place here. In fact, none of the runs achieves an error equal to zero on $\mathcal{I}_T$. However, in each run expressions with a good approximation of the series have been obtained. The program with the best P obtained is the following:

$$f(t) = f(t - 50) + \frac{(-2.35 \cdot (-2.91 \cdot ((f(t - 50) - f(t - 34)) \cdot -2.77)))}{t} \tag{6}$$

Table 2 reports the results, while Fig. 2 shows the behavior on $\mathcal{I}_T$ and $\mathcal{I}_V$ (left) and $\mathcal{I}_P$ (right). A simple analysis of the solution evidences that, although the task difficulty, GP has been able to discover the underlying period Δ. In fact, 50 is close to $\Delta \cdot 3$, while $34 = \Delta \cdot 2$.

Table 2. Results achieved on Mackey–Glass series with $\Delta = 17$

	Best	Average
Generation	1047	864
Nodes	30	49
Fitness F	0.198050	0.271610
Validation V	0.222835	0.270765
Performance Index P	0.231835	0.285495

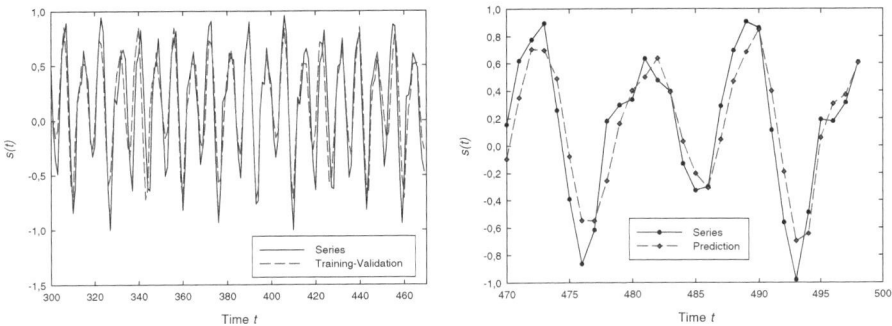

Fig. 2. Behavior of the best program on $\mathcal{I}_T$, $\mathcal{I}_V$ and $\mathcal{I}_P$ (Mackey–Glass series)

4   Conclusion and Future Works

The problem of Inductive Inference of chaotic series has been faced by taking into account the Solomonoff complexity.

Our GP has proved capable of finding the exact expression of the Logistic and Henon series, while pursuing appealing computational strategies. As regards the Mackey–Glass series, GP has been able to discover an expression that, making use of the actual period Δ, provides a good approximation of the data.

An interesting result of the experiments is that the complexity of the series in terms of program length increases from Logistic (17) to Henon series (23) and from this latter to Mackey–Glass series (30), in accordance with what would be supposed by looking at their generating expressions.

Further investigations concern the idea of evaluating the *a priori* probability according to Solomonoff and of using it in the fitness function.

References

1. Solomonoff, R.J.: A formal theory of inductive inference. Information and Control **7** (1964) 1–22, 224–254
2. Kolmogorov, A.N.: Three approaches to the quantitative definition of information. Problems of Information and Transmission **1** (1965) 1–7
3. Falcioni, M., Loreto, V., Vulpiani, A.: Kolmogorov's legacy about entropy, chaos and complexity. Lect. Notes Phys. **608** (2003) 85–108
4. Koza, J.R.: Genetic Programming II: Automatic Discovery of Reusable Programs. MIT Press, Cambridge, Massachusetts (1994)
5. Whigham, P.A.: Grammatical Bias for Evolutionary Learning. PhD thesis, School of Computer Science, University of New South Wales, Australia (1996)
6. Strogatz, S.: Nonlinear Dynamics and Chaos. Perseus Publishing (2000)
7. Hénon, M.: A two–dimensional mapping with a strange attractor. Communications of Mathematical Physics **50** (1976) 69–77
8. Mackey, M.C., Glass, L.: Oscillations and chaos in physiological control systems. Science (1977) 287

Evaluation of Particle Swarm Optimization Effectiveness in Classification

I. De Falco[1], A. Della Cioppa[2], and E. Tarantino[1]

[1] Institute of High Performance Computing and Networking,
National Research Council of Italy (ICAR–CNR),
Via P. Castellino 111, 80131 Naples – Italy
{ivanoe.defalco, ernesto.tarantino}@na.icar.cnr.it
[2] Natural Computation Lab - DIIIE,
University of Salerno, Via Ponte Don Melillo 1,
84084 Fisciano (SA) – Italy
adellacioppa@unisa.it

Abstract. Particle Swarm Optimization (PSO) is a heuristic optimization technique showing relationship with Evolutionary Algorithms and strongly based on the concept of swarm. It is used in this paper to face the problem of classification of instances in multiclass databases. Only a few papers exist in literature in which PSO is tested on this problem and there are no papers showing a thorough comparison for it against a wide set of techniques typically used in the field. Therefore in this paper PSO performance is compared on nine typical test databases against those of nine classification techniques widely used for classification purposes. PSO is used to find the optimal positions of class centroids in the database attribute space, via the examples contained in the training set. Performance of a run, instead, is computed as the percentage of instances of testing set which are incorrectly classified by the best individual achieved in the run. Results show the effectiveness of PSO, which turns out to be the best on three out of the nine challenged problems.

Keywords: Particle Swarm Optimization, Classification.

1 Introduction

In this paper we examine the ability of Particle Swarm Optimization (PSO) [1] [2], heuristic technique for search of optimal solutions based on the concept of swarm, to efficiently face classification [3] [4] of multiclass database instances.

There exist in literature a few papers in which PSO is compared, respectively, against a Genetic Algorithm and a Tree Induction algorithm [5] for classification in databases and against K-means [6] for clustering, yet there are no papers showing a wide comparison against a wide set of techniques typical for this problem. Therefore in this paper we wish to evaluate PSO efficiency in classification by facing nine typical test databases and by comparing results against those achieved by nine well–known classification techniques.

I. Bloch, A. Petrosino, and A.G.B. Tettamanzi (Eds.): WILF 2005, LNAI 3849, pp. 164–171, 2006.

Paper structure is as follows: Section 2 describes PSO basic scheme, while Section 3 illustrates our PSO version applied to classification problem. Section 4 reports on the database faced, the results achieved, the comparison against nine typical classification techniques and some *a posteriori* considerations. Finally Section 5 contains our conclusions and future works.

2   Particle Swarm Optimization

PSO reveals itself very effective in facing multivariable problems in which any variable takes on real values. It has roots in two methodologies. Its links to Artificial Life in general, and with bird flocks, fish schools and swarm theory in particular are very evident. Nonetheless, PSO is also tied to Evolutionary Computation, namely to Genetic Algorithms (GA) [7] and to Evolutionary Programming [8].

PSO is based on a swarm of n individuals called *particles*, each representing a solution to a problem with N dimensions. Its genotype consists of $2 \cdot N$ parameters, the first N representing the coordinates of particle position, and the latter N its velocity components in the N–dimensional problem space. From the evolutionary point of view, a particle moves with an adaptable velocity within the search space and retains in its own memory the best position it ever reached. The parameters are changed when going from an iteration to the next one as described below.

Velocity $\boldsymbol{v}_i(t+1)$ of i-th particle at next step $t+1$ is a linear combination of current velocity $\boldsymbol{v}_i(t)$ of i-th particle at time t, of the difference between the position $\boldsymbol{b}_i(t)$ of the best solution found up to this time by i-th particle and current position $\boldsymbol{p}_i$ of i-th particle, and of the difference between best position ever found in the population $\boldsymbol{b}_g(t)$ and that of i-th particle $\boldsymbol{p}_i(t)$:

$$\boldsymbol{v}_i(t+1) = w \cdot \boldsymbol{v}_i(t) + c_1 \cdot U(0,1) \otimes (\boldsymbol{b}_i(t) - \boldsymbol{p}_i(t)) + c_2 \cdot U(0,1) \otimes (\boldsymbol{b}_g(t) - \boldsymbol{p}_i(t)) \quad (1)$$

where $\otimes$ denotes point–wise vector multiplication, $U(0,1)$ is a function that returns a vector whose positions are randomly generated by a uniform distribution in $[0,1]$, c_1 is the *cognitive parameter*, c_2 is the *social parameter*, and w is the *inertia factor* whose range is $[0.0, 1.0]$. Velocity values must be within a range defined by two parameters $v_{\min}$ and $v_{\max}$.

An improvement to original PSO is in w not being kept constant during execution; rather, starting from a maximal value $w_{\max}$, it is linearly decremented as the number of iterations increases down to a minimal value $w_{\min}$ as follows [9]:

$$w(t) = w_{\max} - (w_{\max} - w_{\min}) \cdot \frac{t}{T_{\max}} \quad (2)$$

where t and $T_{\max}$ are the current and the maximum allowed number of iterations respectively.

The position of each particle at next step is then evaluated as the sum of its current position and of the velocity obtained by eq. (1):

$$\boldsymbol{p}_i(t+1) = \boldsymbol{p}_i(t) + \boldsymbol{v}_i(t+1) \quad (3)$$

These operations are repeated for $T_{\max}$ iterations or until some other stopping criterion gets verified. A typical convergence criterion is constituted by achievement of a minimal desired value of error with respect to optimal solution (of course, where this minimal fitness value is known *a priori*).

PSO pseudocode is the following:

Algorithm 1. PSO Algorithm

---

begin
    for each particle
        initialize particle position and velocity
    while (maximal number of iterations is not reached) **do**
        for each particle
            calculate fitness value $\psi_i(t)$
            if ($\psi_i(t)$ is better than best fitness value $\psi(\boldsymbol{b}_i(t))$ in particle history)
                then $\psi(\boldsymbol{b}_i(t)) = \psi_i(t)$ and take current particle as new $\boldsymbol{b}_i(t)$
        if necessary update the global best particle during execution $\boldsymbol{b}_g(t)$
        for each particle
            calculate particle velocity based on eq. (1)
            update particle position based on eq. (3)
        update the inertia factor based on eq. (2)
end

It should be noted that the algorithm is nonelitist.

3 PSO Applied to Classification

Given a database with C classes and N parameters, classification problem can be seen as that of finding the optimal positions of C centroids in an N-dimensional space, i.e. that of determining for any centroid its N coordinates, each of which can take on, in general, real values. With these premises, the i-th individual of the population is encoded as it follows:

$$(\boldsymbol{p}_i^{\,1}, \ldots, \boldsymbol{p}_i^{\,C}, \boldsymbol{v}_i^{\,1}, \ldots, \boldsymbol{v}_i^{\,C}) \tag{4}$$

where the position of the j–th centroid is constituted by N real numbers representing its N coordinates in the problem space:

$$\boldsymbol{p}_i^{\,j} = \{p_{1,i}^j, \ldots, p_{N,i}^j\} \tag{5}$$

and similarly the velocity of the j-th centroid is made up of N real numbers representing its N velocity components in the problem space:

$$\boldsymbol{v}_i^{\,j} = \{v_{1,i}^j, \ldots, v_{N,i}^j\} \tag{6}$$

Then, any individual in the population consists of $2 \cdot C \cdot N$ components, each of which is represented by a real value.

The fitness function ψ is computed as the sum on all training set instances of euclidean distance in N-dimensional space between generic instance $\boldsymbol{x}_j$ and the centroid of the class $\mathbf{CL}$ it belongs to according to database $(\boldsymbol{p}_i^{\mathbf{CL}_{\text{known}}(\boldsymbol{x}_j)})$. This sum is divided by D_{Train}, which is the number of instances composing the training set. In symbols, i-th individual fitness is given by:

$$\psi(i) = \frac{1}{D_{\text{Train}}} \cdot \sum_{j=1}^{D_{\text{Train}}} d\left(\boldsymbol{x}_j, \boldsymbol{p}_i^{\mathbf{CL}_{\text{known}}(\boldsymbol{x}_j)}\right) \tag{7}$$

When computing distance, any of its components in the N–dimensional space is normalized with respect to the maximal range in the dimension, and the sum of distance components is divided by N. With this choice, any distance can range within $[0.0, 1.0]$, and so can ψ. Given the chosen fitness function, the problem becomes a typical minimization problem.

Performance of a run, instead, is computed as the percentage $\%err$ of instances of testing set which are incorrectly classified by the best individual (in terms of fitness) achieved in the run.

4   Experiments and Results

PSO is used to face a set of nine databases well known in literature taken from UCI database repository [10], and its results are compared against those provided by nine classical classification techniques.

The databases faced and their features are listed in alphabetical order in Table 1. In it for each database total instance number (D), number of classes into which it is divided (C) and number of parameters composing each instance (N) are reported.

We have chosen the classification techniques used for the comparison within the Waikato Environment for Knowledge Analysis (WEKA) system release 3.4 [11]. Namely, we have taken into account the following ones: MultiLayer Perceptron Artificial Neural Network (MLP)[12], Radial Basis Function Artificial Neural Network (RBF)[13], KStar [14], Bagging [15], MultiBoostAB [16], Naive Bayes Tree (NBTree)[17], Ripple Down Rule (Ridor) [18] and Voting Feature Interval (VFI) [19].

Parameter values used for any technique are those set as default in WEKA.

PSO parameters have been set as follows: $n = 50$, $T_{\max} = 1000$, $v_{\max} = 0.05$, $v_{\min} = -0.05$, $c_1 = 2.0$, $c_2 = 2.0$, $w_{\max} = 0.9$, $w_{\min} = 0.4$.

Table 1. Properties of examined databases

	Card	Diabetes	Glass	Heart	Horse	Iris	Wdbc	Wdbc-I	Wine
D	690	768	214	303	364	150	569	699	178
C	2	2	6	2	3	3	2	2	3
N	51	8	9	35	58	4	30	9	13

Table 2. Achieved results in terms of %*err*

	PSO	BAYES NET	MLP ANN	RBF	KSTAR	BAG- GING	MULTI BOOST	NB TREE	RIDOR	VFI
Card	22.84	12.13	13.81	*43.29*	19.18	**10.68**	12.71	16.18	12.65	16.47
Diabetes	**22.55**	25.52	29.16	*39.16*	34.05	26.87	27.08	25.52	29.31	34.37
Glass	41.69	29.62	28.51	44.44	**17.58**	25.36	*53.70*	24.07	31.66	41.11
Heart	**17.46**	18.42	19.46	*45.25*	26.70	20.25	18.42	22.36	22.89	18.42
Horse	40.98	30.76	32.19	38.46	35.71	**30.32**	38.46	31.86	31.86	*41.75*
Iris	2.63	2.63	**0.00**	*9.99*	0.52	0.26	2.63	2.63	0.52	**0.00**
Wdbc	5.73	4.19	2.93	*20.27*	**2.44**	4.47	5.59	7.69	6.36	7.34
Wdbc-I	**2.87**	3.42	5.25	*8.17*	4.57	3.93	5.14	5.71	5.48	5.71
Wine	4.44	**0.00**	2.22	6.43	1.11	4.21	*11.11*	**0.00**	5.77	**0.00**

Each database has been split into a training set, made up by the former 75% of the database instances, and a testing one, constituted by the remaining 25%.

Table 2 shows the results achieved by the 10 techniques on each of the nine databases with respect to %*err*. For any problem the best value obtained among all the techniques is reported in bold, whereas the worst one is in italic. PSO results are averaged over 10 runs differing in the starting seed provided in input to random number generator. MLP, RBF, Bagging, MultiBoostAB and Ridor are based on a starting seed so also for them 10 runs have been carried out by varying this value. Bayes Net, KStar and VFI, instead, do not depend on any starting seed, so ten runs have been carried out as a function of a parameter typical of the technique (*alpha* for Bayes Net, *globalBlend* for KStar and *bias* for VFI). NBTree, finally, depends neither on initial seed nor on any parameter, so only one run has been performed on any database.

PSO execution times vary with database sizes, and range from a minimum of six seconds per run (for Iris) up to a little more than one minute (for Card) on a pc with 1.6–GHz Centrino processor. Thus times are comparable with those of the other technique, which range from 2-3 seconds up to 4-5 minutes.

It can be seen from results shown in Table 2 that PSO is never the worst technique, and is the best one in three cases (Wdbc-I, Heart and Diabetes). Particularly, PSO shows good performance when the problem has two classes: in fact PSO is the best for three out of the five such problems. As regards the other two two-class problems, instead, performance is not brilliant. It can be noted that If we take into account the product $P = D \cdot N$ we can see that the two-class problems are ordered in increasing order as it follows: Diabetes (6144), Wdbc-I (6291), Heart (10105), WDBC (17070), Card (35190). On them PSO gets ranked in positions 1, 1, 1, 6, 9, respectively. So, some relationship between P and PSO performance might exist.

As for the three- and the six–class problems PSO performance is less striking. For example, on two easy problems like Iris and Wine it does not reach 0%, as several techniques do. Also on Horse and Glass PSO is in the second half of the rank. The three-class problems can be ordered based on P value, obtaining: Iris

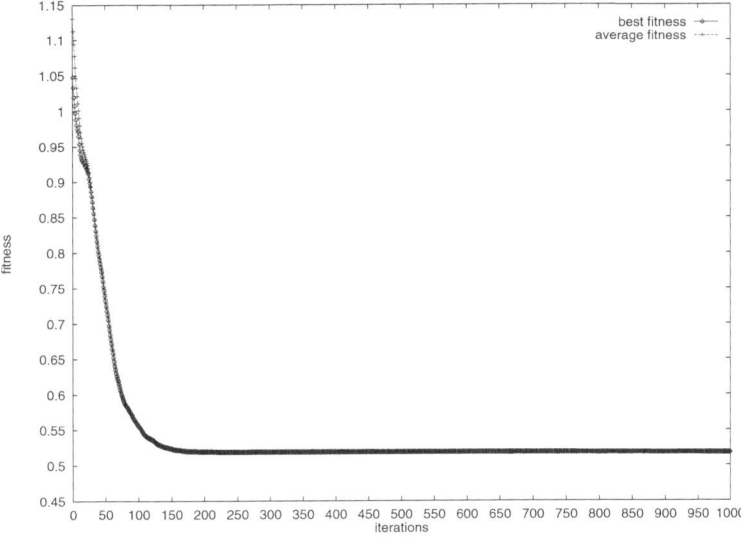

Fig. 1. Typical behavior of best and average fitness as a function of the number of iterations

(600), Wine (2314), Horse (21112). On them PSO gets ranks 6, 7, 9, respectively. Also in this case this might seem to imply some relation between P and PSO performance. For Glass $P = 1926$, and PSO rank is 8.

This leads us to suppose that PSO efficiency shows some limitation as the number of classes increases. For a given number of classes, as P increases so does the difficulty for PSO. This shall be a research issue for our next papers.

From an evolutionary point of view, in Fig. 1 we report the behavior of a typical run in terms of best individual fitness and average fitness in the population as a function of the number of iterations. The execution shown is the run number 1 carried out on database Card.

As it can be seen, PSO shows a first phase of about 100 iterations in which fitness decrease is very strong, starting from 1.0477 for the best and 1.1301 for the average, and reaching about 0.5557 for the best and 0.5574 for the average. A second phase follows, lasting about 80 iterations, in which decrease is slower, and the two values tend to become closer and closer, until they reach 0.5199 and 0.5207 respectively. From now on decrease in fitness values is slower and slower, and those two values become more and more similar. Finally, at iteration 1000 the two values are 0.5182 and 0.5184.

As described above for a specific run and for a specific database is actually true for all runs carried out and for all databases, and is probably a consequence of the good setting of parameters chosen, which allows a very fast decrease in the first iterations, differently from standard settings such as those related to minimal and maximal values for velocities.

5   Conclusions and Future Works

By considering the results achieved on nine typical databases, it can be concluded that PSO can be suitably used to face classification problem. For some test databases, it turns out to be better than nine techniques widely used in this field. Execution times are of the same order of magnitude as those of the nine techniques used.

Future works will aim to shed light on its quality and limitations in this field, and to find possible ways to overcome those limitations. It can be noted that PSO model used is the standard one, which does not allow to use techniques such as fitness sharing in Evolutionary Algorithms. We plan to work on this issue to improve PSO performance.

Firstly, we plan to endow PSO with fitness sharing, aiming to investigate whether this helps in improving performance. Moreover we intend to add to PSO some mutation mechanisms similar to those present in Evolutionary Algorithms. With the same aim we will implement an elitist PSO, in which the velocity of each individual is updated by taking into account the best element found in all iterations performed up to the current one, rather than that present in current iteration only.

References

1. Eberhart, R.C., Kennedy, J.: A new optimizer using particle swarm theory. In: Proceedings of the Sixth International Symposium on Micro Machine and Human Science, Nagoya, Japan, IEEE Press, Piscataway, NJ (1995) 39–43
2. Kennedy, J., Eberhart, R.C.: Particle swarm optimization. In: Proceedings of the IEEE International Conference on Neural Networks IV, IEEE Press, Piscataway, NJ (1995) 1942–1948
3. Han, J., Kamber, M.: Data Mining: Concept and Techniques. Morgan Kaufmann (2001)
4. Hand, D.J., Mannila, H., Smyth, P.: Principles of Data Mining. The MIT Press (2001)
5. Sousa, T., Silva, A., Neves, A.: Particle swarm based data mining algorithms for classification tasks. Parallel Computing **30** (2004) 767–783
6. van der Merwe, D.W., Engelbrecht, A.P.: Data clustering using particle swarm optimization. In: Proceedings of the IEEE Congress on Evolutionary Computation, IEEE Press, Piscataway, NJ (2003)
7. Goldberg, D.E.: Genetic Algorithms in Search, Optimization and Machine Learning. Addison–Wesley, Reading, Massachusetts (1989)
8. Fogel, L.J., Marsh, A.J., Walsh, M.J.: Artificial Intelligence through Simulated Evolution. Wiley & Sons, New York (1966)
9. Shi, Y., Eberhart, R.C.: A modified particle swarm optimizer. In: Proceedings of the IEEE International Conference on Evolutionary Computation, IEEE Press, Piscataway, NJ (1998) 69–73
10. Blake, C.L., Merz, C.J.: UCI Repository of Machine Learning Databases, University of California, Irvine. http://www.ics.uci.edu/~mlearn/MLRepository.html (1998)

11. Witten, I.H., Frank, E.: Data Mining: Practical Machine Learning Tool and Technique with Java Implementation. Morgan Kaufmann, San Francisco (2000)
12. Rumelhart, D.E., Hinton, G.E., Williams, R.J.: Learning representation by back–propagation errors. Nature **323** (1986) 533–536
13. Hassoun, M.H.: Fundamentals of Artificial Neural Networks. The MIT Press, Cambridge, Massachusetts (1995)
14. Cleary, J.G., Trigg, L.E.: K^*: An instance–based learner using an entropic distance measure. In: Proceedings of the 12th International Conference on Machine Learning. (1995) 108–114
15. Breiman, L.: Bagging predictors. Machine Learning **24** (1996) 123–140
16. Webb, G.I.: Multiboosting: a technique for combining boosting and wagging. Machine Learning **40** (2000) 159–196
17. Kohavi, R.: Scaling up the accuracy of naive-bayes classifiers: a decision tree hybrid. In: Proceedings of the Second Internarional Conference on Knowledge Discovery and Data Mining. (1996) 202–207
18. Compton, P., Jansen, R.: Knowledge in context: a strategy for expert system maintenance. In: Proceedings of Artificial Intelligence, Berlin, Springer–Verlag (1988) 292–306
19. Demiroz, G., Guvenir, A.: Classification by voting feature intervals. In: Proceedings of the European Conference on Machine Learning. (1997) 85–92

Identification of Takagi-Sugeno Fuzzy Systems Based on Multi-objective Genetic Algorithms

Marco Cococcioni, Pierluigi Guasqui, Beatrice Lazzerini,
and Francesco Marcelloni

Dipartimento di Ingegneria dell'Informazione, Elettronica, Informatica,
Telecomunicazioni, University of Pisa, Via Diotisalvi, 2, 56122 Pisa, Italy
{m.cococcioni, b.lazzerini, f.marcelloni}@iet.unipi.it

Abstract. In this paper we exploit multi-objective genetic algorithms to identify Takagi-Sugeno (TS) fuzzy systems that show simultaneously high accuracy and low complexity. Using this approach, we approximate the Pareto optimal front by first identifying TS models with different structures (i.e., different number of rules and input variables), and then performing a local optimization of these models using an ANFIS learning approach. The results obtained allow determining a posteriori the optimal TS system for the specific application. Main features of our approach are selection of the input variables and automatic determination of the number of rules.

1 Introduction

Takagi-Sugeno (TS) fuzzy systems [1] have been widely used in regression, classification and control problems. Genetic algorithms (GA's) have been frequently applied to their identification. Typically, after having decided the system structure, e.g., based on either uniform partitioning or clustering of the input space, GA's are used to fine tune the system parameters [2]. In this paper we adopt a different approach: first we use GA's just to assess the TS system structure, then we resort to a local search strategy to fine tune the system parameters, using an ANFIS learning strategy [3]. So doing, we identify different TS structures, with different numbers of rules and input variables. In order to compare these structures with each other based not only on their accuracy, but also on their complexity, we use a Multi-Objective Genetic Algorithm (MOGA), which provides an approximation of the Pareto optimal front, consisting of a set of non-dominated solutions. The user can then select a posteriori, from a variety of Pareto optimal solutions, the solution with the best compromise between desired accuracy and complexity level for the specific application.

2 TS Fuzzy System

Let $X = \{X_1, \ldots, X_F\}$ be the set of input variables. Let U_f, $f = 1..F$, be the universe of variable X_f and let $P_f = \left\{ A_{f,1}, \ldots, A_{f,T_f} \right\}$, $f = 1..F$, be a fuzzy partition of U_f. The generic TS fuzzy rule used in this paper is defined as:

I. Bloch, A. Petrosino, and A.G.B. Tettamanzi (Eds.): WILF 2005, LNAI 3849, pp. 172–177, 2006.
© Springer-Verlag Berlin Heidelberg 2006

$$R_m: \textbf{IF } X_1 \textbf{ is } A_{1,j_{m,1}} \textbf{ and } \dots \textbf{ and } X_F \textbf{ is } A_{F,j_{m,F}} \textbf{ THEN } y_m = f_m(X_1, \dots, X_F) \tag{1}$$

where X_f represents a generic input variable, $A_{f,j_{m,f}}$ identifies which fuzzy set among the T_f fuzzy sets defined on the universe U_f has been selected for X_f in rule R_m, and $f_m(X_1, \dots, X_F)$ is a real function, typically expressed as $f_m(X_1, \dots, X_F) = p_{m,1}X_1 + \dots + p_{m,F}X_F + p_{m,F+1}$ and $p_{m,1}, \dots, p_{m,F+1} \in \Re$.

The TS fuzzy system has been implemented by using a variant of the Adaptive Network-based Fuzzy Inference System (ANFIS) model proposed by Jang [3]. The neurons of the first layer of an ANFIS network represent the membership functions associated with each linguistic term $A_{f,j_{m,f}}$ and compute the membership degree $A_{f,j_{m,f}}(x_f)$ of the input value x_f to fuzzy set $A_{f,j_{m,f}}$. The first layer contains $T = \sum_{f=1}^{F} T_f$ neurons, one for each possible fuzzy set defined on the universe U_f. In the classical approach, the input universes are uniformly partitioned. The neurons of the second layer model the antecedents of the TS rules and compute the activation degrees $w_m = \prod_{f=1}^{F} A_{f,j_{m,f}}(x_f)$ by implementing the *and* operator through the product. The number M of neurons is equal to the number of rules which can be generated by combining all possible fuzzy sets. By simple mathematical considerations, we obtain $M = \prod_{f=1}^{F} T_f$. The neurons of the third layer normalize the activation degrees by computing $v_m = \dfrac{w_m}{\sum_{m=1}^{M} w_m}$. The neurons of the fourth layer represent functions f_m and compute the output of each rule as $y_m = f_m(x_1, \dots, x_F)$. The neuron of the fifth layer computes the output of the TS system by summing the rule outputs as follows: $y = \sum_{m=1}^{M} y_m$.

The learning phase of ANFIS tunes simultaneously the antecedent and consequent parameters using a hybrid learning method: once the antecedent parameters have been computed by adopting the gradient method, the consequent parameters are determined by the Kalman filter. Typically, the ANFIS structure is a priori determined, using either uniform partitioning or clustering of the input space. In the former case, all the first layer nodes are linked to all the second layer nodes, while in the latter case fewer connections are needed since only rules corresponding to clusters are taken into account. In this paper we explore a new approach. We do not search for a unique optimal TS structure, but for a set of optimal structures characterized by different numbers of rules. This allows us to select the TS structure which best balances the accuracy rate on the training set and the system complexity. To this aim, we exploit a multi-objective genetic algorithm. The algorithm explores all the space of possible solutions and computes the optimal solution for a few different TS structures, as explained next.

2.1 The Multi-objective Genetic Algorithm Approach

First, we fix the number T_f of fuzzy sets which partition each input linguistic variable X_f (note that the number of fuzzy sets can be different from an input variable to another). Second, we create a uniform partition $P_f = \{ A_{f,1}, \ldots, A_{f,T_f} \}$, $f = 1..F$, for each variable X_f. Third, we generate all the possible rules in the form (1) considering all the possible fuzzy sets in the F partitions. The number of these rules is $M = \prod_{f=1}^{F} T_f$. The rules in (1) assume that all the input variables are used to define the region of the input space determined by the antecedent of the rule. Actually, some input variables could be irrelevant or even misleading. It is therefore desirable to be able to select the input variables that do contribute to define the input region. To this aim, we introduce a further fuzzy set, denoted $A_{f,0}$, for each partition: $A_{f,0}$ has a membership function equal to 1 on the whole universe. This means that the statement X_f is $A_{f,0}$ does not affect the computation of the activation degree. In other words, for the specific rule, the variable X_f is not taken into account. The terms $A_{f,0}$ allow us to generate rules including only a subset of the input variables, thus implicitly performing a sort of feature selection. The number of rules which can be produced is now $M = \prod_{f=1}^{F} (T_f + 1) - 1$ (note that we have excluded the rule that combines only $A_{f,0}$ terms). The M rules represent the solution space. A TS system will be composed of a subset of the possible rules. To identify the structure of a TS system, therefore, means to identify the subset of rules, and consequently of input variables, which contribute to generate the TS system.

To explore the solution space through a multi-objective genetic algorithm, we describe each possible solution through a chromosome composed of M bits: each bit is associated with one of the possible M rules. If the jth bit is set to 1, then the corresponding jth rule is part of the TS system; otherwise, it is not. An input variable which is not included in at least one rule is eliminated. In this way we perform a selection of the inputs. Furthermore, our approach can determine the number of rules automatically. The TS structure determined by the chromosome is used to build an ANFIS network: as described above, the rules of the TS system are used to initialize the membership function parameters in the neurons of the first layer and to set the connections between the neurons of the first and the second layers. The ANFIS learning tunes the membership function parameters of the antecedents and computes the consequent parameters by minimizing the mean square error between the desired output and the TS system output. As we require that both the system accuracy and complexity are considered, we use a multi-objective genetic algorithm.

Multi-objective genetic algorithms solve optimization problems characterized by multiple objectives. Each chromosome is associated with a vector of N fitness values, where each value typically expresses the fulfillment degree of a different objective. Since the different objectives are often conflicting, typically there does not exist a unique solution able to optimize all the objectives. On the contrary, there exists a trade-off surface, denoted as the Pareto optimal front, which contains all the optimal solutions. To compare different solutions the concept of dominance is introduced. A solution x associated with a fitness vector u dominates a solution y associated with a fitness vector v if and only if, $\forall i \in \{1,...,N\}$, $u_i \leq v_i$ and $\exists i \in \{1,...,N\} : u_i < v_i$, where u_i and v_i are the ith element of vectors u and v, respectively.

To generate the Pareto front, we used the Pareto Archived Evolutionary Strategy (PAES), proposed in [4]. PAES uses a mutation operator to generate new candidate solutions and an archive of non-dominated solutions, which form the Pareto optimal front. At each iteration, a new solution, denoted by m, is generated from the current solution, denoted by c, by using the mutation operator which randomly adds or deletes a rule. The solutions m and c are compared with each other: the dominated one is eliminated while the dominating one, if not dominated by any solution contained in the archive, is inserted into the archive. The possible archived solutions dominated by the newly inserted solution are discarded.

PAES terminates after a given number Z of iterations or when no solution is inserted into the archive for a number G of iterations, with $G < Z$. On PAES termination, the archive contains the set of solutions which compose the Pareto optimal front. Each solution is a TS system. The exploration of the space can be restricted to a subspace of interest by forcing constraints on the minimum and maximum number of rules and the minimum and maximum number of terms in the antecedent of the rules. This allows reducing the learning time of ANFIS and consequently shortening the execution of PAES.

3   Experimental Results

We tested our approach on two well-known datasets, namely Miles per Gallon (MPG) and Pima Indians Diabetes (Pima), obtained from the UCI Machine Learning Repository database (available via anonymous ftp at ftp.ics.uci.edu). As regards the MPG dataset, the goal is to predict the automobile fuel consumption on the basis of eight characteristics. After removing patterns with missing values, the data set was reduced to 392 entries. This dataset was divided into a training set and a test set, both made of 196 patterns. We used only six characteristics, namely number of cylinders, displacement, horsepower, weight, acceleration, model year as inputs, and mpg as output. We did not use origin and car name characteristics, since they are not numerical. As regards the Pima dataset, the goal is to predict whether a Pima Indian individual is diabetes positive or negative, on a basis of eight attributes. It is composed of 768 samples, 500 belonging to the positive class (65.10%) and the remaining 268 to the negative class (34.90%). We randomly split them into a training set of 500 patterns and a test set of 268, with the same class distribution.

We performed 2000 PAES iterations on the MPG dataset, with 100 ANFIS (local) iterations, a uniform partition with 2 fuzzy sets for each input variable, a maximum number of 7 rules, and membership functions of generalized bellman type. The time required was 42 minutes and 3 seconds on a 2Ghz Pentium IV with 1Gbyte RAM. The results are shown in Figure 1, where Rules, Inputs and Terms specify the

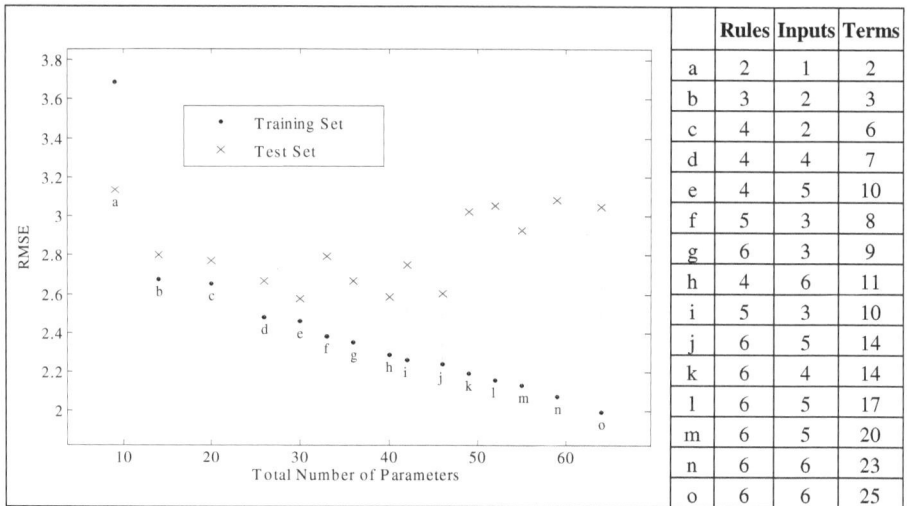

Fig. 1. Approximated Pareto front on training and test sets (MPG dataset)

numbers of rules, input variables and linguistics terms actually used in the TS system. In particular, in the solution denoted c, we obtained better results than in [5], using the same number of inputs (2) and the same number of ANFIS iterations (100).

Figure 2 reports the results obtained running PAES for 2500 iterations on the Pima dataset, with 50 ANFIS (local) iterations, a uniform partition with 3 fuzzy sets for each input variable, a maximum number of 10 rules, and membership functions of generalized bellman type. The classification into classes 1 and 0, for positive and negative samples, respectively, has been performed by rounding the TS system output. The time required was 42 minutes and 25 seconds on the same machine used in

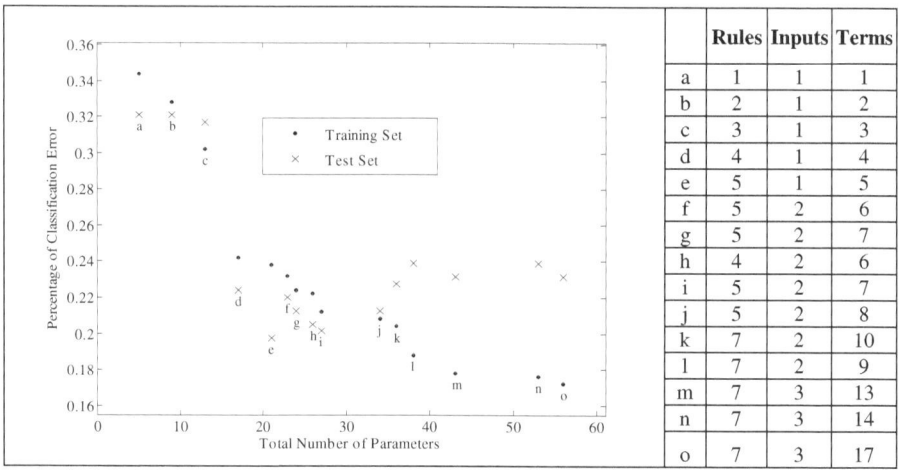

Fig. 2. Approximated Pareto front on training and test sets (Pima dataset)

the first experiment. We observe that our classification performance is comparable to that obtained in [6], where a 2-input, 3-rule TS fuzzy system was employed, for a total of 4 linguistic terms: 24.2% classification error on the training set and 23% on the test set.

The choice of the TS system to be used among all possible alternatives in the Pareto front is demanded to the user, who has to balance the optimisation of the objectives with his/her specific requirements. Obviously, the user has to consider that the increase in the number of parameters leads to improve accuracy on the training set, but, as shown in Figures 1 and 2, may decrease accuracy on the test set (well-known overfitting problem). The choice of the TS system could be automated by using both a training set and a validation set. The Pareto front is generated by applying PAES to the training set. Then, for each TS system in the Pareto front, the accuracy on the validation set is computed. The optimal choice might be the TS system which guarantees the best accuracy on the validation set.

4 Conclusions

In this paper we have applied a Multi-Objective Genetic Algorithm to approximate the Pareto optimal front when the objectives are the TS complexity and accuracy. During the genetic exploration, a local optimization has been performed using the ANFIS local search strategy. We obtain an approximated Pareto front, which allows us to choose a posteriori the best compromise between accuracy and complexity. Our approach automatically selects the input variables by eliminating those that do not contribute to define the input space.

References

1. Takagi,T., Sugeno, M.: Fuzzy identification of systems and its application to modeling and control, IEEE Transaction on System, Man, Cybernetics 15 (1985), 116-132
2. Setnes, M., Roubos, H.: GA-fuzzy modeling and classification: complexity and performance, IEEE Transactions on Fuzzy Systems 8(5) (2000), 509-522
3. Jang, J.S.R.: ANFIS: Adaptive-Network-based Fuzzy Inference Systems, IEEE Trans. on Systems, Man and Cybernetics 23(3) (1993), 665-685
4. Knowles, J.D., Corne, D.W.: Approximating the nondominated front using the Pareto archived evolution strategy. Evolutionary Computation 8(2) (2000), 149-172
5. Jang, J.S.R.: Input Selection for ANFIS Learning, in Proceedings of the Fifth IEEE International Conference on Fuzzy Systems (1996), vol. 2, 1493-1499
6. Chang X., Lilly, J.H.: Evolutionary Design of a Fuzzy Classifier From Data. IEEE Trans. on Systems, Man, and Cybernetics—part b: cybernetics 34(4) (2004), 1894-1906

Genetic Programming and Neural Networks Feedback Linearization for Modeling and Controlling Complex Pharmacogenomic Systems

Alexandru Floares

Oncological Institute Cluj-Napoca, Str. Republicii, Nr. 34-36,
Cluj-Napoca, cod 400015, Romania
alexandru.floares@iocn.ro

Abstract. Modern pharmacology, combining pharmacokinetic, pharmacodynamic, and pharmacogenomic data, is dealing with high dimensional, nonlinear, stiff systems. Mathematical modeling of these systems is very difficult, but important for understanding them. At least as important is to adequately control them through inputs - drugs' dosage regimens. Genetic programming (GP) and neural networks (NN) are alternative techniques for these tasks. We use GP to automatically write the model structure in C++ and optimize the model's constants. This gives insights into the subjacent molecular mechanisms. We also show that NN feedback linearization (FBL) can adequately control these systems, with or without a mathematical model. The drug dosage regimen will determine the output of the system to track very well a therapeutic objective. To our knowledge, this is the first time when a very large class of complex pharmacological problems are formulated and solved in terms of GP modeling and NN modeling and control.

1 Introduction

In the last years, detailed data about the complex molecular interactions between drugs and organism become available. The differential genes expression, induced by drugs, can be investigated by microarray techniques [1]. This pharmacokinetic (PK), pharmacodynamic (PD), and pharmacogenomic (PG) data allow conceptual and mathematical models building. This models are important for *understanding* pharmacological systems. Unfortunately, modeling this very high dimensional nonlinear stiff control systems is a difficult task. At least as important is to be able to adequately *control* them through inputs, which are drugs with different dosage regimens, even with a limited understanding - a gray or black box approach. The final goal is optimizing and individualizing medical therapy in the presence of various degrees of knowledge and uncertainty. Mathematical modeling requires detailed knowledge of the mechanisms involved and the estimation of numerous parameters; it is a tedious, expensive, and time consuming process. This probably explains the law impact of this approach on clinical practice, even for simple models.

I. Bloch, A. Petrosino, and A.G.B. Tettamanzi (Eds.): WILF 2005, LNAI 3849, pp. 178–187, 2006.

We investigate genetic programming [2] and neural networks [3] as alternative techniques for this tasks. In our approach, genetic programming uses information, about the distribution and elimination of the drug (pharmacokinetics), and about cellular and molecular processes's inputs and outputs (pharmacodynamics), to automatically and simultaneously write the mathematical model structure, and calibrate and optimize the model's constants. The end result is a bug free computer model written in computer programming languages like C++ or Java. These models provide a quantitative descriptions of drug(s) pharmacogenomics. Also, because the vast diversity of genes are usually regulated by drugs via a limited array of mechanisms, these mechanisms can be identified from the resulted mathematical models.

This is a reverse engineering methodology, based on the mappings from the computer programs to the mathematical models, and from the mathematical models to the conceptual models describing the molecular mechanisms. It gives insights into the subjacent molecular mechanisms of complex pharmacogenomic systems starting from data. On the same time, this approach address a very difficult pharmacological problem: treatment individualization. If individual pharmacological or pharmacogenomic data are used, the resulted models are individual pharmacological models; the values of the pharmacological parameters are individual values not mean values.

Neural networks are nonlinear, adaptive, independent of modeling assumptions, fault tolerant, universal, and operate in real time. Feedback linearization is one of the most important nonlinear control design strategy [4]. It may results in linearizations which are valid for large operating regions of the system, as opposed to a local Jacobian linearization. In a previous study, we obtained the best published result in a cancer chemotherapy problem using neural networks feedback linearization [5]. This motivates this study and the use of multilayer perceptrons (MLP), instead of other possible neural networks like radial basis functions or dynamical neural networks [4]. In our protocol, neural networks feedback linearization can be applied to complex pharmacological systems, with or without the aid of a conceptual or mathematical model. The established drug dosage regimen will determine the output of the pharmacological system to track very well a therapeutic objective, modeled as a reference signal. To the best of authors knowledge, this is the first time when a very large class of complex pharmacological problems are formulated and solved in terms of genetic programming modeling and neural networks modeling and control.

2 Pharmacological Systems

2.1 Pharmacogenomic Data

For illustrating the proposed methods, we use published microarray pharmacogenomic data [7]. Forty-three male rats weighing 225 to 250 g received a single intravenous bolus dose of 50 mg/kg methylprednisolone (MPL), a synthetic corticoid. Rats were sacrificed and liver, an important action site for corticoids, excised at 17 time points over 72 hours. Four untreated rats were sacrificed at

0 hours as controls. RNAs from individual livers were used to investigate 8000 genes with Affymetrix GeneChips. Cluster analysis revealed six temporal patterns consisting of 197 responsive probes representing 143 genes.

2.2 Pharmacological Mathematical Models

Pharmacokinetics, the relationship between time and plasma concentration, can be simply described as what the body does to the drug. The clinical interpretation of pharmacokinetic results requires another set of information, the relationship between plasma concentrations (or doses) and effect, or pharmacodynamics. This can be described as what the drug does to the body.

Pharmacological models, combining pharmacokinetic and pharmacodynamic models of pharmacogenomic data, are usually nonlinear, high dimensional, stiff control systems. Our investigations show that the vast majority of pharmacological systems, irrespective of drugs formulation and routes of administration, are *affine*. For affine systems, the control input, represented by the drugs with their dosage regimens for pharmacological systems, appears linearly in the state equation:

$$\dot{x} = f(x) + \sum_{j=1}^{m} g_j(x)u_j$$

$$y_i = h_i(x) \tag{1}$$

where $x = [x_1, \dots, x_n] \in \Re^n$ is the state vector, $f(x)$, $g_1(x), \dots, g_m(x)$ are differentiable vector fields, and $h_1(x), \dots, h_p(x)$ are smooth functions, all defined on an open set of $\Re^n$; the system has m inputs $\{u_1, \dots, u_m\}$ and p outputs $\{y_1, \dots, y_m\}$. This is a key aspect of the neural networks feedback linearization approach to pharmacological systems proposed in this study. Feedback linearization converts a nonlinear system into an equivalent linear system through co-ordinate transformation [4]. Feedback linearization is easier to apply to affine systems and more powerful. It is distinguished from Jacobian linearization offering linearizations valid for a larger operating space. By eliminating nonlinearities, conventional linear control techniques can be applied [4].

Pharmacological Mechanistic Blocks. Pharmacokinetics deals with a mathematical description of the rates of drug movement into, within and exit from the body. Rate processes in the field of pharmacokinetics are usually zero order, first order, or Michaelis-Menten (nonlinear) kinetics. These rate processes can be described mathematically by the following *PK blocks*:

1. Zero order kinetics: $dX/dt = -k$; rate processes are described by the rate constant alone.
2. First order kinetics: $dX/dt = -k_X$, where k is a rate constant and X is the amount or concentration of the drug remaining to be transferred.
3. Michaelis-Menten kinetics: $dX/dt = -(V_m X)/(K_m + X)$, where V_m is a maximum rate and K_m is the Michaelis constant.

A PK model is the algebraic sum of the corresponding PK blocks, resulting in one ore more ordinary differential equations. The PK block corresponding to the drug input is usually zero order or first order - *affine* systems.

A pharmacodynamic model attempts to relate drug concentration, ideally at the site of action of the drug, to some pharmacological effect. The pharmacological effects of the drug can be described mathematically by the following, most used, PD blocks [6]:

1. Linear model: $E = S \cdot Ce + E_0$
2. Log-linear model: $E = S \cdot log(Ce) + E_0$
3. Ordinary E_{max} model: $E = E_0 + E_{max} \cdot C_e/(C_e + EC_{50})$
4. Ordinary inhibition E_{max} model: $E = E_0 - E_{max} \cdot C_e/(C_e + EC_{50})$
5. Sigmoid E_{max} model (Hill): $E = E_0 + E_{max} \cdot C_e^n/(C_e^n + EC_{50}^n)$
6. Sigmoid inhibition E_{max} model (Hill): $E = E_0 - E_{max} \cdot C_e^n/(C_e^n + EC_{50}^n)$

where E is the effect variable, E_0 is the baseline effect, E_{max} is the maximum drug induced effect, EC_{50} is the plasma concentration at 50% of maximal effect, S is the slope of the line relating the effect to the concentration, C_e is the concentration to which the effect is related, and n is the sigmoidicity factor (Hill exponent).

Methylprednisolone Pharmacokinetics and Pharmacodynamics. The PK of the drug is described [7] by the following biexponential equation:

$$C_p = C_1 \cdot \exp(-\lambda_1 \cdot t) + C_2 \cdot \exp(-\lambda_2 \cdot t) \tag{2}$$

where C_p is the plasma concentration of the drug, $C_1 = 39,130\,\text{ng/ml}$, $C_2 = 12,670\,\text{ng/ml}$, $\lambda_1 = 7.54\,\text{h}^{-1}$, and $\lambda_2 = 1.20\,\text{h}^{-1}$ are the coefficients for intercepts and slopes. Note that (2), apparently not composed by the above PK blocks, is just the integrated form of an equation following the usual pattern.

The cellular mechanisms of the corticosteroids pharmacogenomics [7] are briefly described. Unbounded methylprednisolone in blood freely diffuse into the cytoplasm of liver cells and quickly binds to the cytosolic receptor and activate it. The activated drug-receptor complex rapidly translocates in the nucleus were it binds to the glucocorticoid responsive element in the target DNA and alter rates of transcription of target genes. Binding of the activated drug-receptor complex results in decreased transcription and reduced levels of receptor mRNA. The mRNA translocates to the cytoplasm were is translated to protein. This further decreases the free receptor cytosolic density. The drug-receptor complex in nucleus may dissociate from the target DNA and return to the cytosol. Part of the receptors may be degraded, whereas the rest may be recycled.

A PD model, describing the drug receptor dynamics in rat liver, after drug administration, was proposed in [7]:

$$\frac{dmRNA_R}{dt} = k_{s\_Rm} \cdot \left(1 - \frac{DR\_N}{IC\_50_R m + DR\_N}\right) - k_{d\_Rm} \cdot mRNA_R \tag{3}$$

$$\frac{dR}{dt} = k_{s\_R} \cdot mRNA_R + R_f \cdot k_{re} \cdot DR\_N - k_{on} \cdot D \cdot R - k_{d\_R} \cdot R \tag{4}$$

$$\frac{dDR}{dt} = k_{on} \cdot D \cdot R - k_T \cdot DR \tag{5}$$

$$\frac{dDR_-N}{dt} = k_T \cdot DR - k_{re} \cdot DR_-N \tag{6}$$

where symbols represent the plasma concentration of the drug (D), the receptor mRNA ($mRNA_R$), the free cytosolic receptor density (R), cytosolic drug-receptor complex (DR), and drug-receptor complex in nucleus (DR_-N); the rate constants include zero-order rates of receptor mRNA synthesis ($k_{s\_Rm} = 2.90$ fmol/g liver/h) , the first-order rates of receptor mRNA degradation ($k_{d\_Rm} = 0.11$ h^{-1}), receptor synthesis ($k_{s\_R} = 1.19$ h^{-1}) and degradation ($k_{d\_R} = 0.0572$ h^{-1}), translocation of the drug-receptor complex into the nucleus ($k_T = 0.63$ h^{-1}), the second-order rate constant of drug-receptor association ($k_{on} = 0.00329$ l/nmol/h), the concentration of DR_-N at which the synthesis rate of receptor mRNA drops to 50% of its baseline value ($IC_{50\_Rm} = 26.2$ fmol/mg of protein), and $R_f = 0.49$ is the fraction of free receptor being recycled. The baseline were defined in [7] using the following equations: $k_{d_{Rm}} = k_{s_{Rm}}/mRNA_R^0$ and $k_{s_R} = (R^0/mRNA_R^0) \cdot k_{d_R}$, where $mRNA_R^0 = 25.8$ (fmol/g liver) and $R_0 = 540.7$ (fmol/mg protein) are the baseline values of receptor mRNA and free cytosolic receptor density.

A careful examination of the above equations can reveal two very important aspects, for our genetic programming and neural networks feedback linearization approaches:

1. This complex model is an algebraic sum of PK and PD blocks.
2. The system is affine, the input D, the plasma concentration of the drug, appears linearly in the state equations.

Methylprednisolone Pharmacogenomics. Binding of the activated drug-receptor complex to the target DNA induces or represses several genes. We investigate two of the six mathematical models proposed in [7] to describe different patterns of gene expression in rat liver, after methylprednisolone treatment. The first model was selected because is typical and the second one because it is the most complex. In the first model, the synthesis and degradation of target RNA (normalized as ratio to control), without drug administration, was assumed as follows:

$$\frac{dmRNA}{dt} = k_{s\_m} - k_{d\_m} \cdot mRNA \tag{7}$$

The $mRNA$ level was assumed to be at steady-state at time 0 in control animals, yielding the following baseline equation: $k_{s\_m} = k_{d\_m} \cdot mRNA^0$. Baseline level of $mRNA^0$ was fixed to 1 for the most genes.

Model I. The induced production of $mRNA$ was described as follows:

$$\frac{dmRNA}{dt} = k_{s\_m} \cdot (1 + S \cdot DR_-N) - k_{d\_m} \cdot mRNA \tag{8}$$

where the increase of transcription rate k_{s_m} is proportional with DR_-N with the constant of proportionality S.

Model II. mRNA with induced degradation in cytosol and secondarily induced transcription by biosignals (BS) was described as follows:

$$\frac{dmRNA}{dt} = k_{s\_m} \cdot (1 + S_{BSm} \cdot DR\_N) - k_{d\_BSm} \cdot mRNA_{BS} \qquad (9)$$

$$\frac{dBS_r}{dt} = k_{s\_BS} \cdot mRNA_{BS} - k_{d\_BS} \cdot BS_r \qquad (10)$$

$$\frac{dmRNA}{dt} = k_{s\_m} \cdot (1 + S_{m\_s} \cdot BS_r) - k_{d\_m} \cdot (1 + S_{m\_d} \cdot DR) \cdot mRNA \quad (11)$$

where $mRNA_{BS}$ is the mRNA of the regulatory biosignals and BS represents their levels, both normalized as ratio to control. k_{s_BSm} is the rate of BS mRNA synthesis, k_{d_BSm} is the rate of BS mRNA degradation, k_{s_BS} is the rate of mRNA translation to BS, and k_{d_BS} is the rate of BS protein degradation. The stimulation of BS transcription is proportional with DR_N with a proportionality constant S_{BSm}. The stimulation of $mRNA$ synthesis is proportional with BS with a proportionality constant S_{m_s}; this stimulation is also present at baseline condition. The cytosolic $mRNA$ degradation is regulated by DR and S_{m_d} is the corresponding stimulation factor. At time 0 (9), (10), and (11) yield the following baseline equations: $k_{s\_BSm} = k_{d\_BSm} \cdot mRNA_{BS}^0$, $k_{s\_BS} = (BS_r^0/mRNA_{BS}^0) \cdot k_{d\_BS}$, and $k_{s\_m} = k_{d\_m} \cdot mRNA^0/(1 + S_{m\_s} \cdot BS_r^0)$, were $mRNA_{BS}^0$ and BS_r^0 are the baseline values of normalized BS mRNA and protein levels, which were fixed as 1.

Examining the last two models one can see again that the models are algebraic sums of PK and PD blocs, and are affine. The input DR_N, the intranuclear concentration of the drug-receptor complex, appears linearly in the equation.

To conclude this section, one can investigate only the pharmacokinetic of a drug (e.g., (2)), or can combine the pharmacokinetics with the pharmacodynamics ((3) - (6)), or even combine the PK and PD descriptions with the pharmacogenomic description ((8), or (9) - (11)) in a PK-PD-PG model, as we did. The models' complexity will increase but they will still be algebraic sums of PK and PD blocks, and affine. This is why we believe that our approach is very general.

3   Genetic Programming Automatically Modeling

Genetic Programming [2], breeds computer programs to solve problems by executing the following three steps:

1. Generate randomly an initial population of compositions of functions and terminals (see below) appropriate to the problem.
2. Iteratively perform the following substeps (a generation) on the population of programs until the termination criterion has been satisfied:
   (a) Execute each program in the population and assign it a fitness value using the problems fitness measure.

(b) Create a new population of programs by applying the following (most usual) operations to program(s) selected from the population with a probability based on fitness.

   i. Reproduction: Copy the selected program to the new population.
   ii. Crossover: Create a new program for the new population by recombining randomly chosen parts of two selected programs.
   iii. Mutation: Create one new program for the new population by randomly mutating a randomly chosen part of the selected program.

3. Designate the individual program with the best fitness as the runs result. This result may be a solution (or approximate solution) to the problem.

The functions may be standard arithmetic or programming operations, standard mathematical functions, logical functions, or domain-specific functions. The terminals are variables and parameters. The major preparatory steps for GP consist in determining the set of terminals, the set of functions, the fitness measure, the parameters for the run, the method for designating a result, and the criterion for terminating a run.

We use a linear version of steady state genetic programming [8]. In linear GP, computer programs are represented as a single sequence of instructions and data, as opposed to binary tree representations, for example. Linear genetic programming is unrelated to linear programming.

It starts with a population of 500 randomly generated computer programs. Then, it selects at random four programs and measures how well each of them maps the inputs to the output. The inputs, measurable in principle, but obtained by simulating the PK-PD-PG model in our study, are: the plasma concentration of the drug, the receptor mRNA, the free cytosolic receptor, the cytosolic

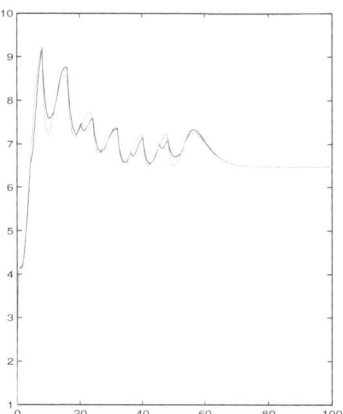 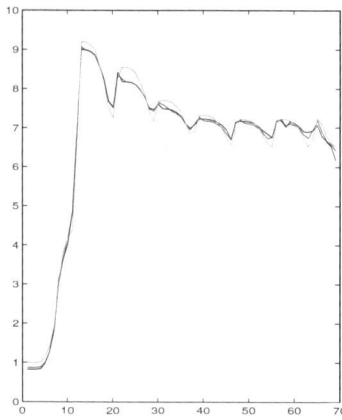

Fig. 1. Left: Neural networks feedback linearization tracking results for the combined pharmacokinetic, pharmacodynamic, and pharmacogenomic model; the resulted drug dosage regimen determine the system to follow closely the desired therapeutic objective. **Right**: The outputs of the combined pharmacokinetic, pharmacodynamic, and pharmacogenomic models generated by genetic programming and the "measured" mRNA.

drug-receptor complex, the drug-receptor complex in nucleus, the mRNA and protein level of intermediate regulators. The output is the mRNA of the corresponding gene. The two best programs win the tournament. They are copied and transformed into two new programs via crossover and mutation with frequency 50% and 95%, respectively. These programs replace the two loser programs from the tournament in the population of programs. GP repeats these steps until it has written a satisfactory program. The performance measure R^2 was 0.99 (the maximum is 1) for test data after 30 minutes and about 20,000,000 programs evaluation on a PC Pentium IV, at 3.2 GHz, and 2Gb RAM (see Fig. 1).

4   Neural Networks Feedback Linearization

Multilayer perceptrons (MLP) have been successfully used in optimizing cancer chemotherapy [5]. Neural networks feedback linearization can be applied to complex pharmacogenomics systems to find adequate drug(s) dosage regimens (see [9] for a more detailed treatment). We use *input-output* feedback linearization [4], in which the output becomes a linear function of a new control input. There are two distinct situations:

1. When a mathematical model exists, neural networks feedback linearization is applied to the input-output data resulting from model simulations.
2. When only experimental data are available, a neural networks model of the system is first identified from the set of input-output data pairs; it consists in two MLP corresponding to the two function f and g in (1).

The input is the plasma concentration of the drug, and the output is the mRNA of the target genes. The proposed methods are applied to both experimental data and to simulation data. In simulation, a random input, between zero and the maximal value of plasma concentration of the drug, is injected into the PK-PD-PG model at random intervals of time. The model structure is the standard Nonlinear Autoregressive-Moving Average (NARMA) model, adapted to the feedback linearization of affine systems - the controller input is *not* contained in the nonlinearity. We want the system output to follow a reference trajectory which is a mathematical formulation of a *therapeutic objective*. To determine the model order, we use the lag-space method [10]. As experimental data we arbitrary select a gene from microarray data [7]. We use regularization and early stopping to avoid over-training, and we start with different random initial conditions to avoid ending in "bad" local minima. The number of hidden layers is one for all neural networks. The number of neurons in the hidden layer, of the two MLPs, is 5 and 3, respectively, for experimental data. For simulated data, both MLPs have 7 neurons. The activation functions are tangent hyperbolic for the hidden layer and linear in the output layer for all NN. The training algorithm is Levenberg-Marquardt. We investigate the prediction errors by cross-validation on a test set. We use the NARMA-L2 [11] version of input-output feedback linearization. Tracking results, of the proposed neural networks feedback linearization methods, applied to the PK-PD-PG model, are shown in Fig. 1.

5 Conclusions

Genetic programming and feedback linearization using neural networks are applied to model and control the most complex nonlinear pharmacological systems. The goal is to develop powerful but easy to use methods for optimizing and individualizing medical therapy, in the presence of variate degrees of knowledge and uncertainty. The genetic programming and neural networks proposed methods can be applied to data, without a mathematical or even a conceptual model.

When a mathematical model exists, the neural networks methods can be applied to simulated data. In both situations the neural networks protocol is the same - identify a neural model and design a neuro-controller. Feedback linearization has a strong theoretical foundation, is simple, and gives a linear control low, valid everywhere in the space of admissible inputs and outputs. As we show, most pharmacological systems are affine. For affine systems feedback linearization is particularly simple to apply and powerful.

Starting from data, a mathematical model, expressed in a programming language like C++ or Java, can be automatically generated, without bugs, with our genetic programming approach. As we show, even very complex pharmacogenomic models are build by summing pharmacokinetic and pharmacodynamic blocks. Knowing this, greatly helps checking if genetic programming is identifying the correct model structures from data, reflecting the molecular reality. This models, giving insights into complex molecular mechanisms involved, are preferred to a black box neural networks model, when understanding is the main goal. investigation. The investigation of differential gene expressions, induced by drugs, when combined with powerful computational intelligence methods, represents a strong foundation for optimizing and individualizing medical therapies. The genetic programming and neural networks feedback linearization protocols allow the elaboration of drug dosage regimens, which determine the pharmacological systems to follow the desired therapeutic objectives, without the aid of mathematical models, which are difficult to build for such complex data.

References

1. Berrar, D., P., Dubitzky, W., Granzow, M., eds.: A Practical Approach to Microarray Data Analysis. Ed: Kluwer Academic Publisher New York 2003
2. Koza, J., R.: Genetic Programming: On the Programming of Computers by Means of Natural Selection. MIT Press: Cambridge MA 1992
3. Bishop, C., M.: Neural Networks for Pattern Recognition. Oxford University Press 1995
4. Garces, F., Becerra, V., M.,Kambhampati, C., and Warwick, K.: Strategies for feedback linearisation: a dynamic neural network approach. Springer series *Advances in Industrial Control* London 2003
5. Floares, A., Floares, C., Cucu, M., Lazar, L.: Adaptive Neural Networks Control of Drug Dosage Regimens in Cancer Chemotherapy. Proceedings of the IJCNN 2003 Portland OR Jully 20-24 pp 154–159 2003
6. Mager, D., D., Wyska, E., Jusko, W., J.: Diversity of Mechanism-Based Pharmacodynamic Models. Drug Metab. Dispos. **31** (2003) 510–518

7. Jin, J., J., Almon, R., R., Dubois, D., C., and Jusko, W., J.: Modeling of Corticoids Pharmacogenomics in Rat Liver Using Gene Microarrays.J. Pharmacol. Exp. Ther. **307** (2003) 93–109
8. Banzhaf, W., Nordin, P., Keller, R., E., Francone, F.D.: Genetic Programming: An Introduction: On the Automatic Evolution of Computer Programs and Its Applications. Morgan Kaufmann 1997
9. Floares, A.: Feedback Linearization Using Neural Networks Applied to Advanced Pharmacodynamic and Pharmacogenomic Systems. Proceedings of the IJCNN 2005 Montreal Canada July 31-August 4 2005
10. Nrgaard, M.:Neural Network Based System Identification toolbox, Version 2. Technical Report **00-E-891**, Department of Automation Technical University of Denmark January 23 2000
11. Narendra, K., S, and Mukhopadhyay, S.: Adaptive Control Using Neural Networks and Approximate Models. IEEE Transactions on Neural Networks **8** 1997 475-485

OR/AND Neurons for Fuzzy Set Connectives Using Ordinal Sums and Genetic Algorithms

Angelo Ciaramella[1], Witold Pedrycz[2,3], and Roberto Tagliaferri[1]

[1] Department of Mathematics and Computer Science, University of Salerno,
via Ponte Don Melillo, I-84084, Fisciano, Salerno
{ciaram, robtag}@unisa.it
[2] Department of Electrical and Computer Engineering,
University of Alberta, Edmonton AB, Canada
[3] System Research Institute, Polish Academy of Sciences,
Warsaw, Poland
{pedrycz}@ee.ualberta.ca

Abstract. The paper introduces a generalization of the fuzzy logic connectives AND and OR. To define the logical connectives different t-norms and t-conorms are used. To generalize the t-norms (t-conorms) the Ordinal Sums are introduced. To learn the parameters of the builded Ordinal Sums and the of weights of the connectives the Genetic Algorithms are applied. Two experiments using both synthetic and benchmark data are made. From one hand, a 2-dimensional classification problem to show the behavior of the approach is considered and on the other hand the Zimmermann-Zysno data set to show the capability of the model is analyzed.

Keywords: Fuzzy Logic Connectives, t-norms (t-conorms), Ordinal Sums, Genetic Algorithms, Experiments with Logic Connectives.

1 Introduction

The semantics of logic operators (logic connectives) in fuzzy sets is enormously rich. Some of the most recent conceptual developments along this line involve uninorms [5, 6, 12] nullnorms [1, 4] and ordinal sums [3] of t-norms, just to name a few pursuits. Logic operators are the crux of the fundamentals of the fuzzy sets or granular computing in general. Quite commonly, more advanced models of logic connectives come with a suite of adjustable parameters whose role is to customize the operator to available experimental data. This flexibility should come hand in hand with a viable method of optimizing these parameters. Subsequently, the optimization should be efficient enough and supported through some architectural realizations of the operators. This tendency is quite visible in case of Neural Network (NN) implementations of *AND* and *OR* operators with weights, AND/OR logic hybrids [6] and ordinal sums [3]. In this paper we focus our attention to study a NN-based model of logical connectives [9, 10]. The logical connectives AND and OR are defined using different t-norms and t-conorms. We generalize these norms by using Ordinal Sums (OSs) [3]. To learn the parameters of the OSs and the weights of the connectives we use Genetic Algorithms (GAs) [3].

I. Bloch, A. Petrosino, and A.G.B. Tettamanzi (Eds.): WILF 2005, LNAI 3849, pp. 188–194, 2006.

2   Logic-Based Neurons and Ordinal Sums

In this section we briefly remind the two basic types of logical-based neurons. A connective can be conveniently regarded as multivariable nonlinear transformations between unit hypercubes, say $[0, 1]^n \rightarrow [0, 1]$. The AND neuron aggregates input signals (membership values) $\mathbf{x} = [x_1, x_2, \ldots, x_n]$ by first combining them individually with the connections (weights) $\mathbf{w} = [w_1, w_2, \ldots, w_n] \in [0, 1]^n$ and afterwards globally ANDing these results (see figure 1a),

$$y = \mathbf{AND}(\mathbf{x}, \mathbf{w}) \tag{1}$$

i.e.,

$$y = \mathbf{T}_{i=1}^n(x_i s w_i) \tag{2}$$

where t- and s-norms (t-conorms) are used to represent the AND and OR operation, respectively.

The structure of the OR neuron is dual to that reported for the AND neuron (see figure 1b), namely,

$$y = \mathbf{OR}(\mathbf{x}, \mathbf{w}) \tag{3}$$

i.e.,

$$y = \mathbf{S}_{i=1}^n(x_i t w_i) \tag{4}$$

On the other hand we have that a *triangular norm* (t-norm for short) is a function $t : [0, 1]^2 \rightarrow [0, 1]$ which is commutative, associative, non-decreasing in both components, and satisfies the boundary condition $t(x, 1) = x$ for each $x \in [0, 1]$. Given a t-norm, its dual t-conorms is defined by the De Morgan's relationship that is $s(x, y) = 1 - t(1 - x, 1 - y)$.

Now we also mark that for every natural number $n \geq 2$ the n-ary extension $\mathbf{t}_{i=1}^n : [0, 1]^n \rightarrow [0, 1]$ of a t-norm is recursively defined as [3, 7]

$$\mathbf{t}_{i=1}^{n+1}(x_1, x_2, \ldots, x_{n+1}) = \begin{cases} t(x_1, x_2) & \text{if } n = 1 \\ t(\mathbf{t}_{i=1}^n(x_1, x_2, \ldots, x_n), x_{n+1}), & \text{if } n > 1 \end{cases} \tag{5}$$

Finally we stress that an important way to construct new t-norms from given ones is that of an ordinal sum (OS) [3, 11, 7]:
let $\mathcal{I} = \{]\alpha_k, \beta_k[\}_{k \in K}$ be a non-empty countable family of pairwise disjoint open

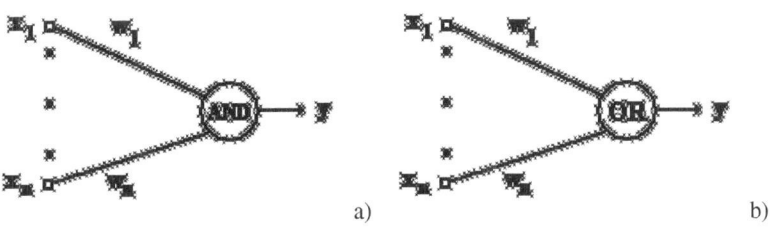

a)   b)

Fig. 1. AND/OR neurons: a) AND neuron of equation 1; b) OR neuron of equation 2

subintervals of $[0, 1]$ and let $\mathcal{F} = \{t_k\}_{k \in K}$ be a family of corresponding t-norms, then the ordinal sum $\{\langle \alpha_k, \beta_k, t_k \rangle\}_{k \in K}$ is the function $\mathbf{T} : [0, 1]^2 \to [0, 1]$ defined by

$$\mathbf{T}(x, y, \mathcal{F}, \mathcal{I}) = \begin{cases} \alpha_k + (\beta_k - \alpha_k) \cdot t_k\left(\frac{x - \alpha_k}{\beta_k - \alpha_k}, \frac{y - \alpha_k}{\beta_k - \alpha_k}\right) & \text{if } x, y \in [\alpha_k, \beta_k] \\ \min(x, y) & \text{otherwise} \end{cases} \quad (6)$$

which is always a t-norm. We also remark that ordinal sums of t-conorms are defined in the same way as OSs of t-norms, only replacing min by max [7].

3   Ordinals Sums and GAs

In the previous section we showed how to construct new t-norms from given ones using OSs. In this section we show how to apply OSs to define the t-norms and t-conorms of the equation 2 (or 3) and how to learn the parameters using Genetic Algorithms (GAs). In this case the equation 2 becomes

$$y = \widehat{\mathbf{T}}_{i=1}^n (x_i \widehat{s} w_i) \quad (7)$$

where $\widehat{t}$- and $\widehat{s}$-norms are used to represent AND and OR operation, respectively, and are obtained by using OSs. In details for a t-norm (for t-conorms is equivalent) we have

$$\widehat{T} = \mathbf{T}(x_i, w_i, \mathcal{F}, \mathcal{I}) \quad (8)$$

where from equation 6 we have that $\mathcal{F} = \{t_k\}_{k \in n}$ and $\mathcal{I} = [\alpha_k, \beta_k]_{k \in m}$ [3]. In the same way we formulate the equation 4

$$y = \widehat{\mathbf{S}}_{i=1}^n (x_i \widehat{t} w_i). \quad (9)$$

In figure 2 we show a possible mapping obtained using an OS and considering two variables x and y. The parameter m denotes the number of segments $s_k = \beta_k - \alpha_k$ (see equation 6).

Now we have to mark that the GA is a stochastic global search method that mimics the metaphor of natural biological evolution [3, 2, 8].

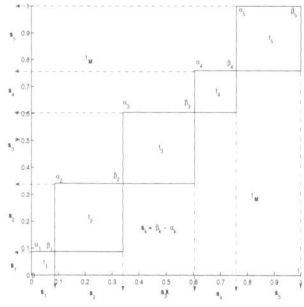

Fig. 2. Example of OSs with five t-norms

From one hand one of the main problems that we have to apply a GA is to define the structure of a chromosome. In our case, if we define m to be the number of intervals of the OS, we mark that in equation 7 (or equation 9) the parameters to learn are the t-norms $\tilde{t}_i$ ($i = 1, \ldots, m$) and the t-conorms $\tilde{s}_i$, the size of the subintervals $\tilde{v}_i^t$ of the t-norms and the subintervals $\tilde{v}_i^s$ of the t-conorms, respectively. Moreover we have also to learn the weights $\mathbf{w} = [w_1, w_2, \ldots, w_K]$ of the connections (where K is the number of the input neurons). Then to determine the best parameters we define the structure of a single chromosome in the following manner

$$\boxed{\tilde{t}_1}\boxed{\tilde{t}_2}\ldots\boxed{\tilde{t}_m}\boxed{\tilde{v}_1^t}\boxed{\tilde{v}_2^t}\ldots\boxed{\tilde{v}_m^t}\boxed{\tilde{s}_1}\boxed{\tilde{s}_2}\ldots\boxed{\tilde{s}_m}\boxed{\tilde{v}_1^s}\boxed{\tilde{v}_2^s}\ldots\boxed{\tilde{v}_m^s}\boxed{w_1}\boxed{w_2}\ldots\boxed{w_K}$$

where the values are defined in the $[0, 1]$ range.

If we denote n to be the number of known t-norms (for the t-conorms the following procedure is the same) then from the values of the chromosomes we associate a known t-norm (t_k with $k = 1, 2, \ldots, n$) in this way

$$\tilde{t}_j \in \left[\frac{k-1}{n}, \frac{k}{n}\right] \rightarrow t_k \tag{10}$$

where $j \in 1, 2, ..., m$.

Moreover, to determine the size of the subintervals we normalize each value in this way

$$v_i^t = \frac{\tilde{v}_i^t}{\sum_{j=1}^m \tilde{v}_j^t} \tag{11}$$

On the other hand another problem is to define the fitness function. In our experiments the fitness function is defined by using a 2-norm ($\|.\|_2$). If we suppose to have two n-dimensional vectors $\mathbf{x} = [x_1, x_2, \ldots, x_n]$ and $\mathbf{y} = [y_1, y_2, \ldots, y_n]$, respectively, then the fitness function using a 2-norm becomes (Sum-of-Squares Error (SSE))

$$J = \min \frac{1}{2}\|\mathbf{o} - \mathbf{t}\|_2^2 \tag{12}$$

where $\mathbf{t} = [t^1, t^2, \ldots, t^n]$ is a n target vector with values t^i and $\mathbf{o} = [o^1, o^2, \ldots, o^n]$ is a vector where o^i is the objective function estimated as follows

$$o^i = \mathbf{T}(x_i, y_i, \mathcal{F}, \mathcal{I}). \tag{13}$$

From equation 10 we have that $\mathcal{F} = \{t_k\}_{k \in n}$ and $\mathcal{I} = [\alpha_k, \beta_k]_{k \in m}$. We mark that $\alpha_{k+1} = \beta_k = \alpha_k + s_k$ and $\alpha_1 = 0$. For more details on the other operators and parameters needed for a GA process see [3, 2, 8]

4   Experimental Results

In this section we show some results obtained applying the approach described in section 3. We use the following t-norms (and the respective dual t-conorms):

$$\begin{aligned}
t_{\mathbf{M}}(x, y) &= \min(x, y) \\
t_{\mathbf{P}}(x, y) &= xy \\
t_{\mathbf{L}}(x, y) &= \max(0, x + y - 1) \\
t_{\mathbf{W}}(x, y) &= \begin{cases} \min(x, y), & \text{if } \max(x, y) = 1 \\ 0, & \text{otherwise} \end{cases}
\end{aligned} \tag{14}$$

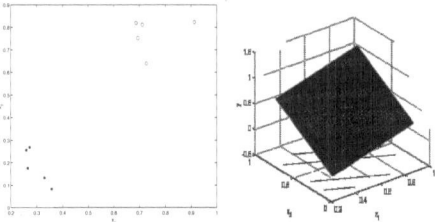

Fig. 3. Data set and classification: a) two classes with Gaussian distributions; b) SLP decision boundary

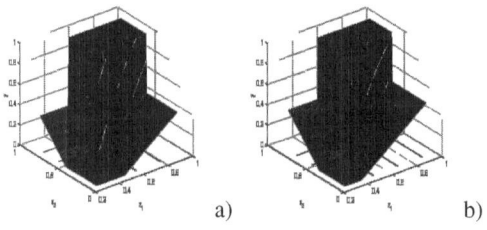

a)   b)

Fig. 4. AND neuron decision boundary: a) one norm; b) two norms

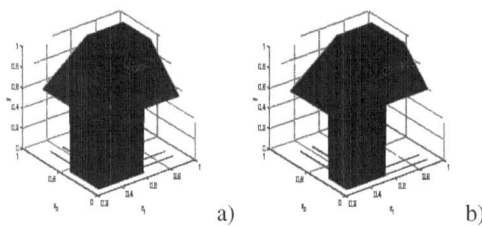

a)   b)

Fig. 5. OR neuron decision boundary: a) one norm; b) two norms

Aim of the first experiment is to explain the behavior of the proposed connectives to accomplish classification. We consider a 2-dimensional data set composed by two separated classes having both Gaussian distributions (fig. 3a). We also compare the AND and OR neurons with a Single Layer Neural Network (SLNN) with a linear output. In figure 3b we plot the decision boundary obtained applying the SLNN. In this case we obtain a SSE of 0.04. On the other hand in figure 4 we plot the decision boundaries obtained using the AND connective. In detail in figure 4a we plot the decision boundary obtained using equation 7 by using one norm and in figure 4b by using two norms.

Moreover in figure 5 we show the same experiment but using the connective described by equation 9. We have to note that both the SSEs are closed to 0 but we also note that using more than one norm in the OS we obtain a better result. This is clearer in the next experiment.

In the second experiment we study a data set coming from Zimmermann and Zysno [13]. It consists of 24 pairs of membership values $A(x)$ and $B(x)$ along with the experimentally collected results of their aggregation (fig. 6). Also in this case we apply the AND and OR neurons varying the number of t-norms and t-conorms. In figure 7 we

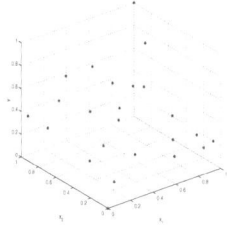

Fig. 6. Data set: Zimmermann and Zysno data set

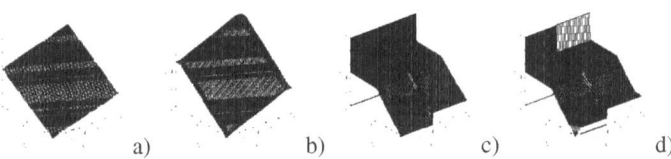

Fig. 7. AND neuron decision boundary: a) one norm (SSE 0.0950); b) two norms (SSE 0.0888); c) three norms (SSE 0.0950); d) four norms (SSE 0.090)

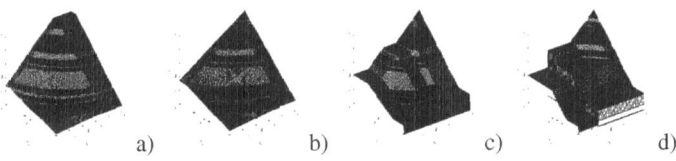

Fig. 8. OR neuron decision boundary: a) one norm (SSE 0.0558); b) two norms (SSE 0.0341); c) three norms (SSE 0.0361); d) four norms (SSE 0.035)

plot the results obtained by using an AND neuron. In these figures from a to d we plot the results obtained using one, two, three and four norms, respectively. We also obtain that the SSEs are of 0.0950, of 0.0988, of 0.0950 and of 0.090, respectively. In figure 8 we plot the results obtained by using an OR neuron. In this case the SSEs are of 0.0558, 0.0341, 0.0361 and 0.035, respectively.

We conclude that using the t-norms and t-conorms to build the connectives we achieve a more complex decision boundary. The decision boundary can be automatically defined by the proposed approach directly from the data. This is also confirmed by the different experimental results that we obtained using other data sets. We also have to note that about the Zimmermann-Zysno data set by using more than one t-norm (t-conorms) we achieve better results in the case of the OR connective.

5   Conclusion

In this paper we introduced a NN-based model of logical connectives. The logical connectives AND and OR have been defined using different t-norms and t-conorms. These norms have been generalized using Ordinal Sums (OSs). To learn the parameters of the OSs and the weights of the connectives we used GAs.

From the experiments that we made and from the results that we presented in the paper we noted that using OSs we achieve a better result than using one norm both on classification tasks and on the approximation of the Zimmermann-Zysno data set.

We stress that using GAs the learning process permits automatically to discover the norms and the weights of the connections from the data.

We also mark that the connectives obtained using different $t-$norms (and/or $t-$conorms) permit to obtain a more complex decision boundary than a single neuron. Could be noted that a similar boundary, using NNs, can be obtain only using more than one layer.

Moreover we stress that the logical connectives also permit to generalize the inference systems in a fuzzy logic framework.

In the next future the authors focus their attention to validate and to generalize the model to obtain a two layer NN for a complex inference system and to use GAs to generalize the uninorms.

References

1. T. Calvo, R. Mesier, Continuous Generate Associative Aggregation Operators, Fuzzy Sets and Systems, 126, 2002, 191-197
2. A. Chipperfield, P. Fleming, H. Pohlheim, C. Fonseca, Genetics Algorithms Toolbox, Dept. of Automatic Control and Systems Engineering, University of Sheffield
3. A. Ciaramella, W. Pedrycz, R. Tagliaferri, The Genetic Development of Ordinal Sums, Fuzzy Sets and Systems, vol. 151, Page(s): 303-325, 2005
4. P. Drygas, A Characterizaion of Nullnorms, Fuzzy Sets and Systems, 145, 2004, 455-461
5. J.C. Fodor, R. R. Yager, A. Rybalov, Structure of Uninorms, Int. J. of Uncertainty, Fuzziness and Knowledge-Based Systems, 5, 1997, 411-427
6. K. Hirota, W. Pedrycz, OR/AND Neuron in Modeling Fuzzy Set Connectives, IEEE Trans. on Fuzzy Systems, 2, 1994, 151-161
7. E. P. Klement, R. Mesiar, E. Pap, Triangular Norms, Kluwer Academic Publishers, Dordrecht 2001
8. C.-T. Lin, C.S.G. Lee, Neural Fuzzy Systems, Prentince Hall PTR, Upper Saddle River, NJ 07458, 1996
9. W. Pedrycz, A. F. Rocha, Fuzzy-Set Based Models of Neurons and Knowledge-Based Networks, IEEE Transactions on Fuzzy Systems, Vol. 1, NO. 4, 1993
10. K. Hirota, W. Pedrycz, OR/AND Neuron in Modeling Fuzzy Set Connectives, IEEE Transactions on Fuzzy Systems, Vol. 2, NO. 2, 1994
11. S. Sessa, R. Tagliaferri, G. Longo, A. Ciaramella, A. Staiano, Fuzzy similarities in stars/galaxies classification Systems, Man and Cybernetics, 2002 IEEE International Conference on , Volume: 2, Oct. 6-9, 2002 Page(s): 494 -496
12. R. R. Yager, Uninorms in Fuzzy Systems Modeling, Fuzzy Sets and Systems, 122, 2001, 167-175
13. H. H. Zimmermann and P. Zysno, Latent connectives in human decision making, Fuzzy Sets and Systems, 1980, Vol. 4, pp. 37-51

Intelligent Track Analysis on Navy Platforms Using Soft Computing

Robert Richards, Richard Stottler, Ben Ball, and Coskun Tasoluk

Stottler Henke Associates, Inc.,
951 Mariner's Island Blvd, Suite 360,
San Mateo, CA 94404
{Richards, Stottler, Bball, Tasoluk}@stottlerhenke.com
www.stottlerhenke.com

Abstract. We have developed and continue to enhance automated intelligent software that performs the tasks and decision making which now occurs by the personnel manning watch stations in the Combat Direction Center (CDC) and Task Force Combat Center (TFCC), on-board aircraft carriers and other Navy ships. Integrating information from various sources in a combat station is a complex task; disparate sources of information from radars, sonars, and other sensors are obtained by watch station surveillance guards, who must interpret it and relay it up the chain of command. The *Intelligent Identification Software Module* (IISM) alleviates some of the burden placed on battle commanders by automating tasks like management of historical data, disambiguating multiple track targets, assessing threat levels of targets, and rejecting improbable data. We have knowledge engineered current CDC/TFCC experts and designed IISM using C++ and SimBionic, a visual AI development tool. IISM uses multiple soft computing techniques including Baysian inference and fuzzy reasoning. IISM is interfaced to the Advanced Battle Station (ABS) for use on many US Navy sea vessels.

1 Introduction

The Combat Direction Center (CDC) and Task Force Combat Center (TFCC) on-board aircraft carriers and other ships must be manned with dozens of highly trained technical and tactical personnel [1]. The reason for this is the complexity of the weapon systems and associated information, as shown by the high-level organization of it in Figure 1. The combat areas consist of people; computers; and displays; and the arrows (in the figure) roughly correspond to information flow between combat areas and from sensors, to combat areas and from combat areas to weapons/countermeasures. CDC/TFCC operation is complicated by a large number of sensors, weapons and countermeasures. These operations will only become more complicated as additional sensors, weapons, and even war-fighting areas are added. Furthermore, through the Cooperative Engagement Capability (CEC), each ship can use the sensors and weapons on other ships thus adding additional combat areas, sensors, and weapons.

A naval commander must make complex decisions based on limited or noisy information. In partially observable and adversarial environments it is vital to keep

I. Bloch, A. Petrosino, and A.G.B. Tettamanzi (Eds.): WILF 2005, LNAI 3849, pp. 195–204, 2006.
© Springer-Verlag Berlin Heidelberg 2006

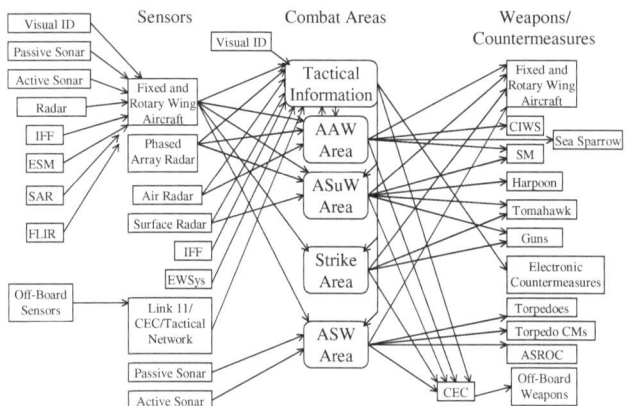

Fig. 1. Weapon System High Level Overview

track of an approximate model of the world that simultaneously maintains multiple hypotheses about the world state [2]. These hypotheses facilitate reasonable decisions to take in response to the hostile environment.

To ameliorate the complexity of these systems, Stottler Henke has developed the *Intelligent Identification Software Module* (IISM) that performs the tasks and decision making which now occurs by the human manning that watch station, such as tracking objects that merge and later split up, maintaining history of possible tracks for an object, assessing threat level, rejecting "insane" data, and handling errors.

IISM is interfaced to the Advanced Battle Station (ABS) for use on many US Navy ships. Given tracking data and time stamps from the Advanced Battle Station (ABS), IISM updates the history list of tracking and identification data, rejects nonsense tracks, compares recent history to past patterns of activity, alerts the commander when necessary, and provides customizable identifications of targets as well as the threat level of each target. IISM is also capable of correcting errors and recovering snap-shot and history data after unforeseen catastrophes.

We have knowledge engineered current CDC/TFCC experts and determined that the cognitive processes being utilized were reproducible with Artificial Intelligence techniques [3]. We determined the types of tasks performed and the knowledge required for those tasks. A breadth of positions was important to keep the representation schema truly general. We designed the general CDC/TFCC knowledge representation schema and then an intelligent CDC/TFCC equipment control, monitoring, processing, and fusion system. From knowledge engineering and the schema, we designed and implemented IISM using C++ and SimBionic, a visual AI development tool that can help in the development of fuzzy, Bayesian and other AI techniques.

2   IISM Input / Output Description and Functional Overview

Human tactical decision making in warfare scenarios can be described with the simplified diagram shown in Figure 2. Imperfect information about the current state

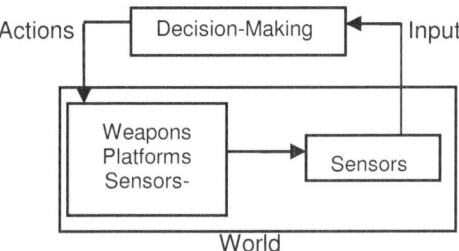

Fig. 2. Human Tactical Decision-Making

of the world is gathered by a diverse set of sensors. These sensors can be in several modes, may be off ship, and may be human in nature. The human decision-maker receives the sensor data through a communication or perception processes. Based on that information he makes decisions to take actions that affect the objects in the world over which he has direct control. These might include CDC/TFCC display systems, airborne platforms, weapon systems, communications, and sensors.

On a highly conceptual level, IISM's task can be viewed as a classification problem of the threat level assigned to individual entities, e.g. ships, present in the scenario. Maintaining a consistent and reasonably approximate model of several entities' attributes that are only partially perceivable implies the task of track handling and analysis. The latter is exploited in IISM to: 1) determine the identity of an entity (or some degree of certainty about it), 2) perform path analysis of entities and (3) infer abstract conclusions regarding the behavior of entities on the basis of their movement over time. Stated another way, both positive and negative evidence is tracked to form multiple, possibly competing hypotheses. Conclusions about these hypotheses are inferred for tracks through the process of elimination reasoning.

IISM stores and reasons about incoming track data in a flexible and customizable manner as defined by the control logic defined in SimBionic (see below). During this processing, IISM checks the quality of incoming messages, it updates its history of vessel movements (*tracks*) and IDs and performs threat assessment of units. This functionality is presently performed by trained watch-standing personnel aboard ships. It requires reasoning about whether the perceptions align with the internal model of the world and how *insane* (i.e. misaligned) perceptions are treated.

Insane and noisy data handling

Insane data can arise through an incorrect model or faulty perceptions, and special care must be taken in order to extract hints to potential threats instead of discarding them just like incorrect perceptions are discarded. The IISM reasoning functionality is performed in three subsequent steps in IISM's *Insanity Checker*: (1) *Threat processing* marks a unit as a potential threat in case insane perceptions are indicating this. (2) *Data Neglect Checking* takes account of an erroneous internal model caused by sensor noise and updates the model with the insane update. (3) *Inconsistency with ID checking* keeps track of harmless, but questionable/suspicious pieces of information and thus allows reasoning about temporally dispersed perceptions.

Track Hypothesis Handling

Instead of keeping a flat organization of unit ID hypotheses, IISM uses a hierarchical approach to refine an ID hypothesis as needed, such as in the case of determining the exact type of the enemy's unit. IISM assigns each hypothesis a particular *certainty level* that describes its reliability. When we get new data we use a *Bayesian network* update to keep track of the proper certainties for each track hypothesis. When the certainty for one of multiple hypothesis of a track is changed, or when a new hypothesis for a track arrives, an update algorithm is called on that track. This update algorithm uses the hyperbolic arctangent adjustment algorithm on each certainty to propagate the change made by the additional information. This algorithm runs through every hypothesis that is related to the changed one, updating each certainty according to Bayesian rules. These rules update the certainties based on the prior values and how closely they are related to the other related certainties.

Example Situation

Figure 3, shows an area around the Persian Gulf and provides an idea of how cluttered the environment being monitored and assessed can be.

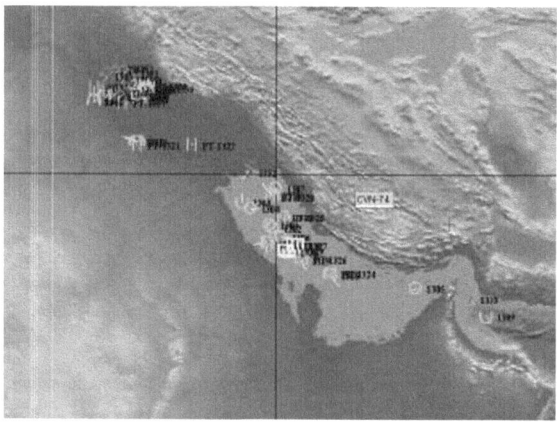

Fig. 3. Example of Density of Contacts that Need Monitoring and Assessing

Let's examine a situation where two surface tracks (track 1 & 2) are first detected, both traveling at a high speed (50 knots). At this point, IISM would already inference a subset of platform types based on their speed. Later these two tracks split up. Track 2 later merges with track 3, which had previously been IDed (identified) as an Iranian Houdong Fast Patrol Boat. These tracks (2 & 3) soon split up; at this point IISM does NOT know which of the tracks (2 or 3) is the formerly identified Iranian Houdong Fast Patrol Boat. Therefore, IISM will keep both sets of past information and use new information to improve its hypothesis on what each boat is. As can been seen even with this simple scenario, the situation is very fluid and multiple hypotheses must be tracked and re-evaluated as new information is obtained.

3 SimBionic

SimBionic is a visual framework that simplifies the authoring of simulated behaviors or algorithms. SimBionic's framework consists of a canvas depicting algorithms as a finite state machine (FSM) graph, a palette of geometric objects and glyphs, and a dictionary of actions and predicates.

SimBionic employs four programming constructs; 1) actions, which define all the different actions the algorithms can perform; 2) algorithms (also referred to as behaviors) that string together actions and conditional logic; 3) predicates, which set the conditions under which each action and algorithm will happen; and 4) connectors, which control the order in which conditions are evaluated, and actions and algorithms take place.

These four constructs allow one to create algorithms that range from simple sequences to complex conditional logic. Via SimBionic's authoring canvas, see Figure 4 (left image), users can visually create algorithms by drawing actions and invoke algorithms (represented as rectangles) and conditions (represented as ovals) to interact in both simple and complex combinations via connectors (represented as arrow-shaped lines with priority numbers). This canvas also allows users to assign arbitrary expressions and comments to these elements.

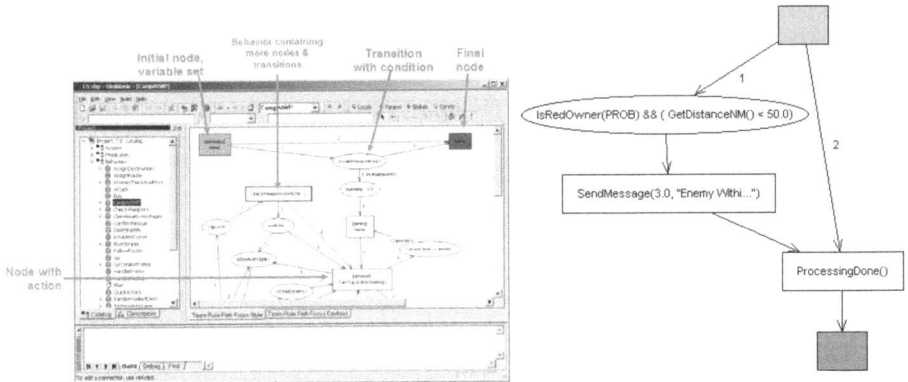

Fig. 4. SimBionic Authoring Environment & Trigger_NearByEnemy Behavior

SimBionic extends the usual notion of finite state machines by making it possible for states to refer to other finite state machines hierarchically, to define modular algorithms that can be combined powerfully. SimBionic software also provides four extensions that increase the power and expressiveness of the basic engine: global and local variables, interrupt transitions, "blackboards" for sharing knowledge among finite state machines, and polymorphic indexing for run-time selection of algorithms.

IISM uses the SimBionic visual AI code generator platform to instantiate intelligent modules that track target paths, assess threat, and identify targets. For example, the Trigger_NearByEnemy behaviour found in IISM, see Figure 4 (right image), is a schema for interacting with the possible enemy labelled RED, within some predefined distance. This behaviour is called when tracking data of the target are consistent with type RED and calculated "distance" from own ship. It invokes an action to contact the target by messaging. Other more complicated behaviours are invoked for identifying targets as friend or foe, for tracking specific targets over time, and for rejecting nonsense/insane data.

4 IISM Detailed Capabilities

IISM has been implemented using C++, and SimBionic. SimBionic can output its behaviors as C++ code for fast execution. IISM utilizes this facility to create a fast executing AI-based solution. Not all of the major capabilities or requirements utilize SimBionic, so listed first are those major capabilities or requirements that do not exploit SimBionic, and then those that do are described.

Intelligent Tactical Memory
One of the important functions that humans currently provide in the CDC/TFCC is that of intelligent memory and IISM mimics this capability. This memory includes all track attributes (position, velocity, ID information, etc.) along with a time stamp for each. Current ship systems do not keep, in a readily recalled format, the trajectory and ID history of each track. IISM fulfills this purpose.

System Independence
If tactical decision systems go down, IISM will continue to remember (and update from other sources if possible) the current tactical picture. This memory function is important for rebuilding the tactical picture. IISM is set up to take inputs from multiple sources.

IISM Reliability
IISM is required to be very robust, never crashing and able to run around the clock without requiring reboots. To handle the cases of hardware failure, IISM constantly backs up its memory to disk and automatically restore it upon start up.

Human Computer Interaction (HCI)
Most of the HCI occurs through the Advanced Battle Station (ABS). This way watch station personnel do not need to learn anything new, the information will appear in the same manner as if the current human decision makers had provided the information.

4.1 SimBionic Supported Capabilities

SimBionic is used to support IISM's core capabilities of automating the task of intelligent track analysis. The track's position and velocity with historical information, if any, regarding position, velocity, proximity and other interactions with other platforms is analyzed by IISM to estimate the probability of hostile intentions of and assess the threat posed by the track. Whenever a track significantly changes its velocity, analysis is made to determine if the maneuver warrants a change in the current ID estimate. Considerations include existing ship and air lanes, motion toward or away from blue forces or the assets that they are protecting, whether tracks appear to be cooperating, and attacks. For example, consider two tracks proceeding together at high speed. One breaks off and mingles with local fishing traffic. Later the other attacks. IISM will warn the watch stander about the other track. If the attack track has merged with other tracks, IISM will notify the user of which ones are possible enemy. IISM can reason from process of elimination as the non-enemy tracks are IDed to identify the remaining possibilities.

For example, the Track Id Processing Behavior (TIPB) is a hierarchal decision tree to classify the track into one of the ID categories (BLUE, RED, GRAY, WHITE) with a given certainty level by analyzing current information as well as historical information of the track, see Figure 5 (left image). TIPB has 3 top-level behaviors: Surface Track Behavior for analyzing surface tracks, Air Track Behavior for analyzing air tracks and Undersea Behavior for analyzing undersea tracks. When IISM receives new updates for the track it runs through TIPB.

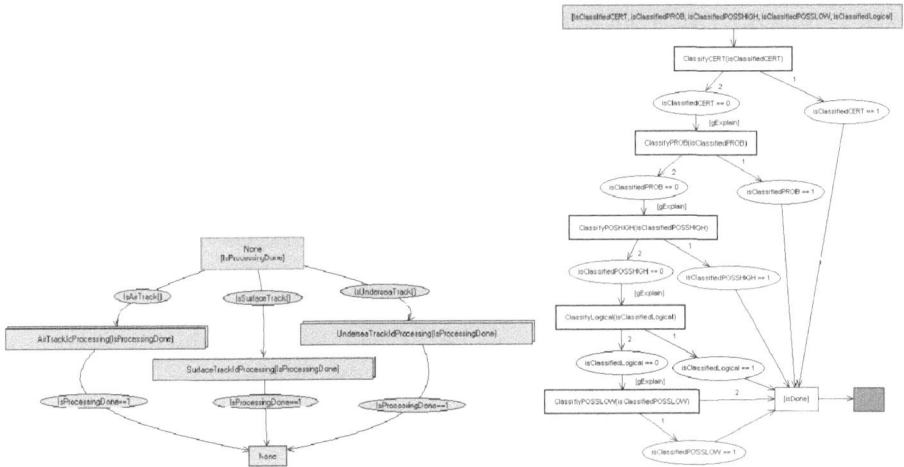

Fig. 5. Track Id Processing Behavior& Surface Track Id Processing

Now looking at the Surface Track Behavior, see Figure 5 (right image), it consists of five behaviors:

- ClassifyCERT
- ClassifyPROB
- ClassifyPOSSHIGH
- ClassifyLogical
- ClassifyPOSSLOW

The analysis of the information starts with ClassifyCERT and goes through ClassifyPOSSLOW if the track cannot be classified by any of the behaviors.

The following details some of the reasoning techniques used to perform the intelligent track analysis.

Track History Maintenance
Memory is also used to correlate previous tracks with new track information. A complete track history is kept, which allows IISM (or a human operator) to quickly determine if the track's ID is ambiguous because of a track merge or ID swap. Several mistakes, during naval exercises, caused by merges and swaps resulted in the targeting of several neutral, and even blue, platforms. Such mistakes during exercises cause commanders to limit their own options during future exercises or real missions. They are much less likely to use a weapon like the Harpoon, since they lack faith in their own ID picture. Although these problems are rare during random or benign scenarios (tracks don't normally pass that close to each other), a real adversary will go out of his way to try to create them. E.g., a terrorist attacking platforms under US protection would try to mingle, possibly several different times, with commercial platforms, such as fishing boats and merchant traffic. IISM has algorithms implemented with SimBionic that will handle the most complex set of merge/split scenarios (e.g. platforms merging with several different platforms and each other at separate times) logically and correctly. These algorithms already outperform humans in their ability to determine the possible IDs of tracks involved in several merges.

Historical Comparison
A track's history is kept in varying levels of detail, depending upon its age. IISM will remember all tactical data (to different levels of detail, minutes, hours, days, months, or even years before) and compare the current data, events, and situation to the recent or distant past. IISM will retrieve tracks similar to the current one and make recommendations accordingly.

Multiple Competing Hypotheses for ID
IISM keeps simultaneous competing hypothesis for each track as to the type/hull of the platform and its country of ownership. It will track both positive and negative evidence and reach both positive and negative conclusions. IISM explicitly keeps track of all possible hypotheses and the associated likelihoods for each track. Initially, a track can be anything, but incoming evidence impacts the certainties of each hypothesis. Positive ID information, such as a good visual ID, eliminates the competing hypotheses until the track is involved with a merge, at which time the resulting tracks each contain all the hypotheses of both tracks that merged.

Hierarchy of possible ID values
For both dimensions of ID information, IISM will include a hierarchy (from general to specific) of possible ID values. E.g.:

- Blue – UK, Combatant – frigate – FFG-7 – Specific platform; or
- White – Merchant, Cargo Carrier – Ship Class – Specific Hull

ID is often hierarchical with the goal of determining the most precise value that is worthwhile. Thus while an ID of White Merchant might be adequate, a Red Combatant may need to be IDed more precisely, perhaps as Chinese Houdong Fast Patrol Boat. These hierarchical symbols interact with the competing hypotheses described above. Thus, if the only competing hypotheses for a track are Gray Destroyer and Red fast patrol boat, and information is received that it has a speed greater than is possible for a destroyer, then IISM will conclude it is red.

Sanity Checking

When new data is received, before the track information is updated, the new data is compared to the recent history to make sure it makes sense and is at least physically possible. Any inconsistencies are reported, and to the degree practical, automatically resolved. This sanity checking function occurs for red, blue, gray, and white forces. IISM compares the current position/velocity to the last reported position for that track and determines if it is physically possible, given the platform type. If not, it determines if it is most likely a spurious data point, that the assigned track type is wrong, that a completely different platform as been assigned the same track number, or that the reported position of a friendly track is incorrect. It then recommends the appropriate action.

Fuzzy Reasoning

Classify Logical of the surface track behavior is an example of the fuzzy reasoning used by IISM. It will analyze the trajectory of the track to try and classify what kind of platform it is. Please refer to Figure 6.

In this behavior, first the turnRadius and Weight of the track is estimated based on the history of the trajectory. Next these numbers are converted into one of three fuzzy values, representing heavy, light, or middle weights, and small, middle, and large turn radii. The reason we use fuzzy values for the calculations is because this algorithm now becomes much more robust in the presence of noise or other negative

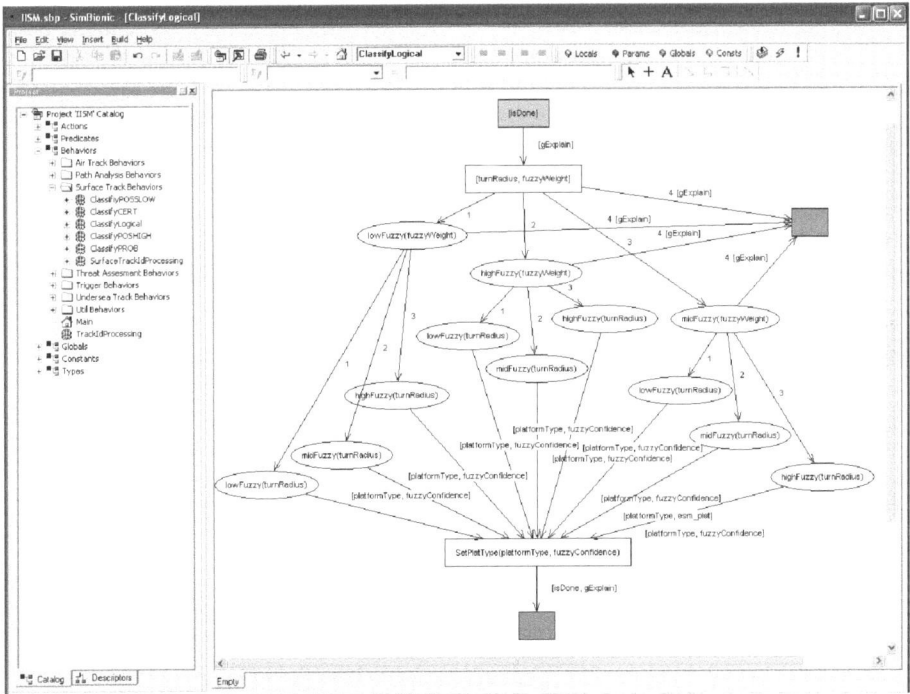

Fig. 6. Classify Logical Fuzzy Reasoning Behavior

factors. Finally, the platform type is recommended with various fuzzyConfidence levels depending on the fuzzy values. For example, if we have a low weight and high turn radius, we are PROB small light platform, and similarly if we are high weight and low turn radius we are POSHIGH large platform. The reason the large is only poshigh while the small is prob is because a large ship cannot move quickly, but a small ship can, thus we are more confident a ship is small when it moves quickly than that a ship is large when it moves slowly. This kind of intuitive reasoning is only possible via fuzzy reasoning.

Process of Elimination Reasoning
IISM employs logic and the process of elimination in making ID decisions. For example, IISM may know one combatant is out in a particular area where several other tracks are present. Even though every track seems to have low probability of being a combatant based on their behavior, a higher probability bias is used since one of them must be the combatant. The process of elimination is used to determine the most likely tracks to investigate first.

5 Conclusion

IISM is an AI module that alleviates the burdens placed on battle commanders by tracking sometimes ambiguous target signals, storing and handling past target data, assessing threat levels of targets, filtering out insane data, as well as robustly recovering from crashes and errors. IISM's rule-based logic is used to compute track IDs, estimate threats, and notify users of alert conditions; its probabilistic hypothetical reasoning system keeps track of multiple track hypotheses based on the fusion of evidence from multiple sources, and uses statistical algorithms to find correlations between track movements. IISM is a seamless enhancement to the current Advanced Battle Station, providing enhanced reasoning without the need for any user to learn a new system. By applying multiple soft computing techniques including Bayesian inference and fuzzy reasoning, as well as other AI techniques, including polymorphic finite-state-machines, IISM is performing as well as or better than Navy personnel.

References

1. Navy Warfare Development Command, "Sea-Based Theater Air and Missile Defense: A 21st-Century Warfighting Concept", http://www.ndcweb.navy.mil.
2. Hutchins, S. G., Technical Report 1718, Principles for Intelligent Decision Aiding.
3. Salas, E., Cannon-Bowers, J. A., & Johnston, J. H., "How can you turn a team of experts into an expert team?: Emerging training strategies". In C. Zsambok & G. Klein (Eds.), Naturalistic decision making (pp359-370).

Software Implementation of Fuzzy Controller with Conditionally Firing Rules, and Experimental Comparisons

Corrado Manara[1], Paolo Amato[2], Antonio Di Nola[1],
Maria Linawaty[3], and Immacolata Pedaci[3]

[1] Dipartimento di Matematica e Informatica,
Università degli Studi di Salerno,
Via Ponte don Melillo, 84084 Fisciano (SA), Italy
{cmanara, adinola}@unisa.it
[2] STMicroelectronics, FTM R&D,
Via C. Olivetti 2, 20041 Agrate Brianza (MI), Italy
paolo.amato@st.com
[3] STMicroelectronics, PST R&D,
Via Remo De Feo 1, 80022 Arzano (NA), Italy
{maria.linawaty, immacolata.pedaci}@st.com

Abstract. In this work we present a MATLAB implementation of a fuzzy controller with Conditionally Firing Rules (CFR). The performance of Mamdani-Assilian, Takagi-Sugeno-Kang and CFR inferences are compared and analyzed on two test examples.

1 Introduction

A fuzzy controller with *Conditionally Firing Rules* (CFR) was introduced in [8] as a generalization of Mamdani-Assilian (MA) controller [6], using an inference rule that is not compositional [11]. This extension was motivated by the effort of optimally fitting the behaviour of the controller with respect to the meaning of the fuzzy rule base, since MA controllers represent the rule base only roughly, with some distortion.

Both theoretical and practical arguments support the expectation that CFR controller allows a better performance without changing the rule base [8, 7, 9]. Thus CFR may improve the quality of control while keeping the same computational complexity as MA inference. Moreover, in [1] it is proved that the action of CFR inference can be implemented by integrating hardware circuits designed for MA controllers with crisp input variables.

However, in tackling real-world problems several issues should be taken into account and MA controllers are not the only possible choice. For instance, the efficiency of Takagi-Sugeno-Kang (TSK) inference [10] is witnessed by its use in many control applications. Consequently, it is not easy to have prior knowledge on which fuzzy inference system (MA, TSK, or CFR) could have the best performance on a given problem. To quickly test different inferences on the same

I. Bloch, A. Petrosino, and A.G.B. Tettamanzi (Eds.): WILF 2005, LNAI 3849, pp. 205–214, 2006.

problem, here we extend the MATLAB Fuzzy Logic Toolbox including CFR among its inference options.

In this work we discuss some practical issues related to this software implementation, and then we analyze the performance of the three fuzzy inference systems on two different test problems.

2   MATLAB Implementation of CFR Inference

In this section we describe a MATLAB implementation of CFR inference, extending the preexistent Fuzzy Logic Toolbox.

Any reference to MATLAB and its toolboxes can find exhaustive explanation at `www.mathworks.com`. Details of the software implementation and the simulations are freely available on request to the authors. The reader is referred to [8] for all the theoretical issues related to the following algorithm and a detailed treatment of the background information.

2.1   The Controller with Conditionally Firing Rules

Let X and Y denote the input and the output space of a controller, respectively. For a set Z, we denote by $\mathcal{F}(Z)$ the set of all fuzzy subsets of Z. The *support* of a fuzzy set $A \in \mathcal{F}(Z)$ is $\operatorname{Supp} A = \{z \in Z : A(z) > 0\}$. A fuzzy set is called *convex* if all its α-cuts are convex sets. For fuzzy subsets $A_1, \ldots, A_n$ of the same universe, Z, their *convex hull* is the smallest (w.r.t. the pointwise ordering) convex fuzzy set $C \in \mathcal{F}(Z)$ satisfying $A_i(z) \leq C(z)$ for all $z \in Z$, $i \in \{1, \ldots, n\}$. A fuzzy set is called *normal* if it attains the value 1 at some point. For $x \in Z$, let $\chi_x \in \mathcal{F}(Z)$ denote a singleton (crisp value), i.e., $\chi_x(z) = 1$ if $z = x$ and $\chi_x(z) = 0$ if $z \neq x$, for all $z \in Z$.

The expert's knowledge may be expressed by a rule base of the form

$$\textbf{if } x \in X_i \textbf{ then } y \in Y_i \,,$$

where $X_i \in \mathcal{F}(X)$ are *antecedents* and $Y_i \in \mathcal{F}(Y)$ are *consequents*, $i \in \{1, \ldots, n\}$ (see [3]).

The desirable properties a fuzzy controller should satisfy from a mathematical point of view are described in [8]. In our present context, when only crisp inputs are considered, those desiderata can be specialized as follows:

[Int1] Let $x^* \in X$ be a crisp input. If $X_i(x^*) = 1$ for a fixed i, then the output coincides with the respective consequent, i.e., $Y^* = Y_i$.

[Int2] For each crisp input $x^* \in X$, the output Y^* is not contained in all consequents, i.e., $Y^* \not\leq \min_{j \leq n} Y_j$.

[Int3] For each crisp input $x^* \in X$, the output Y^* belongs to the convex hull of consequents Y_i of all rules such that $x^* \in \operatorname{Supp}(X_i)$ (i.e., all firing rules).

According to [8], in rather typical situations many fuzzy controllers do not admit to satisfy [Int1], [Int2] and [Int3] simultaneously. This is the principal motivation for the introduction of CFR controller as a generalization of MA controller.

An effective calculation of the output of a MA controller is given by

$$Y^*(y) = \max_{i \leq n} T\left(\mathcal{D}(\chi_{x^*}, X_i), Y_i(y)\right), \tag{1}$$

where T is a fixed t-norm modelling a fuzzy conjunction [4], and

$$\mathcal{D}(\chi_{x^*}, X_i) = \sup_{x \in X} T(\chi_{x^*}(x), X_i(x)) = X_i(x^*) \tag{2}$$

is the *degree of overlapping* of χ_{x^*} and X_i.

MA controller is generalized to CFR controller by rescaling the membership degrees in both input and output spaces and replacing the degree of overlapping $\mathcal{D}(\chi_{x^*}, X_i)$ in (1) by the *degree of conditional firing* of the ith rule. Using the rescaled membership functions $\chi_{x^*} \circ \varrho$ and $X_j \circ \varrho$ for a proper choice of the increasing bijection $\varrho : [0, 1] \to [0, 1]$, the latter is defined as

$$\mathcal{C}_i = \frac{\mathcal{D}\left(\chi_{x^*} \circ \varrho, X_i \circ \varrho\right)}{\max_{j \leq n} \mathcal{D}\left(\chi_{x^*} \circ \varrho, X_j \circ \varrho\right)} = \frac{\varrho(X_i(x^*))}{\max_{j \leq n} \varrho(X_j(x^*))}. \tag{3}$$

Therefore, we have the following:

Theorem 1. [8–Theorem 3] *Let* $\Theta = (X_i, Y_i)_{i=1}^n$ *be a rule base satisfying the following properties:*

[C1] *"normality": each X_i is normal,*
[C2] *"covering of antecedents":* $\inf_{x \in X} \max_{i \leq n} X_i(x) > 0$,
[C3] *"significance of consequents":* $\forall i \in \{1, \ldots, n\} : Y_i \not\leq \min_{j \neq i} Y_j$.

Let $c < 1$ and let $\varrho\colon [0, 1] \to [0, 1]$ be any increasing bijection satisfying

[C4] *"disjointness of antecedents":* $\mathcal{D}(X_i \circ \varrho, X_j \circ \varrho) \leq c$ *whenever $i \neq j$.*

Then, for any increasing bijection $\sigma\colon [0, 1] \to [c, 1]$, the input-output correspondence $f\colon \mathcal{F}(X) \to \mathcal{F}(Y)$ of the CFR controller satisfies properties [Int1], [Int2] *and* [Int3].

2.2 The Algorithm

The input of the algorithm is:

- A rule base Θ satisfying properties [C1], [C2] and [C3] of Theorem 1.
- An increasing bijection $\varrho : [0, 1] \to [0, 1]$ and a constant $0 \leq c < 1$, satisfying property [C4] of Theorem 1.
- An increasing bijection $\sigma : [0, 1] \to [c, 1]$. To extend the inverse of σ to the whole interval $[0, 1]$, we use its *pseudoinverse* $\delta(t)$ that is equal to $\sigma^{-1}(t)$ if $t \geq c$ and to 0 otherwise.
- A t-norm $\otimes$.
- A crisp value $x^* \in X$.

The output is a membership function $Y^* \in \mathcal{F}(Y)$, which will be defuzzified by one of the classical methods available in the Fuzzy Logic Toolbox. The implemented algorithm [8] can be summarized as follows:

1. Rescale the membership degrees in the input space by ϱ, that is $\varrho(X_i(x))$, $x \in X$, $i \in \{1, \ldots, n\}$.
2. Rescale the membership degrees in the output space by σ, that is $\sigma(Y_i(y))$, $y \in Y$, $i \in \{1, \ldots, n\}$.
3. Calculate the degree of overlapping $\mathcal{D}_i = \varrho(X_i(x^*))$, for every $i \in \{1, \ldots, n\}$.
4. Calculate the degree of conditional firing, $\mathcal{C}_i(z) = \mathcal{D}_i / \max_{j \leq n} \mathcal{D}_j$, for every $i \in \{1, \ldots, n\}$.
5. Calculate the output $Y^*(y) = \delta(\max_{i \leq n}(\mathcal{C}_i(z) \otimes \sigma(Y_i(y))))$.

To keep the numerical complexity of the changes of scale as low as possible, we set $\varrho(t) = t^r$ with $r \in \mathbb{R}^+$ and $\sigma(t) = (1 - c) \cdot t + c$ with $c \in [0, 1[$. The optimal values of r and c are determined by an optimization algorithm, in this case PSOA [5].

2.3   The Graphical Interface

By the introduction of some MATLAB m-functions and the rewriting of some C-language routines used inside the Fuzzy Logic Toolbox, the Fuzzy Inference System (FIS) Editor is extended to include CFR inference as a new option in the choice of the inference type.

After its selection, the CFR structure can be imported from the disk or from the MATLAB workspace, or can even be created from scratch: the fields to be filled are now the same of the standard FIS structure plus the new fields *Rho constant* and *Sigma constant*, while a t-norm can be selected from the field *AND operator*. Moreover, the control surface and the graphs of the inserted CFR structure can be generated and depicted. Appropriate *warning dialog boxes* appear when the inserted values do not satisfy the criteria mentioned in the previous subsection.

3   Examples and Comparison with Other Fuzzy Controllers

In this section we analyze the performance of MA, TSK and CFR inference on two classical examples: the *ball on beam* and the *inverted pendulum*.

In all the previous works [8, 1, 7, 9], CFR was always compared with MA only, while in this work the comparison is extended to TSK as well. The three inference systems process the same rule base in different ways but, whereas MA and CFR share also the same antecedent and consequent membership functions, the consequents of TSK need to be properly defined to get a meaningful comparison. Here the membership constant functions of TSK consequents are determined as the centroids of the output membership functions of the original MA and CFR systems.

The examples are developed inside the MATLAB Simulink framework. In particular, we have modified the C S-functions in the Fuzzy Logic Toolbox to process and simulate CFR inference model and to reduce the simulation time. In both examples, we chose to use the *minimum* t-norm and the *centroid* defuzzification method. For the sake of readability, we collect and depict all the simulation results in Appendix A.

3.1 Ball on Beam

The goal of this control problem is to achieve and stabilize a desired position of a ball on a plate by changing the angle of the plate.

The system accepts two inputs (position and velocity of the ball) and produces one output (angle of the plate). The input and output membership functions for MA and CFR and the rule base are fixed beforehand and are shown in Fig. 1.

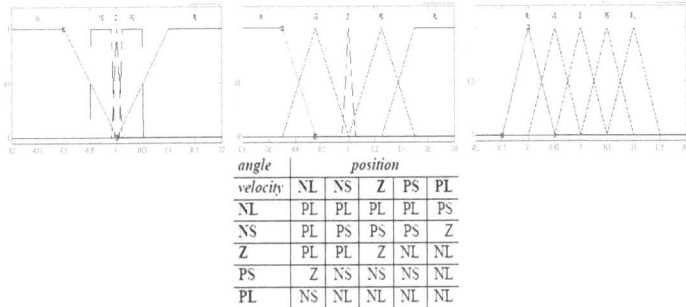

angle	position				
velocity	NL	NS	Z	PS	PL
NL	PL	PL	PL	PL	PS
NS	PL	PS	PS	PS	Z
Z	PL	PL	Z	NL	NL
PS	Z	NS	NS	NS	NL
PL	NS	NL	NL	NL	NL

Fig. 1. Ball on beam: membership functions for variables position, velocity, angle, and the corresponding rule base

The results of our simulations are depicted in Fig. 3, 4, 5 and 6. The main parameter we consider is the transient time: by comparing the three inference systems we obtain a transient time of 83 seconds for MA, 70 for TSK and 81 for CFR.

Note that TSK inference shows the best transient time but it is obtained through a velocity overshoot (while in MA and CFR the velocity varies monotonically and smoothly). Moreover, the equilibrium state is characterized by oscillations of higher frequency and amplitude. All these conditions could not be physically feasible in real problems, or imply higher energy consumption and mechanical stress.

In this simple example, apart from a smaller transient time, CFR does not show any greater improvement than MA, which is instead more evident in the next (more complex) example.

3.2 Inverted Pendulum

The inverted pendulum consists of a pole hinged on a cart moving on a track. The controller's goal is to avoid the pole falling by exerting force on the left or

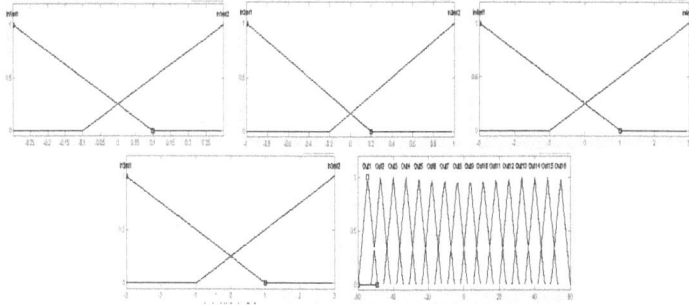

Fig. 2. Inverted pendulum: membership functions for variables angle, angular velocity, cart position, cart velocity and force

on the right of the cart, without moving it outside the track. The system accepts four inputs (pendulum angle, angular velocity, cart position, cart velocity) and produces one output (the force applied to the cart in order to balance the pendulum). Also in this case the membership functions for MA and CFR are fixed beforehand, for both input and output space and are shown in Fig. 2. The rule base is composed by $2^4 = 16$ rules, that is, by all possible combinations, since there are 16 output membership functions.

The performance of the inverted pendulum is mainly measured by the result of pendulum angle (Fig. 7) that, in order to keep a balanced rest state, has to be equal or close to zero.

From Fig. 7 and 10 it is possible to notice that the transient time of MA controller is equal to 7 seconds, while those of TSK and CFR are both equal to 3 seconds. Anyway, once the equilibrium state is reached, CFR controller exhibits a much more stable behavior, since oscillations amplitude is lower compared to MA and TSK (see Fig. 8).

Figure 9 shows that MA keeps the cart continuously moving around the equilibrium point, while TSK and CFR tend to stop it. In addition, CFR exhibits the lowest frequency of oscillation.

Finally, Fig. 11 shows that using CFR inference the cart velocity varies monotonically and smoothly after its transient time, while it oscillates when using MA and TSK controllers.

4   Conclusions

In this work we have presented the first simulation results of a MATLAB implementation of CFR controller, which integrates CFR inference in the Fuzzy Logic Toolbox, thus giving the chance to easily use it in comparison with the built-in inference systems (MA and TSK).

The application of this software on two examples (ball on beam and inverted pendulum) have been shown and discussed. The advantages of CFR, especially in terms of smoother responses, tend to emerge with the increase of the complexity of the examples. Though the results of these simulations are not yet statistically significant, this work gives some indications that could be confirmed by further experiments on more complex examples and by the software optimization.

Acknowledgments. We thank Antonella Valerio, who gave a valuable contribution to the development of this software during a stage at STMicroelectronics, and Francesco Pirozzi (STMicroelectronics) for discussions on this work.

References

1. Amato, P., Di Nola, A., Navara M.: Reformulation of controller with conditionally firing rules. In Proc. Int. Conf. Computational Intelligence for Modelling, Control and Automation, M. Mohammadian (ed.) (2003) 140–151
2. Amato, P., Di Nola, A., Navara, M.: Mathematical aspects of fuzzy control. Proc. Int. Workshop on Fuzzy Logic and Applications (2003) 1-6
3. Driankov, D., Hellendoorn, H., Reinfrank, M.: An Introduction to Fuzzy Control. Springer, Heidelberg (1993)
4. Gottwald, S.: Fuzzy Sets and Fuzzy Logic. Vieweg, Braunschweig (1993)
5. Kennedy, J., Eberhart, R. C.: Particle swarm optimisation. Proc. IEEE Int. Conf. on Neural Networks (1995) IEEE Press 1942–1948
6. Mamdani, E. H., Assilian S.: An experiment in linguistic synthesis of fuzzy controllers. Int. J. Man-Mach. Stud. **7** (1975) 1–13
7. Moser, B., Navara M.: Conditionally firing rules extend the possibilities of fuzzy controllers. In Proc. Int. Conf. Computational Intelligence for Modelling, Control and Automation, M. Mohammadian (ed.) (1999) 242-245
8. Moser, B., Navara M.: Fuzzy controllers with conditionally firing rules: IEEE Trans. Fuzzy Systems **10** (2002) 340–348
9. Navara, M., Št'astný, J.: Properties of fuzzy controller with conditionally firing rules. In Intelligent Technologies, Theory and Applications, P. Sinčák, J. Vaščák, V. Kvasnička, J. Pospíchal (Eds) IOS Press (2002) 111–116
10. Takagi, T., Sugeno, M.: Fuzzy identification of systems and its applications to modeling and control. IEEE Transactions on Systems, Man and Cybernetics, **15** (1985) 116–132
11. Zadeh, L. A.: Outline of a new approach to the analysis of complex systems and decision processes. IEEE Trans. Syst., Man, Cybern. **3** (1973) 28-44

A   Simulation Results

A.1   Ball on Beam

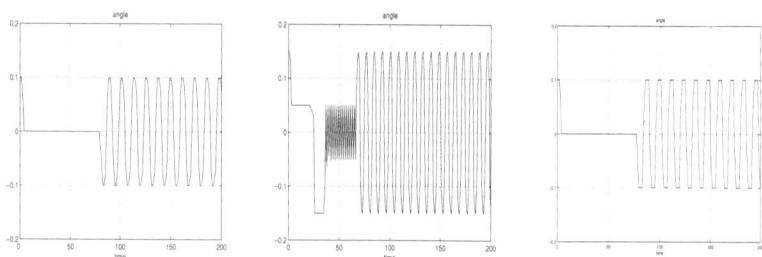

Fig. 3. Angle: simulation results using MA, TSK and CFR inference

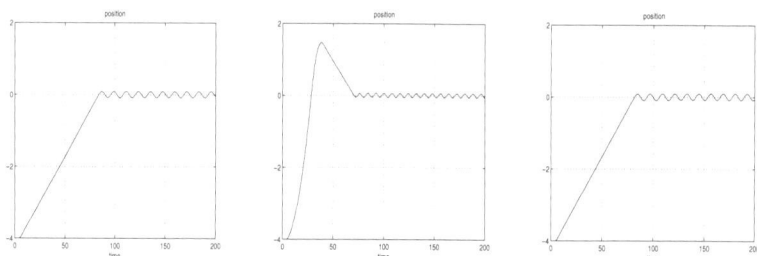

Fig. 4. Position: simulation results using MA, TSK and CFR inference

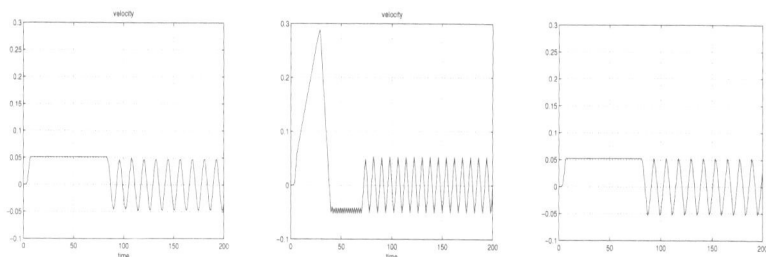

Fig. 5. Velocity: simulation results using MA, TSK and CFR inference

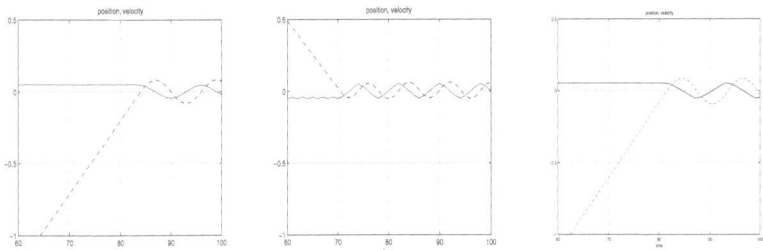

Fig. 6. Position (dashed line) and velocity around the transient time point, for MA, TSK and CFR systems

A.2 Inverted Pendulum

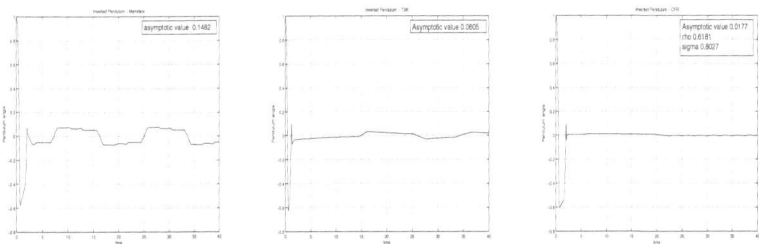

Fig. 7. Pendulum angle: simulation results using MA, TSK and CFR inference

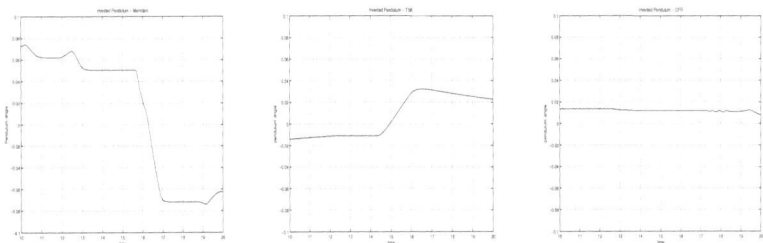

Fig. 8. Pendulum angle: oscillations amplitude (zoom) of MA, TSK and CFR inference

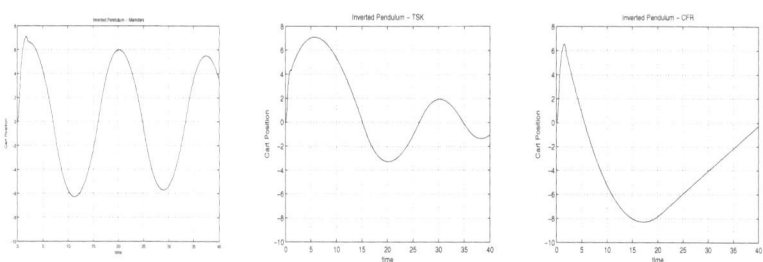

Fig. 9. Cart position: simulation results using MA, TSK and CFR inference

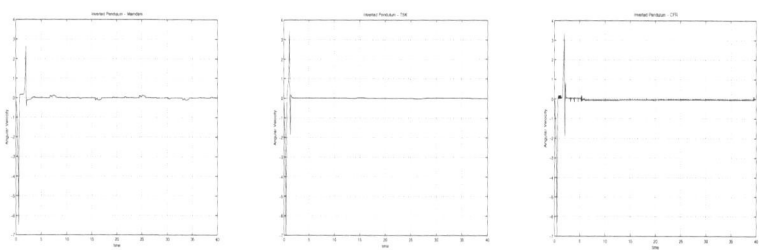

Fig. 10. Angular velocity: simulation results using MA, TSK and CFR inference

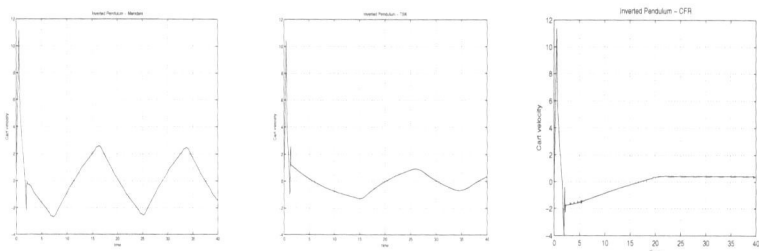

Fig. 11. Cart velocity: simulation results using MA, TSK and CFR inference

Adaptive Feature Selection for Classification of Microscope Images

Ralf Tautenhahn[1], Alexander Ihlow[2], and Udo Seiffert[2]

[1] Leibniz Institute of Plant Biochemistry,
Weinberg 3, D-06120 Halle (Saale), Germany
ralf.tautenhahn@ipb-halle.de
[2] Leibniz Institute of Plant Genetics and
Crop Plant Research (IPK) Gatersleben,
Corrensstr. 3, D-06466 Gatersleben, Germany
{ihlow, seiffert}@ipk-gatersleben.de
http://mue.bic-gh.de

Abstract. For high-throughput screening of genetically modified plant cells, a system for the automatic analysis of huge collections of microscope images is needed to decide whether the cells are infected with fungi or not. To study the potential of feature based classification for this application, we compare different classifiers (kNN, SVM, MLP, LVQ) combined with several feature reduction techniques (PCA, LDA, Mutual Information, Fisher Discriminant Ratio, Recursive Feature Elimination). We achieve a significantly higher classification accuracy using a reduced feature vector instead of the full length feature vector.

1 Introduction

Recent biomolecular methods produce large amounts of raw data exceeding all limitations of currently used manual or semiautomatic analysis. To study resistance mechanisms of crop plants against fungi a high-throughput screening of genetically modified cells is performed and the desired automated process should be able to analyse an immense number of microscope images without human interaction. An overview of computerized cell image analysis can be found in [1]. Automated classification of cell images – from a medical point of view – has been documented in e.g. [20,15,18] and the recognition of plankton images from an underwater video microscope system has been described in [16].

This paper focuses on a feature based classification of biological objects which have been previously segmented in high-resolution microscope images. The biological relevant object [14] to be automatically detected is a so called *haustorium* – a complex object consisting of a "waist" with "fingers" (see Fig. 1 for some typical samples). In the underlying processing pipeline, regions of interest containing relevant biological cells (more precisely, genetically transformed cells characterized by a greenish blue dye) are extracted from the acquired images [9]. Next, these individual transformed cells are checked for potential *haustoria*, using advanced image segmentation methods [11,10]. This step leads to a rather

I. Bloch, A. Petrosino, and A.G.B. Tettamanzi (Eds.): WILF 2005, LNAI 3849, pp. 215–222, 2006.

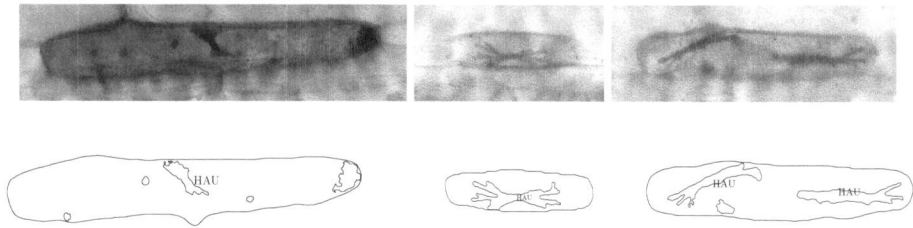

Fig. 1. Three different regions of interest containing biologically relevant cells extracted from original microscope images. The segmented objects inside those cells have to be classified into *haustoria* (marked by "HAU" in the sketches) and other objects. As can clearly be seen, the contrast may be rather poor and the objects differ very much in colour, shape, size and orientation.

large number of objects which might be either *haustoria* or similar image structures being any other objects. Because this segmentation does not provide a sufficiently correct recognition of *haustoria*, classification has to be done to distinguish between real *haustoria* and *similar* objects.

Since the objects stand out only slightly against the background, the object recognition has to be rather sophisticated. Furthermore, the objects differ in colour, shape, size, and orientation (see Fig. 1). Thus, for example, template based approaches or any solution requiring model assumptions or a-priori knowledge will not be suitable. A common and very flexible approach is to extract a number of features from labeled examples for all different object classes (here: *haustorium* or not) from the image and to perform training and classification subsequently. Since the impact of particular features often depends strongly on the subsequent classification method – a fact that is often highly underestimated, both the feature selection and the classification have to be considered together. By means of the above mentioned quite challenging real-world application of *haustoria* recognition, this paper investigates a number of common statistical and neural network based classification methods in conjunction with several common feature selection algorithms and comes up with some expected results but also some maybe unanticipated ones.

2 Feature Generation

A total number of 38 features is generated, characterizing shape as well as colour and texture. An overview of the features is given in table 1. During the segmentation procedure described in [11], a contrast enhancement is done using the morphological top-hat operations. Features can be extracted from the original or the enhanced images: the average colour values of every object were measured in RGB and HSV from both image versions and texture features were also calculated for original and enhanced image.

From the objects curvature [12] the normalized multiscale bending energy NMBE can be calculated. This measure is 1 for a circle and larger for every other,

Table 1. Overview of features from different categories which were generated for classification purposes. The texture and colour features were calculated from both the original and the morphological contrast enhanced image.

Category	Feature	Number	Comment
Simple geometric	Area	1	
	Roundness metric	1	$R = \frac{4\pi F}{U^2}$
Shape	Hu-Moments	7	[8]
	Granlund-Descriptors	7	[6]
	NMBE	1	[3]
Texture	Contrast	2	[5]
	Correlation	2	
	Energy	2	
	Homogeneity	2	
Colour	RGB	3	
	RGB (enhanced)	3	
	HSV	3	
	HSV (enhanced)	3	
Other	CSAT	1	see text
	Total	38	

more 'twisted' object, independent of its size. Before calculating the curvature, the contour is smoothed using a Gaussian function with $\sigma = 1.5$. We constructed another feature, CSAT, which is calculated from the enhanced image by counting each object's pixels with saturation value $= 1$.

3   Feature Selection

Dimensionality Reduction with PCA and LDA
Principal component analysis (PCA) and linear discriminant analysis (LDA) are two common techniques for feature reduction. While the PCA provides axes with maximal variance, the aim of the LDA is to find vectors which maximize the separability of predefined classes. More precisely, a vector d is obtained such that the ratio of the between-class variance to the within-class variance is maximized. This criterion C can be expressed as

$$C = \frac{d^T B d}{d^T W d} ,$$

with B being the between-class covariance matrix and W the within-class covariance matrix. The best discriminant vector d_1 is provided by

$$W^{-1} B d_1 = \lambda d_1 ,$$

where d_1 is the eigenvector of $W^{-1}B$ associated with the largest eigenvalue. It is well known as the *Fisher linear discriminant*. However, if K classes were defined, at most $K-1$ eigenvectors exist. To obtain an orthogonal set of more than $K-1$ vectors, a method proposed in [4] was applied.

Three different techniques for feature selection were used: the recursive feature elimination as a method of measuring the influence of features on the weight vector of the classifier, the mutual information to quantify correlation between several features and classes as well as the Fisher's discriminant ratio to rate individual features.

Recursive Feature Elimination
The RFE [7] is a different version of the *Sequential Backward Selection* [17]. It can be performed with classifiers which rely on minimizing a cost function of a weight vector $\mathbf{w}$, e.g. $\gamma(\mathbf{w}) = \frac{1}{2}\mathbf{w}^T\mathbf{w}$ for a support vector machine. The idea is to quantify the influence of the feature i by measuring the absolute value of the weight w_i. The process consists of the following steps:

- Train the classifier (optimize the weight vector $\mathbf{w}$ with respect to $\gamma(\mathbf{w})$).
- Compute the ranking criteria $c_i = (w_i)^2$ for all i.
- Remove the feature j with smallest ranking criterion c_j.

The result of this algorithm is a feature ranking, but the top ranked (most recently eliminated) features are not necessarily the ones that are individually most relevant, only their combination in terms of a feature vector allows an assessment of their relevance [7].

Mutual Information
Mutual information $MI(X,Y)$ is a measure of relative entropy between the joint probability $p(x,y)$ of two random variables X,Y and the product of their marginal probabilities $p(x)p(y)$ [2]:

$$MI(X,Y) = \sum_{x,y} p(x,y) \log \frac{p(x,y)}{p(x)p(y)}.$$

In the context of classification the mutual information for features v_i and classes ω_j is given as:

$$MI(v_i,\omega_j) = p(v_i,\omega_j) \log \frac{p(v_i,\omega_j)}{p(v_i)p(\omega_j)}$$

To evaluate the feature v_i, the MI-values for all classes ω_j weighted with their priors $p(\omega_j)$ are summarized:

$$MI(v_i) = \sum_{\omega_j \in \Omega} p(\omega_j)\, MI(v_i,\omega_j)\,.$$

Fisher Discriminant Ratio
The FDR can be used to quantify the separability capabilities of individual features [17]. For the two class case, the FDR of feature v is given as

$$FDR(v) = \frac{(\mu_{v1} - \mu_{v2})^2}{\sigma_{v1}^2 + \sigma_{v2}^2},$$

where μ_{v1} is the mean and σ_{v1} the variance of class 1 and μ_{v2} the mean and σ_{v2} the variance of class 2 corresponding to the feature v.

4   Results

After applying the mentioned selection techniques, feature rankings can be calculated. The rankings reflect the diversity of the selection methods. In our experiments the RFE rates the colour features very high, whereas the Mutual Information tends to place form attributes on top of the list.

Table 2. Classification accuracies (with standard deviations) measured using the full sized feature vector. Most classifiers show similar, moderate performance. LVQ fails classification if the feature vector is used with full length.

Classifier	KNN3	kNN5	kNN7	SVM RBF2	SVM POLY3	MLP	LVQ
Classification accuracy	0.90 ± 0.05	0.90 ± 0.05	0.90 ± 0.05	0.89 ± 0.05	0.86 ± 0.06	0.88 ± 0.05	0.56 ± 0.08

Table 3. Classification accuracies achieved with reduced dimensionality. The combinations of feature reduction techniques and classifiers with best results are shown. All classifiers show an improved accuracy compared to classification using feature vectors with full length. The values of SVM-RBF2, SVM-POLY3 and LVQ are significantly increased.

Classifier	kNN3	kNN5	kNN7	SVM RBF2	SVM POLY3	MLP	LVQ
Reduction method	FDR	RFE	RFE	RFE	LDA	LDA	LDA
Dimensionality	18	22	21	19	5	36	7
Classification accuracy	0.92 ± 0.05	0.93 ± 0.04	0.93 ± 0.04	**0.95** ± 0.03	**0.92** ± 0.04	0.90 ± 0.05	**0.91** ± 0.05

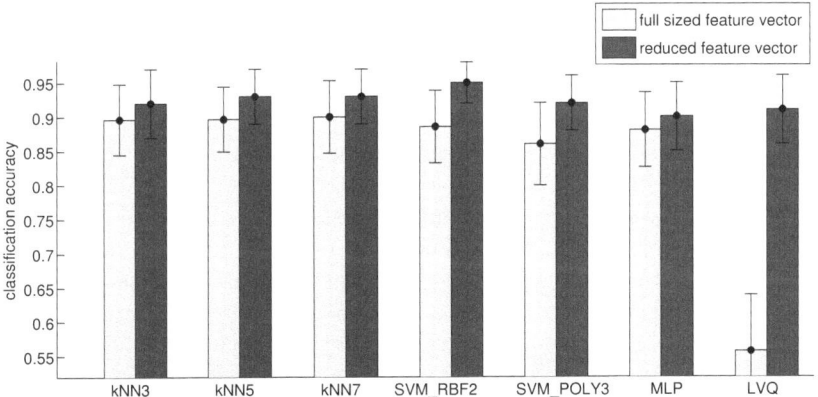

Fig. 2. Comparison of classification accuracies achieved with reduced feature vectors and feature vectors with full size. All classifiers benefit from feature reduction.

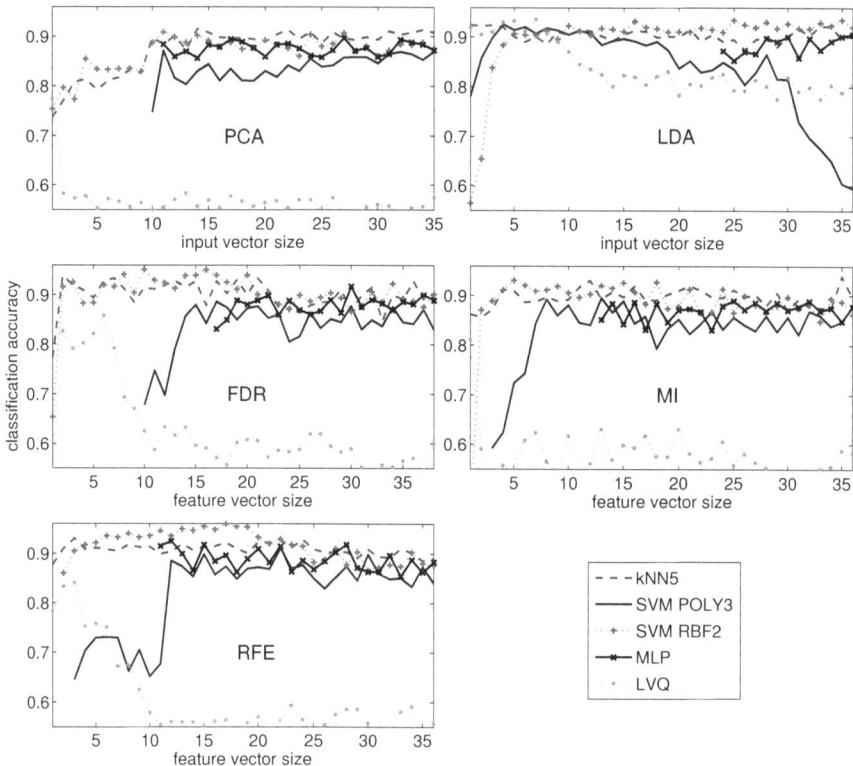

Fig. 3. Comparison of different feature reduction techniques combined with several classifiers. SVM-POLY3 and LVQ achieve their greatest values with a low dimensional feature vector, calculated with LDA, while the kNN5 classifier and the SVM-RBF reach their maxima with a medium sized feature vector, containing features obtained by recursive feature elimination. Some combinations could not be calculated due to bad convergence. The kNN3 and kNN7 classifier behave similar like kNN5 and are not drawn for clearness reasons.

To get an impression of the performance on our dataset we use different classifiers: a k-nearest-neighbor classifier ($k = \{3, 5, 7\}$), a multilayer perceptron with two hidden layers (12 neurons in the first and 3 in the second hidden layer), learning vector quantization (16 neurons in the hidden layer) and support vector machines with polynomial ($n = 3$) kernel and also with a radial base function ($\sigma = 2$) [19]. The specified parameters are the result of preselection and optimization.

Our sample set consists of 364 annotated micrographs of single plant cells. It was split into training- and test sets using 10-fold cross validation. To compare the results of several classification results on one sample set, the *corrected re-sampled t-test* [13], which takes into account the variability due to the choice of the training sets, is used.

Table 2 shows the results using the feature vector of full length (38 features). The classification accuracies of kNN, SVM-RBF2 and the MLP are similar in the range between 0.88 and 0.90. LVQ shows very poor performance.

The situation changes considerably when the feature reduction algorithms are applied. The achieved classification accuracies are shown in figure 3 as a function of the feature vector size. The classifiers respond differently to the reduction techniques: LVQ and the SVM with POLY3 kernel show great improvements with LDA-transformed input data. The accuracy of the kNN classifier and the SVM with RBF kernel can be enhanced using the RFE-selected features[1].

5   Conclusion

For the automatic classification of microscope images of plant cells we studied the influence of feature selection and -reduction techniques on several classifiers. Using reduced feature vectors the classification accuracy of learning vector quantization, a support vector machine with a radial base function and also with a polynomial kernel could be significantly improved compared to the classification accuracy achieved with a feature vector of full length. In our tests, the highest accuracy (95%) was obtained by a support vector machine with RBF-kernel in conjunction with recursive feature elimination.

Acknowledgements. We thank Patrick Schweizer and his staff for their expertise in interpretation of LM-micrographs. This work was supported by the German Federal Ministry of Education and Research (BMBF) under grant 0312706A.

References

1. Ewert Bengtsson. Computerized cell image analysis: Past, present, and future. In *SCIA*, pages 395–407, 2003.
2. Thomas M. Cover and Joy A. Thomas. *Elements of Information Theory.* Wiley, New York, 1991.
3. Luciano da Fontoura Costa and Roberto Marcondes Cesar Jr. *Shape Analysis and Classification.* CRC Press, 2001.
4. J. Duchene and S. Leclercq. An optimal transformation for discriminant and principal component analysis. *IEEE Transactions on Pattern Analysis and Machine Intelligence (PAMI)*, 10(6):978–983, 1988.
5. Rafael C. Gonzalez and Richard E. Woods. *Digital Image Processing.* Addison-Wesley, Reading, Massachusetts, 1993.
6. Gösta H. Granlund. Fourier preprocessing for hand print character recognition. *IEEE Transactions on Computers*, C-21(3):195–201, March 1972.
7. Isabelle Guyon, Jason Weston, Stephen Barnhill, and Vladimir Vapnik. Gene selection for cancer classification using support vector machines. *Machine Learning*, 46(1-3):389–422, 2002.

---

[1] The RFE-feature ranking is computed with the SVM-RBF2, but also evaluated with other classifiers.

8. Ming-Kuei Hu. Visual pattern recognition by moment invariants. *IRE Transactions on Information Theory*, 8(2):179–187, February 1962.
9. Alexander Ihlow and Udo Seiffert. Microscope color image segmentation for resistance analysis of barley cells against powdery mildew. In *9. Workshop "Farbbildverarbeitung"*, ZBS Zentrum für Bild- und Signalverarbeitung e.V. Ilmenau, Report Nr. 3/2003, pages 59–66, Ostfildern-Nellingen, Germany, October 2003.
10. Alexander Ihlow and Udo Seiffert. Automating microscope colour image analysis using the Expectation Maximisation algorithm. In *Pattern Recognition: 26th DAGM Symposium*, pages 536–543, Tübingen, Germany, September 2004. Springer.
11. Alexander Ihlow and Udo Seiffert. Haustoria segmentation in microscope images of barley cells. In *10. Workshop 'Farbbildverarbeitung'*, pages 119–126, Koblenz, October 2004. Der andere Verlag.
12. Farzin Mokhtarian and Alan Mackworth. Scale-based description and recognition of planar curves and two-dimensional shapes. *IEEE Transactions on Pattern Analysis and Machine Intelligence (PAMI)*, 8(1):34–43, January 1986.
13. Claude Nadeau and Yoshua Bengio. Inference for the generalization error, 2000. Advances in Neural Information Processing Systems 12, MIT Press.
14. Patrick Schweizer, Jana Pokorny, Olaf Abderhalden, and Robert Dudler. A transient assay system for the functional assessment of defense-related genes in wheat. *Molecular Plant-Microbe Interactions*, 12(8):647–654, 1999.
15. W. Nick Street, William H. Wolberg, and Olvi L. Mangasarian. Nuclear feature extraction for breast tumor diagnosis. *Biomedical Image Processing and Biomedical Visualization*, 1905:861–870, 1993.
16. Xiaoou Tang, W. Kenneth Stewart, Luc Vincent, He Huang, Marty Marra, Scott M. Gallager, and Cabell S. Davis. Automatic plankton image recognition. *Artificial Intelligence Review*, 12(1-3):177–199, February 1998.
17. Sergios Theodoridis and Konstantinos Koutroumbas. *Pattern Recognition*. Elsevier Academic Press, San Diego, 2003.
18. Jean-Philippe Thiran and Benoît M. Macq. Morphological feature extraction for the classification of digital images of cancerous tissues. *IEEE Transactions on Biomedical Engineering*, 43(10):1011–1020, October 1996.
19. Jason Weston, Andre Elisseeff, Gökhan Baklr, and Fabian Sinz. The SPIDER: object-orientated machine learning library. http://www.kyb.tuebingen.mpg.de/bs/people/spider/index.html.
20. William H. Wolberg, W. Nick Street, and Olvi L. Mangasarian. Breast cytology diagnosis via digital image analysis. *Analytical and Quantitative Cytology and Histology*, 15(6):396–404, 1993.

Genetic Algorithm Against Cancer*

F. Pappalardo[1,2], E. Mastriani[2], P.-L. Lollini[3], and S. Motta[2]

[1] Faculty of Pharmacy, University of Catania,
V.le A. Doria, 6, I-95125 Catania, Italy
Tel.: +39 095 7384071; Fax.: +39 095 7384278
[2] Department of Mathematics and Computer Science, University of Catania,
V.le A. Doria, 6, I-95125 Catania, Italy
Tel.: +39 0957383073; Fax.: +39 0957337046
francesco@dmi.unict.it, mastriani@dmi.unict.it, motta@dmi.unict.it
[3] Sezione di Cancerologia, Dipartimento di Patologia Sperimentale
and Centro Interdipartimentale di Ricerche sul Cancro "Giorgio Prodi",
University of Bologna,
Viale Filopanti 22, I-40126 Bologna, Italy
Tel.: +39 051241110; Fax.: +39 051242169
pierluigi.lollini@unibo.it

Abstract. We present an evolutionary approach to the search for effective vaccination schedules using mathematical computerized model as a fitness evaluator. Our study is based on our previous model that simulates the Cancer - Immune System competition activated by a tumor vaccine. The model reproduces pre-clinical results obtained for an immunoprevention cancer vaccine (Triplex) for mammary carcinoma on HER-2/neu mice. A complete prevention of mammary carcinoma was obtained *in vivo* using a Chronic vaccination schedule. Our genetic algorithm found complete immunoprevention with a much lighter vaccination schedule. The number of injections required is roughly one third of those used in Chronic schedule.

1 Introduction

Immunoprevention of mammary carcinoma in HER-2/neu transgenic mice was attempted using various immunological strategies, including cytokines, non-specific stimulators of the immune response, and HER-2/neu specific vaccines made of DNA, proteins, peptides, or whole cells. Most approaches achieved a delay of mammary carcinogenesis, but a complete prevention of tumor onset was not attained, particularly in the most aggressive tumor models.

A new vaccine, called Triplex, was proposed in [6]. The vaccine combined three different stimuli for the immune system. The first was p185neu, protein

* F.P. and S.M. acknowledge partial support from University of Catania research grant and MIUR (PRIN 2004: *Problemi matematici delle teorie cinetiche*). This work has been done while F.P. is research fellow of the Faculty of Pharmacy of Universiy of Catania. P.-L.L. acknowledges financial support from the University of Bologna, the Department of Experimental Pathology ("Pallotti" fund) and MIUR.

I. Bloch, A. Petrosino, and A.G.B. Tettamanzi (Eds.): WILF 2005, LNAI 3849, pp. 223–228, 2006.
© Springer-Verlag Berlin Heidelberg 2006

product of HER-2/neu, which in this system is at same time the oncogene driving carcinogenesis and the target antigen. p185neu was combined with the two non specific adjuvants, allogeneic class I major histocompatibility complex (MHCI) glycoproteins and interleukin 12 (IL-12). Allogeneic MHCI molecules stimulate a relatively large fraction of all T cell clones, up to 10% of the available repertoire. IL-12 is a cytokine normally produced by antigen presenting cells (APC) such as dendritic cells (DC) to stimulate T helper cells and other cells of the immune system, such as natural killer cells (NK) [2].

A complete prevention of mammary carcinogenesis with the Triplex vaccine was obtained when vaccination cycles started at 6 weeks of age and continued for the entire duration of the experiment, at least one year (chronic vaccination). One vaccination cycle consisted of four intraperitoneal administrations of non-replicating (mitomycin-treated) vaccine cells over two weeks followed by two weeks of rest [4]. Other tested vaccination schedules were unable to prevent mammary carcinogenesis. Triplex is an immunoprevention vaccine: *in vivo* experiments have shown that the vaccine is no more effective after the solid tumor formation.

The question whether the chronic protocol is the minimal vaccination protocol yielding complete protection from tumor onset, or whether a lower number of vaccination cycles would provide a similar degree of protection, is still an open question.

Finding an answer to this question via a biological solution would be too expensive in time and money as it would require an enormous number of experiments, each lasting at least one year.

For this reason we developed an accurate model [9] of immune system responses to vaccination. We performed *in silico* experiments considering a large population of individual mice. Each individual mouse is characterized by a sequence of uniformly distributed random numbers which will determine the probabilistic events. As showed in [9], comparison with *in vivo* experiments shows excellent agreement.

Our evolutionary approach uses the model and its computer implementation described above as a fitness evaluator to find a schedule which controls the growth of cancer cells by a minimal number of vaccine injections.

The paper is organized as follows. In section 2, we describe in depth our algorithm. Section 3 will provide computational results and, in Section 4, we will give conclusions and final remarks.

2   The Genetic Algorithm

Evolutionary Algorithms have been applied with satisfactory results to a very long list of hard combinatorial problems. A complete description or enumeration of such results is, per se, a hard problem. The interested reader can found extensively review in [3, 7].

The approach we present in this paper differs from "standard" GA applications as we use a *simulator* to compute a parameter of the fitness function. To our best knowledge, very few applications [1] use a complete simulator in a genetic algorithm and no applications of this type exist in cancer immunoprevention.

Each chromosome in the population represents a vaccine schedule. It is a binary string of 1200 bits, in which each gene represents a timestep in which is possible to inoculate a vaccine injection. One timestep is equal to 8 hours of mouse's life. If the i-th gene is expressed (respectively not expressed), i.e. the i-th bit is set to 1 (respectively the i-th bit is set to 0), then a vaccination has to be administered at timestep i (respectively no vaccination has to be administered at timestep i). We decide to set a population formed by 40 chromosomes.

The selection operator used is *tournament selection* [5]. Reproduction uses uniform crossover.

Mutation acts as follows: first a gene subject to mutation is chosen with probability $p = 1/g_n$, where g_n represents the number of chromosome's genes; second the chosen gene is setted to 1 with probability $p = 1/p_s$ and to 0 with probability $p = 1 - (1/p_s)$, where p_s represents the population size.

Elitism is used on two specific elements of the population: *i)* best fitness element; *ii)* second best fitness element.

Results showed in [9], fully justify the usage of the SimTriplex simulator for the definition of a good fitness function. SimTriplex simulator computes the main biological entities of the cancer - immune system competition. If the number of cancer cells overcomes 10^5, then the simulator recognizes the solid tumor formation (carcinogenesis) and simulation ends at the reached time. We will refer to this time as mice survival time. An effective vaccination must reach a mice survival time of 1200 timesteps, i.e. 400 days.

In setting up a fitness function we must take into account two fundamental and competing requirements: *i)* any schedule must be an effective one, i.e. the mice survival time must reach 400 days; *ii)* the best schedules must have a minimal cardinality, i.e. they must provide mice survival with the minimum number of vaccine injections.

Any evolutionary approach which just takes into consideration the first requirement, will produce populations of individuals *very rich* of ones, thus, not minimal. If instead, we take into consideration just the second requirement, we will have populations of individuals full of zeroes, and thus very likely we will obtain a non effective schedule. Any fitness function therefore must be, at least, a two-variables function of type $f(n, s, \dots)$ where n is the number of injections, and s is the number of timesteps survived by the mouse. Also, f must satisfy the following two properties:

$$f(n, s, \dots) < f(n, s', \dots) \text{ if } s > s' \tag{1}$$

$$f(n, s, \dots) > f(n', s, \dots) \text{ if } n > n' \tag{2}$$

We restrict ourself the simple case of a two variable function and we chose the following function:

---

[1] The only one we found is reported in http://www.cs.ucl.ac.uk/staff/P.Bentley/ WLBEC1.pdf

$$f(n, s) = \frac{n^2}{s} \qquad (3)$$

which meets the properties (1) and (2). Obviously, the fitness function (3) has to be minimized.

Determining the quality (fitness) of each chromosome involves the use of the SimTriplex simulator which takes a non negligible amount of time. In particular, the simulator returns the timesteps survived by the mouse with the proposed therapy coded by the chromosome.

3   Results

In setting the computer experiments, we randomly chose 10 virtual mice over the 100 of the first sample set we used in [9]. Each run of the GA took about 36 hours on a 686-class PC machine. All the 10 virtual mice gave similar, but obviously not identical, results. They all got complete prevention of mammary carcinoma with 19 vaccine injections, against the 59 required by Chronic vaccination schedule. Figure 1 shows the effect of Chronic vaccination schedule for one of the chosen mice. For the same mouse the GA suggested schedule is shown in figure 2.

This shows, as in Chronic vaccination schedule, an initial burst of cancer cells. This effect is due to the time-lag of vaccine effect [9]. Chronic schedule then controls the growth of cancer cells with regular successive vaccine injections. Our GA was not required to do this, so a second burst of cancer cells appears in GA proposed schedule. This is a reasonable minimum from the point of the view of the GA. One could argue if this is also reasonable for mice safety. In this specific case we notice that both the cancer cells maximum in the GA suggested

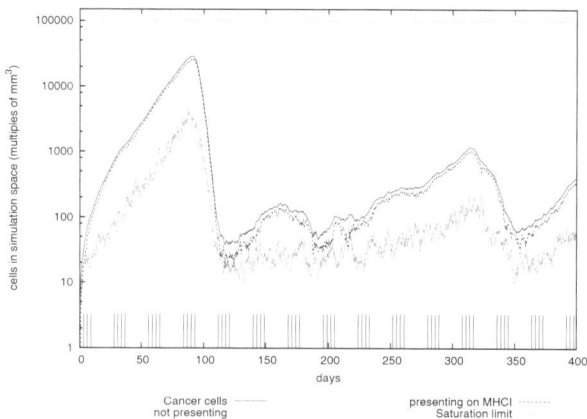

Fig. 1. Cancer cells dynamics. Cancer cells controlled by CHRONIC vaccination schedules. Red ticks above x axis represent the timing of vaccine administration. Saturation limit indicates that a solid tumor is formed, i.e. the number of cancer cells in the simulated space becomes greater than 10^5.

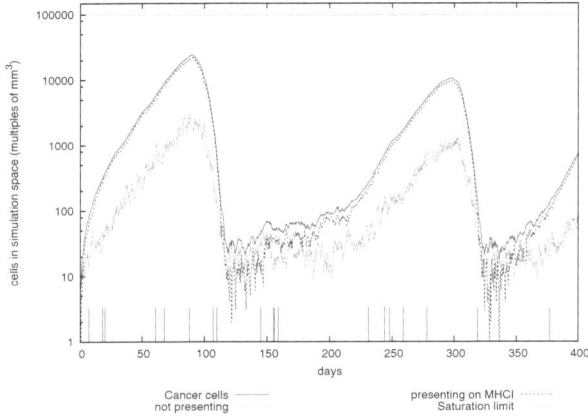

Fig. 2. Cancer cells dynamics. Cancer cells controlled by genetic algorithm proposed vaccination schedules. Red ticks above x axis represent the timing of vaccine administration. Saturation limit indicates that a solid tumor is formed, i.e. the number of cancer cells in the simulated space becomes greater than 10^5.

schedule are lower than the Chronic schedule maximum. So in this case the GA suggested schedule is safer than the Chronic one. However this could not be the case for other mice. This suggests that more requirements should be added to the GA fitness function. Those requirements must obviously be biologically driven.

4   Conclusion and Future Work

We have presented an evolutionary algorithm which turns out to be efficient in finding effetctive therapies for protecting virtual mice from mammary carcinoma.

One of the major novelties of our algorithm, is the usage of a *simulator* to compute a parameter of the fitness function. To our best knowledge, very few applications use a complete simulator in a genetic algorithm and no applications of this type exist in cancer immunoprevention.

Future work will see a GRID/parallel implementation of our GA that can be used to perform a large number of simulated experiments which suggest key *in vivo* experiments on a small set of schedules which take into account biological or clinical constraints.

References

1. Bäck, T., Eiben A., Vink, M.: A superior evolutionary algorithm for 3-SAT. Proceedings of the 7th Annual Conference on Evolutionary Programming, Lecture Notes in Computer Science, Springer (1998) 125-136
2. Colombo, MP., Trinchieri, G.: Interleukin-12 in anti-tumor immunity and immunotherapy. Cytokine Growth Factor Rev., **13:2** (2002) 155-168

3. Corne, D., Dorigo, M., Glover, F.: New Ideas in Optimization. McGraw-Hill, Advanced Topics in Computer Science (1999)
4. De Giovanni, C., Nicoletti, G., Landuzzi, L., Astolfi, A., Croci, S., Comes, A., Ferrini, S., Meazza, R., Iezzi, M., Di Carlo, E., Musiani, P., Cavallo,F., Nanni, P., Lollini, P.-L.: Immunoprevention of HER-2/neu transgenic mammary carcinoma through an interleukin 12-engineered allogeneic cell vaccine. Cancer Res., **64:11** (2004) 4001-4009.
5. Goldberg, D.E.: A comparative analysis of selection schemes used in genetic algorithms. Foundations of Genetic Algorithms, Morgan Kaufmann, G. Rawlins eds. (1991)
6. Lollini, P.-L., De Giovanni, C., Pannellini, T., Cavallo, F., Forni, G., Nanni, P.: Cancer immunoprevention. Future Oncology, **1:1** (2005) 57-66
7. Mitchell, M.: An Introduction to Genetic Algorithm. The MIT Press (1996)
8. Motta, S., Lollini, P.-L., Castiglione, F., Pappalardo, F.: Modelling Vaccination Schedules for a Cancer Immunoprevention Vaccine. Immunome Research, submitted (2005)
9. Pappalardo, F., Lollini, P.-L., Castiglione, F., Motta, S.: Modelling and simulation of cancer immunoprevention vaccine. Bioinformatics, **21:12**, (2005) 2891-2897

Unsupervised Gene Selection and Clustering Using Simulated Annealing

Maurizio Filippone[1], Francesco Masulli[3], and Stefano Rovetta[1]

[1] Dipartimento di Informatica e Scienze dell'Informazione, Università di Genova,
Via Dodecaneso 35, I-16146 Genova, Italy
[2] Dipartimento di Informatica, Università di Pisa, Largo B. Pontecorvo 3, I-56127 Pisa, Italy

Abstract. When applied to genomic data, many popular unsupervised explorative data analysis tools based on clustering algorithms often fail due to their small cardinality and high dimensionality. In this paper we propose a wrapper method for gene selection based on simulated annealing and unsupervised clustering. The proposed approach, even if computationally intensive, permits to select the most relevant features (genes), and to rank their relevance, allowing to improve the results of clustering algorithms.

1 Introduction

Unsupervised explorative data analysis using clustering algorithms provide an useful tool to explore data. In the case of genomic data, that are often characterized by small cardinality and high dimensionality (e.g., in the case of gene expression data obtained from DNA microarrays) this approach can fail, as many clustering algorithms suffer from being applied in high-dimensional spaces (each dimension or feature corresponding in our case to a gene expression data), as clustering algorithms often seek for areas where data is especially dense. Moreover, some (or most) genes are not relevant for the clustering learning task and a gene (feature) selection procedure could highlight the relevant genes and improve the clustering results at the same time [11, 14, 2].

Feature selection algorithms can be broadly divided into two categories [3, 10]: filters and wrappers. Filters evaluate the relevance of each feature (subset) using the data set alone, while wrappers invoke the learning algorithm to evaluate the quality of each feature (subset). Both approaches, filters and wrappers, usually involve combinatorial searches through the space of possible feature subsets. Anyway, wrappers are usually more computationally demanding, but they can be superior in accuracy when compared with filters.

Most of the literature on feature selection pertains to supervised learning, and not much work has been done for feature selection in unsupervised learning [13, 6, 11, 8, 14, 2].

In this paper we propose a wrapper approach to gene selection in clustering of gene expression data. The combinatorial search is performed using the Simulated Annealing (SA) method [9] which is a global search method technique derived from Statistical Mechanics and based on the Metropolis algorithm [12], while the learning algorithm is the Fuzzy C-Means (FCM) that is one of most popular clustering algorithms (for a detailed description of the FCM see [1]).

I. Bloch, A. Petrosino, and A.G.B. Tettamanzi (Eds.): WILF 2005, LNAI 3849, pp. 229–235, 2006.
© Springer-Verlag Berlin Heidelberg 2006

In the next section we describe the proposed SA algorithm for gene selection. In Sect. 3 an evaluation index of gene relevance is presented. Sect. 4 describes the experimental validation of our method on the data set by Golub et al. [7]. Conclusions are presented in Sect. 5.

2   SA for Gene Selection

The method for feature selection we propose makes use of Simulated Annealing (SA) technique [9] that is a global search method technique derived by Statistical Mechanics.

SA is based on the Metropolis algorithm [12] that has been proposed to simulate the behavior and small fluctuations of a system of atoms starting from an initial configuration, by the generation of a sequence of iterations. In the Metropolis algorithm each iteration is composed by a random perturbation of the actual configuration and the computation of the corresponding energy variation (ΔE). If $\Delta E < 0$ the transition is unconditionally accepted, otherwise the transition is accepted with probability given by the Boltzmann distribution:

$$P(\Delta E) = exp\left(\frac{-\Delta E}{KT}\right) \tag{1}$$

where K is the Boltzmann constant and T the temperature.

In SA this approach is generalized to the solution of general optimization problems [9] by using an *ad hoc* selected cost function (*generalized energy*), instead of the physical energy. SA works as a probabilistic hill-climbing procedure searching for the global optimum of the cost function. The temperature T takes the role of a control parameter of the search area (while K is usually set to 1), and is gradually lowered until no further improvements of the cost function are noticed. SA can work in very high-dimensional searches, given enough computational resources.

In Tab. 1, a detailed description of the proposed Simulated Annealing Feature Selection (SAFS) algorithm is presented.

The system state (configuration) is represented by a binary mask $\mathbf{g} = \left(g_1, g_2, \ldots, g_q\right)$, where each bit g_i (with $i = 1, \ldots, q$) corresponds to the selection ($g_i = 1$) / deselection ($g_i = 0$) of a feature (if we want to select a set of s features, at each time only s bits will be set to 1). The initialization of the vector mask $\mathbf{g}$ (Step 2) is done by generating s integer numbers with uniform distribution in the interval $[1, q]$ and setting the corresponding bits to 1 of $\mathbf{g}$ and the remaining ones to 0. A perturbation or move (Step 5c) is done by switching $2 \times r$ bits of $\mathbf{g}$, by randomly selecting r bits set to 0 and r bits set to "1", and flipping their values.

The unsupervised clustering (Steps 3 and 5d) is performed in the sub-space of selected features defined by the vector mask $\mathbf{g}$. After each run of the unsupervised clustering algorithm we can obtain an evaluation of E as a function of either the cost function associated to the clustering algorithm, clustering validation indexes [4], or, when the data set is labeled, the *Representation Error* (RE) defined as the percentage of data points belonging to the same cluster sharing the same label.

The initial value of temperature T is obtained as the average value of ΔE computed over an assigned number p of random perturbations of the mask $\mathbf{g}$.

Table 1. Simulated Annealing Feature Selection (SAFS) Algorithm

---

1. Assign s (number of features to be selected), r (number of bits to be switched for making a move), T (initial temperature of the system), α (cooling parameter), f_{max} (maximum number of iteration at each T), h_{min} (minimum number of success for each T), c (number of clusters), m (fuzzification parameter);
2. Initialize **g** at random (binary mask);
3. Perform unsupervised clustering and evaluate the generalized system energy E;
4. **do**
5. Initialize $f = 0$ (number of iterations), $h=0$ (number of success);
   (a) **do**
   (b) Increment number of iterations f;
   (c) Perturb mask **g**;
   (d) Perform unsupervised clustering and evaluate the generalized system energy E;
   (e) Generate a random number rnd in the interval $[0,1]$;
   (f) **if** $rnd < P(\Delta E)$ **then**
       i. Accept the new **g** mask;
       ii. Increment number of success h;
   (g) **endif**
   (h) **loop until** $h \le h_{min}$ **and** $f \le f_{max}$;
6. update $T = \alpha T$;
7. **loop until** $h > 0$;

SAFS is a very computational intensive algorithm, but it is able to work with every kind of features (e.g., continuous, ordinal, binary, discrete, nominal).

It is worth noting that each time we run the SAFS algorithm we can find a sub-optimal sub-set of s features from the original q. In principle, different independent runs of SAFS can lead to different sub-sets of s features.

3 Ranking Feature Relevance

SA is an algorithm implementing a stochastic time-varying dynamical system where the state vector evolves in the direction of the minima of the generalized energy function.

In our case during the evolution of the SAFS algorithm the bits set in the state vector **g** will be related to the more relevant features (genes) with increasing probability.

The features more relevant in cluster discrimination should appear soon in the set of bits set to 1 and will be more frequent in the following iterations of the algorithm.

In order to estimate the relevance of features, we can include in the SAFS an aging algorithm. To this aim, we can define a vector $\mathbf{r} = (r_1, r_2, \ldots, r_{iq})$. At Step 2 of the SAFS algorithm, we set $r_i = 1/q \; \forall i$. Every time a perturbation is accepted (Step 5.f), according to the Boltzmann distribution, we update **r** using this formula:

$$\mathbf{r} = \gamma\mathbf{r} + \mathbf{g} \tag{2}$$

where γ is the aging constant chosen in the interval $[0,1]$, and then we normalize the vector $\mathbf{r}$ using the following constraint:

$$\sum_{i=1}^{N} r_i = 1 \tag{3}$$

At the end of the SAFS the vector $\mathbf{r}$ tells us how long each feature has belonged to it in the last few successful moves of the algorithm. We give then to vector $\mathbf{r}$ an interpretation as vector of feature relevances.

Table 2. Parameters choice

Meaning	Symbol	Value
Number of random perturbations of $\mathbf{g}$ used to estimate the initial value of T	p	10000
Number of features to be selected	s	20
Number of bits to be switched for making a move	r	3
Cooling parameter	α	0.9
Aging constant	γ	0.98
Maximum number of iteration at each T	f_{max}	10000
Minimum number of success for each T	h_{min}	1000
FCM algorithm repetitions for each move	1	5
Number of clusters	c	2
Fuzzification parameter	m	2

4   Experimental Validation

The method was tested on the publicly available Leukemia data by Golub et al. [7]. The Leukemia problem consists in characterizing two forms of acute leukemia, Acute Lymphoblastic Leukemia (ALL) and Acute Mieloid Leukemia (AML). The original work proposed both a supervised classification task ("class prediction") and an unsupervised characterization task ("class discovery"). Here we obviously focus on the latter, but we exploit the diagnostic information on the type of leukemia to assess the goodness of the clustering obtained.

The data set contains 38 samples for which the expression level of 7129 genes has been measured with the DNA microarray technique (the interesting human genes are 6817, and the other are controls required by the technique). These expression levels have been scaled by a factor of 100. Of these samples, 27 are cases of ALL and 11 are cases of AML. Moreover, it is known that the ALL class is in reality composed of two different diseases, since they are originated from different cell lineages (either T-lineage or B-lineage). In the data set, ALL cases are the first 27 objects and AML cases are the last 11. Therefore, in the presented results, the object identifier can also indicate the class (ALL if id ≤ 27, AML if larger).Using those data (with dimensionality $q = 7129$), Golub et al. [7] selected a set of 50 most relevant genes.

Table 3. The ten most relevant genes found in a run ranked in order of relevance

Name	Description
M33680_at	26-kDa cell surface protein TAPA-1 mRNA
J03801_f_at	LYZ Lysozyme
X04085_rna1_at	Catalase EC1.11.1.6 5'flank and exon 1 mapping to chromosome 11, band p13 (and joined CDS)
S71043_rna1_s_at	Ig alpha 2=immunoglobulin A heavy chain allotype 2 {constant region, germ line} [human, peripheral blood neutrophils, Genomic, 1799 nt]
M19722_at	FGR Gardner-Rasheed feline sarcoma viral $v - fgr$ oncogene homolog
AB002332_at	KIAA0334 gene
M10942_at	Metallothionein-Ie gene (hMT-Ie)
HG2238-HT2321_s_at	Nuclear Mitotic Apparatus Protein 1, Alt. Splice Form 2
S34389_at	HMOX2 Heme oxygenase (decycling) 2
M96956_at	TDGF1 Teratocarcinoma-derived growth factor 1

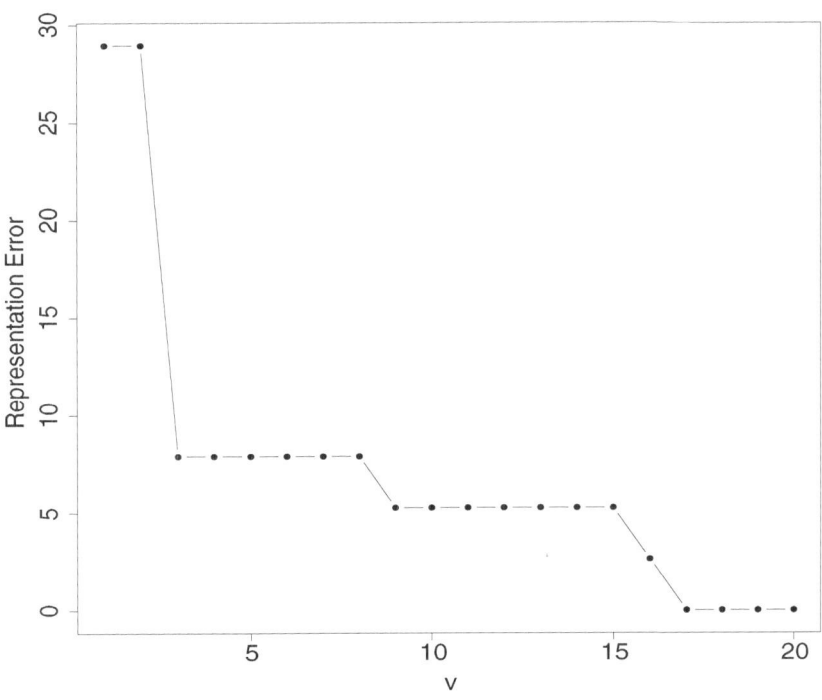

Fig. 1. Representation Error versus the size (v) of gene subsets

We describe here the results obtained using the SAFS algorithm. In the implementation of SAFS we used as the clustering algorithm the Fuzzy C-Means (FCM) [1] that is one of most popular clustering algorithms. Moreover, as the Leukemia data base contains is labeled, we used the *Representation Error* (RE) as an evaluation the generalized energy E.

It is worth noting that the FCM is an unstable algorithm, as his results depend not only from even small perturbations of the data set, but also from the initialization of his parameters (i.e., number of clusters c, clusters centroids y_k and fuzziness parameter m). For this reason at the beginning of the SAFS algorithm (Step 3) and for each perturbation (Step 5c) of SAFS we run the FCM $l = 5$ times and we choose the solution corresponding to the minimum of the generalized energy E.

SAFS has been implemented in R-language (*http://www.r-project.org/*) under Linux operating system. On a Pentium IV 1900 Mhz personal computer a complete running of SAFS least about 10 hours (involving the run of about one million FCMs).

We done 10 independent runs of SAFS using the assumptions in Tab. 2. For each run we obtained a different sets of 20 genes giving a $RE = 0$, containing at the least one gene found by Golub et al. [7]. In Tab. 3, we list the ten must relevant genes found in a run.

Adding the relevance vectors r obtained in the 10 runs, the genes ranked at positions 1, 2, 4 are contained also in the set selected in unsupervised way by Golub et al. [7]. This is a symptom of a strong redundancy in the features of the data set.

In Fig. 1, we show the Representation Error (RE) computed using subset of genes including the v most relevant ones. As shown, at least the 17 most relevant genes must be considered in clustering in order to obtain $RE = 0$.

5   Conclusions

In this paper we have proposed a wrapper method for selecting features based on simulated annealing technique [9] and FCM algorithm [1]. The proposed approach, even if computationally intensive, permits to select the most relevant features (genes), and to rank their relevance, allowing to improve the results of clustering algorithms.

On the 7129-dimensional Leukemia data set by Golub et al. [7] the proposed feature selection method is able to find for each run a subset of 20 genes, that is sufficient to perform FCM clustering algorithm with null Representation Error.

It is worth noting that the proposed algorithm can work with every kind of features (e.g., continuous, ordinal, binary, discrete, nominal). Moreover, the proposed feature selection approach using simulated annealing can be used also with other learning machines, not only for unsupervised clustering, but also for supervised classification, regression, etc.

Acknowledgment

Work funded by the Italian Ministry of Education, University and Research (2004 "Research Projects of Major National Interest", code 2004062740), and the Biopattern EU Network of Excellence.

References

1. J.C. Bezdek, *Pattern Recognition with Fuzzy Objective Function Algorithms*, Plenum Press, New York, 1981.
2. D.R. Bickel, Robust cluster analysis of microarray gene expression data with the number of cluster determined biologically, Bioinformatics, vol. 19, no. 7 pp. 818–824, 2003.
3. A. Blum and P. Langley, Selection of Relevant Features and Examples in Machine Learning, Artificial Intelligence, vol. 97, nos. 1–2, pp. 245–271, 1997.
4. N. Bolshakova and F. Azuaje, Cluster validation techniques for genome expression data Source, Signal Processing, vol.83, pp. 825–833, 2003.
5. R.O. Duda and P.E. Hart, *Pattern Classification and Scene Analysis*, Wiley, New York, 1973.
6. J.G. Dy, C.E. Brodley, A. Kak, L.S. Broderick, and A.M. Aisen, Unsupervised Feature Selection Applied to Content-Based Retrieval of Lung Images, IEEE Trans. Pattern Analysis and Machine Intelligence, vol. 25, no. 3, pp. 373–378, 2003.
7. Golub, T., Slonim, D., Tamayo, P., Huard, C., Gaasenbeek, M., Mesirov, J., Coller, H., Loh, M., Downing, J., Caligiuri, M., Bloomfield, C., Lander, E.: Molecular classification of cancer: Class discovery and class prediction by gene expression monitoring. Science vol. 286, pp. 531–537, 1999.
8. R. Jornsten, B. Yu, Simultaneous gene clustering and subset selection for sample classification via MDL, Bioinformatics, vol. 19, no. 8, pp. 1100–1109, 2003.
9. S. Kirkpatrick, C.D. Gelatt, and M.P. Vecchi. Optimization by simulated annealing. Science, vol. 220, pp.661–680, 1983.
10. R. Kohavi and G. John, Wrappers for Feature Subset Selection, Artificial Intelligence, vol. 97, nos. 1–2, pp. 273–324, 1997.
11. M.H. Law, M.A.T. Figueiredo, and A.K. Jain, Simultaneous Feature Selection in and Clustering Using Mixture Models, IEEE Trans. Pattern Analysis and Machine Intelligence, vol 28, no. 9, 2004.
12. N. Metropolis, A.W. Rosenbluth, M.N. Rosenbluth, A.H. Teller, and E. Teller. Equation of state calculations for fast computing machines. Journal of Chemical Physics, vol. 21, pp. 1087–1092, 1953.
13. P. Mitra and C.A. Murthy, Unsupervised Feature Selection Using Feature Similarity, IEEE Trans. Pattern Analysis and Machine Intelligence, vol. 24, no. 3, pp. 301–312, Mar. 2002.
14. B. Mumey, L. Showe, M. Showe, A Combinatorial Approach to Clustering Gene Expession Data, Bioinformatics, 2003.
15. Q. Wang, Y. Shen, Y. Zhang, and JQ. Zhang, A quantitative method for evaluating the performances of hyperspectral image fusion, IEEE Trans. Instrumentation and Measurement vol. 52, pp. 1041–1047, 2003.

SpecDB: A Database for Storing and Managing Mass Spectrometry Proteomics Data

Mario Cannataro and Pierangelo Veltri

University Magna Græcia of Catanzaro Viale Europa, 88100 Catanzaro, Italy
{cannataro, veltri}@unicz.it

Abstract. Data produced by mass spectrometer (MS) have been using in proteomics experiments to identify proteins or patterns in clinical samples that may be responsible of human diseases. Nevertheless, MS data are affected by errors and different preprocessing techniques have to be applied to manipulate and gathering information from data. Moreover, MS samples contain a huge amount of data requiring an efficient organization both to reduce access time to data, and to allow efficient knowledge extraction. We present the design and the implementation of a database for managing MS data, integrated in a software system for the loading, preprocessing, storing and managing of mass spectra data.

1 Introduction

Mass spectrometry (MS) data are produced by a ionization process on biological samples. Each spectrum contains a large number of (m/z) data, thus that analysis by manual inspection is not feasible. Such problem is much more evident when information has to be obtained by comparing thousands of different spectra, for instance to capture proteins profiles responsible of human diseases. Each spectrum is generated mostly in raw flat file, managed by file system thus that applying data management knowledge can be limited by data format. Moreover, mass-spectrometry based experiments comprise more phases that may require data manipulation. Extracting information from single samples is not a simple task if data are organized in raw data. The problem becomes more complex when experiments involve hundreds of spectra forming huge amount of data. with each spectrum reported in a flat file, often instrument dependent, managed by file systems.

In this paper we are interested in studying the problem of storing and manipulation of spectra data. In particular we focus on designing and implementing of a spectra database for loading, storing and querying spectra obtained by a MALDI TOF spectrometer. We report the results of our study obtained in the contest of a proteomics project for early detection of cancer, which requires the manipulation of huge amount of spectra to cluster data identifying peaks in a spectrum whose profiles classify diseases. The proposed database architecture offers a loading interface to extract data from text file, allows an efficient solution for storing treated spectra in the same database, and offers an API for preprocessing functionalities. Preprocessing such as, noise reduction and calibration is required before using extracting knowledge based algorithms (e.g., data mining). Obviously applying pre-processing using database management framework results in improved efficiency. Pre-processed data are associated to the original sample

I. Bloch, A. Petrosino, and A.G.B. Tettamanzi (Eds.): WILF 2005, LNAI 3849, pp. 236–245, 2006.

taking trace of experiments. Our relational model based database called SpecDB allows to : (i) read spectra data from flat files and loading in the database; (ii) manage spectra data to maintain consistent data, (iii) support preprocessing of spectra data, and finally (iv) allow to store for each spectrum its historical manipulation, adding a temporal dimension to our database. The proposed database is part of MS-analyzer, a system designed and implemented to support proteomics experiments using MALDI spectra data [6]. Our database is organized in such a way that each spectrum is stored in a set of relations each containing portion of the spectrum according to a predefined window of mass/charge values, while a main MS-spectrum relation maintains the organization in single spectrum, improving data management.

Companies supporting biomedical activities, are defining new modules for treatment of mass spectra data (see [2], [10]). Most of them left spectra management to the file system, storing only the files physical locations in the database. For instance [18] stores the physical address of the .txt file. Research communities are working on the idea of creating an MS data repository. The Proteomics Standard Initiative (PSI)[17] is working on the definition of an XML based standard to represent proteomics experiments, mzData format for capturing peak list information, with the aim of uniformly represent different formats into one. In such paper we focus on the storing of MS data in a relational model; however we are working in a framework that includes also data in the XML format [5].

2   Related Works

Recently there has been an increasing interest of computer scientists in studying biological problems, defining environments supporting biological experiments [15, 3], defining methods for biological data integration [14] and designing information extraction methods. Often the results contain lots of data mixed to noise and irrelevant data, that cannot be filtered out without automatic supports.

Most of the companies supporting biomedical activities through new technologies, are defining new modules for treatment of mass spectra data. For instance, the Applied Biosystems company, one of the leader producing mass spectrometers [2], offers LIMS an integrated interface that allows to design and implement an application using workflows. In particular it supports phases of proteomics experiments: sample preparation, protein separation (using for instance 2D gel), data analysis and visualization, modules for protein identifications from spectra. It supports querying publicly available databases for protein identifications. Nevertheless it is not yet available any MS data repository that allows users to create and manipulate local owned spectra database. Data are still managed by file system. Similarly, SampleManager [10] integrates data obtained both by bio-clinical instruments and by Electronic Patient Records, but no database support is given to the MS spectra data.

Most of the existing products offer commercial database management systems to model mass spectrometer experiments, storing meta-information such as experiments, results, laboratory, but spectra management are left to the file system, storing only the physical locations of spectra files in the database. For instance the suite offered by [18] stores the physical address of the file containing raw data. Data are then loaded whenever is necessary, and managed in main memory. Such approach does not allow any data management and indexing, limiting experimental tests that need comparing several data, as for instance required by data mining approaches.

Recently, research communities are working on the idea of creating an MS data repository to store experiments similarly on what offered by publicly available data sets (as PDB [11]). For instance, the Proteomics Standard Initiative (PSI) is working on the definition of an XML based standard format, the mzData, to represent proteomics experiments, where also meta information on mass spectrometer experiments is reported [17]. XML data format allows accessing data of different repositories in a similar way, while value pairs related to spectrum are compressed and contained in a single element. Moreover [17] proposes mzData data format for capturing peak list information, with the aim of uniformly represent different formats into one. Still, no repository propositions have been reported for defining centralized or distributed data storage for mzData. The IBM studied the definition of a platform for MS experimental storing in [13], for monitoring and managing bioinformatics experiments. In such framework, particular interest is given to the mass spectra data management, even coming from MALDI MS or from LC/MS spectrum analysis.

Database community is treating MS data as interesting data sets: [7] presents a labeling technique for adding labels to each spectra about ions expected chemical species, but no attentions have been given for data managements. MS data contains lots of information, that is hard to manage in plain text formats.

3   Mass Spectrometry Data

Mass Spectrometry (MS) is a technique more and more used to identify macromolecules in a compound. The mass spectrometer is an instrument designed to separate gas phase ions according to their m/Z (mass to charge ratio) values. The MassSpectrometry process [1] [12] can be decomposed in four sub-phases: (i) Sample Preparation (e.g. Cell Culture, Tissue, Serum); (ii) Proteins Extractions; (iii) ICAT protocol (optional); and

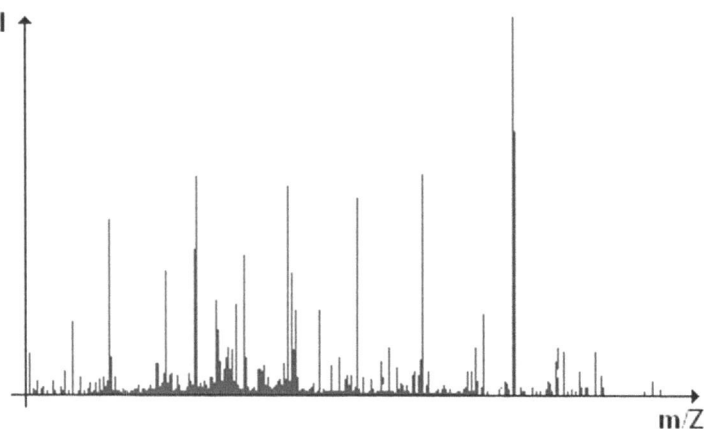

Fig. 1. Mass Spectrum of a Biological Sample

Fig. 2. Spectrum Generation Phases

m/z	intensity
.....	...
799.976004	135.864
800.004478	156.232
800.032953	140.765
800.061429	152.13
800.089905	137.15
800.118381	132.145
800.146858	131.137
800.175336	122.761
800.203814	124.499
800.232292	125.993
...	...

Fig. 3. Mass Spectrum Loaded in a Text Raw Data

(iv) Mass Spectrometry processing. In Figure 2, is reported a simple representation of the spectrum generation phases. Mass Spectrometry output is represented, at a first stage, as a (large) sequence of value pairs, where each pair contains a measured intensity, which depends on the quantity of the detected biomolecules, and a mass to charge ratio (m/z), which depends on the molecular mass of detected biomolecules. The large number of (m/z) data justifies the dimension of each output associated to an analyzed sample. Indeed, MALDI-TOF output are in order of 10 MB of flat files: analysis by manual inspection is not feasible.

Figure 1 shows an example of spectrum, and Figure 3 a snapshot of data representation. In some applications, such as identification of proteins, research has focused on obtaining a list of significant peaks, each of which represents a peptide. Such problems is usually performed by experts that know the instruments: up to 80% of peaks in a spectrum maight be unsignificant in researching interesting peaks. Such preprocessing phase cannot be performed automatically manipulating flat files. The complexity of the problem increases when several spectra have to be compared or need to be preprocessed to be compared. In this simple case manipulating data as flat files is an hard task. In data mining processes, used to extract information from collections of thousands of samples, and that require high input/output rate, managing data using database management facilities is mandatory.

Mass spectra require preprocessing phases. For instance one of the more frequent preprocessing phase consists in aligning a set of spectra with respect to their m/z axes, i.e. peaks are aligned to be comparable. Our approach, as we present in the following, allows to perform preprocessing phases, such as normalization, binning, alignment etc., (see [6] for more details) and allows to store preprocessed data in the database.

4   SpecDB: Data Model and Architecture

Raw mass spectra are not complex data, they are couples of real numbers (raw data). Nevertheless, complexity derives from the fact that each spectrum contains thousands of values, stored in flat files with average space occupancy of 5 Mbytes: problems occur in case of manipulation (e.g., preprocessing) of large set of spectra. Even available data mining tools such as [16] are limited by manipulating large amounts of data. In our proposal, data are reported in a relational database where spectra are stored in a set of relations. The database is realized on a MySQL [8] database management system. Data are loaded through a Java loading interface that dump data from text files in the database.

Main requirements for a spectra database are the following:

- supporting efficient storing and retrieval of data (single spectrum, set of spectra and portions of spectra);
- supporting import/export functions (e.g. loading of raw spectra available in different text files, exporting of spectra in XML mzData data);
- offering query/update functions able to enhance performance of data preprocessing and analysis (e.g. avoiding full main memory processing). Such functions could be offered through a set of specialized APIs, for instance range queries could be used to implement basic preprocessing techniques as aggregation of peaks, such as binning and data reduction.

4.1   Data Model

In Figure 4, we report the conceptual organization of data that have to be contained in SpecDB database. The main entity *MassSpectrum* represents the set of spectra, each represented by a unique identifier, the m/z value and the corresponding intensity value, information on date and experimental details (laboratory, operator etc). The self-reference relation indicate that each spectrum may be associated to other spectra generated by preprocessing it. This allows to populate the database by considering all spectra obtained by experiments or manipulated in preprocessing operation required by data treatments. Treatment entity represents manipulation and preprocessing techniques that have been applied to data. The entity spectrometer gives information on the instruments, while *SampleOrigin* contains information on data origins and biological sample.

Once spectra are loaded, from text files they are loaded into Mass spectrum table (see Figure 5). To otimize data management, values of intensity for each spectrum may be reported in a relation associated to range of m/z axes.

4.2   Architecture and Functions

We integrated our system in MS-Analyzer, an integrated environment to manage MS proteomics experiments. In particular, MS-Analyzer provides an integrated tool able to

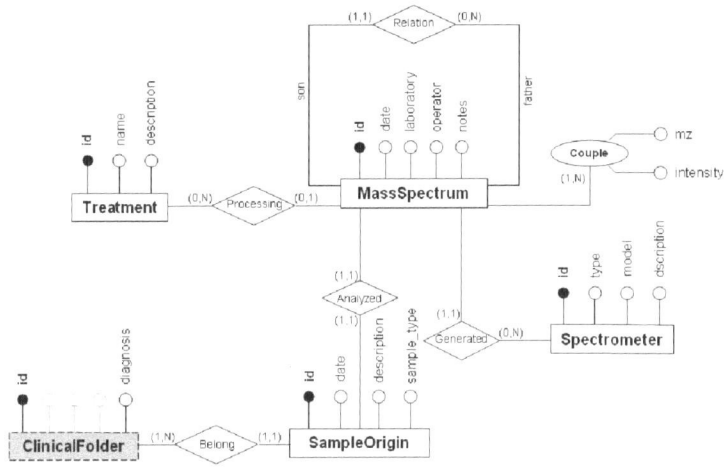

Fig. 4. Conceptual organization of Data

Table: MassSpectrum

id	father	treatment	sample_origin	spectrometer	date	operator	laboratory
1	1	1	1	2	18.12.04	op1	unicz
2	1	2	-	-	18.12.04	op2	unicz
3	2	6	-	-	18.12.04	op2	unicz
4	3	8	-	-	18.12.04	op2	unicz
5	5	1	2	2	18.12.04	op1	unicz
6	6	2	3	1	18.12.04	op1	NCI
7	6	6	-	-	18.12.04	op1	unicz
8	5	2	-	-	18.12.04	op1	unicz

Fig. 5. Mass_Spectrum Relations with some Spectra

monitor and manage each phase of a proteomics experiments that uses mass spectrometer: sample collection (e.g. serum, tissue, cell line), sample preparation and storing (e.g. ICAT protocol), mass spectrometry processing, data preprocessing, through data mining algorithms. The target is to provide support in extracting information form thousands of spectra associated to a population presenting genetic attitude in developing breast cancer [4]. Figure 6 reports the architecture of MS-Analyzer. MS-Analyzer provides the following functions:

1. MS proteomic data acquisition, getting raw spectra produced by different Mass Spectrometers. Spectra are locally stored in file system and using the SpecDB loading function, raw data are dumped into SpecDB.
2. MS proteomic data pre-processing that gets spectra from SpecDB, and applies different pre-processing techniques. Preprocessed data are also dumped in SpecDB.
3. MS proteomic data preparation, that gets MS preprocessed spectra and prepare them to be given in input to different kind of data mining tools. E.g. WEKA data mining tool requires that MS spectra being organized in a unique file containing a metadata header (ARFF).

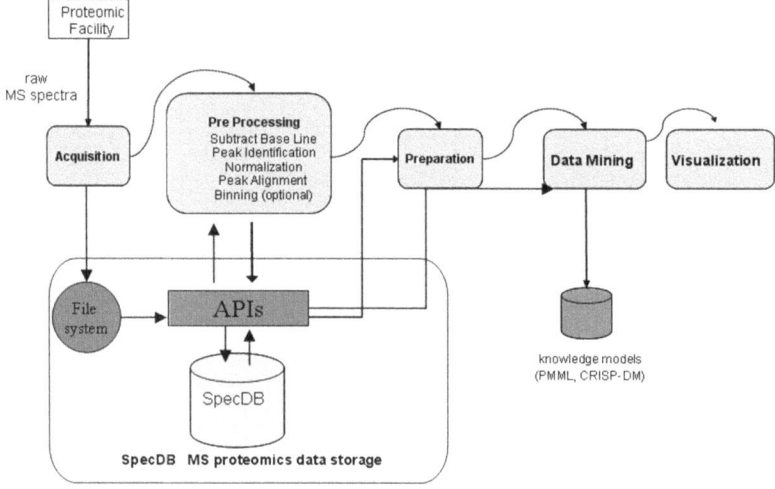

Fig. 6. MS-Analyzer Tool

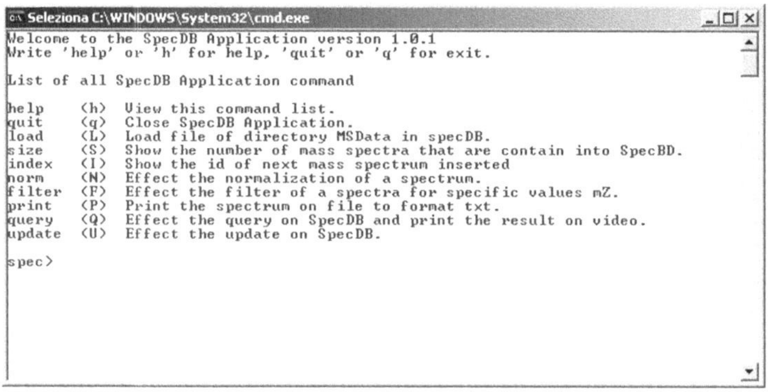

Fig. 7. SpecDB textual Interface

4. Data Mining analysis: this phase allows to select different data mining tasks (e.g. classification, clustering, pattern analysis), different data mining algorithms and tools (e.g. Q5, C5, K-means, ect.), loading data from SpecDB and to give them as input to the chosen data mining tools.

Preprocessing are relevant processes whenever is necessary to prepare spectra data for information extraction (e.g. Subtract Base Line, Peak identification, Normalization of intensities, Peak Alignment, Binning), some of which can be optional, and stores data in an efficient manner into a Pre-processed Spectra Repository (PSR). Pre-processing can be applied to one spectrum or to many spectra. In latter case use of SpecDB improves performances than using files. The idea is to provide a set of database APIs that allow efficient preprocessing, in such a way, preprocessing can be thought as a basic database function.The current version presents a simple textual interface as the one presented in Figure 7 used to perform access and preprocessing functions on MS data.

5   Experiments

We tested the SpecDB database by loading 280 MB of raw data (text files) containing 500 spectra (see [9]). The average time of loading has been of 1675 milliseconds, and have been loaded almost 20 thousands of records. First simple experiment shows that loading data from flat files is linear in number of loaded spectra (see Figure 8).

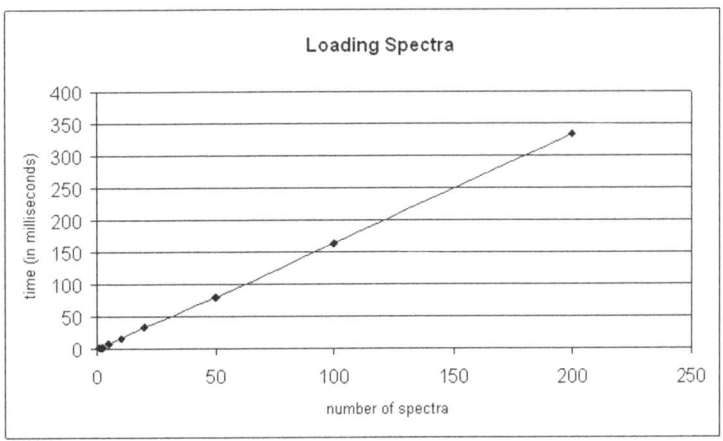

Fig. 8. Loading Spectra in SpecDB

We report a simple script for querying data using a filter, creating a new spectrum.

```
FilterSpectrum(MassSpectrum ms, Double mz[]){
  /*create a new spectrum*/
Q1=insert into Mass_Spectrum (id,father,treatment,...)
values (ms.id(),ms.father(),ms.treatment(),...)
 for(int i=0; i<mz.length; i++){
  /*select (mz, intensity) */
Q2=select * from Couple
    where mass_spectrum='ms.father()' and mz='mz[i]'
  while(next){
  /*insert Q2 results*/
  Q3=insert into Couple (mass_spectrum, mz, intensity)
  values('id','Q2.mz','Q2.intensity')
  }//end while
}//end for
}//end filterSpectrum
```

Such simple method is useful when is required to compare portion of spectra filtered with respect to m/z windows, or when comparing spectra from similar biological samples. The filtering operation has an execution time that depends on the m/Z values used as filter parameters. Time filtering is in the range comprised between 40 ms for single value and 300 ms for more than 10 filter values.

We now report a simple experiment on normalization performed directly using text interface of Figure 7. The normalization of spectra data consists in reducing peak values with respect to the maximum peak. The script in pseudo code that can be directly used on SpecDB, is as follows.

```
Normalization(Mass_Spectrum ms){
 /*select max intensity*/
Q1=select max(intensity) from Couple
    where mass_spectrum=ms.getFather()
 /*create New Spectrum*/
Q2=insert into Mass_Spectrum (id,father,treatment,...)
values(ms.getId(),ms.getFather(),ms.getTreatment(),..)
 /*Normalize*/
Q3=select * from Couple where mass_spectrum=ms.getFather()
while(next()){
  Q4=insert into Couple (mass_spectrum, mz, intensity)
  values(ms.getId(),Q3.mz,(Q3.intensity/maxIntensity))
  }//end while
}//end Normalizzation
```

We remind that once data are preprocessed, they are loaded into the database as a new spectrum keeping trace of the original spectrum (id Father in the relation of Figure 4). Preprocessing is much more efficient using SpecDB than using single flat files. Normalization procedure is performed in 1400 milliseconds, on one spectrum of 15.000 average records. Moreover such normalization may be realized for several spectra considering the normalization with respect to the maximum value of peak for all spectra. We plan to report complete comparison of performance measures of preprocessing by using SpecDB and file system based data.

6   Conclusion

We propose a relational data base system, SpecDB, to store data and to support preprocessing functions. SpecDB is part of MS-Analyzer, an integrated platform for monitoring and managing proteomics experiments from sample preparation to MS data generation, storing and preprocessing. We are improving the architecture of SpecDB architecture to manage mzData, the XML based format for spectra, and we are improving the API for further functions.

Acknowledgments. This work is partially supported by the COFIN 2003 project titled *Hereditary breast cancer: genetics and proteomics*, funded by MIUR.

References

1. Ruedi Aebersold and Matthias Mann.   Mass spectrometry-based proteomics. *Nature*, 422:198–207, 13 March 2003 2003.
2. Applied Biosystem. Applied biosystem. http://www.cmpharm.ucsf.edu/ nomi/nnpredict.htm.

3. Carl-Ivar Branden and John Tooze. *Introduction to Protein Structure*. Garland Publishing, Inc., New York, USA, second edition, 1998.

4. M. Cannataro, G. Cuda, M. Gaspari, P. H. Guzzi, T. Mazza, P. Tagliaferri, G. Tradigo, and P. Veltri. Mass Spectrometry Data Analysis for Early Detection of Inherited Breast Cancer. In *WIRN 2004*, 2004.

5. M. Cannataro, P. Guzzi, T. Mazza, A. Oliverio, G. Tradigo, and P. Veltri. An Architecture for Managing Mass Spectrometry Data for a Distributed Proteomics Laboratory. In *Heterogeneous and Distributed Information Retrieval, 28th conf. on Research and Development Information Retrieval (SIGIR)*, 2005.

6. M. Cannataro, P. Guzzi, T. Mazza, G. Tradigo, and P. Veltri. On the Preprocessing of Mass Spectrometry Proteomics Data. In *WIRN*, 2005.

7. Lei Chen, Zheng Huang, and Raghu Ramakrishnan. Cost-Based Labeling of Groups of Mass Spectra. In *SIGMOD 2004*, 2004.

8. Paul DuBois. *MySQL*. 0-7357-1212-3. Sams Developer's Library, second edition, Agust 2004.

9. Petricoin EF Ardekani AM Hitt BA Levine PJ Fusaro VA Steinberg SM Mills GB Simone C Fishman DA Kohn EC and Liotta. Use of proteomic patterns in serum to identify ovarian cancer . *Lancet*, 359:572–577, 2002.

10. Thermo electrono corporation. Sample manager. http://www.thermo.com/com/cda/home.

11. Research Collaborator for Structural Bioinformatics.   Protein data bank, 2004. http://www.rcsb.org/pdb/.

12. Gary L. Glish and Richard W. Vachet. The basic of mass spectrometry in the twenty-first century. *Nature Reviews*, 2:140–150, February 2003 2003.

13. IBM.   Proteomiq   and   ibm   life   sciences   framework. http://www.proteomesystems.com/uploads/Proteomepartnerbrief.pdf.

14. Zoé Lacroix and Terence, editors. *Bioinformatics: Managing Scientific Data*. Learn how, 2001.

15. Janusz M.Bujnicki. *Practical Bioinformatics*. 0933-1891. Springer, Berlin, frist edition, 2004.

16. University of Waikato. Weka data mining tool. http://www.cs.waikato.ac.nz/ml/weka/.

17. Proteomics Standard Initiative PSI. Psi. http://psidev.sourceforge.net/ms/.

18. Nonlinear Bioinformatics Solutions. Non linear phoretyx. http://www.nonlinear.com/.

NEC for Gene Expression Analysis

R. Amato [1], A. Ciaramella [2], N. Deniskina [1,3], C. Del Mondo [1], D. di Bernardo [4],
C. Donalek[1,5], G. Longo [1,6,7], G. Mangano [6,1,8], G. Miele [1,6], G. Raiconi [2,6],
A. Staiano [1,2], and R. Tagliaferri [2,6]

[1] Dipartimento di Scienze Fisiche, University of Naples "Federico II", Naples, Italy
[2] Dipartimento di Matematica e Informatica, University of Salerno, Fisciano, Salerno, Italy
[3] Institute of Information Transmission Problems, Russian Academy of Sciences,
Moscow, Russia
[4] Telethon Institute for Genetics and Medicine, Naples, Italy
[5] Department of Astronomy, California Institute of Technology, Pasadena CA, USA
[6] INFN - Istituto Nazionale Fisica Nucleare - Sezione di Napoli, Naples, Italy
[7] INAF - Istituto Nazionale di Astrofisica - Sezione di Napoli, Naples, Italy
[8] Department of Physics, Syracuse University, Syracuse NY, USA

Abstract. Aim of this work is to apply a novel comprehensive data mining
machine learning tool to preprocess and to interpret gene expression data. Furthermore, some visualization facilities are provided. The data mining framework
consists of two main parts: preprocessing and clustering-agglomerating phases.
To the first phase belong a noise filtering procedure and a non-linear PCA Neural Network for feature extraction. The second phase is used to accomplish an
unsupervised clustering based on a hierarchy of two approaches: a Probabilistic
Principal Surfaces to obtain the rough regions of interesting points and a Fisher-
Negentropy information based approach to agglomerate the regions previously
found in order to discover substructures present in the data. Experiments on gene
microarray data are made. Several experiments are shown varying the threshold,
needed by the agglomerative clustering, to understand the structure of the analyzed data set.

1 Introduction

Scientists have been successful in cataloguing genes through genome sequencing
projects, and they can now generate vast quantities of gene expression data using microarrays. However, due to the sheer size of the data sets involved, and to complexity
of the problems to be tackled, the biological community has so far had less success
in understanding how genes and proteins are connected and how they operate within
networks. Such challenges call for novel approaches to data mining and understanding
heavily relying on artificial intelligence tools.

In this work we propose a new approach, based on a solid mathematical formalism,
able to cluster noisy gene expression data with missing data points. The method, which
provides a graphical user friendly interface for several data visualization options, represents an automatic procedure to discover, with no a priori assumptions, the number of
clusters present in the data.

I. Bloch, A. Petrosino, and A.G.B. Tettamanzi (Eds.): WILF 2005, LNAI 3849, pp. 246–251, 2006.

The method was tested against the microarray data set of cell cycle in yeast Saccharomyces Cerevisiae made publicly available by [6]. The Spellman data set consists of four synchronization experiments (alpha factor arrest, elutriation and arrest of CDC15 and CDC28 temperature-sensitive mutants) which were performed for a total of 73 microarrays during cell cycle. A detailed description of the data set can be found in [6].

The work is presented as follows: in Section 2 we introduce the methods for the preprocessing step, while in Section 3 the NEC approach is detailed. Section 4 describes the experimental results and, finally, conclusions, in Section 5 close the paper.

2   Preprocessing Phase

Microarray data have a very noisy nature, and thus preprocessing plays a fundamental role. We used a two step preprocessing phase: a preliminary noisy gene filtering, followed by a non-linear PCA features extraction.

The first step of the preprocessing phase is the filtering of the genes with high noise–to–signal ratio. This requires a reliable noise estimation which was achieved by exploiting the periodicity of the input data set [1]. From our analysis we have that the 0.5 threshold was found to be a reasonable value for the noise–to–signal ratio. It is worth observing, as a further a posteriori check of the validity of our rejection method, that most of the genes used in the Spellman analysis pass our test.

The second step of preprocessing is the features extraction process that is based on a non-linear PCA method which allows to obtain the eigenvectors from unevenly sampled data. This approach is based on the STIMA algorithm described in [4,7,8].

3   NEC Approach

The preprocessed data represent the input to the NEC algorithm. As mentioned before, our clustering model is composed by two phases hierarchically organized: a preliminary unsupervised clustering approach and an agglomerative method.

I Phase: Clustering Algorithm

The approach that we consider at the first stage is the Probabilistic Principal Surfaces (PPS) method: a latent variable model which has been shown to be very effective for data mining purposes [1,3]. PPS defines a non-linear, parametric mapping $y(x; W)$ from a Q-dimensional latent space ($x \in R^Q$) to a D-dimensional data space ($t \in R^D$), where normally $Q < D$. The mapping $y(x; W)$ (defined continuous and differentiable) maps every point in the latent space to a point into the data space.

Since the latent space is Q-dimensional, these points will be confined to a Q-dimensional manifold non-linearly embedded into the D-dimensional data space. Substantially, the PPS approach builds a constrained mixture of Gaussians and the EM algorithm can be used to estimate the parameters of the model.

If $Q = 3$ is chosen, a spherical manifold [1] can be constructed using a PPS with nodes arranged regularly on the surface of a sphere in R^3 latent space, with the latent basis functions evenly distributed on the sphere at a lower density. After a PPS model is fitted to the data, several visualization possibilities are available for analyzing the data points [1].

II Phase: Agglomerative Information

The information that we use to merge clusters (rough regions) is the Fisher's discriminant and Negentropy: on one hand, the Fisher's linear discriminant is a classification method that projects high-dimensional data onto a line and performs classification in this one-dimensional space [2]. The projection maximizes the distance between the means of the two classes while minimizing the variance within each class. We have to note that the Fisher discriminant aims to achieve an optimal linear dimensionality reduction. It is therefore not strictly a discriminant itself, but it can easily be used to construct a discriminant.

The Fisher criterion for two classes is given by

$$J_F(\mathbf{w}) = \frac{\mathbf{w}^{\mathrm{T}}\mathbf{S}_{\mathrm{B}}\mathbf{w}}{\mathbf{w}^{\mathrm{T}}\mathbf{S}_{\mathrm{W}}\mathbf{w}} \tag{1}$$

where $\mathbf{S}_{\mathrm{B}}$ is the between-class covariance matrix and $\mathbf{S}_{\mathrm{W}}$ is the total within-class covariance matrix.

On the other hand, the definition of Negentropy J_N is given by

$$J_N(\mathbf{x}) = H(\mathbf{x}_{\mathrm{Gauss}}) - H(\mathbf{x}), \tag{2}$$

where $\mathbf{x}_{\mathrm{Gauss}}$ is a Gaussian random vector of the same covariance matrix as $\mathbf{x}$ and $H(.)$ is the differential entropy. We note that the Negentropy can also be interpreted as a measure of non-Gaussianity.

The classic method of approximate Negentropy is using higher-cumulants, through the polynomial density expansions [5].

However, such cumulant-based methods sometimes provide a rather poor approximation of the entropy.

A special approximation is obtained if one uses two functions G^1 and G^2, which are chosen so that G^1 is *odd* and G^2 is *even*. Such a system of two functions can measure the two most important features of non-Gaussian 1-D distributions. The odd function measures the asymmetry, and the even function measures the dimension of bimodality vs. peck at zero, closely related to sub- vs. supergaussianity. Classically, these features have been measured by skewness and kurtosis, which correspond to $G^1(x) = x^3$ and $G^2(x) = x^4$.

Then the Negentropy approximation of equation 2

$$J_N(\mathbf{x}) \propto k_1 E\{G^1(\mathbf{x})\}^2 + k_2 (E\{G^2(\mathbf{x})\} - E\{G^2(v)\})^2 \tag{3}$$

where v is a Gaussian variable of zero mean and unit variance (i.e. standardized), the variable $\mathbf{x}$ is assumed to have also zero mean and unit variance and k_1 and k_2 are positive constants. We note that choosing the functions G^i that do not grow too fast, one obtains more robust estimators. See [5] for more details on this kind of functions.

NEC Algorithm

In order to agglomerate the regions found by PPS, we combine the Fisher's discriminant and the Negentropy information in a measure that we call $J_{\mathbf{NEC}}$:

$$J_{\mathbf{NEC}}(\mathbf{X}) = \alpha_F J_F(\mathbf{w}) + \alpha_N J_N(\mathbf{X}) \tag{4}$$

where α_F and α_N are two defined constants and $\mathbf{w}$ is the Fisher's direction.

Algorithm 1. NEC Algorithm

 1: **Begin initialize** dt, $\hat{c} = c$, $D_i \leftarrow X_i$, $i = 1, \ldots, c$
 2: **for** i = 1 to ĉ-1 **do**
 3:    **for** j = 1 to ĉ **do**
 4:       calculate the Fisher's direction between D_i and D_j and project the data on it
 5:       calculate the $J_{\mathbf{NEC}}$ information
 6:       **if** $J_{\mathbf{NEC}} < dt$ **then**
 7:          merge the clusters D_i and D_j and recursively all the clusters previously selected
 8:       **end if**
 9:    **end for**
10: **end for**
11: **return** $\tilde{c}$ new clusters
12: **End**

The only *a priori* information is a dissimilarity threshold (dt in the following). We suppose to have c multi-dimensional regions X_i with $i = 1, \ldots, c$ that have been defined by the clustering approach. The NEC algorithm is described in Algorithm 1.

4 Experimental Results on Microarray Data

In this section we show the results obtained using our approach on a microarray data set of cell cycle in yeast Saccharomyces Cerevisiae [6,1]. After the noisy gene filtering, the non-linear PCA is used to extract the first eigenvector (8-dimensional) of the auto-correlation matrix of the overall 2445 genes that passed the filtering procedure, for each of the 4 microarray experiment, so obtaining 32 features. Then a PPS model with 98 latent variables is trained on the 2445×32 preprocessed data set. After the completion of the training phase we projected the data into the latent space for visualization purposes. Figure 1a shows the projection of genes which belong to some relevant clusters (clusters found after the agglomerating phase).

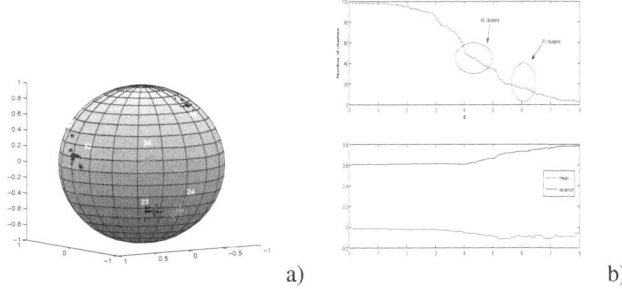

a)   b)

Fig. 1. Experimental results: a) Projections of genes belonging to some relevant clusters (clusters found after the agglomerating phase): front view. A similar image can be produced for the back view; b) plateaus obtained varying the threshold dt (up) and clusters variance and mean varying the threshold dt (down).

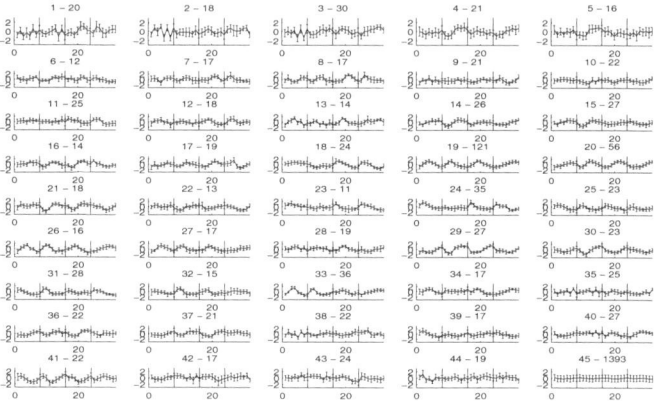

Fig. 2. 45 clusters error bars plot showing the prototype behavior and standard deviation. On the horizontal axes is reported, for each of the 4 microarray experiments, the 8-dim. parameter vector extracted by nonlinear PCA. For each subplot the cluster number and the number of genes in it contained are reported.

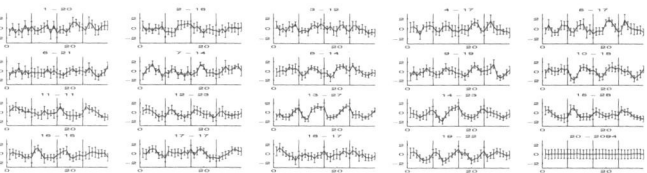

Fig. 3. 20 cluster error bars plot showing the prototype behavior and standard deviation. On the horizontal axes is reported, for each of the 4 microarray experiments, the 8-dim. parameter vector extracted by nonlinear PCA. For each subplot the cluster number and the number of genes in it contained are reported.

The structure of the data set can be deeply analyzed using the NEC approach by plotting the number of clusters found by the algorithm varying the threshold (dt). We note that in this case the threshold is chosen with a step of 0.05. From figure 1b it is worth noting that, even though the data set is noisy, we are able to discover two main substructures corresponding to the plateaus on 45 and 20 clusters, respectively. In this way we identify the points of stability to understand the structure of the data set and the meaningful number of clusters.

For each discovered cluster, we computed a prototype defined as the mean of the data points within it, and the standard deviation with respect to the prototypes. The resulting error bars plots are shown in figure 2 and figure 3 for 45 and 20 clusters, respectively.

From the two figures we note that we mainly have one cluster with noise and all the other with a structured shape. We also note, from the confusion matrices calculated between both the agglomerations, that the NEC algorithm is stable since merge only regions with high noise.

5   Conclusion

We have presented an unsupervised tool for preprocessing, visualization and clustering. The clustering procedure is based on a novel approach hierarchically conceived, i.e. a hierarchy of two unsupervised clustering algorithms, called by us NEC. The first algorithm is based on a Probabilistic Principal Surfaces approach and the second one on an agglomerative clustering based on both Fisher and Negentropy information.

We note that since the NEC clustering does not use non-parametric estimation of the distribution it is very fast and very stable. In fact, in many cases the non-parametric estimation could give worse results depending on the values assigned to the parameters.

Moreover, we stress that applying the approach on microarray data of cell cycle we could obtain, in an unsupervised and simple way, biological information from the data. Furthermore, it is highlighted that using the NEC approach, varying the agglomerative threshold we can study the structure of the data set and to choose the right number of clusters.

In the next future the authors will focus their attention on the validation of the found clusters from a biological point of view, i.e. using biological p-values of our clusters by using the GOTerm Finder tool from the Saccharomyces Cerevisiae database and to apply the approach on different biological data sets.

References

1. R. Amato, A. Ciaramella, N. Deniskina, C. Del Mondo, D. di Bernarndo, C. Donalek, G. Longo, G. Mangano, G. Miele, G. Raiconi, A. Staiano, R. Tagliaferri, A Novel Approach to Gene Expression Clustering, submitted to Bioinformatics Journal;
2. C.M. Bishop, Neural Networks for Pattern Recognition, Oxford University PRESS, 1995;
3. A. Ciaramella, G. Longo, A. Staiano, R. Tagliaferri, NEC: a Hierarchical Agglomerative Clustering Based on Fisher and Negentropy Information, accepted for pubblication on Lecture Notes in Computer Science, Proceedings of WIRN 2005, LNCS PRESS;
4. A. Ciaramella, C. Bongardo, H. D. Aller, M. F. Aller, G. De Zotti, A. Lähteenmaki, G. Longo, L. Milano, R. Tagliaferri, H. Teräsranta, M. Tornikoski, S. Urpo, A Multifrequency Analysis of Radio Variability of Blazars *Astronomy & Astrophysics Journal* **419** 485-500, 2004;
5. A. Hyvärinen, J. Karhunen, E. Oja, Independent Component Analysis, John Wiley & Sons, 2001;
6. P.T. Spellman, G. Sherlock, M.Q. Zhang, V.R. Iyer, K. Anders, M.B. Eisen, P.O. Brown, D. Botstein, B. Futcher, Comprehensive identification of cell cycle-regulated genes of the yeast saccharomyces cerevisiae by microarray hybridization *Molecular Biology of the Cell* **9** 3273-3297, 1998;
7. R. Tagliaferri, A. Ciaramella, L. Milano, F. Barone, G. Longo, Spectral Analysis of Stellar Light Curves by Means of Neural Networks *Astronomy & Astrophysics Suppl. Series* **137** 391-405, 1999;
8. R. Tagliaferri, N. Pelosi, A. Ciaramella, G. Longo, L. Milano, F. Barone, Soft Computing Methodologies for Spectral Analysis in Cyclostratigraphy *Computer and Geosciences* **27** 535-548, 2001.

Active Learning with Wavelets for Microarray Data*

D. Vogiatzis and N. Tsapatsoulis

Dept. of Computer Science, University of Cyprus,
Kallipoleos 75, Nicosia CY-1678, Cyprus
Tel.: +357-22892749; Fax.: +357-22892701
{dimitrv, nicolast}@cs.ucy.ac.cy

Abstract. In Supervised Learning it is assumed that is straightforward to obtained labeled data. However, in reality labeled data can be scarce or expensive to obtain. Active Learning (AL) is a way to deal with the above problem by asking for the labels of the most "informative" data points. We propose a novel AL method based on wavelet analysis, which pertains especially to the large number of dimensions (i.e. examined genes) of microarray experiments. DNA Microarray expression experiments permit the systematic study of the correlation of the expression of thousands of genes. We have applied our method on such data sets with encouraging results. In particular we studied data sets concerning: Small Round Blue Cell Tumours (4 types), Leukemia (2 types) and Lung Cancer (2 types).

Keywords: active learning, wavelets, microarray data.

1   Introduction to Active Learning

The idea that a large set of labeled data is available for training a classifier under a supervised regime is often wrong. Data labeling can be a time consuming and expensive process. For instance in the microarray experiments in biology, a large data set is produced which represents the expression of hundreds or thousands genes (i.e. production of RNA) under different experimental conditions (see [1] for an overview of microarray experiments), labeling exhaustively all the experimental samples can be very expensive or simply impossible. An alternative would be to start with a small set of labeled data, which may be easy to obtain, then to train a classifier with supervised learning. At this point a query mechanism pro-actively asks for the labels of some of the unlabeled data; whence the name active learning. The query implements a strategy to discover the labels of the most "informative" data points.

---

* The work presented in this paper has been undertaken in the framework of the OPTOPOIHSH project (PLHRO/0104/04 - Development of knowledge-based Visual Attention models for Perceptual Video Coding) funded by the Cyprus Research Promotion Foundation, Framework Programme for Research and Technological Development 2003-2005.

I. Bloch, A. Petrosino, and A.G.B. Tettamanzi (Eds.): WILF 2005, LNAI 3849, pp. 252–258, 2006.

The concept of Active Learning is hardly new, an important contribution is in [2], where optimal data selection techniques for feedforward neural networks are discussed. In addition the authors show how the same techniques can be used for mixtures of Gaussians and locally weighted regression. An information based approach for active data selection is presented in [3]. In particular three different techniques for maximising the information gain are tested on an interpolation problem. In yet another approach, the geometry of the learning space is derived by computing the Voronoi tessellation, and the queries request the labels of data points at the borders of Voronoi regions [4]. The concept of active learning has also been realised in the context of Support Vector Machines for text classification in [5]. The method is based on selecting for labeling, data points that reduce the version space (the hyperplanes that separate the data) as much as possible. In sect. 2 we discuss about the use of wavelet analysis in bioinformatics, then in sect. 3 we present our proposed algorithm on wavelet based active learning. In sect. 4 we present the experimental setting and the results. In sect. 5 we elaborate on certain choices we made and especially, in the use of wavelets. Finally, in sect. 6 we present conclusions and future work.

2   Wavelets as Data Mining in Bioinformatics

Wavelets have been widely used in signal processing for more than 20 years. Moreover, their usefulness has also been proved in the domain of data mining [6] as well as in the biomedical domain [7]. The main advantage of the wavelets with respect to the Fourier transforms, is that they allow the localisation of a signal in both the time and frequency domains. From the point of view of mathematics, a function can be represented as an infinite series expansion in terms of a dilated and translated version of a basis function called the *mother wavelet*. For practical purposes, we can use the discrete wavelet transform, which removes some of the redundancy found in the continuous transform. In the experiments that we consider we have a small number vectors (microarray experiments), where the dimensions of each vector (genes involved) are an order of magnitude greater than the number of vectors (see [8] for overview of computational methods for microarray data). In particular a data set from a microarray experiment has the form of a two dimensional matrix $\mathbf{X}$. Let, $\underline{x}_i$ be a row vector of $\mathbf{X}$, and $\underline{x}_i=(x_{i,1}, x_{i,2} \ldots x_{i,L})$. The index i refers to the microarray experiment or time step and its maximum value is relatively small, up to 100. On the other hand each dimension represents the expression level (i.e. the RNA that has been produced) of a specific gene in a specific experiment. A characteristic property of microarray which differentiates them from most of the other data is that the number of samples is small when compared to the number of dimensions. Because of this property each sample can be considered as a time series.

3   Wavelet Based Active Learning

It is presumed that we have a pool of labeled and unlabeled multidimensional data, and that the number of dimensions is far greater than the number of data.

The purpose is to reduce the error of a classifier (built with a labeled training set), by selectively asking for the labels of the unlabeled data.

We propose an algorithm which implements a query strategy based on wavelet analysis, and it can be informally stated as: find two data items, one from the testing set of the classifier and the other from the pool of unlabeled data, such that their distance is minimal and the item from the testing set is from the worst performing class (in terms of the Root Mean Square Error). Perform that computation on multiple levels of wavelet analysis, which possibly returns multiple candidates from the unlabeled pool. Then apply a voting scheme and choose the best unlabeled data item from the pool. Next, we present formally the algorithm:

1. Train the classifier with the labeled data.
2. Perform 1D, discrete wavelet transform on each labeled and unlabeled data item, from scale S_1 up to $S_{log_2(L)}$, where L is the number of dimensions. Here greater index in scale denotes lower resolution of wavelet analysis and S_1 denotes the original signal.
3. *Active Learning step:* Form a query to ask for the label of an unlabeled datum: Let class k is the class with the worst classification error and let $\mathbf{X_k}$ be the set of data classified, by the current classifier, to class k. Let also $\underline{x}_{k,i} \in \mathcal{R}^L$ be the i-th signal in $\mathbf{X_k}$. By $\underline{x}_{k,i}^{S_r}$ we denote signal $\underline{x}_{k,i}$ at scale S_r, $r \in \{0, .., log_2(L)\}$. If $\mathbf{U}$ is the set of unlabeled data and $\underline{u}_j \in \mathcal{R}^L$, is the j-th signal in $\mathbf{U}$, represented at scale S_r by $\underline{u}_j^{S_r}$, then the following criterion is used to select the next candidate datum $\underline{u}_\xi$ at scale S_r:

$$\xi^{S_r} = \arg \min_j \{\min_i \|\underline{u}_j^{S_r} - \underline{x}_{k,i}^{S_r}\|_2\} \tag{1}$$

where $\underline{u}_j^{S_r}$ denotes that the comparison is made at scale S_r. It is possible (though not necessary) that $\underline{u}_\xi^{S_r} \neq \underline{u}_\xi^{S_l}$ $r, l \in \{0, .., log_2(L)\}$, $r \neq l$.
4. The previous step is performed for all S_r, which produces a series of

$$\underline{u}_\xi^{S_1} \ldots \underline{u}_\xi^{S_{log_2(L)}} \tag{2}$$

5. Assign weights to all $\underline{u}_\xi^{S_r}$, where each weight is reversely proportional to the euclidean distance of $\underline{u}_\xi^{S_r}$ from its closest $\underline{x}_{k,i}^{S_r}$
6. All $\underline{u}_\xi^{S_1} \ldots \underline{u}_\xi^{S_{log_2(L)}}$ vote according to a weighted majority voting scheme.
7. The label of winner from the previous voting $\underline{u}_\xi^{S_r}$ is requested, and the datum is entered into the training data set, and the classifier is retrained with the updated training set.
8. The algorithm terminates, when the user decides that the overall classifier's performance is good enough, or when there can be no more labels.

The rationale of the algorithm, is to query for labels to the data that are closest to the worst performing data in terms of the classification error. Asking at multiple scales, implies that the similarity that it is sought must be present at multiple resolutions; and thus more impervious to noise.

4   Experiments

We have used datasets from 3 labeled microarray experiments. The first data set was obtained from "The Microarray Project cDNA Library" (http://research. nhgri.nih.gov/microarray/Supplement/). The second and third data sets were obtained from the Gene Expression Datasets collection (http://sdmc.lit.org.sg/ GEDatasets).

The first data set is about Small Round Blue-Cell tumours (SRBCT), investigated with cDNA microarrays containing 2308 genes, over a series of 63 experiments. The 63 samples included tumour biopsy material and cell lines from 4 different types: 23 Ewing's sarcoma (EWS), 20 rhabdomyo sarcoma (RMS), 12 neuroblastoma (NB) and 8 Burkitt's lymphoma (BL). There are also available 20 samples (6 EWS, 3 BL, 6 NB and 5 RMS) for testing [9].

The provenance of the second data set stems is also from oligonucleotide microarrays, with a view of distinguishing between acute lymphoblastics leukemia (ALL) and acute meyeloid leukemia (AML). The training data set consisted of 38 bone morrow samples (27 ALL, 11 AML) from 7130 human genes. The test data set consisted of 34 samples (20 ALL, 14 AML) [10].

The third data set also stems from a microarray experiment and consists of lung malignant pleural mesothylioma (MPM) and adenocarcinoma (ADCA) samples [11]. The training set consists of 32 samples (16 MPM and 16 ADCA) each class) from 12534 human genes. The test set consists of 149 samples (15 MPM and 134 ADCA).

He have split the data sets into training, pool and testing subsets. The first is used for training the classifier. Under the active learning regime a query asks for the labels of specific data from the pool. The testing subset is used for independent control (see Table 1). The best datum (according to the query asked) receives a label and it is subsequently integrated into the training data set. The results reported in Fig. 1 depict the testing set error and are averages

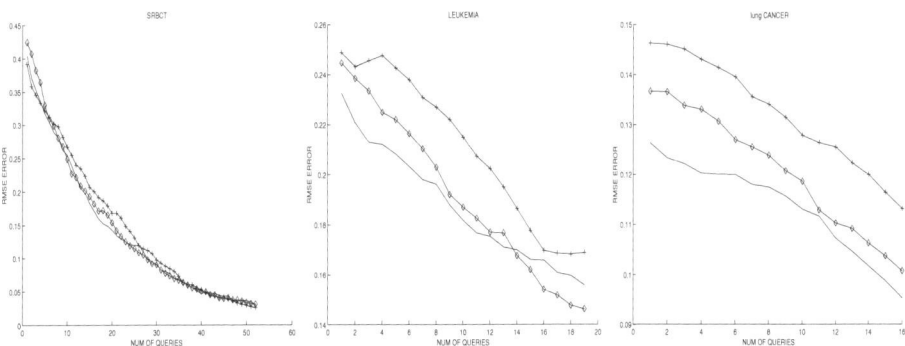

Fig. 1. $\diamond$: (diamond line) random choice of datum, __: (solid line) wavelet weighted majority voting, +-+ (cross line) minimum distance at the largest scale

Table 1. Data Sets characteristics and Performance

Description					Cumulative Query Error		
Data set	Training	Pool	Test	# dimensions	Random	Distance	Wavelet
SRBC	13	52	20	2308	7.4655	7.7540	7.2860
Leukemia	19	19	34	7130	3.6385	4.0050	3.5556
Lung	16	16	117	12534	1.9301	2.1144	1.8165

over 100 experiments. In each experiment the whole data set is shuffled while the number of training, pool and test sets remains the same.

As a classifier, we have used a Support Vector Machine (with a polynomial kernel of degree 3), primarily for practical reasons; our smallest data set has 2308 dimensions and training a multilayered perceptron would take a prohibitively long time. Training on all data sets always resulted in learning 100% of the training sample.

The experimental setting aims to compare the reduction of the classification error on the testing set by comparing three different query types. Each query type progressively asks for the labels of all data in the pool. Next, we present the three query types (the Wavalet Distance Query is what we proposed):

Random Query. For comparison only: The next datum $\underline{u}_\xi$ to be included in the training data set is selected from the pool at random.

Distance Query. The selected datum (from the pool) is the one which minimises: $\|\underline{u}_\xi - \underline{x}_{k,i}\|$, without applying any wavelet transformation. k denotes the category where the classifier has the highest Root Mean Square Error (RMSE), i is the index of the datum $\underline{x}$, and $\underline{x}$ is selected from the testing subset.

Wavelet Distance Query. The proposed method: Apply the criterion of distance query at all scales of wavelet analysis. The winner from each scale votes according to a weight that is reversely proportional to the minimum distance. We have applied a 1D discreet wavelet (Daubechies family order 2) transform up to scale $log_2(L)$, where L is the number of dimensions and $\underline{x}$ is selected from the testing subset.

The results are summarised in Fig. 1 and in Table 1 where the cumulative query error is reported over all pool data for all three data sets (lower values are better). For all data sets the wavelet based active learning method outperformed the two other methods.

All experiments were carried out on the Matlab 6.5 platform, with the OSU SVM classifier and the Wavelet toolboxes.

5 Discussion

The signals we consider are made of the expression levels of genes. The task we are addressing is that of supervised learning, by active (on the part of classifier) selection of the training samples. The signal is not assumed to be stationary, thus the frequencies extend over limited regions of the signal. The wavelet transform offers the best trade off between localisation and extraction of frequencies. Let us assume that the wavelet transform was not applied, and the best datum was selected from the pool for labeling, based on the proximity to another datum of known label. Proximity, can be defined in terms of euclidean/mahalanobis etc. distance. This would impose a certain structure on data belonging in the same category (i.e. categories are spheres/ellipsoids) which of course is not necessarily the case. Naturally, data of the same category must have something in common, but this common property has to been "mined". With this work, we advocate that the intrinsic properties of each datum are hidden in its frequencies and their location. This analysis has to proceed at different scales, because higher frequencies tend to be shorter in duration than lower frequencies, subsequently lower frequencies can be resolved in frequency.

To enforce the claim that wavelet analysis is necessary to detect the most similar object; i.e. it is better to look at multiple scales than at a single scale, it is interesting to observe the data of Fig. 1, where the selection based on minimum distance at largest scale is even worse than selecting at random.

Principal Component Analysis (PCA) is widely used in statistics and machine learning. In many cases the data dimensions are highly correlated with each other; PCA can transform the dataset in such a way that the new dimensions are uncorrelated, then the new dimensions with the lowest variance could be discarded. The assumption of PCA is that the intrinsic data dimensionality is lower that the original dimensionality. In the problem domain we have considered each dimension is the expression level of a gene, and whereas some of the dimensions are expected to be related there is no guarantee that they are linearly related. Concomitant to that is that if the data distribution is not gaussian like, PCA being a linear transformation will not be of much use. Therefore, a better choice in the vein of PCA is to implement and experiment with non-linear extensions of PCA.

6   Conclusions and Future Work

We have designed a method for active learning with a weighted wavelet based voting scheme and we evaluated the method on three datasets from microarray experiments with encouraging results —the wavelet distance query outperformed all other methods. It is interesting to observe that the simple distance query performs on average worse than random choice.

We intent to apply our proposed method on a larger number of microarray data sets to test experimentally its validity. In particular the Gene Expression Datasets collection contains a large number of microarray data sets (see http://sdmc.lit.org.sg/GEDatasets).

An important issue is the choice of the unlabeled data item to be included in the training set, in our case we have devised a weighted voting scheme, however there are also other good choices such as considering only the scale with the lowest entropy which have to be investigated.

We also need to investigate the role of the wavelet. In particular we have use the Daubechies order 2 (db2)—an orthogonal wavelet. What would be the result of a biorthogonal wavelet? It remains to test it experimentally. However, in the experiments we conducted the Haar wavelet underperformed when compared to db2. Moreover, the characteristics of the microarray experiments, might lead us to design a new wavelet to fit the problem of active learning. Another important issue is that the order of the components in the data vectors is not important, they represent expression levels of genes, therefore a rearrangement of the components might lead to improved results.

Finally, we have not taken advantage of the characteristics of the Support Vector Machine in particular we have not considered the separating hyperplanes that are produced. This could suggest a way of improving the active learning query.

References

1. Bergeron, B.: Chapter 6. In: Bioinformatics Computing. Prentice Hall (2004) 222–231
2. Cohn, D.A., Ghahramani, Z., Jordan, M.I.: Active learning with statistical models. In Tesauro, G., Touretzky, D., Leen, T., eds.: Advances in Neural Information Processing Systems. Volume 7., The MIT Press (1995) 705–712
3. MacKay, D.: Information-based objective functions for active data selection. Neural Computation 4 (1992) 590–604
4. Hasenjäger, M., Ritter, H.: Active learning with local models. Neural Processing Letters 7 (1998) 107–117
5. Tong, S., Koller, D.: Support vector machine active learning with applications to text classification. In Langley, P., ed.: Proceedings of ICML-00, 17th International Conference on Machine Learning, Stanford, US, Morgan Kaufmann Publishers, San Francisco, US (2000) 999–1006
6. Li, T., Li, Q., Zhu, S., Ogihara, M.: A survey on wavelet applications in data mining. SIGKDD Explorations 4 (2003) 49–68
7. Liò, P.: Wavelets in bioinformatics and computational biology: state of art and perspectives. Bioinformatics 19 (2003) 2–9
8. Quackenbush, J.: Computational Analysis of Microarray Data. Nature Reviews 2 (2001) 418–428
9. Khan, J., Wei, J., Ringer, M., Saal, L., Ladanyi, M., Westermann, F., Berthold, F., Schwab, M., Antonescu, C., Peterson, C., Meltzer, P.: Classification and diagnostic prediction of cancers using gene expression profiling and artificial neural network. Nature Medicine 7 (2001) 673–679
10. Golub, T.R., Slonim, D.K., Tamayo, P., Huard, C., Gaasenbeek, M., Mesirov, J.P., Coller, H., Loh, M.L., Downing, J.R., Caligiuri, M.A., Bloomfield, C.D., Lander, E.: Molecular Classification of Cancer: Class Discovery and Class Prediction by Gene Expression Monitoring. Science (1999)
11. Gordon, G., Jensen, R., Hsiao, L., Gullans, S., Blumenstock, J., Ramaswamy, S., Richard, W., Sugarbaker, D., Bueno, R.: Translation of microarray data into clinically relevant cancer diagnostic tests using gene expression ratios in lung cancer and mesothelioma. Cancer Research (2002) 4963–4967

Semi-supervised *Fuzzy c-Means* Clustering of Biological Data

M. Ceccarelli and A. Maratea

Research Centre On Software Technology-RCOST,
University of Sannio, Via Traiano 11, 82100 Benevento, Italy
{ceccarelli, amaratea}@unisannio.it

Abstract. Semi-supervised methods use a small amount of labeled data as a guide to unsupervised techniques. Recent literature shows better performance of these methods with respect to totally unsupervised ones even with a small amount of side-information This fact suggests that the use of semi-supervised methods may be useful especially in very difficult and noisy tasks where little *a priori* information is available. This is the case of biological datasets' classification. The two more frequently used paradigms to include side-information into clustering are *Constrained Clustering* and *Metric Learning*. In this paper we use a *Metric Learning* approach as a preliminary step to fuzzy clustering and we show that *Semi-Supervised Fuzzy Clustering* (SSFC) can be an effective tool for classification of biological datasets. We used three real biological datasets and a generalized version of the Partition Entropy index to validate our results. In all cases tested the metric learning step produced a better highlight of the datasets' clustering structure.

Keywords: Semi-Supervised Learning, Fuzzy Clustering, Adaptive Metric, Validity Index.

1 Introduction

In Computational Biology, recent techniques as *Microarray Chips* produce a wealth of data that need to be analyzed and interpreted. In such experiments, the level of mRNA expression of thousands of genes in a cell is simultaneously measured in various experimental conditions. The result is usually presented in form of a matrix, whose columns are the various experimental conditions (time evolution, case/control, etc) and whose rows are the genes fragments spotted on the chips. Pattern Analysis and Machine Learning methods are extensively used to gain insights into biological phenomena and to extract genetic information coded in the DNA chips [5], [6],[9], [10], [14]. Thanks to the application of automatic classification methods, successful results in the understanding of genes roles and interactions have been reached , although there is no literature's agreement on a general method that would work outside the tested datasets. Due to the complexity of the underlying phenomena, the results of functional genomics experiments are very hard to be validated without the aid of a well

I. Bloch, A. Petrosino, and A.G.B. Tettamanzi (Eds.): WILF 2005, LNAI 3849, pp. 259–266, 2006.
© Springer-Verlag Berlin Heidelberg 2006

trained expert of the field. Hence we believe that a fundamental step towards the availability of new and more powerful tools to analyze this kind of data is the inclusion in automatic procedures of *a priori* knowledge, supplied by the field's experts.

In many biological experiments, often there are genes that are biologically known to be involved in the same process under certain conditions. Using conventional unsupervised classification techniques this information is lost, producing poor results due to the algorithms' inability to recognize genes' correlations that are evident for an expert of the field. Such an expert may be able to indicate explicitly at least a reduced list of genes known to be "similar" and some others genes known to be "dissimilar" in a given experiment. A way to use this *a priori* substantive knowledge is tweaking the metric, as a preliminary step to clustering, in such a way that genes declared to be "similar" fall into the same cluster and genes declared to be "dissimilar" fall into different ones.

Often a metric is suitable for some data and completely unsuitable for other, or worse it's suitable for some variables or experiments and completely inadequate for others on the same data. One solution could be to manually "adapt" the metric to the data or to automatically "learn" from data a metric that respects some constraints. If constraints are formulated by a user in a way that expresses his substantive knowledge of the phenomenon under study, the metric can be thought as representing a way to include an *a priori* information into the analysis performed. This can be done as a preliminary step to many classification and clustering algorithms.

The two more frequently used paradigms to include side-information into clustering are *Constrained Clustering* [12],[16] and *Metric Learning* [1], [3], [4], [7], [13], [15]. In the former case the objective function of a clustering algorithm is modified to include a penalty for wrongly classified points, while in the latter a suitable metric that makes similar points be closer and dissimilar points be farther away is learned prior to clustering. Among previous works in metric learning methods, in [15] a Mahalanobis distance is learned using convex optimization. This is a very effective approach, although severely limited in application to real data by the computational complexity $O(d^6)$. In [7] the metric is learned considering pairs of samples belonging to the same class and the computational complexity is reduced to $O(d^3)$ showing similar the. In [3] a k-means family algorithm that joins metric learning and clustering in the same step is proposed. It is more general with respect to [7] and [15] because it considers a different metric for each cluster and so allows for clusters of different shape. In the Semi-Supervised Fuzzy Clustering (SSFC) here proposed, we use a two step approach, separating metric learning from clustering. We also use a different metric for each cluster and we generalize further to fuzzy clustering. Optimization in the learning step in this case is done through an heuristic.

In this paper we show the efficacy and the benefits of learning a metric that respects some user-defined constraints as a preliminary step to clustering, using three real biological dataset and the well known fuzzy c-means algorithm. Results are validated compared to fuzzy c-means with classical euclidean metric and through a generalized version of the Partition Entropy index [2]. Membership distribution shows a clear improvement in detecting the clustering structure for all tested cases.

2   Method

2.1   Fuzzy c-Means with Euclidean Metric

Fuzzy Clustering is a partition-optimization technique that aims to group data based on their similarity in a non-exclusive manner, that is permitting each sample to belong to more than one group. The strength of each sample's belonging to each group is measured through a function, called *membership* that has values between 0 and 1 and that sums to 1 on all clusters. Values closer to 1 indicate a stronger belonging of that sample to that cluster. There are various algorithms with which grouping can be performed and one of the most used is the fuzzy c-means [2].

Main known limitations of the fuzzy c-means are that:

- it can remain trapped in local optima;
- the number of clusters and the amount of fuzziness are free parameters;
- all clusters are of hyperspherical form;
- it produces in every case a grouping, even if the data have no clustering structure.

Its objective function is:

$$J_b = \sum_i^c \sum_j^N d(x_j, m_i)\mu_{ij}^b \tag{1}$$

where μ_{ij} are the membership, m_i are the cluster centroids and b is the overlap parameter.

2.2   Fuzzy c-Means with Learned Metric

The algorithm used is a two steps approach: In the first step we use the *a priori* information to "tweak" the metric d_{A_i}

$$d_{A_i}(x, y) = [(x - y)^T A_i (x - y)]^{\frac{1}{2}} \tag{2}$$

To gain more generality and more flexibility, we considered a different matrix for each cluster A_i $i = 1, ..., c$. To define a criterion for the metric we demand that samples declared to be "similar" have small squared distance and samples declared to be "dissimilar" have high squared distance. Let us call $\mathcal{S}_i$ and $\mathcal{D}_i$ respectively the sets of similar points and the set of dissimilar points in the i-th cluster, then we pose a set of c constrained problems [15]:

$$\min_{A_i} \sum_{(x_k, x_l) \in \mathcal{S}_i} ||x_k - x_l||_{A_i}^2 \qquad (3)$$

$$\text{subject to} \sum_{(x_k, x_l) \in \mathcal{D}_i} ||x_k - x_l||_{A_i} \geq K, \quad \text{and } A_i \geq 0. \qquad (4)$$

where $i = 1 \ldots c$ and $K > 0$ is an arbitrary constant, the constraint term captures the notion of between-cluster dissimilarity whereas the functional to minimized captures the notion of within-cluster similarity. Here, as in [15], we use $K = 1$. It can be easily shown that the problems (4) are equivalent, under the assumption of diagonal matrices A_i, to the minimization of the following c convex functionals:

$$g(A_i) = \sum_{(x_i, x_j) \in S} ||x_i - x_j||_{A_i}^2 - log(\sum_{(x_i, x_j) \in D} ||x_i - x_j||_{A_i}) \qquad (5)$$

The Netwon-Raphson method has been adopted in [15], it leads to an $O(d^6)$ algorithm, where d is the dimension of data. In our case, the problem is even more complex as we have a set of c such problems, moreover in microarray experiments, the data dimension can reach several thousands, therefore here we adopted a stochastic search based minimization algorithm based on the well-known Simulated Annealing (SA) [11] method.

Then we calculate the fuzzy c-means clustering with the more general distance metrics calculated in the previous step.

$$J_b^* = \sum_i^c \sum_j^N d_{A_i}(x_j, m_i)\mu_{ij}^b \qquad (6)$$

The convergence of the fuzzy c-means algorithm is independent from the change in the distance function if the distances are all positive and the prototypes are calculated according to the minimization of the objective function [2].

2.3   Cluster Validity Measures

In a well defined fuzzy clustering the first memberships should be much higher than the others, reflecting scarce ambiguity and good model's matching to the data structure. Considering the list of sorted memberships for each sample it is obvious that a more pronounced asymmetry towards higher values indicates a better defined clustering [8]. There are various possibilities to express quantitatively this fact. If we assimilate membership to probabilities it is possible to use the Entropy Index as a quantitative measure of asymmetry. The mean value on the whole dataset of the Entropy Index is known as Partition Entropy [2] and has the following form:

$$PE = \frac{1}{N} \sum_i^c \sum_j^N \mu_{ij} \log_a \mu_{ij} \qquad (7)$$

where N is the number of points in the dataset, μ_{ij} are the membership and c is the clusters' number. Lower values indicate more asymmetric partitions.

This is a well-known index based only on memberships' values that shows an increasing trend with the number of clusters, mainly because the membership tends to spread over clusters. One way to enhance index's sensitivity to higher memberships' values and to avoid this limitation is to raise the memberships to the power p and to normalize the new values so that they sum to one. The higher the power, the higher index's sensitivity.

$$PEm_p = \frac{1}{N} \sum_i^c \sum_j^N \hat{\mu}_{ij}^p \log_a \hat{\mu}_{ij}^p \qquad (8)$$

Where $\hat{\mu}_{ij}^p$ are memberships raised to power p and normalized.

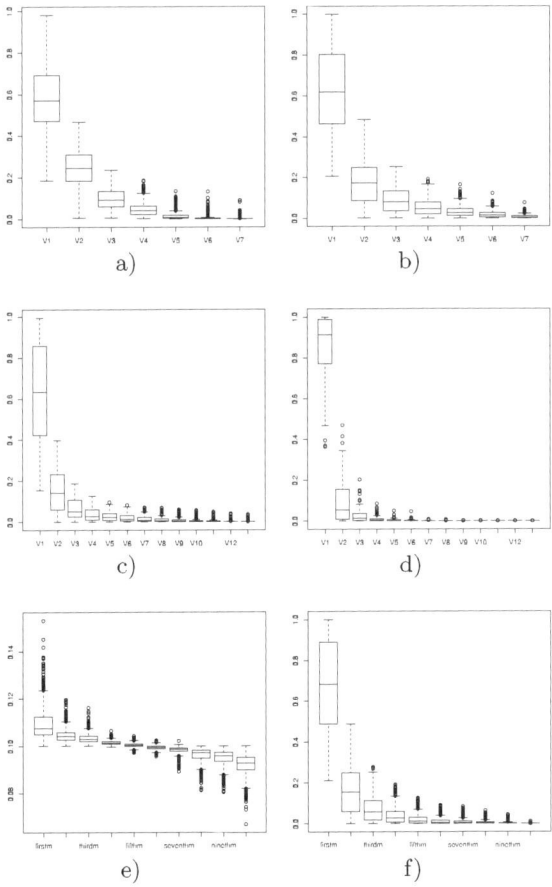

Fig. 1. *Boxplots* of ordered memberships of the three datasets analyzed. On the left data are clustered with conventional FCM and on the right data are clustered with our algorithm. Figures a) and b) are referred to the Sporulation dataset; figures c) and d) to the Rat dataset; figures e) and f) to the Yeast dataset.

3   Results

The three dataset used are:

– The Yeast dataset, downloaded from the UCI Machine Learning repository. It has 1484 rows and 10 columns of attributes that are a series of

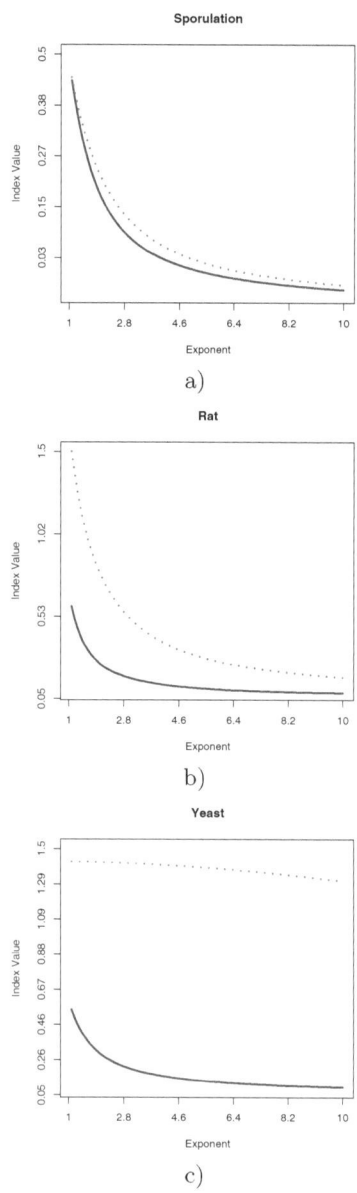

Fig. 2. Plots of PEm Validity Index for conventional fuzzy c-means (dotted line) and
for SSFC (solid line) on the three datasets analyzed, in function of power p

measurements to establish the localization site of proteins. Last column is the localization site.

- The Rat dataset is the data set of Wen, Fuhrman, Michaels, Carr, Smith, Barker and Somogyi that measures the mRNA expression levels of 112 genes during rat central nervous system development from embryonic through postnatal to adult stage (9 stages). Last column is the functional classification of genes.

- The Sporulation dataset is the Spellmann dataset of Yeast expression levels. It has 6118 rows and 86 columns. We selected only the 477 rows that we know to be Cell-cycle regulated and the 7 columns of g/r ratios.

Boxplots of sorted memberships for all dataset tested are shown in figure 1, in the case of conventional fuzzy c-means and in the case of SSFC. As we can see, in all cases our modified fuzzy c-means algorithm produces more asymmetric memberships. We can see that the ordered memberships are much more asymmetric in the case of the "learned" metric to prove a much more evident clustering structure on the data. This is particularly evident for the Yeast dataset (1 e) and f)) an it's a bit more attenuated for the Rat dataset (1 c) and d)). It is less evident in the Sporulation dataset (1 a) and b)) but however there is still an improvement.

In all cases tested we have plotted the PEm indexes' values versus the power p from 1 to 10 in 0.1 steps, comparing the two algorithms, classical fuzzy c-means and SSFS (Fig.2). As we can see the power p affects index performances and makes it much more sensible to dataset clustering structure. Specifically, in the Sporulation case, classical index PE fails in highlight the distribution improvement due to the metric learning, while starting from the power of 2 the modified index PEm reaches the task. In all tested cases, the index PEm reflects distribution improvement due to metric learning and seems to be more reliable and sensible with respect to the original index PE.

4 Conclusions and Future Work

We have shown the efficacy of using side information in unsupervised techniques. Learning a metric that respects some user-defined constraints as a preliminary step to clustering improves clearly its performance, extending utilization possibilities to more difficult tasks without substantial changes to the technique. We used three real biological dataset to validate our method and the well known fuzzy c-means algorithm. In all tested cases there is an advantage in using side information that can be quantified through a generalized version of the Partition Entropy index. Membership distribution shows a clear improvement in detecting the clustering structure for all tested cases. Future work is in joining metric learning to the clustering step and in studying more complex distance functions.

References

1. Bar-Hillel A., Hertz T., Shental N. and Weinshall D.(2005) "Learning a Mahalanobis Metric from Equivalence Constraints", *Journal of Machine Learning Research*, vol.6, pp. 937-965.
2. Bezdek J. C. (1981) *Pattern Recognition with Fuzzy Objective Function Algorithms*, Plenum Press, NY.
3. Bilenko M., Basu S., Mooney R.J. (2004) "Integrating Constraints and Metric Learning in Semi-Supervised Clustering", *Proceedings of the 21st ICML*, pp. 81-88.
4. Chang H. and Yeung D.Y. (2004) "Locally Linear Metric Adaptation for Semi-Supervised Clustering", *Proceedings of the 21st ICML*, pp. 153-160.
5. R.J.Cho, M.J.Campbell, E.A.Winzeler, L.Steinmetz, A.Conway, L.Wodicka, T.G.Wolfsberg, A.E.Gabrielian, D.Landsman, D.J.Lockhart, R.W.Davis (1998) "A Genome-Wide Transcriptional Analysis of the Mitotic Cell Cycle", *Molecular Cell*, vol. 2, pp. 65-73.
6. S.Chu, J. DeRisi, M.Eisen, J. Mulholland, D.Botstein, P.O. Brown, I.Herskowitz (1998) "The Transcriptional Program of Sporulation in Budding Yeast", *Science*, vol. 282, pp.699-705.
7. De Bie T., Momma M., Cristianini N. (2003) "Efficiently Learning the Metric With Side-Information", *Lecture Notes in Artificial Intelligence*, vol. 2842, pp.175-189.
8. Dembl D.,Kastner P. (2003) "Fuzzy C-means Method for Clustering Microarray Data", *Bioinformatics*, vol. 19, pp. 973-980.
9. M.B.Eisen, P.T.Spellman, P.O.Brown, D.Botstein (1998) "Cluster Analysis and Display of Genome-Wide Expression Patterns", *PNAS* vol. 95, pp. 14863-14868.
10. T.R. Golub, D.K. Slonim, P.Tamayo, C.Huard, M.Gaasenbeek, J.P. Mesirov, H.Coller, M.L.Loh, J.R. Downing, M.A.Caligiuri, C.D.Bloomfield, E.S.Lander (1999) "Molecular Classification of Cancer: Class Discovery and Class Prediction by Gene Expression Monitoring", *Science*, vol. 286, pp.531-537.
11. Kirkpatrick S., C.D. Gelatt Jr and M.P. Vecchi (1983),"Optimization by Simulated Annealing", *Science*, vol. 220, pp. 671- 680.
12. Shental N., Bar-Hillel A., Hertz T. and Weinshall D.(2003) "Computing Gaussian Mixture Models with EM using Equivalence Constraints", *Proceedings of Neural Information Processing Systems 2003*, vol.16.
13. Schultz M. and Joachims T. (2003) "Learning a Distance Metric From Relative Comparisons", *Proceedings of Neural Information Processing Systems 2003*, vol. 16.
14. P.T.Spellman, G.Sherlock, M.Q.Zhang, V.R.Iyer, K.Anders, M.B.Eisen, P.O.Brown, D.Botstein, B. Futcher (1998)"Comprehensive Identification of Cell Cycle-Regulated Genes of the Yeast Saccaromyces Cervesiae by Microarray Hybridization" *Molecular Biology of the Cell*, vol. 9, pp. 3273-3297.
15. Xing E. P., Ng A. Y., Jordan M. I., Russell S. (2002) "Distance Metric Learning, With Application to Clustering With Side-Information", *Advances in Neural Information Processing Systems*, vol. 15.
16. Zhengdong L., Leen T. (2004) "Semi-supervised Learning with Penalized Probabilistic Clustering" *Proceedings of Neural Information Processing Systems 2004*, vol. 17.

Comparison of Gene Identification Based on Artificial Neural Network Pre-processing with k-Means Cluster and Principal Component Analysis

Leif E. Peterson[1] and Matthew A. Coleman[2]

[1] Baylor College of Medicine, Houston, TX 77030, USA
peterson@bcm.tmc.edu
http://www.chipst2c.org
[2] Lawrence Livermore National Laboratory, Livermore, CA 94550, USA

Abstract. A combination of gene ranking, dimensional reduction, and recursive feature elimination (RFE) using a BP-MLP artificial neural network (ANN) was used to select genes for DNA microarray classification. Use of k-means cluster analysis for dimensional reduction and maximum sensitivity for RFE resulted in 64-gene models with fewer invariant and correlated features when compared with PCA and mimimum error. In conclusion, k-means cluster analysis and sensitivity may be better suited for classifying diseases for which gene expression is more strongly influenced by pathway heterogeneity.

1 Introduction

Artificial neural networks (ANNs) have been applied to DNA microarray data through several approaches. Tarca et al used ANNs to normalize cDNA microarray data and demonstrated a reduction in both intensity-dependent bias and spatial-dependent bias[1]. The agreement between regulatory motifs and functional classes of *Saccharomyces cerevisiae* genes in clusters based on Euclidean distance, correlation, and mutual information was found to be lower than ANN-derived clusters[2]. Using expression data for cardiovascular disease, Tham et al reported that an ANN approach provided promising prediction results[3]. The remaining clinical papers on ANNs focused on diagnostic classification of several types of cancer such as leukemia, lymphoma, lung cancer, prostate cancer, and various neurological malignancies [4-9].

The goal of this paper was to assess correlation and differential expression among features identified through a combination of methods involving gene ranking, dimensional reduction, and recursive feature elimination (RFE). Comparisons are provided describing the amount of between-gene correlation in 64-gene models as a function of dimensional reduction and RFE methods. Also provided is the proportion of genes among the 64-gene models with significant between-class differential expression.

I. Bloch, A. Petrosino, and A.G.B. Tettamanzi (Eds.): WILF 2005, LNAI 3849, pp. 267–276, 2006.

2 Methods

2.1 Simulated Data Set

Let the matrix $\mathbf{E}$ of gene expression profiles have dimension $G \times A$, where G is the number of genes ($g = 1, 2, \ldots, G$) and A is the number of microarrays ($a = 1, 2, \ldots, A$). A simulated data set with 400 genes and 20 arrays was generated with expression values distributed normally as $N(\mu, \sigma^2)$ as shown in Table 1. Symmetry in the simulated expression values was preserved among the 2 classes in order to prevent bias in the sensitivity for a particular class.

Table 1. Description of 400 simulated genes for 20 arrays

	Simulated genes			
	Class A (10 arrays)		Class B (10 arrays)	
# Genes	Odd arrays	Even arrays	Odd arrays	Even arrays
40	N(0,1)	N(0,1)	N(0,1)	N(0,1)
20	N(5,1)	N(5,1)	N(0,1)	N(0,1)
20	N(0,1)	N(0,1)	N(5,1)	N(5,1)
20	N(5,1)	N(5,1)	N(-5,1)	N(-5,1)
20	N(-5,1)	N(-5,1)	N(5,1)	N(5,1)
20	N(5,1)	N(0,1)	N(0,1)	N(0,1)
20	N(0,1)	N(0,1)	N(5,1)	N(0,1)
20	N(5,1)	N(-5,1)	N(0,1)	N(0,1)
20	N(0,1)	N(0,1)	N(5,1)	N(-5,1)
20	N(5,1)	N(-5,1)	N(5,1)	N(0,1)
20	N(5,1)	N(0,1)	N(5,1)	N(-5,1)
20	N(5,1)	N(-5,1)	N(5,1)	N(-5,1)
20	N(2.5,1)	N(0,1)	N(0,1)	N(0,1)
20	N(0,1)	N(0,1)	N(2.5,1)	N(0,1)
20	N(2.5,1)	N(-2.5,1)	N(0,1)	N(0,1)
20	N(0,1)	N(0,1)	N(2.5,1)	N(-2.5,1)
20	N(2.5,1)	N(-2.5,1)	N(2.5,1)	N(0,1)
20	N(2.5,1)	N(0,1)	N(2.5,1)	N(-2.5,1)
20	N(2.5,1)	N(-2.5,1)	N(2.5,1)	N(-2.5,1)

2.2 Empirical Data Sets

We used two empirical data sets available in the public domain. The first was published by Hedenfalk et al [10] on *BRCA1* and *BRCA2* mutations with 3170 genes and 15 arrays comprising 2 classes (7 arrays for *BRCA1* and 8 arrays for *BRCA2*). The second was published by Khan et al [9] on childhood small round blue-cell tumors (SRBCT) with 2308 genes and 63 arrays comprising 4 classes (23 arrays for EWS-Ewing Sarcoma, 8 arrays for BL-Burkitt lymphoma, 12 arrays for NB-neuroblastoma, and 20 arrays for RMS-rhabdomyosarcoma).

2.3 Gene Ranking

We applied non-parametric independent k-sample statistical tests and ranked genes based on their significance level. For the 2-class simulated and Hedenfalk et al data sets, we applied the Mann-Whitney test to rank genes based on significance. The Mann-Whitney test approximates the Gini diversity index commonly used for feature selection[11]. While all of the 400 simulated genes were used, we applied a cutoff criterion of $p \leq 0.2$ for the 3170 original genes in the Hedenfalk et al data set and identified 967 genes. For the 4-class Khan et al data set with 2308 original genes, we used the independent k-sample Kruskal-Wallis ANOVA test to rank genes and applied a cutoff $p \leq 0.01$. This led to 898 gene expression profiles. Tail probabilities for parametric t-tests applied to the Hedenfalk et al data resulted in 1191 genes with $p \leq 0.20$ and 920 genes for F-tests applied to the Khan et al data for which $p \leq 0.01$. Parametric test results were not used for gene ranking, but stored for bookkeeping.

2.4   Dimension Reduction with K-Means Cluster Analysis

In addition to gene ranking and use of p-value cutoffs, we applied k-means clustering and principal components analysis (PCA) for dimension reduction in order to minimize effects from the curse of dimensionality[12]. For k-means clustering, let k $(k=1,2,...,K)$ be the the kth cluster of a clustering, and K the total number of clusters. The optimal value of K is determined by cycling through values of $K = 2, 3, ..., \sqrt{G}$. This is performed as follows. For K clusters, the total *within-cluster sum-of-squares* is

$$SSW(K) = \sum_{k=1}^{K} \sum_{g=1}^{G_k} \|\mathbf{x}_{gk} - \mathbf{m}_k\|, \tag{1}$$

where $\mathbf{x}_{gk}$ is the row vector containing expression values for gene g in cluster k over the A arrays and $\mathbf{m}_k$ is the mean vector for G_k genes in cluster k, and $\|.\|$ is the Euclidean distance. For the same K clusters, the smallest *between-cluster distance* is

$$d(K) = \min_{1 \leq k < l \leq K} \|\mathbf{m}_k - \mathbf{m}_l\|, \tag{2}$$

and the score function for a set of K clusters is

$$S_K = \frac{d(K)}{SSW(K)}. \tag{3}$$

After evaluating the score function S_K for values of K ranging from 2 to $\sqrt{G}$, the optimal value of K is

$$K_{opt} = \max_{2 \leq K \leq \sqrt{G}} \{S_K\}. \tag{4}$$

Once K_{opt} is determined, the k-means algorithm is rerun using K_{opt} clusters. K-means clustering results in a $A \times K$ $\mathbf{M}$ matrix of *k-means centers*. For each gene, determine the k-means score which maps the gene back to the center k as

$$z_{gk} = \frac{\|\mathbf{x}_g - \mathbf{m}_k\| - \mu_k}{\sigma_k} \qquad k = 1, 2, \ldots, K, \tag{5}$$

where $\mathbf{x}_g$ is the standardized expression vector for gene g, $\mathbf{m}_k$ is the mean vector for center k, $\|\mathbf{x}_g - \mathbf{m}_k\|$ is the Euclidean distance between expression for gene g and center k, and μ_k and σ_k are the average and standard deviation of distances $\|\mathbf{x}_g - \mathbf{m}_k\|$ between all genes and center k. This was repeated for each cluster center to yield a $G \times K$ $\mathbf{Z}$ matrix of *k-means scores*. Since the scores are standard normal distributed, the bulk of scores will be centered around zero and genes with the smallest or greatest distance from the cluster center will yield greater scores. For the simulated data set, we identified 12 centers (i.e., K_{opt}=12). For the Hedenfalk et al breast cancer data, we identified 30 centers, whereas for the Khan et al SRBCT data, we identified 28 centers.

2.5 Dimension Reduction with PCA

During PCA, the top 10 eigenvalues were extracted from the $G \times G$ correlation matrix $\mathbf{R}$. The array by component $(A \times P)$ $\mathbf{F}$ matrix of *PC scores* is determined with the matrix of standardized expression values (standardized with mean and s.d. over the genes) as follows

$$\underset{A \times P}{\mathbf{F}} = \underset{A \times G}{\mathbf{Z}} \underset{G \times P}{\mathbf{W}}. \tag{6}$$

In order to map genes back to the arrays via the principal components, the matrix of *PC score coefficients* was obtained using the matrix operation

$$\underset{G \times P}{\mathbf{W}} = \underset{G \times P}{\mathbf{L}} \underset{P \times P}{(\mathbf{L}'\mathbf{L})^{-1}}, \tag{7}$$

where $\mathbf{L}$ is the loading matrix reflecting the correlation between each gene expression profile and the extracted PC scores. The top 10 PC's were always extracted from the gene by gene correlation matrix and used for training. Orthogonal rotations were not performed.

2.6 ANN Training During Recursive Feature Elimination

Recursive feature elimination (RFE) was based on a BP-MLP ANN with one hidden layer. The ANNChip computer program (http://www.chipst2c.org) was used for RFE and included 8-fold cross-validation and leave-one-out testing where arrays were randomly assigned to 8 validation groups. Each validation group was selected singly resulting in a single ANN model in which the remaining 7/8 of arrays were used for training. During leave-one-out testing, array 1 was left out of models 1-8, array 2 left out of models 9-16, etc., so that each array was left out during 8 models. Table 2 summarizes the input data sets with the number of samples and genes, the reduction methods and derived matrices used to feed the ANN during RFE, and the total number of models used based on 8-fold cross-validation with leave-one-out testing.

Selection of Genes Based on Maximum Sensitivity. In order to gauge the influence of each gene on the classification, target outputs $\hat{t}_c^g$ for each gene were calculated during the last sweep of every model using the last known weights

Table 2. ANN training during recursive feature elimination (RFE) and ANN model summary for 8-fold cross-validation with leave-one-out testing

Data	Reduction method	Samples	Genes	Training matrix[a]	Network size	ANN models[b]
Simulated	k-means	20	400	$\mathbf{M}_{20 \times 12}$	12-5-2	160
	PCA	20	400	$\mathbf{F}_{20 \times 10}$	10-4-2	160
Hedenfalk et al	k-means	15	967	$\mathbf{M}_{15 \times 30}$	30-12-2	120
	PCA	15	967	$\mathbf{F}_{15 \times 10}$	10-4-2	120
Khan et al	k-means	63	898	$\mathbf{M}_{63 \times 28}$	28-11-4	504
	PCA	63	898	$\mathbf{F}_{63 \times 10}$	10-4-2	504

[a] Matrices from reduction results were used for training the ANN during RFE.
[b] Number of models equal to 8 validation groups times number of samples.

and setting the input nodes x_i equal to either the $1 \times K$ row vector of k-means scores $\mathbf{z}_g$ for each gene or the $1 \times P$ row vector $\mathbf{w}_g$ of PC score coefficients for each gene. It warrants noting that the ANN was not retrained here, but rather gene-specific values of $\hat{t}_c^g$ were determined by applying the last known weights to gene-specific row vectors of $\mathbf{Z}$ or $\mathbf{W}$, which map the genes back to the original $\mathbf{M}$ and $\mathbf{F}$ matrices used for training. The average gene-class-specific sensitivity of each gene was then determined as

$$S_c^g = \frac{1}{n} \sum_{i=1}^{n} \frac{\partial \hat{t}_c^g}{\partial x_i} \tag{8}$$

where g is the gene, c is the class, n is the number of input nodes based on $n = K$ and $\mathbf{x} = \mathbf{z}_g$ if the ANN was trained with k-means centers based on $\mathbf{M}$, or $n = P$ and $\mathbf{x} = \mathbf{w}_g$ if the ANN was trained with PC scores based on $\mathbf{F}$. The partial derivative $\partial \hat{t}_c^g / \partial x_i$ is determined via the chain rule, by first differentiating $\hat{t}_c^g$ w.r.t. hidden layer outputs, v_j, and then input row values, x_i, given by

$$\begin{aligned}
\frac{\partial \hat{t}_c^g}{\partial x_i} &= \sum_j \frac{\partial \hat{t}_c^g}{\partial v_j} \frac{\partial v_j}{\partial x_i} \\
&= \sum_j \frac{d\hat{t}_c^g}{dy_c} \frac{\partial y_c}{\partial v_j} \frac{dv_j}{du_j} \frac{\partial u_j}{\partial x_i} \\
&= \sum_j \left(\frac{\exp(y_c) \left(\sum_l \exp(y_l) - \exp(y_c) \right)}{\left(\sum_l \exp(y_l) \right)^2} w_{jc}^{ho} \frac{e^{-u_j}}{(1+e^{-u_j})^2} w_{ij}^{ih} \right)
\end{aligned} \tag{9}$$

Class-specific sensitivities for each gene were summed over all models and then sorted in descending order. Genes at the top of the sort were selected as the best predictors based on gene-class-specific sensitivity. The list of genes was divided equally into genes with the greatest sensitivity for discriminating each class.

For example, for a list of 8 genes and 2 classes, the 4 genes with the greatest sensitivity for discriminating class 1 were used along with the 4 genes with the greatest sensitivity for discriminating class 2.

Selection of Genes Based on Minimum Error. In addition to RFE based on sensitivity, we also calculated the gene-class-specific mean square error during the last sweep, $E_c^g = 0.5(\hat{t}_c^g - t_c)^2$, using the recomputed values of $\hat{t}_c^g$ described above. Analogously, we derived lists of genes for which each class was represented equally by genes having the lowest gene-class-specific MSE.

2.7 Generating Lists of Selected Genes

A modular approach was employed for generating the list of genes identified during RFE. Lists were divided uniformly into genes that best discriminated each outcome class, depending on whether the selection criterion was minimum gene-class-specific MSE or maximum gene-class-specific sensitivity. The total number of genes in a list was based on powers of 2 multiplied by the number of classes, such that the list was uniformly loaded with genes that best discriminated each class.

2.8 ANN Training with Selected Gene Expression Profiles

After recursive feature identification, we trained the ANN models with the actual standardized values of expression for the identified genes. For example, for 64 genes (features) and 2 outcome classes a 64-26-2 network was employed, where the number of hidden nodes is equal to 40% of the number of input

Table 3. ANN training input using standardized expression profiles of genes selected during recursive feature elimination (RFE). 8-fold cross-validation with leave-one-out testing used.

Data	Reduction method	RFE method	Samples	Genes(n)	Training matrix[a]	Network size	ANN models[b]
Simulated	k-means	$\min(E_c^g)$	20	2,4,8,16,32,64	$\mathbf{E}_{20 \times n}$	n-0.4n-2	160
		$\max(S_c^g)$	20	2,4,8,16,32,64	$\mathbf{E}_{20 \times n}$	n-0.4n-2	160
	PCA	$\min(E_c^g)$	20	2,4,8,16,32,64	$\mathbf{E}_{20 \times n}$	n-0.4n-2	160
		$\max(S_c^g)$	20	2,4,8,16,32,64	$\mathbf{E}_{20 \times n}$	n-0.4n-2	160
Hedenfalk et al	k-means	$\min(E_c^g)$	15	2,4,8,16,32,64	$\mathbf{E}_{15 \times n}$	n-0.4n-2	120
		$\max(S_c^g)$	15	2,4,8,16,32,64	$\mathbf{E}_{15 \times n}$	n-0.4n-2	120
	PCA	$\min(E_c^g)$	15	2,4,8,16,32,64	$\mathbf{E}_{15 \times n}$	n-0.4n-2	120
		$\max(S_c^g)$	15	2,4,8,16,32,64	$\mathbf{E}_{15 \times n}$	n-0.4n-2	120
Khan et al	k-means	$\min(E_c^g)$	63	4,8,16,32,64	$\mathbf{E}_{63 \times n}$	n-0.4n-4	504
		$\max(S_c^g)$	63	4,8,16,32,64	$\mathbf{E}_{63 \times n}$	n-0.4n-4	504
	PCA	$\min(E_c^g)$	63	4,8,16,32,64	$\mathbf{E}_{63 \times n}$	n-0.4n-4	504
		$\max(S_c^g)$	63	4,8,16,32,64	$\mathbf{E}_{63 \times n}$	n-0.4n-4	504

[a] Training matrix of standardized gene expression $\mathbf{E}$ based on genes sorted by RFE method.
[b] Number of models equal to 8 validation groups times number of samples.

nodes. During runs with actual gene expression profiles, we assessed accuracy, the proportion of between-gene correlation coefficients that were significant ($p \leq 0.01$ and $p \leq 0.05$), and the proportion of genes in the list that had significant parametric test statistics ($p \leq 0.05$) during the original gene ranking calculations. Table 3 lists the gene expression data used for training the ANN.

3   Results and Discussion

The choice for using non-parametric Mann-Whitney U and Kruskal-Wallis tests for gene ranking should have a minimal effect on the observed results. Li et al assessed the effect of 8 different feature selection statistics on SVM outcome and determined that for more than 150 microarray-based genes the variation in performance was small[13]. Figure 1 illustrates the average and standard deviation of sensitivity for different types of simulated expression profiles based on dimensional reduction by k-means clustering and PCA. K-means resulted in near-zero values of sensitivity for genes with lower within-class variation of expression, such as N(5;1)|N(0;1), N(0;1)|N(5;1), N(5;1)|N(-5;1), and N(-5;1)|N(5;1). However, for PCA genes N(0;1)|N(5,1) and N(-5;1)|N(5;1) showed large negative values of sensitivity. For these genes, PCA inflated sensitivity that was not detected by k-means. One can notice in Figure 1 that, for the remainder of genes with larger differential entropy, k-means resulted in greater sensitivity when compared with PCA. Another disadvantage of PCA is that orthogonal projections may have nothing to do with class discrimination. Moreover, the bulk of data including noise and outlier patterns that often load on the lower components (>3)

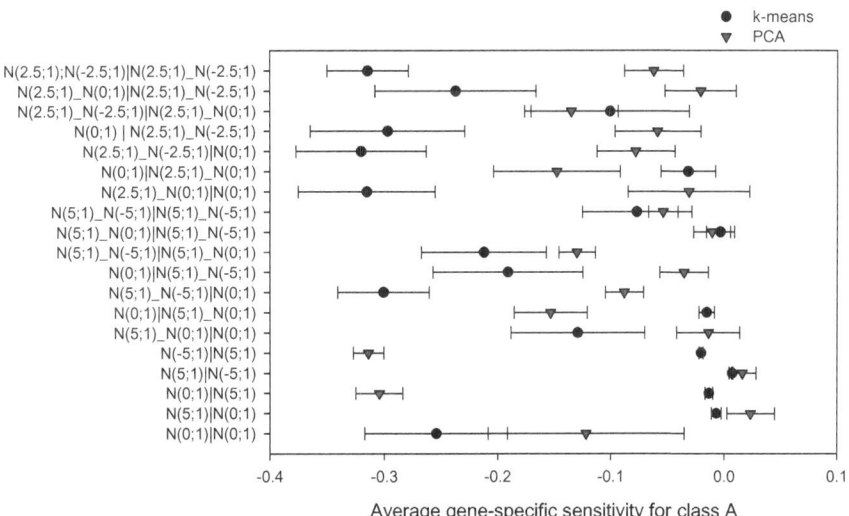

Fig. 1. Average and standard deviation of sensitivity, S_c^g, for class A

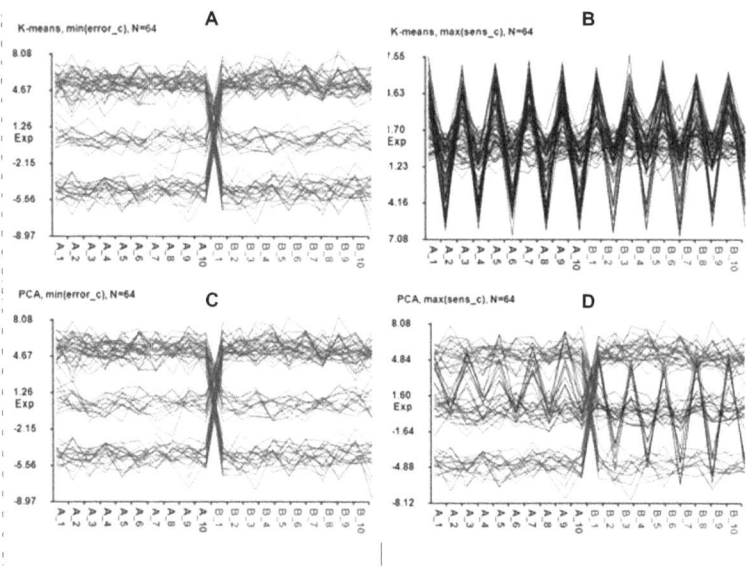

Fig. 2. Expression profiles of top 64 simulated genes identified after gene reduction with k-means or PCA followed by recursive feature elimination (RFE). (A) Dimension reduction with k-means cluster analysis and RFE based on minimum MSE. (B) K-means and maximum sensitivity. (C) PCA and minimum MSE. (D) PCA and maximum sensitivity. Red lines denote genes with significantly different ($p \leq 0.05$) expression levels between class A and B. Blue lines represent genes for which differential expression is not significantly different at the 0.05 level.

contribute less to discerning the classes. The advantage of PCA dimensional reduction is that more genes with lower within-class variation will be identified, driving up the accuracy.

Table 4 shows that, at the 0.01 level of significance, RFE based on k-means dimensional reduction along with maximum sensitivity resulted in the least amount of between-gene correlation for all 3 data sets and the least proportion of significantly differentially expressed genes for the 2-class data sets. In Figure 2B, one can visualize for simulated expression profiles that for k-means and sensitivity only 17% (0.172 from Table 4, row 2) of genes in a 64-gene model had significantly different expression. On the other hand, PCA with sensitivity (Figure 2D) resulted in 81% of the genes in a 64-gene model that had significantly different expression. The smaller proportion of significant between-gene correlation of expression due to k-means and sensitivity can also be noticed in Figure 3 for all data sets, and in particular for the Hedenfalk et al breast cancer data for which the proportion of significant ($p \leq 0.01$) between-gene correlation coefficients was 0.138.

Genes that are strongly differentially expressed and correlated may be co-regulated by shared upstream signaling molecules. A classifier based on such genes may have greater misclassification when pathway heterogeneity is important for classification. We have shown that, for the data considered, an ANN

Table 4. Recursive feature elimination (RFE) results for 64-gene ANN models including classification accuracy, proportion of significant positive between-gene correlation, and proportion of parametric tests significant among the 64 genes during original gene ranking. ANNs were fed with actual standardized expression values for the 64 genes after they were identified with RFE.

Data set	Data reduction	RFE method	Accuracy	$r > 0$ $p \leq 0.01^a$	$r > 0$ $p \leq 0.05$	Signif genes[b]
Simulated	K-means	$\min(E_c^g)$	1.000	0.492	0.492	1.000
		$\max(S_c^g)$	0.976	0.328	0.467	0.172
	PCA	$\min(E_c^g)$	1.000	0.492	0.492	1.000
		$\max(S_c^g)$	1.000	0.341	0.371	0.813
Hedenfalk et al	K-means	$\min(E_c^g)$	1.000	0.322	0.404	0.969
		$\max(S_c^g)$	0.962	0.138	0.266	0.515
	PCA	$\min(E_c^g)$	0.967	0.197	0.325	0.703
		$\max(S_c^g)$	0.900	0.177	0.284	0.641
Khan et al	K-means	$\min(E_c^g)$	0.963	0.235	0.293	1.000
		$\max(S_c^g)$	0.960	0.200	0.270	1.000
	PCA	$\min(E_c^g)$	0.996	0.239	0.295	1.000
		$\max(S_c^g)$	0.998	0.301	0.359	1.000

[a] Proportion of 2016 between-gene correlation coefficients (i.e., $n(n-1)/2$) for 64 gene expression profiles with $p \leq 0.01$.
[b] Proportion of 64 genes with significant parametric test (t-test or F-test) during original gene ranking.

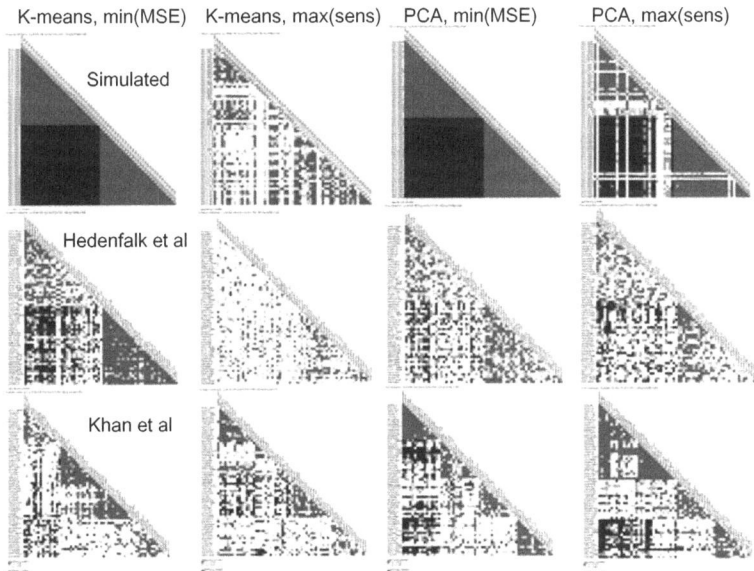

Fig. 3. Plot of significant between-gene correlation for 64-gene models. ANNs trained with standardized expression profiles for 64 genes identified during recursive feature elimination (RFE). Red denotes significant ($p \leq 0.01$) positive correlation, whereas blue signifies significant negative correlation.

classifier based on k-means dimensional reduction and sensitivity for RFE can result in accuracy levels exceeding 90% with fewer invariant and correlated features. K-means cluster analysis coupled with sensitivity for RFE may increase detection of patients with pathway heterogeneity, which may not be tackled as well by RFE with minimum error or dimensional reduction with PCA.

References

1. Tarca, A.L., Cooke, J.E., Mackay, J.: A robust neural networks approach for spatial and intensity dependent normalization of cDNA microarray data. Bioinformatics **21** (2005) 2674-2683
2. Sawa, T., Ohno-Machado, L.: A neural network-based similarity index for clustering DNA microarray data. Comput. Biol. Med. **33** (2003) 1-15
3. Tham, C.K., Heng, C.K., Chin, W.C.: Predicting risk of coronary artery disease from DNA microarray-based genotyping using neural networks and other statistical analysis tool. J. Bioinform. Comput. Biol. **1** (2003) 521-539
4. Ando, T., Suguro, M., Hanai, T., et al.: Fuzzy neural network applied to gene expression profiling for predicting the prognosis of diffuse large B-cell lymphoma. Jpn. J. Cancer Res. **93** (2002) 1207-1212
5. Bicciato, S., Pandin, M., Didone, G., Di Bello, C.: Pattern identification and classification in gene expression data using an autoassociative neural network model. Biotechnol. Bioeng. **81** (2003) 594-606
6. O'Neill, M.C., Song, L.: Neural network analysis of lymphoma microarray data: prognosis and diagnosis near-perfect. BMC Bioinformatics **4** (2003) 13
7. Liu, B., Cui, Q., Jiang, T., Ma, S.: A combinational feature selection and ensemble neural network method for classification of gene expression data. BMC Bioinformatics. **5** (2004) 136
8. Wei, J.S., Greer, B.T., Westermann, F., et al.: Prediction of clinical outcome using gene expression profiling and artificial neural networks for patients with neuroblastoma. Cancer Res. **64** (2004) 6883-6891
9. Khan, J., Wei, J.S., Ringner, M., et al.: Classification and diagnostic prediction of cancers using gene expression profiling and artificial neural networks. Nature Med. **7** (2001) 673-679
10. Hedenfalk, I., Duggan, D., Chen, Y., et al.: Gene-expression profiles in hereditary breast cancer. N. Engl. J. Med. **344** (2001) 539-548
11. Hand, D.J., Mannila, H., Smyth, P.: Principles of Data Mining (Adaptive Computation and Machine Learning). Cambridge(MA): MIT Press, 2001
12. Hastie, T., Tibshirani, R., Friedman, J.: The Elements of Statistical Learning: Data Mining, Inference, and Prediction. New York(NY): Springer, 2005
13. Li, T., Zhang, C., Ogihara, M.: A comparative study of feature selection and multiclass classification methods for tissue classification based on gene expression. Bioinformatics **20** (2004) 2429-2437

Biological Specifications for a Synthetic Gene Expression Data Generation Model

Francesca Ruffino[1], Marco Muselli[2], and Giorgio Valentini[1]

[1] DSI - Dipartimento di Scienze dell'Informazione,
Università degli Studi di Milano, 20135 Milano, Italy
{ruffino, valentini}@dsi.unimi.it
[2] IEIIT, Istituto di Elettronica, Ingegneria dell'Informazione,
e delle Telecomunicazioni, Consiglio Nazionale delle Ricerche, Genova, Italy
muselli@ice.ge.cnr.it

Abstract. An open problem in gene expression data analysis is the evaluation of the performance of gene selection methods applied to discover biologically relevant sets of genes. The problem is difficult, as the entire set of genes involved in specific biological processes is usually unknown or only partially known, making unfeasible a correct comparison between different gene selection methods. The natural solution to this problem consists in developing an artificial model to generate gene expression data, in order to know in advance the set of biologically relevant genes. The models proposed in the literature, even if useful for a preliminary evaluation of gene selection methods, did not explicitly consider the biological characteristics of gene expression data. The main aim of this work is to individuate the main biological characteristics that need to be considered to design a model for validating gene selection methods based on the analysis of DNA microarray data.

1 Introduction

Analysis of gene expression data may be performed at different levels, ranging from the analysis of differential expression of genes, to unsupervised and supervised analysis of sets of genes and tissues [1].

An important related problem is to determine the subset of genes involved in the biological process under examination. Such problem is generally referred to as gene selection and several statistic and machine learning techniques have been proposed in literature to face with it [2–4].

Unfortunately, the entire set of genes involved in specific biological processes is usually unknown or only partially known, and as a consequence the evaluation of the real effectiveness of gene selection methods is very difficult and in many cases unfeasible.

Several models have been proposed to simulate gene expression data, in order to make available synthetic gene expression data for classification, clustering and gene selection problems [5, 6]. Even if these models may be in principle helpful to test gene selection methods, their main limitation consists in a drastic

I. Bloch, A. Petrosino, and A.G.B. Tettamanzi (Eds.): WILF 2005, LNAI 3849, pp. 277–283, 2006.

modelling simplification, without sufficiently taking into account the biological characteristics of gene expression data.

In this paper we address the problem of the analysis of the specifications needed to properly model the biological characteristics of gene expression data. In particular, the main concerns of this work are the relationships between the biological and modelling issues involved in the design of a flexible tool to generate synthetic gene expression data. To this end we performed an analysis of the gene expression literature to individuate structural commonalities in gene expression data. The design and the implementation of an artificial model will allow us to properly evaluate the performance of clustering and gene selection methods, as all subsets of the simulated genes involved in simulated biological processes will be known in advance.

2   Profiles and Expression Signatures

The main goal of gene selection methods is to find sets of genes significantly related to a specific functional status (e.g. diseased vs. healthy). In the biomolecular literature sets of biologically relevant and differentially expressed genes are named *expression signatures* [7–11].

To our knowledge the term *expression signature* has been introduced by Alizadeh et al. [7], to characterize gene expression patterns found by gene expression profiling. More precisely this term refers to a group of genes coordinately expressed in a given set of specimens and in a specific physiological or pathophysiological condition.

The correlation between the mRNA levels of the genes is due to the underlying regulatory system, by which the same set of transcription factors and binding sites may be directly or indirectly shared by the genes belonging to the same expression signature. Hence a gene expression signature indicates a cluster of coordinately expressed genes, whose coordination reveals the fact that they participate to the same biological process (and hence they are controlled by the same set of regulation factors); indeed they are usually named by either the cell type in which their component genes are expressed, or by the biological process in which their component genes are known to function.

From this standpoint the overall *expression profile* of a patient can be interpreted as a collection of gene expression signatures that reveal different biological features of the analyzed sample [7].

2.1   Gene Expression Signatures in Human Diseases

Expression signatures has been mainly discovered and analyzed in gene expression profiles of diseases. For instance, the expression profiling of B-cell malignancies through hierarchical clustering, revealed expression signatures related to cell-proliferation, to lymph-nodes, T-cells, Germinal Centre B-cells (GCB) and others [7].

Independent Component Analysis performed on gene expression data from ovarian cancer tissues found gene expression signatures representing potential pathophysiological processes in ovarian tissue samples [8]. Expression profiling of rhabdomyosarcoma (RMS), the most common soft tissue sarcoma in children, identified two signatures associated with metastatic RMS, responsible for most of the fatal outcome of this disease [11], while two way hierarchical clustering analysis identified several expression signatures expressed in different types of bladder carcinoma [9].

Expression signatures have been also identified in species other than humans and in contexts not related to tumoral differentiation. For instance comparative functional genomics based on shared patterns of regulations across orthologous genes identified shared expression signatures of aging in orthologous genes of *D. melanogaster* and *C. elegans* [10].

Summarizing, *expression profiles* and *expression signatures* seem to be well-established biological structures that characterize gene expression data.

2.2   Characteristics of Gene Expression Signatures

In this section we discuss the main characteristics of gene expression signatures.

Differential Expression and Co-expression. Differential expression analysis of single genes, even if it may be useful to identify specific genes involved in biological processes [12], cannot capture the complexity of tightly regulated processes, crucial for the proper functioning of a cell.

Correlations between gene expression levels have been observed [13, 7], reflecting the fact that in most biological processes genes are co-regulated. As recently observed, not all changes in co-regulation are manifested by up or down regulation of individual genes, and we need to explicitly consider interactions between genes to discover patterns in the data [14].

Hence, we need sets of co-regulated genes, that is expression signatures, to reveal functional relationships between genes.

Gene Expression Signatures as a Whole Rather Than Single Genes Contain Predictive Information. Many times is the signature taken as a whole that seems to contain predictive information for a biologically meaningful identification of tissue samples. For instance, it was found an expression signature of 8 upergulated and 9 downregulated genes associated with metastasis in different types of adenocarcinoma: none of these genes represents a marker, but it is the signature as a whole that represents a "collective marker" of tumor metastasis [15].

In other works [15, 14] it has been shown that in some cases relevant differences are subtle at the level of individual genes but coordinate in gene expression groups.

Genes May Belong to Different Gene Expression Signatures at the Same Time. Many genes may be involved in a number of distinct behaviours, depending on the specific conditions of the tissue. From this standpoint they may belong to different expression signatures [16]. Indeed each gene may be influenced by several transcription factors, each of which influences several genes [8]. Moreover many underlying conditions in a given sample may concur to define a gene expression signature (e.g. tumorigenesis, angiogenesis, apoptosis) [17].

Expression Signatures May be Independent of Clinical Parameters.
An expression signature of 153 genes can be used to correctly classify hepatocellular carcinoma (HCC) intra-hepatic metastasis from metastatic-free HCC [18]. This expression signature, that embeds high predictive information, has been shown to be independent of tumor size, tumor encapsulation and patient age, and also very similar to that of their corresponding metastases.

Several other works showed that a bio-molecular characterization of tumours can discover different subtypes of malignancies, not detectable with traditional morphological and histopathological features (see e.g. [7, 2]).

Different Gene Expression Profiles May Share Signatures and May Differ Only for Few Signatures. It has been shown that gene expression signatures may be shared and partially expressed in different gene expression profiles [7, 15, 18].

For instance, it has been shown that Diffuse Large B-Cell Lymphoma (DLBCL) subgroups (GCB-like and activated B-like DLBCL) share most of the expression signatures but they differ mainly for two signatures (GCB and activated B-cell signatures) partially expressed respectively in germinal centre B-cell and activated peripheral blood B cell [7].

Moreover, hierarchical clustering, in the space of a 128 genes signature of metastatic adenocarcinoma nodules of diverse origin, showed two clusters of primary tumors that were highly correlated with metastatic ones: this fact, together with a differential overall survival in primary adenocarcinoma tumors showed that this gene expression signature is present in subpopulation of primary tumors [15].

Hence gene expression profiles of functionally different tissues may share expression signature and differ only for a subset of expression signatures. These expression signatures may be also partially expressed (that is, not all the genes belonging to the expression signature are over-expressed or under-expressed), reflecting functional alterations in diseased patients.

3   Biological and Modelling Issues

In light of the characteristics of gene expression signatures (Sect. 2.2), in this section we discuss the relationships between the biological and modelling issues we need to consider to design an artificial model for gene expression data synthesis. Schematically, we identified the following main items:

1. Expression profiles may be characterized as a set of gene expression signatures. A set of gene expression signatures defines a *functional group* of samples. The model should allow us to define expression profiles in terms of expression signatures, with a large flexibility with respect to the number and gene composition of the synthetic expression signatures.

2. Expression signatures are interpreted in the literature as a set of coexpressed genes. These genes may be overexpressed and underexpressed with respect to the other genes and with respect to a particular condition. Accordingly, in the model, each expression signature should be defined as a set of overexpressed or underexpressed genes, that is genes with gene expression levels above or below a given threshold. The model should define a signature *active* if its genes are coordinately over(under)expressed.

3. Expression signatures may be defined either by the overall available knowledge about bio-molecular processes (e.g. by Gene Ontology categories) or may be discovered through statistical and machine learning methods. The model should permit to define arbitrary signatures, in order to allow us a large range of applications in different biological contexts.

4. Genes may belong to different signatures at the same time. As a consequence the model should allow us to assign the same gene to different signatures.

5. The model should permit to select from few few units to few hundreds of genes for each gene expression signatures, as the number of genes within a signature usually vary within this range.

6. Apart from technical variation (that in principle should be detected and canceled by proper design and implementation of bio-technological experiments and suitable pre-processing procedures [19]), gene expression is biologically variable also within functional classes (conditions) [20]. The model should reproduce the variation of gene expression data. Variation of single genes may be simulated sampling from a predefined distribution. Our preliminary analysis of gene expression data showed that gene expression values are close to be normally distributed, but it would be useful to analyze a larger number of gene expression data to properly evaluate this item.

7. Not always expression signatures show large variations of gene expression levels: some signatures may present modest but coordinate variation. The model should be sufficiently flexible to allow small variations of coexpressed genes, and to this end it should include tunable parameters of the gene distributions.

8. Not all the genes within a signature may be expressed in all samples. Moreover gene expression variation between individuals may introduce variation into expression signatures. The model should permit to introduce flexibility in the number of genes that can be underexpressed or overexpressed, as well as to introduce individual variability within a functional group.

9. Different expression profiles may differ only for few signatures, that is different functional groups may share the same (or very similar) expression signatures. The model should allow to define an expression profile as a set of signatures and to define other functional groups in terms of subsets of previously defined signatures, eventually modifying or adding new signatures.

10. Some signatures may be only partially expressed within a particular expression profile. The model should be sufficiently flexible to allow us to define an expression profile in several ways: (a) a set of active signatures; (b) a set of randomly active signatures; (c) a set of randomly active signatures with a set of "mandatory" active signatures.

4   Conclusions

In this paper we analyzed the biological issues underlying the modelling of an artificial system for simulating gene expression data.

We identified the expression signatures as a major common biological structure in gene expression data and we provided the biological specifications to develop an artificial model for gene expression data synthesis.

The next step of this works consists in developing and implementing a biologically motivated gene expression data generation model, to properly evaluate the performance of gene selection methods.

References

1. Baldi, P., Hatfield, G.: DNA Microarrays and Gene Expression. Cambridge University Press, Cambridge, UK (2002)
2. Golub, T., et al.: Molecular Classification of Cancer: Class Discovery and Class Prediction by Gene Expression Monitoring. Science **286** (1999) 531–537
3. Guyon, I., Weston, J., Barnhill, S., Vapnik, V.: Gene Selection for Cancer Classification using Support Vector Machines. Machine Learning **46** (2002) 389–422
4. Muselli, M.: Gene selection through Switched Neural Networks. In: NETTAB-2003, Workshop on Bioinformatics for Microarrays, Bologna, Italy (2003)
5. Weston, J. et al.: Use of the zero-norm with linear models and kernels methods. Journal of Machine Learning Research **3** (2003) 1439–1461
6. Dudoit, S., Fridlyand, J.: Bagging to improve the accuracy of a clustering procedure. Bioinformatics **19** (2003) 1090–1099
7. Alizadeh, A. et al.: Distinct types of diffuse large B-cell lymphoma identified by gene expression profiling. Nature **403** (2000) 503–511
8. Martoglio, A., Miskin, J., Smith, S., MacKay, D.: A decomposition model to track gene expression signatures: preview on observer-independent classification of ovarian cancer. Bioinformatics **18** (2002) 1617–1624
9. Dyrskjøt, L. et al.: Identifying distinct classes of bladder carcinoma using microarrays. Nature Genetics **33** (2003) 90–96
10. McCarroll, S. et al.: Comparing genomic expression patterns across species identifies shared transcriptional profile in aging. Nature Genetics **36** (2004) 197–204
11. Yu, Y. et al.: Expression profiling identifies the cytoskeletal organizer ezrin and the developmental homoprotein Six-1 as key metastatic regulators. Nature Medicine **10** (2004) 175–181
12. Cui, X., Churchill, G.: Statistical tests for differential expression in cDNA microarray experiments. Genome Biology **4** (2003)
13. Eisen, M., Spellman, P., Brown, P., Botstein, D.: Cluster analysis and display of genome-wide expression patterns. PNAS **95** (1998) 14863–14868
14. Kotska, D., Spang, R.: Finding disease specific alterations in the co–expression of genes. Bioinformatics **20** (2004) i194–i199
15. Ramaswamy, S., Ross, K., Lander, E., Golub, T.: A molecular signature of metastasis in primary solid tumors. Nature Genetics **33** (2003) 49–54
16. Gasch, P., Eisen, M.: Exploring the conditional regulation of yeast gene expression through fuzzy k-means clustering. Genome Biology **3** (2002)
17. Ihmels, J., Bergmann, S., Barkai, N.: Defining transcription modules using large-scale gene expression data. Bioinformatics (2004)

18. Ye, Q. et al.: Predicting hepatitis b virus-positive metastatic hepatocellular carcinomas using gene expression profiling and supervised machine learning. Nature Medicine **9** (2003) 416–423
19. Chen, J., et al.: Analysis of variance components in gene expression data. Bioinformatics **20** (2004) 1436–1446
20. Cheung, V., et al.: Natural variation in human gene expression assessed in lymphoblastoid cells. Nature Genetics **33** (2003) 422–425

Semisupervised Profiling of Gene Expressions and Clinical Data

Silvano Paoli, Giuseppe Jurman, Davide Albanese,
Stefano Merler, and Cesare Furlanello

ITC-irst - Trento, Italy
{silpaoli, jurman, albanese, merler, furlan}@itc.it
http://biodcv.itc.it

Abstract. We present an application of BioDCV, a computational environment for semisupervised profiling with Support Vector Machines, aimed at detecting outliers and deriving informative subtypes of patients with respect to pathological features. First, a sample-tracking curve is extracted for each sample as a by-product of the profiling process. The curves are then clustered according to a distance derived from Dynamic Time Warping. The procedure allows identification of noisy cases, whose removal is shown to improve predictive accuracy and the stability of derived gene profiles. After removal of outliers, the semisupervised process is repeated and subgroups of patients are specified. The procedure is demonstrated through the analysis of a liver cancer dataset of 213 samples described by 1993 genes and by pathological features.

Keywords: statistical learning, semisupervised classification, feature selection, Support Vector Machines, functional genomics, DNA microarray.

1 Introduction

BioDCV is a software set-up for the predictive molecular profiling of gene expression data. It implements complete validation schemes on distributed computing resources such as clusters and virtual GRID facilities [1]. Complete validation is a methodology for correctly assessing predictive accuracy in gene expression studies. It requires intensive resampling and replication of the classification processes in order to control for selection bias [2].

Here we apply the BioDCV system to liver cancer profiling, considering a relatively large dataset endowed with a description of some pathological features [3]. This study is based on the method recently introduced in [4] for semisupervised pattern discovery from functional genomics data. Given a basic classification task and a complete validation scheme, we showed that subtyping and outlier detection may be derived from a sample-by-sample analysis of errors at different feature sets and for different resamplings.

In the case of the liver cancer dataset [3], the basic task was the discrimination of patients from control cases. Answering this biological question is a relatively easy problem on this dataset since estimated predictive error is close to 3%.

I. Bloch, A. Petrosino, and A.G.B. Tettamanzi (Eds.): WILF 2005, LNAI 3849, pp. 284–289, 2006.
© Springer-Verlag Berlin Heidelberg 2006

Questions such as subtyping for response to treatment are typically much more difficult to answer. Semisupervised learning has been proposed in particular for predicting survival [5]. The dataset of the Sese study includes more than 100 positive samples of different age, sex and previous exposure to diseases or physiological states potentially correlated to liver cancer. The availability of several covariate features of pathological relevance provides a challenging opportunity for subtype discovery methods, with applications for the search of biomarkers in a very interesting class of studies. In this study, we apply semisupervised profiling to remove selected patterns, as proposed in [6,7]. Both predictive accuracy as well as the stability of the resulting gene signature are improved. Moreover, we analyze the subtypes derived according to the expression data in conjunction with the available pathological information.

The structure of the paper is as follows: We first describe the main characteristics of the BioDCV system and the semisupervised approach in Section 2. Data and original classification task are discussed in Section 3. Outlier detection and list stability are discussed in Section 4, with comments in Section 5.

2   Semisupervised Profiling with BioDCV

The BioDCV system [1] implements the E-RFE complete validation setup developed at ITC-irst for predictive profiling of gene expression data [8]. BioDCV is portable from single workstations to local Linux clusters and virtual GRID facilities. It is written in C and interfaced with the SQLite database management library (http://www.sqlite.org). Note that learning, tuning and evaluation tasks may be replicated for up to a few millions of models in complete validation setup. We use SQLite to support concurrent access and transactions and manage results and parameters in a distributed environment during this high-throughput process. The main engine of BioDCV is the libml library, a C toolbox for learning problems which includes the Support Vector Machine (SVM) classifiers applied in this study. BioDCV also runs within the Egrid (http://www.egrid.it) computational infrastructure, based on Globus/EDG/LCG2 middleware and integrated as an independent virtual organization within Grid.it, the INFN production grid. Part of the computation described in this paper was performed on a local computing facility (an Open Mosix cluster of 26 bi-processor units and one data server).

The semisupervised procedure implemented in BioDCV is based on an analysis of the effect of the feature selection and ranking process for each individual sample. Given a complete validation setup (such as the one described in [8]), for each sample s we define the sample-tracking profile as the function of the number of features k as $E_s(k) = W(s,k)/N(s)$, where $N(s)$ is the number of runs in which s belongs to the test set and $W(s,k)$ is the number of runs in which s belongs to the test set and it is wrongly classified when the model is built with k features. The sequences $E_s(k)$ may be studied as an estimate of the classification error as a function of the size of the feature set. Sample-Tracking curves of easy-to-classify points quickly reach zero, while curves not far from the

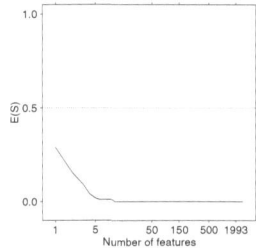 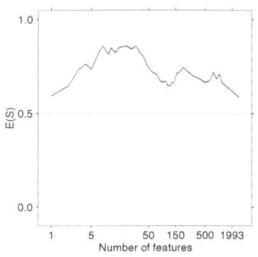

Fig. 1. Examples of sample-tracking profiles from the liver cancer dataset: a easy-to-classify point (left) and an outlier (right)

no-information error rate should correspond to hard-to-classify points. A profile lying systematically above the no-information error rate indicates a typical outlier behaviour. A complete description and examples of the sample-tracking procedure may be found in [4]. Examples for the liver cancer dataset used in this study are shown in Fig. 1.

The response to the supervised classification task is then used to drive a secondary unsupervised pattern discovery process. Similar classification responses may be aggregated by a hierarchical clustering technique. In this study we adopted Dynamic Time Warping (DTW) as the distance, with weight configuration $(1, 2, 1)$. The use of DTW for the analysis of gene expression time series was recently introduced in [9]. This distance is more suited than Euclidean metric in curves comparison, because it takes morphology into account instead of just evaluating the pointwise distance of the vectors [10].

3   Data and Classification Task

The liver dataset originally analysed in [3] consists of 213 cases described by 1993 genes. The classification problem consists in the discrimination between the 107 samples extracted from tumors (liver cancer patients) and the 106 control samples. Several binary features of interest for the pathology are available: sex, age> 65, positive to Hepatitis virus-type B, or to virus-type C, presence of cirrhosis, Child score (A or B). Linear SVMs are used for classification (regularization parameter $C = 100$) and the feature ranking method is E–RFE (reverting to standard RFE in the last 100 steps). Results (average test set error ATE and confidence intervals) are reported in the columns in the left block of the table included in Fig. 2. On the MPBA cluster, the computing time of a complete validation experiment with 400 replicated runs is 216 minutes, plus 91 minutes on a single unit for the semisupervised analysis (a single non-parallelizable process).

4   Outlier Detection and List Stability

The sample-tracking curves were aggregated with respect to DTW distance by hierarchical clustering with the `hclust` algorithm (average link) from the `stats`

Feat.	Complete dataset ATE	CI	Shaved dataset ATE	CI
1	27.7	(26.9,28.6)	24.3	(23.6,25.0)
2	24.3	(23.5,25.3)	20.4	(19.5,21.6)
3	21.8	(21.0,22.8)	15.4	(14.6,16.4)
4	18.4	(17.8,19.1)	11.4	(10.9,11.9)
5	16.6	(16.1,17.3)	9.9	(9.4,10.4)
10	12.7	(12.3,13.1)	5.9	(5.6,6.2)
20	8.9	(8.6,9.3)	3.4	(3.2,3.7)
50	5.8	(5.5,6.1)	1.8	(1.6,2.0)
100	4.8	(4.5,5.0)	1.5	(1.3,1.6)
1000	3.1	(2.9,3.4)	1.5	(1.4,1.7)
1993	3.2	(3.0,3.5)	1.5	(1.3,1.7)

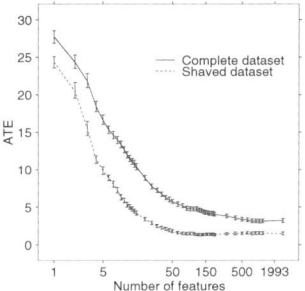

Fig. 2. Predictive error (ATE: average test error) for the liver cancer dataset on the complete dataset of 213 samples (solid line) and on the shaved dataset of 198 samples after outlier removal (dashed line). Bars indicate bootstrap studentized confidence intervals (.95 level).

package in the R system [11]. The analysis of the sample-tracking profiles detected 15 outliers. The BioDCV system was applied again after removing the outliers. The improved ATE scores on the shaved dataset are reported in the right block of the table in Fig. 2.

For stability analysis, we compared the ranked gene lists before and after the removal of outliers. For the two cases, the genes ranked in the top h positions in the 400 lists were listed and ordered for number of extractions ($Exts$). After outlier removal, the number of genes extracted at least once ($Exts$ greater than 1) is smaller. The best genes are extracted more frequently, and they have lower position means and standard deviations. A comparison for the best 10 genes is detailed in Table 1 for $h = 20$: $Exts$ resulted greater than 1 for 330 genes instead of 427. The resulting groups of samples were then analyzed in terms of their pathological features. Clustering of the liver cancer cases is displayed in Fig. 3. The analysis of clinical data for the cluster-1 subgroup is provided in Fig. 4 (left) in comparison with the aggregated values of all the 98 liver cancer cases. All subjects in cluster-1 are positive to Virus C and negative to Virus B.

Table 1. Multiplicity of extraction (Exts.), position mean and standard deviation (S.D.) of the best 10 genes extracted in the top $h = 20$ positions before and after removing outliers

Pos.	h = 20 - Complete Gene	Exts.	Mean	S.D.	Pos.	h = 20 - Shaved Gene	Exts.	Mean	S.D.
1	GS201	260	9	5.5	1	GS3244	356	5.6	4.4
2	GS1324	212	8.4	5.8	2	GS6094	336	5.9	4.3
3	GS1686	204	6.2	6.1	3	GS201	276	8.9	4.8
4	GS3244	198	8.7	5.9	4	GS1686	264	5.8	5.2
5	GS6094	194	8.7	6.1	5	GS10759	248	8.6	5.1
6	GS11601	194	9.7	5.9	6	GS1710	235	10.1	5.4
7	GS11954	187	7.5	5.9	7	GS11954	232	7.2	5.5
8	GS3097	163	11.3	4.9	8	GS2954	232	9.6	4.9
9	GS2375	152	10.8	4.8	9	GS1324	214	10.4	6
10	GS10424	145	11.2	5.2	10	GS2303	212	7.5	6.2

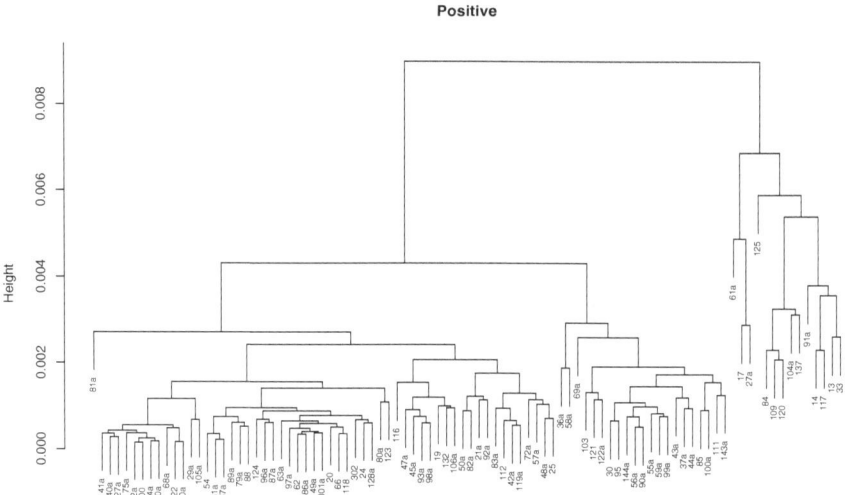

Fig. 3. DTW based clustering of the sample-tracking profiles of the positive samples in the shaved liver cancer dataset

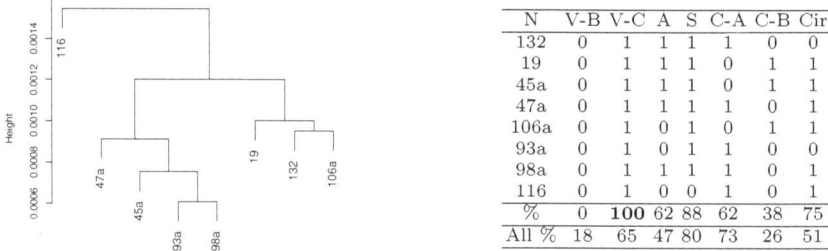

N	V-B	V-C	A	S	C-A	C-B	Cir
132	0	1	1	1	1	0	0
19	0	1	1	1	0	1	1
45a	0	1	1	1	0	1	1
47a	0	1	1	1	1	0	1
106a	0	1	0	1	0	1	1
93a	0	1	0	1	1	0	0
98a	0	1	1	1	1	0	1
116	0	1	0	0	1	0	1
%	0	**100**	62	88	62	38	75
All %	18	65	47	80	73	26	51

Fig. 4. DTW based clustering of the sample-tracking profiles of the cluster-1 subgroup (left) and description of clinical features of the corresponding samples (right): name (N), virus-type B (V-B) and C (V-C), age> 65 (A), sex (S), child score A (C-A) and B (C-B), and presence of cirrhosis (Cir). The last two lines show the percentage of values equal to 1 in the cluster-1 subgroup and in all the positive samples.

They belong mostly to the elderly group. Note that they are all males but for sample 116. This subject is detected as a singleton by the DTW-based clustering focused on this subgroup (Fig. 4, right).

5   Conclusions

With BioDCV and the concept of sample-tracking profiles, the high-throughput structure of complete validation schemes for gene profiling may be reused to support new models of semisupervised analysis. The availability of covariate features of pathological relevance provides a challenging opportunity for subtype discovery methods, with applications for the search of biomarkers in a very interesting class of studies.

Acknowledgments

The research was partially financed by the AIRC-BICG grant. GJ is supported by the FUPAT post-graduate project "Algorithms and software environments for microarray gene expression experiments". We thank S. Dzerowsky for the helpful indication of this application. We particularly thank the EGrid Project at ITCP Trieste for guidance in developing the Grid implementation of BioDCV.

References

1. Albanese, D.: BioDCV: a distributed computing system for the complete validation of gene profiles. Master's thesis, University of Trento (2005)
2. Simon, R., Radmacher, M., Dobbin, K., McShane, L.: Pitfalls in the use of DNA microarray data for diagnostic and prognostic classification. J Natl Cancer Inst **95** (2003) 14–18
3. Sese, J., Kurokawa, Y., Monden, M., Kato, K., Morishita, S.: Constrained clusters of gene expression profiles with pathological features. Bioinformatics **20** (2004) 3137–3145
4. Furlanello, C., Serafini, M., Merler, S., Jurman, G.: Semisupervised learning for molecular profiling. IEEE/ACM Transactions on Computational Biology and Bioinformatics **2** (2005) 110–118
5. Bair, E., Tibshirani, R.: Semi-supervised methods to predict patient survival from gene expression data. PLoS Biol **2** (2004) DOI: 10.1371/journal.pbio.0020108
6. Merler, S., Caprile, B., Furlanello, C.: Bias-variance control via hard points shaving. International Journal of Pattern Recognition and Artificial Intelligence **18** (2004) 891–903
7. Li, L., Pratap, A., Lin, H., Abu-Mostafa, Y.: Generalization by data categorization. In: The 16th ECML and the 9th PKDD. (2005) In press.
8. Furlanello, C., Serafini, M., Merler, S., Jurman, G.: Entropy-based gene ranking without selection bias for the predictive classification of microarray data. BMC Bioinformatics (2003) 54
9. Aach, J., Church, G.: Aligning gene expression time series with time warping algorithms. Bioinformatics **17** (2001) 495–508
10. Furlanello, C., Merler, S., Jurman, G.: Combining feature selection and DTW for time-varying functional genomics. Technical Report T05-05-01, ITC-irst (2005)
11. R Development Core Team: R: A Language and Environment for Statistical Computing. R Foundation for Statistical Computing, Vienna, Austria. (2005)

Local Metric Adaptation for Soft Nearest Prototype Classification to Classify Proteomic Data

F.-M. Schleif[1,3,*], T. Villmann[2], and B. Hammer[4]

[1] Univ. Leipzig, Dept. of Math. and Comp. Science, Leipzig, Germany
[2] Univ. Leipzig, Clinic for Psychotherapy, Leipzig, Germany
[3] Bruker Daltonik GmbH, Permoserstrasse 15, D-04318 Leipzig, Germany
Tel.: +49 341 24 31-408; Fax.: +49 341 24 31-404
schleif@informatik.uni-leipzig.de
[4] Clausthal Univ. of Technology, Dept. of Comp. Science

Abstract. We propose a new method for the construction of nearest prototype classifiers which is based on a Gaussian mixture approach interpreted as an annealed version of Learning Vector Quantization. Thereby we allow the adaptation of the underling metric which is useful in proteomic research. The algorithm performs a gradient descent on a cost function adapted from soft nearest prototype classification. We investigate the properties of the algorithm and assess its performance on two clinical cancer data sets. Results show that the algorithm performs reliable with respect to alternative state of the art classifiers.

Keywords: classification, learning vector quantization, metric adaptation, mass spectrometry, proteomic profiling.

1 Introduction

During last years proteomic[1] profiling based on mass spectrometry (MS) became an important tool for studying cancer at the protein and peptide level in a high throughput manner. MS based serum profiling is under development as a potential diagnostic tool to distinguish patients with cancer from normal subjects. Reliable classification methods which can cope with typically high dimensional characteristic profiles constitute a crucial part of the system. Thereby, a good generalization ability and interpretability of the results are highly desirable.

KOHONEN'S Learning Vector Quantization (LVQ) belongs to the class of supervised learning algorithms for nearest prototype classification (NPC) [2]. It relies on a set of prototype vectors (also called codebook vectors) which are adapted by the algorithm according to their respective classes. Thus, it forms a very intuitive local classification method with very good generalization ability

---

[*] Corresponding author.
[1] Proteome - is an ensemble of protein forms expressed in a biological sample at a given point in time [1].

I. Bloch, A. Petrosino, and A.G.B. Tettamanzi (Eds.): WILF 2005, LNAI 3849, pp. 290–296, 2006.
© Springer-Verlag Berlin Heidelberg 2006

also for high dimensional data [3] which constitutes an ideal candidate for an automatic and robust classification tool for high throughput proteomic patterns.

However, original LVQ is only heuristically motivated and shows instable behavior for overlapping classes. Recently a new method, Soft Nearest Prototype Classification (SNPC), has been proposed by SEO ET AL. [4] based on the formulation as a Gaussian mixture approach which yields soft assignments of data. We adapt this algorithm by introducing local and global relevance learning to SNPC and apply it to profiling of mass spectrometric data in cancer research. The approach is well suited to deal with high dimensional data focusing on optimal class separability. Further, it is capable to determine relevance profiles of the input which can be used for identification of relevant data dimensions. Interestingly, strong dimensionality independent large margin generalization bounds hold for this method. In addition, the relevance weights may be taken as indicators for biomarkers in the underlying biological data. We demonstrate the abilities of SNPC-R for two cancer data sets: the Wisconsin Breast Cancer (WBC)[5] and a leukemia data set (LEUK) provided by [6].

Now, we introduce the concept of LVQ first. The SNPC is reviewed in section 2 followed by the generalization to Self Adapted Soft Nearest Prototype Classification with Relevance Learning (SNPC-R). Subsequently, we report the application results in comparison to other well known state of the art methods as Supervised Relevance Neural GAS (SRNG) as introduced in [7] and Support Vector Machines (SVM) as introduced in [8]. We conclude by a short discussion of the method and show the benefits of the metric adaptation.

2   Soft Nearest Prototype Classification

Usual learning vector quantization is mainly influenced by the standard algorithms LVQ1...LVQ3 introduced by KOHONEN [2]. Several derivatives have been developed to ensure faster convergence, a better adaptation of the receptive fields to optimum Bayesian decision, or an adaptation for complex data structures, to name just a few [9, 10, 4]. Standard LVQ does not possess a cost function in the continuous case; it is based on the heuristic to minimize misclassifications using Hebbian learning. The first version of learning vector quantization based on a cost function, which formally assesses the misclassifications, is the Generalized LVQ (GLVQ) [11]. We will use GLVQ resp. its extension SRNG as introduced in [7] for comparison in this article.

Now, we introduce the basic notation for LVQ schemes. Inputs are denoted by $\mathbf{v}$ with label $c_{\mathbf{v}} \in \mathcal{L}$. Assume $\mathcal{L}$ is the set of labels (classes) with $\#\mathcal{L} = N_{\mathcal{L}}$ and $V \subseteq \mathbb{R}^{D_V}$ a finite set of inputs $\mathbf{v}$. LVQ uses a fixed number of prototypes (weight vectors, codebook vectors) for each class. Let $\mathbf{W} = \{\mathbf{w_r}\}$ be the set of all codebook vectors and $c_{\mathbf{r}}$ be the class label of $\mathbf{w_r}$. Furthermore, let $\mathbf{W}_c = \{\mathbf{w_r}|c_{\mathbf{r}} = c\}$ be the subset of prototypes assigned to class $c \in \mathcal{L}$. The classification of vector quantization is implemented by the map Ψ as a winner-take-all rule, i.e. a stimulus vector $\mathbf{v} \in V$ is mapped onto that neuron $\mathbf{s} \in A$ the pointer $\mathbf{w}_s$ of which is closest to the presented vector $\mathbf{v}$,

$$\Psi_{\mathcal{V} \to \mathcal{A}}^{\lambda^{\mathbf{r}}} : \mathbf{v} \mapsto \mathbf{s}\left(\mathbf{v}\right) = \operatorname*{argmin}_{\mathbf{r} \in A} d^{\lambda^{\mathbf{r}}}\left(\mathbf{v}, \mathbf{w_r}\right) \tag{2.1}$$

with $d\left(\mathbf{v}, \mathbf{w}\right)$ being an arbitrary distance measure, usually the standard euclidean metric. The neuron $\mathbf{s}$ is called winner or best matching unit. The subset of the input space $\Omega_{\mathbf{r}} = \{\mathbf{v} \in V : \mathbf{r} = \Psi_{V \to A}\left(\mathbf{v}\right)\}$ which is mapped to a particular neuron $\mathbf{r}$ according to (2.1), forms the (masked) receptive field of that neuron. Standard LVQ training adapts the prototypes such that for each class $c \in \mathcal{L}$, the corresponding codebook vectors $\mathbf{W}_c$ represent the class as accurately as possible, i.e. the set of points in any given class $V_c = \{\mathbf{v} \in V | c_{\mathbf{v}} = c\}$, and the union $\mathcal{U}_c = \bigcup_{\mathbf{r}|_{\mathbf{w_r} \in \mathbf{W}_c}} \Omega_{\mathbf{r}}$ of receptive fields of the corresponding prototypes should differ as little as possible. This is either achieved by heuristics as for LVQ1...LVQ3 [2], or by the optimization of a cost function related to the mismatches as for GLVQ [11] and SRNG as introduced in [7].

Soft Nearest Prototype Classification (SNPC) has been proposed as alternative stable NPC learning scheme. It introduces soft assignments for data vectors to the prototypes which have a statistical interpretation as normalized Gaussians. In the original SNPC as provided in [4] one considers as the cost function

$$E\left(\mathcal{S}, \mathcal{W}\right) = \frac{1}{N_{\mathcal{S}}} \sum_{k=1}^{N_{\mathcal{S}}} \sum_{\mathbf{r}} u_{\tau}\left(\mathbf{r}|\mathbf{v}_k\right)\left(1 - \alpha_{\mathbf{r}, c_{\mathbf{v}_k}}\right) \tag{2.2}$$

with $\mathcal{S} = \{(\mathbf{v}, c_{\mathbf{v}})\}$ the set of all input pairs, $N_{\mathcal{S}} = \#\mathcal{S}$, and $\mathcal{W} = \{(\mathbf{w_r}, c_{\mathbf{r}})\}$ whereby $c_{\mathbf{r}}$ is the class label of $\mathbf{w_r}$, as before. The value $\alpha_{\mathbf{r}, c_{\mathbf{v}_k}}$ equals one if $c_{\mathbf{v}_k} = c_{\mathbf{r}}$. $u_{\tau}\left(\mathbf{r}|\mathbf{v}_k\right)$ is the probability that the input vector $\mathbf{v}_k$ is assigned to the prototype $\mathbf{r}$. A crisp *winner-takes-all* mapping (2.1) would yield $u_{\tau}\left(\mathbf{r}|\mathbf{v}_k\right) = \delta\left(\mathbf{r} = \mathbf{s}\left(\mathbf{v}_k\right)\right)$.

In order to minimize (2.2) in [4] the variables $u_{\tau}\left(\mathbf{r}|\mathbf{v}_k\right)$ are taken as soft assignment probabilities. This allows a gradient descent on the cost function (2.2). As proposed in [4], the probabilities are chosen as normalized Gaussians

$$u_{\tau}\left(\mathbf{r}|\mathbf{v}_k\right) = \frac{\exp\left(-\frac{d(\mathbf{v}_k, \mathbf{w_r})}{2\tau^2}\right)}{\sum_{\mathbf{r}'} \exp\left(-\frac{d(\mathbf{v}_k, \mathbf{w}_{\mathbf{r}'})}{2\tau^2}\right)} \tag{2.3}$$

whereby d is the distance measure used in (2.1). Then the cost function (2.2) can be rewritten as

$$E_{soft}\left(\mathcal{S}, \mathcal{W}\right) = \frac{1}{N_{\mathcal{S}}} \sum_{k=1}^{N_{\mathcal{S}}} lc\left(\left(\mathbf{v}_k, c_{\mathbf{v}_k}\right), \mathbf{W}\right) \tag{2.4}$$

with local costs

$$lc\left(\left(\mathbf{v}_k, c_{\mathbf{v}_k}\right), \mathbf{W}\right) = \sum_{\mathbf{r}} u_{\tau}\left(\mathbf{r}|\mathbf{v}_k\right)\left(1 - \alpha_{\mathbf{r}, c_{\mathbf{v}_k}}\right) \tag{2.5}$$

i.e., the local error is the sum of the assignment probabilities $\alpha_{\mathbf{r}, c_{\mathbf{v}_k}}$ to all prototypes of an incorrect class, and, hence, $lc\left(\left(\mathbf{v}_k, c_{\mathbf{v}_k}\right), \mathbf{W}\right) \leq 1$. Because the local costs $lc\left(\left(\mathbf{v}_k, c_{\mathbf{v}_k}\right), \mathbf{W}\right)$ are continuous and bounded, the cost function (2.4) can be minimized by stochastic gradient descent using the derivative of the local costs:

$$\triangle \mathbf{w_r} = \begin{cases} \frac{1}{2\tau^2} u_\tau\left(\mathbf{r}|\mathbf{v}_k\right) \cdot lc\left(\left(\mathbf{v}_k, c_{\mathbf{v}_k}\right), \mathbf{W}\right) \cdot \frac{\partial d_\mathbf{r}}{\partial \mathbf{w_r}} & \text{if } c_{\mathbf{v}_k} = c_\mathbf{r} \\[2mm] -\frac{1}{2\tau^2} u_\tau\left(\mathbf{r}|\mathbf{v}_k\right) \cdot \left(1 - lc\left(\left(\mathbf{v}_k, c_{\mathbf{v}_k}\right), \mathbf{W}\right)\right) \cdot \frac{\partial d_\mathbf{r}}{\partial \mathbf{w_r}} & \text{if } c_{\mathbf{v}_k} \neq c_\mathbf{r} \end{cases} \qquad (2.6)$$

where

$$\frac{\partial lc}{\partial \mathbf{w_r}} = -u_\tau\left(\mathbf{r}|\mathbf{v}_k\right)\left(\left(1 - \alpha_{\mathbf{r}, c_{\mathbf{v}_k}}\right) - lc\left(\left(\mathbf{v}_k, c_{\mathbf{v}_k}\right), \mathbf{W}\right)\right) \cdot \frac{\partial d_\mathbf{r}}{\partial \mathbf{w_r}} \qquad (2.7)$$

This leads to the learning rule

$$\mathbf{w_r} = \mathbf{w_r} - \epsilon\left(t\right) \cdot \triangle \mathbf{w_r} \qquad (2.8)$$

with learning rate $\epsilon\left(t\right)$ fulfilling $\sum_{t=0}^{\infty} \epsilon\left(t\right) = \infty$ and $\sum_{t=0}^{\infty}\left(\epsilon\left(t\right)\right)^2 < \infty$ as usual. All prototypes are adapted in this scheme according to the soft assignments. Note that for small bandwidth τ, the learning rule is similar to LVQ2.1.

A window rule like for standard LVQ2.1 can be derived for SNPC, too, which is necessary for numerical stabilization [2],[4]. The update is restricted to all weights for which the local value $\eta_\mathbf{r} = lc\left(\left(\mathbf{v}_k, c_{\mathbf{v}_k}\right), \mathbf{W}\right) \cdot \left(1 - lc\left(\left(\mathbf{v}_k, c_{\mathbf{v}_k}\right), \mathbf{W}\right)\right)$ is less than a threshold value η with $0 \ll \eta < 0.25$.

3   Relevance Learning for SNPC

Like all NPC algorithms, SNPC heavily relies on the metric d, usually the standard euclidean metric. For high dimensional data as occur in proteomic patterns, this choice is not adequate since noise present in the data set accumulates and likely disrupts the classification. Thus, a focus on the (priorly not known) relevant parts of the inputs, the biomarkers, would be much more suited. Relevance learning as introduced in [12] offers the opportunity to learn metric parameters which account for the different relevance of input dimensions during training. In analogy to the above learning approaches we include this relevance learning idea into SNPC leading to SNPC-R. Instead of the standard metric $d\left(\mathbf{v}_k, \mathbf{w_r}\right)$ a metric incorporating adaptive relevance factors $d^\lambda\left(\mathbf{v}_k, \mathbf{w_r}\right)$ is included into the soft assignments (2.3), whereby the component λ_k of λ is usually chosen as weighting parameter for input dimension k. The relevance parameters λ can be adjusted according to the given training data, taking the derivative of the cost function, i.e. $\frac{\partial lc\left(\left(\mathbf{v}_k, c_{\mathbf{v}_k}\right), \mathbf{W}\right)}{\partial \lambda}$ using the local cost (2.5):

$$\frac{\partial lc\left(\left(\mathbf{v}_k, c_{\mathbf{v}_k}\right), \mathbf{W}\right)}{\partial \lambda_j} = \frac{\partial}{\partial \lambda_j}\left[\frac{\sum_\mathbf{r} \exp\left(-\frac{d^\lambda\left(\mathbf{v}_k, \mathbf{w_r}\right)}{2\tau^2}\right) \cdot \left(\left(1 - \alpha_{\mathbf{r}, c_{\mathbf{v}_k}}\right)\right)}{\sum_{\mathbf{r}'} \exp\left(-\frac{d^\lambda\left(\mathbf{v}_k, \mathbf{w_{r'}}\right)}{2\tau^2}\right)}\right] \qquad (3.1)$$

$$= -\frac{1}{2\tau^2}\sum_\mathbf{r} u_\tau\left(\mathbf{r}|\mathbf{v}_k\right) \cdot \frac{\partial d_\mathbf{r}^\lambda}{\partial \lambda_j} \cdot \left(\left(1 - \alpha_{\mathbf{r}, c_{\mathbf{v}_k}} - lc\left(\left(\mathbf{v}_k, c_{\mathbf{v}_k}\right), \mathbf{W}\right)\right)\right) \qquad (3.2)$$

with subsequent normalization of the λ_k.

We would like to emphasize that SNPC-R can also be used with *individual* metric parameters λ^r for each prototype $\mathbf{w_r}$ or with a classwise metric shared within prototypes with the same class label c_r as it is done here, referred as localized SNPC-R (LSNPC-R). If the metric is shared by all prototypes, LSNPC-R is reduced to SNPC-R. The respective adjusting of the relevance parameters λ can easily be determined in complete analogy to (3.2).

It has been pointed out in [3] that NPC classification schemes which are based on the euclidean metric can be interpreted as large margin algorithms for which dimensionality independent generalization bounds can be derived. Instead of the dimensionality of data, the so-called hypothesis margin, i.e. the distance, the hypothesis can be altered without changing the classification on the training set, serves as a parameter of the generalization bound. This result has been extended to NPC schemes with *adaptive* diagonal metric in [7]. This fact is quite remarkable, since D_V new parameters, D_V being the input dimension, are added this way, still, the bound is independent of D_V. This result can even be transferred to the setting of *individual* metric parameters λ^r for each prototype or class such that a generally good generalization ability of this method can be expected. Despite from the fact that (possibly local) relevance factors allow a larger flexibility of the approach without decreasing the generalization ability, they are of particular interest for proteomic pattern analysis because they indicate potentially semantically meaningful positions.

4   Experiments and Applications

In the following we give experimental results for the application of both SNPC and SNPC-R for two data sets in comparison to SRNG and SVM. Thereby, for SNPC, the usual Euclidean distance is applied whereas in case of SNPC-R, the weighted Euclidean metric as distance measure $d^\lambda(\mathbf{v}, \mathbf{w}) = \sum \lambda_i (v_i - w_i)^2$ is used whereby the factors λ are global (SNPC-R) or attached to the singular classes (LSNPC-R). This choice allows a direct interpretation of the relevance parameters as a weighting of importance of the spectral bands for cancer detection, which may give a hint for potential biomarkers. The WBC data set consists of 100 training samples and 469 test data, whereby for the training samples exactly half the data set is to cancer state. The spectra are given as 30-dimensional vectors. For a detailed description of the data including facts about preprocessing we refer to [5]. The second data set LEUK was generated by [6]. It was obtained by spectral analysis of blood plasma of patients suffering from cancer and control probands. A mass range between 2 to 20kDa was used. The spectra were first processed using the standardized workflow as given in [13]. After preprocessing the spectra are obtained as 145-dimensional vectors. The data set consists of 74 cancer and 80 control samples.

For classification, we use 8 prototypes for WBC data and 4 prototypes for LEUK data. The classification results are given in Tab. 1. Clearly, metric adaptation significantly improves the classification accuracy. The relevance profiles are depicted in Fig. 1. High relevance values refer to greater importance of the

Table 1. Classification accuracy for the two cancer data sets

	SNPC		SNPC-R		LSNPC-R		SRNG		SVM	
	train	test	train	test	train	test	train	test	train	test
WBC	98%	85%	95%	91.4%	94%	92%	98%	86%	98%	96%
LEUK	94%	100%	96%	100%	96%	100%	98%	100%	100%	≈ 100%

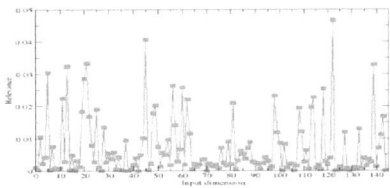

Fig. 1. Relevance profiles for the WBC (left) and LEUK (right) data set using SNPC-R

respective spectral bands for classification accuracy and, therefore, hints for potential biomarkers. We see that SNPC-R is capable to generate a suitable classification model which leads to prediction rates above 91%. The results are better than those obtained by ordinary SNPC. The results are reliable in comparison with SVM and SRNG. Besides the good prediction rates obtained from SNPC-R we get additional information from the relevance profiles. For metrics per class we get specific knowledge on important input dimensions per class.

5   Conclusion

We extended the usual SNPC by relevance learning as one kind of metric adaptation. We derived an adaptation dynamic for metric adaptation and prototype adjustment according to a gradient descent on a cost function. This cost function is obtained by appropriate modification of the SNPC. As demonstrated, this new soft nearest prototype classification with relevance learning can be efficiently applied to the classification of proteomic data and leads to results which are competitive to results as reported by alternative state of the art algorithms.

Acknowledgment. The authors are grateful to U. Clauss and J. Decker both Bruker Daltonik GmbH Leipzig for support by preprocessing of the prostate cancer data set.

References

1. Binz, P., Appel, R.: Mass spectrometry-based proteomics: current status and potential use in clinical chemistry. Clin. Chem. Lab. Med. **41** (2003) 1540–1551
2. Kohonen, T.: Self-Organizing Maps. Volume 30 of Springer Series in Information Sciences. Springer, Berlin, Heidelberg (1995) (2nd Ext. Ed. 1997).

3. Crammer, K., Gilad-Bachrach, R., A.Navot, A.Tishby: Margin analysis of the lvq algorithm. In: Proc. NIPS 2002. (2002)
4. Seo, S., Bode, M., Obermayer, K.: Soft nearest prototype classification. IEEE Transaction on Neural Networks **14** (2003) 390–398
5. Blake, C., Merz., C.: UCI rep. of mach. learn. databases. (1998) available at: http://www.ics.uci.edu/ mlearn/MLRepository.html.
6. MHH Hannover, I.S., Daltonik, B.: internal results on leukaemia (2004)
7. Hammer, B., Strickert, M., Villmann, T.: Supervised neural gas with general similarity measure. Neural Processing Letters **21** (2005) 21–44
8. Vapnik, V.: Statistical Learning Theory. Wiley, New York (1998)
9. Hammer, B., Villmann, T.: Mathematical aspects of neural networks. In Verleysen, M., ed.: Proc. of Europ. Symp. on Art. Neural Netw., Brussels (2003) 59–72
10. Kohonen, T., Kaski, S., Lappalainen, H.: Self-organized formation of various invariant-feature filters in the adaptive-subspace SOM. Neural Computation **9** (1997) 1321–1344
11. Sato, A.S., Yamada, K.: Generalized learning vector quantization. In Tesauro, G., Touretzky, D., Leen, T., eds.: Advances in Neural Information Processing Systems. Volume 7. MIT Press (1995) 423–429
12. Hammer, B., Villmann, T.: Generalized relevance learning vector quantization. Neural Netw. **15** (2002) 1059–1068
13. Adam, B., Qu, Y., Davis, J., Ward, M., Clements, M., Cazares, L., Semmes, O., Schellhammer, P., Yasui, Y., Feng, Z., Wright, G.: Serum protein finger printing coupled with a pattern-matching algorithm distinguishes prostate cancer from benign prostate hyperplasia and healthy men. Cancer Research **62** (2002) 3609–3614

Learning Bayesian Classifiers from Gene-Expression MicroArray Data

Andrea Bosin[1], Nicoletta Dessì[1], Diego Liberati[2], and Barbara Pes[1]

[1] Università degli Studi di Cagliari, Dipartimento di Matematica e Informatica,
Via Ospedale 72, 09124 Cagliari
andrea.bosin@dsf.unica.it,
{dessi, pes}@unica.it
[2] IEIIT CNR c/o Politecnico di Milano, Piazza da Vinci 32, I-20133 Milano
liberati@elet.polimi.it

Abstract. Computing methods that allow the efficient and accurate processing of experimentally gathered data play a crucial role in biological research. The aim of this paper is to present a supervised learning strategy which combines concepts stemming from coding theory and Bayesian networks for classifying and predicting pathological conditions based on gene expression data collected from micro-arrays. Specifically, we propose the adoption of the Minimum Description Length (MDL) principle as a useful heuristic for ranking and selecting relevant features. Our approach has been successfully applied to the Acute Leukemia dataset and compared with different methods proposed by other researchers.

Keywords: Bayesian Classifiers, Gene-Expression Data Analysis, Feature Selection, MDL.

1 Introduction

In the growing field of bio-informatics, new technologies, like the so-called Micro-arrays [1], provide thousands of gene expression data on a single cell in a simple and fast-integrated way. In contrast with data sets from other fields, a typical Micro-array data-set has a large number of genes (>5000) and a small number of samples (<100) requiring a large amount of computational effort.

Since little is currently known about how gene expression values differentiate a pathological condition from a normal one, one could take advantage from the statistical power of such a number of gene expression data, but the amount of information makes the inference computationally intractable by classic statistical procedures. However, we can expect that not all the genes will carry relevant information. Thus, the process of selecting the most important ones is a useful technique for promoting faster learning of classification models whose essence is to deal efficiently with the automatic classification of new subjects. If such a learning is obtained directly by the data, without using in some way any parallel information not included in them, the procedure is deemed to be unsupervised. In contrast, a supervised learning approach analyzes patient's gene expression measurements and the related diagnosis provided by a pathologist that assigns each particular subject to already defined pathological classes.

I. Bloch, A. Petrosino, and A.G.B. Tettamanzi (Eds.): WILF 2005, LNAI 3849, pp. 297–304, 2006.
© Springer-Verlag Berlin Heidelberg 2006

The aim of this paper is to present a supervised learning strategy which combines concepts stemming from coding theory and Bayesian networks for classifying and predicting pathological conditions based on gene expression data collected from mi-cro-arrays. Specifically, we propose the adoption of the Minimum Description Length (MDL) principle [2] as a useful heuristic for ranking and selecting relevant features. We first apply MDL principle for assigning weights to each feature, based on a pro-cedure that is independent from the classification task. Then, the feature selection is accomplished by progressively choosing the features step by step. At each step, a Bayesian classifier is built on the selected subset of features and new features are entered one by one, until no improvement in accuracy results.

Our approach has been applied to the Acute Leukemia dataset [3] and compared with different methods proposed in other works [4] [5].

The paper is organized as follows. Section 2 illustrates the feature selection heuris-tics and provides some insight on Bayesian classifiers. Our experimental results are presented in Section 3. A comparison with other learning approaches is discussed in Section 4. Section 5 briefly presents some related works about the correlation between genes and Acute Leukemia. Finally, conclusions and future work are outlined in Section 6.

2 The Learning Strategy

We propose a learning strategy that looks for a trade-off between a high predictive accuracy of the classifier and a low cardinality of the selected feature subset. Our central hypothesis is that a good feature subset contains features that are highly corre-lated with the class to be predicted, yet uncorrelated with each other.

Generally, methods for feature selection can be classified as wrappers and filters. Wrappers are tuned to a specific learning algorithm since they conduct the search for a good subset of features using the classifier itself to evaluate the merit of the subset. Filters only look at intrinsic characteristics of the data and rank features leaving the final choice to the user. Wrappers give good results, but, in practice, may be too slow on datasets containing many features. While more practical, filters don't select explic-itly the set of features because they result unable in handle both redundant and irrele-vant features.

Based on information theory, the MDL principle [2] provides our operational sup-port in that it states that the best theory to infer from training data is the one that minimizes the length (i.e. the complexity) of the theory itself and the length of the data encoded with respect to it. In particular, MDL can be employed as a criteria to judge the quality of a classification model.

The motivation underlying the MDL method is to find a compact encoding of the training data $(u_1, \ldots, u_N)$. To this end we adopt the following MDL measure [7]:

$$\mathrm{MDL} = \log_2(m) - \sum_{i=1}^{N} \log_2 \left(p(u_i) \right) \qquad (1)$$

where m is the number of potential candidate models and $p(u_i)$ is the model predicted probability assigned to the training instance u_i. The first term in Equation 1 represents how many bits we need to encode the specific model (i.e. its length), and the second term measures how many bits are needed to describe the data based on the probability distribution associated to the model.

This approach can be applied to address the problem of feature selection, by considering each feature as a simple predictive model of the target class. As described in [6], we rank each feature according to its description length, that reflects the strength of its correlation with the target. In this context, the MDL measure is given by [10]:

$$MDL = \sum_j \log_2 \binom{N_j + C - 1}{C - 1} - \sum_j \sum_{i=1}^{N_j} \log_2(p_{ji}) \qquad (2)$$

where N_j is the number of training instances with the j-th value of the given feature, C is the number of target class values, and p_{ji} is the probability of the value of the target class taken by the i-th training instance with the j-th value of the given feature (estimated from the distribution of target values in the training data). As in the general case (Eq. 1), the first term expresses the encoding length, where we have one sub-model for each value of the feature, while the second gives the number of bits needed to describe the data, based on the probability distribution of the target value associated to each sub-model.

However, once all features have been ordered by rank, no a priori criterion is available to choose the cut-off point beyond which features can be discarded. To circumvent this drawback, we adopt a wrapper approach that starts with building a classifier on the set of the n-top ranked features. Then, we sequentially add a new feature to this set, and build a new classifier until no improvement in accuracy is achieved.

Our approach compares two different classifiers derived from Bayesian Networks, i.e. the Naïve Bayes (NB) and the Adaptive Bayesian Network (ABN).

NB is a very simple Bayesian network consisting of a special node (i.e. the target class C) that is parent of all other nodes (i.e. the features or attributes, $A_1, ..., A_n$) that are assumed to be conditionally independent, given the value of the class (Fig. 1). The NB network can be "quantified" against a training dataset of pre-classified instances, i.e. we can compute the probability associated to a specific value of each attribute, given the value of the class label. Then, any new instance can be easily classified making use of the Bayes rule. Despite its strong independence assumption is clearly

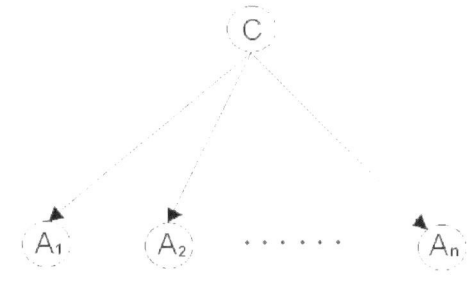

Fig. 1. The structure of a Naive Bayes network

unrealistic in several application domains, NB has been shown to be competitive with more complex state-of-the-art classifiers [7][8][9].

In the last years, a lot of research has focused on improving NB classifiers by relaxing their full independence assumption. One of the most interesting approaches is based on the idea of adding correlation arcs between the attributes of a NB classifier. On these "augmenting arcs" are imposed specific structural constraints [7] [8], in order to maintain computational simplicity on learning. The algorithm we evaluate, the Adaptive Bayesian Network (ABN) [10], is a greedy variant, based on MDL, of the approach proposed in [8].

In brief, the steps needed to build an ABN classifier are the following. First, the attributes (predictors) are ranked according to their MDL importance. Then, the network is initialized to NB on the top k ranked predictors $(A_1, A_2,..., A_k)$, that are treated as conditionally independent. Next, the algorithm attempts to extend NB by constructing a set of tree-like multi-dimensional features.

Feature construction proceeds as follows. The top ranked predictor is stated as a seed feature, and the predictor that most improves feature predictive accuracy, if any, is added to the seed. Further predictors are added in such a way to form a tree structure, until the accuracy does not improve. Using the next available top ranked predictor as a seed, the algorithm attempts to construct additional features in the same manner. The process is interrupted when the overall predictive accuracy cannot be further improved or after some pre-selected number of steps.

The resulting network structure consists of a set of conditionally independent multi-attribute features, and the target class probabilities are estimated by the product of feature probabilities. Interestingly, each multi-dimensional feature can be expressed in terms of a set of if-then rules enabling users to easily understand the basis of model predictions.

3   Experimental Study Cases

We tested the proposed approach on the Acute Leukemia dataset [3], which consists of 72 samples (38 training samples, 34 testing samples) each described by 7129 attributes (genes for which a quantitative expression level was measured). The pathological classes (targets) to be predicted are AML (acute myeloid leukemia) and ALL (acute lymphoblastic leukemia).

As a preprocessing step, we ranked all the 7129 attributes using MDL scoring. Table 1 shows the top 20 attributes.

Then, we performed two series of experiments for learning both NB and ABN classifiers on the 38 samples of the training set. We started with the set of the first 10 predictors of Table 1, and added the following attributes, one by one, feeding them to both NB and ABN classifiers, until no further improvement in accuracy was achieved.

Their predictive accuracy has been evaluated by the rate of correct predictions on the 34 samples of the test set. Table 2 shows the accuracy of NB and ABN classifiers with 10, 15 and 20 predictors. Note that the classifiers built with 15 and 20 predictors result in the same accuracy. Table 3 shows the confusion matrices and, interestingly, the only errors of both classifiers consist in the misclassification of AML samples (i.e. 3 samples for NB and 2 for ABN).

Table 1. MDL top-ranked genes

1. APLP2 (L09209_s_at)	11. PPGB (M22960_at)
2. LYN (M16038_at)	12. Azurocidin (M96326_rna1_at)
3. p62 (U46751_at)	13. Rhesus (HG627-HT5097_s_at)
4. CD36 (M98399_s_at)	14. FAH (M55150_at)
5. Zyxin (X95735_at)	15. CST3 (M27891_at)
6. Adipsin (M84526_at)	16. LGALS3 (M57710_at)
7. LTC4S (U50136_rna1_at)	17. CAB3b (L27584_s_at)
8. CD33 (M23197_at)	18. ME491 (X62654_rna1_at)
9. LEPR (Y12670_at)	19. SNRPN (J04615_at)
10. CHRNA7 (X70297_at)	20. CTSD (M63138_at)

Table 2. Overall accuracy of NB and ABN classifiers (10, 15, 20 predictors)

	10	15	20
NB	0.853	0.912	0.912
ABN	0.882	0.941	0.941

Table 3. Confusion matrix for NB and ABN classifiers with 15 predictors

	NB			ABN		
	ALL	AML	rate	ALL	AML	rate
ALL	20	0	100%	20	0	100%
AML	3	11	78.6%	2	12	85.7%

4   Discussion

In this section, we compare our results with different approaches [4][5] in analyzing the Acute Leukemia dataset. Specifically, in [4] a supervised approach is proposed, where a subset of 50 informative genes (25 most highly correlated with AML and 25 most highly correlated with ALL) has been selected. In [5] an unsupervised approach is adopted, where a first subset of 178 statistically relevant genes has been considered, then PDDP clustering and K-means algorithms have been applied to the 72 samples, resulting in two clusters (AML and ALL). A final gene-pruning step results in only 7 significant genes out of 178 .

The following observations can be made.

- Both [4] and [5] carried out the selection of a limited number of genes by statistical methods, where the choice of a more or less arbitrary cut-off threshold is a major drawback.

- Interestingly, 12 of the genes shown in Table 1 are among the 25 genes most highly correlated with AML in [4]. No genes in Table 1 are among the 25 most highly correlated with ALL in [4]. In Table 4 we evaluate the MDL ranking order for each AML predictor reported in [4]. The ranking is high for most genes, suggesting

Table 4. MDL ranking of AML predictors proposed in [4]

Gene	MDL rank	Gene	MDL rank	Gene	MDL rank
M55150	14	X17042	54	M57710	16
X95735	5	Y00787	53	M69043	35
U50136	7	M96326	12	M81695	52
M16038	2	U46751	3	X85116	33
U82759	28	M80254	39	M19045	38
M23197	8	L08246	25	M83652	41
M84526	6	M62762	104	X04085	47
Y12670	9	M28130	20		
M27891	15	M63138	20		

that both MDL measure and high statistical correlation with AML give similar results in detecting genes whose informative content is the highest. The situation is different for ALL, since our MDL scoring gives quite a low ranking (only 2 genes in the first 50 positions, 5 in the first 100) of the 25 genes most highly correlated with ALL found in [4].

- Conversely, only 2 of the 7 significant predictors in [5] are among the 15 genes sufficient in our approach (Table 5). A plausible explanation is that in our approach, as well as in [4], features are ranked based on their correlation with the target class, while in [5] feature selection is carried out through a principal component analysis, designed to select the linear combinations of variables with higher intersubject covariances.

- Since the number of predictive variables is large relative to the available sample, the unsupervised learning approaches may be affected by over-fitting. To account for this, we have trained and tested both NB and ABN classifiers using the 7 predictors reported in [5]. The resulting accuracy on the test set is 88% for both NB and ABN. Comparison with Table 2 indicates that for both NB and ABN a larger number of top ranked genes does improve accuracy, while, if such a number needs to be reduced,

Table 5. MDL ranking of predictors proposed in [5]

Gene	MDL rank
M11147	70
M19507	159
M27891	15
M96326	12
Y00433	1202
Y00787	53
Z19554	1321

NB benefits from external gene selection. This result also seems to suggest that over-fitting is not a major concern in [5] and that, on the contrary, different subsets of genes can be used to build classifiers of similar accuracy.

- Finally, the number of correct predictions on the test dataset is 32 out of 34 for our ABN classifier with 15 predictors, to be compared with 29 out of 34 obtained in [4], and with 70 out of 72 reported in [5] (all samples).

5 Related Work

Several authors have reported many genes statistically correlated to AML/ALL, i.e. cathepsin D [11][12], integrins [12], zyxin [4]: cathepsin D is a lysosomal enzyme functionally correlated to Cystatin C, an inhibitor of cysteine proteinases, mostly cathepsins; integrins associate with the actin cytoskeleton near the adhesion plaques; zyxin, 5th in MDL ranking, is a phosphoprotein that is concentrated at adhesion plaques and along the actin filament bundles near where they insert at the adhesion plaques [13]. Thus, integrins, zyxin and vimentine might be different actors or aspects in the process we aim to identify. In particular, CST3 Cystatin C, the most significant gene in [5], here ranked as the 15th, was identified as discriminant also in [14] [11] [15] on the basis of the same database via different methodological approaches.

6 Conclusions

A learning strategy which combines concepts stemming from coding theory and Bayesian networks has been successfully applied in predicting ALL-AML leukemia distinction based on gene expression data collected from micro-arrays. Our experimental results confirm that gene expression data combined with powerful learning algorithms can lead to excellent diagnostic models, even with very modest sample size. In particular, by a selection heuristic based on MDL, we identify a subset of predictive genes which differently overlaps with other subsets presented in literature. This supports the hypothesis that even disjoint subsets of genes can provide evidence of the same pathological conditions. An interesting problem not addressed here is the correlation among genes in the same and different subsets and we plan to investigate it in a future work.

References

[1] Hardimann G., Microarray methods and applications: Nuts & bolts. DNA Press, 2003.
[2] Barron A., Rissanen J., Yu B., The minimum description length principle in coding and modelling, IEEE Transactions on Information Theory, 44: 2743-2760, 1998.
[3] http://sdmc.lit.org.sg/GEDatasets/Datasets.html
[4] Golub T.R., et al., Molecular Classification of Cancer: Class Discovery and Class Prediction by Gene Expression Monitoring, Science, Vol. 286, 531-537, 1999.
[5] Liberati D., Bittanti S., Garatti S.: Unsupervised Mining of Genes Classifying Leukemia, Enciclopedia of Data Warehousing and Mining, Wang J. ed., Idea Group, Hershey, PA, USA, 2005.

[6] Kononenko I., On biases in estimatine multi-valued attributes, IJCAI95, 1034-1040, 1995.

[7] Friedman N., Geiger D., Goldszmidt M., Bayesian Network Classifiers, Machine Learning, 29: 131-161, 1997.

[8] Keogh E., Pazzani M.J., Learning the structure of augmented Bayesian classifiers, International Journal on Artificial Intelligence Tools, Vol. 11, No. 4, 587-601, 2002.

[9] Cheng G., Greiner R., Comparing Bayesian Network Classifiers, Proceedings of the Fifteenth Conference on Uncertainty in Artificial Intelligence, Morgan Kaufmann Publishers, Inc., San Francisco, 1999.

[10] Yarmus J.S., ABN: A Fast, Greedy Bayesian Network Classifier, 2003. http://otn.oracle.com/products/bi/pdf/adaptive_bayes_net.pdf

[11] Chow ML, Moler EJ, Mian IS (2001). Identifying marker genes in transcription profiling data using a mixture of feature relevance experts. Physiol Genomics 5: 99-111.

[12] Moos PJ, Raetz EA, Carlson MA, Szabo A, Smith FE, Willman C, Wei Q, Hunger SP, Carroll WL (2002). Identification of gene expression profiles that segregate patients with childhood leukaemia, Clinical Cancer Research 8: 3118-3130.

[13] Crawford AW and Beckerle MC (1991). Purification and characterization of zyxin, an 82000-Dalton component of adherents junctions, The Journal of Biological Chemistry 266 (9): 5847-5853

[14] Li X, Rao SQ, Wang YD, Gong BS (2004). Gene mining: a novel and powerful ensemble decision approach to hunting for disease genes using microarray expression profiling, Nucleic Acids Research 32: 2685-2694

[15] Valdes et al. (2004), Seventeenth International Conference on Industrial & Engineering Applications on Artificial Intelligence & Expert Systems.

On the Evaluation of Images Complexity:
A Fuzzy Approach

Maurizio Cardaci[1,2], Vito Di Gesù[1,3],
Maria Petrou[4], and Marco Elio Tabacchi[3]

[1] Università di Palermo, C.I.T.C., Italy
[2] Università di Palermo, DDP, Italy
[3] Università di Palermo, DMA, Italy
[4] University of Surrey, UK

Abstract. The inherently multidimensional problem of evaluating the complexity of an image is of a certain relevance in both computer science and cognitive psychology. Computer scientists usually analyze spatial dimensions, to deal with automatic vision problems, such as feature-extraction. Psychologists seem more interested in the temporal dimension of complexity, to explore attentional models. Is it possible, by merging both approaches, to define an more general index of visual complexity? We have defined a fuzzy mathematical model of visual complexity, using a specific entropy function; results obtained by applying this model to pictorial images have a strong correlation with ones from an experiment with human subjects based on variation of subjective temporal estimations associated with changes in visual attentional load, which is also described herein.

Keywords: Fuzzy sets, image analysis, complexity, entropy, mental clock, internal clock.

1 Introduction

The problem of evaluating the complexity of an image is of a certain relevance to both cognitive and computer science studies, although in broader contexts the general problem of visual complexity measurement is ill-defined. The evaluation of visual complexity is useful in understanding relations among different levels of the recognition process and it is also of interest to real applications such as image compression and information theory.

The Computer Science approach to visual complexity is generally space-based: local feature extraction and selection plus global statistical parameter estimation are employed to quantify complexity from the point of view of a rational agent. Nevertheless, complexity is not only relevant to the stimulus' spatial properties, but, as an emerging factor affecting the human perceiver's cognitive operations, it can also involve the temporal dimension. The recent Mental Clock Model [1], developing an intuition by Ornstein [2], relies on the simple hypothesis that the subjective passing duration is affected by a hypothetical internal clock which tends to modify its speed according to the attentional load of the current task.

I. Bloch, A. Petrosino, and A.G.B. Tettamanzi (Eds.): WILF 2005, LNAI 3849, pp. 305–311, 2006.

These biases in the subjective time evaluation allow us to indirectly determine the visual complexity of stimuli, by comparing (under the same experimental conditions) their perceived durations.

Is it possible to define a fuzzy model of complexity that encompasses both approaches, leading to the calculation of a true visual index of complexity?

We have defined such a fuzzy model, based on local and global spatial features of the image and a definition of entropy. We chose an entropic measure because information theory and several branches of statistics have been proven to be powerful tools in quantifying the infinitesimal differences between two probability density functions. Entropic distances have been used successfully for image comparison and object matching problems in query by content applications, showing their ability to grasp the pictorial visual content [3]. For each image we computed its fuzzy index of complexity using the adopted model.

The results from this fuzzy model were compared, using proven data analysis techniques, with the ones obtained by an experiment on subjective estimate of the perceived time, performed while the subjects were exposed to pictorial stimuli of increasing complexity.

2   The Entropic Model of Visual Complexity

To create a mathematical model of the visual complexity based on spatial parameters we have reviewed many of the local and global features from literature. Global features are suited to derive single values from the general properties of an image. Local features are needed to take into account classical verbal explanations for the meaning of complexity: many versus few, curved and/or detailed versus linear and planar, complex textures versus flat areas. Using local features also helps reducing ambiguities in results.

2.1   Local Features Extraction

Points of interest may be identified by using local operators. We chose two well-known local features: the image edges [4, 5], and the local symmetries computed by the *Discrete Symmetry Transform* (*DST*) [6]. Edge detectors highlight image zones with abrupt changes in luminosity level, associated with surface discontinuity. The rationale is straightforward: the more edges, the more objects (or the more surfaces), and a greater perceived complexity. *DST* extracts zones of the image in which the local gray levels show a high degree of radial symmetry (where the degree of locality depends on the radius of the local window used). It is interesting to note that points of interest detected by *DST* appear to be related with points to which shifts of gaze are directed performed by humans watching the same image. Apart from the natural attraction of symmetry, this also means that the more the points of interest in an image, the more complex the image is perceived as. More specifically, *DST* computes local symmetries of an image based on a measure of *axial moments* of a body around its center of gravity. In the image case, the pixels inside a circular window are considered as point masses, with their mass expressed by their gray value g. Details of the algorithm can be found in [6]

2.2   An Entropic Measure of Complexity

We are now interested in a global algorithm that can output a single value for each filtered images, while preserving its class of complexity. We decided to investigate the usefulness for this task of the fuzzy entropic distance functions detailed in [3]. There are plenty of reasons for considering these functions among many others usually employed in this kind of task: first, a soft computing approach using fuzzy values seems appropriate when we are trying to describe a situation where binary logic is too strict. As for the entropic distance function, we can reformulate our main question from *"How complex is this image?"* to *"What is the distance of this image from the simplest possible image in the defined feature space?"*. This approach leads to the use of standard distance functions, which respect the usual properties of identity, symmetry and triangular inequality, augmented by entropic functions. We chose the following functions:

$$G_0(\eta) = -\frac{1}{\log(2)} \times (\eta \log(\eta) + (1 - \eta) \log(1 - \eta). \ [7]$$
$$G_1(\eta) = \frac{2\sqrt{e}}{e-1} \left(\eta e^{1-\eta} - \eta e^{\eta-1} \right), \ G_2(\eta) = 4\eta(1 - \eta)$$

where, $\eta = \frac{1}{n} \sum_{i=1}^{n} |h_i|$, and h_i are the gray levels of the image pixels normalized in the range [0,1]. It can be easily shown that $G_j (j = 0, 1, 2)$ satisfies the properties of a distance function, and it takes values in the interval [0, 1]. For each G_j function, the first 15% of input values is mapped to more then half the range of output values. Most of the image would have a complexity index in the first quartile, so we expected to obtain a better classification through the expansion of exactly this range of input values. We considered three complexity classes of images, three different filters, and three entropy distance functions. In Table 1 we show the mean value for each category. It is easy to see that all choices of functions and filters give results in line with our expectations, with a slightly better performance obtained by the use of function G_2 and the symmetry filter without thresholding. Note that the reported values of entropy are normalized in the interval [0, 1] with the same procedure described in Section 3.

Table 1. Results from filtering and applying entropy functions to all images in our pool, and then taking mean values for each category

	G_0			G_1			G_2		
	I	II	III	I	II	III	I	II	III
Edge detection	1	0.60	0	1	0.39	0	1	0.48	0
Local Symmetry	1	0.45	0	1	0.39	0	1	0.49	0
Thresholded Symmetry	1	0.42	0	1	0.39	0	1	0.34	0

3   The Experiment on Perceived Time

We devised an experiment to demonstrate that a subjective measure of the perceived time can be used as an indirect measure of the complexity of an image. We asked a number of volunteers to observe some images on a computer screen, and recorded their perceived duration of the observation.

We run two sets of experiments, one in University of Surrey's CVSSP group (50 individuals) and the other in the Dipartimento di Psicologia dell'Università di Palermo (15 individuals). In order to minimize the cultural bias, all experimental subjects had university backgrounds. Participants were part of the staff and undergraduate students, on a volunteering basis, without any knowledge of our research's aims. Privacy of the subjects was taken care of according to the Italian law on personal data; only initials, age and gender were recorded for each subject.

Fig. 1. Examples of test images, classified by intuitive complexity: high complexity (top); medium complexity (middle); low complexity (bottom)

The experiments were held in a dim light room to reduce visual distraction, giving time to the participant for darkness adaptation. All the usual ergonomic precautions, such as using a quasi-soundproof room, were taken, and the subject was allowed to choose their own preferred position and visual angle. The images were presented full screen. The software used was home-made using the multimedia programming environment Macromedia Dreamweaver 2004 MX on an Apple Macintosh computer with a TFT LCD monitor. The chosen images were computer scans of paintings, divided in three categories representing different levels of visual complexity, based on the presence or absence of certain classes of features and cue points. Figure 1 shows examples of painting used in this study.

Each image were presented for a fixed period of time (90 secs.), with no temporal clues; the experiment also had a *controlled* design in order to minimize side effects: lights dimmed and uniform, subject alone in a soundproof room. The subject was alerted to focus their attention on the contents of the displayed images. The images used for the experiments were chosen according to the intuitive hypothesis that the complexity of a scene increases with the number of objects and their relative position, and with its overall structure [8]. The chosen images were paintings, divided in three categories representing different levels of visual

Table 2. Mean and Normalized Time Estimation

	Class I	Class II	Class III
$\hat{\mu}_T$	61.74	73.38	85.15
$\hat{\sigma}_T$	33.05	38.09	41.14
n_T	0	0.49	1

complexity. Here the estimate time perceived by each subject is reported. We consider it as a subjective measures of complexity for the three categories of images introduced above. In the following it will be denoted as $\{(t_T^{(i)})|i = \text{I,II,III}\}$. The sample mean value and the variance of the perceived time $(\hat{\mu}_T, \hat{\sigma}_T)$ are reported in Table 2. It is evident that the mean perceived time decreases with the complexity of the image. In order to compare the perceived time with the objective measure of complexity, and to highlight our interest in relative differences between estimations made by the same subject when watching different images, the following normalized time measures are introduced:

$$n_T^{(i)} = \frac{t_T^{(i)} - \hat{\mu}^{min}}{\hat{\mu}^{max} - \hat{\mu}^{min}} \quad \text{for } i = \text{I,II,III}$$

where $\hat{\mu}^{min} = \min_{i=I,II,III}\{\hat{\mu}_T^{(i)}\}$ and $\hat{\mu}^{max} = \max_{i=I,II,III}\{\hat{\mu}_T^{(i)}\}$, and $0 \leq n_T^{(i)} \leq 1$.

The proposed normalization allows us a better comparison with the results obtained from the mathematical model, carried out in the next section. In this context, 0 and 1 have no strict numerical significance, but should be interpreted more like subjective degrees of complexity, which suits best with our fuzzy model.

Results are in agreement with our model of time perception: complex images (category I) produce shorter time estimations than images in category II and the same is true for categories II and III.

4 Comparison of Measures and Data Validation

As shown by comparing the entries of Tables 1 and 2, our experimental data match those of the mathematical model. In fact, images with a high entropic complexity index generate, on average, a shorter estimation of the perceived time. Therefore, category I has the shortest evaluations and category III the longest. The strong anti-correlation between the entropic measure of complexity and the mental clock is shown in Figure 2. The values of Gs are the averages of those in Table 1.

We carried out a strict validation of the results using proven data analysis methods in order to ascertain the relation between data and model, minimizing the effects derived from the use of mean values and the cardinality of the dataset. To verify that a correlation between the experimental data and the mathematical results exists, we calculated the coefficient of correlation between the results

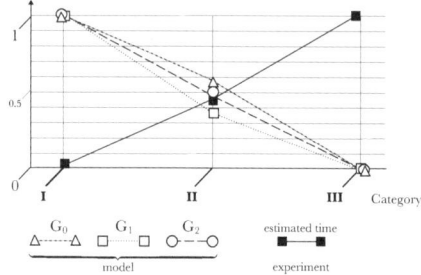

Fig. 2. Anti-correlation between the estimate of the mental clock and the measure of complexity via entropy functions

from the experimental data and the complements of the entropic measures of complexity using Spearmans' ρ. Even in the worst case, the probability of data and model sequences being correlated is more than 0.98. To confirm that the correlation is not due to the size of the data-set, we carried out many non-parametric bootstrap tests, using 10,000 virtual sets. In each test the difference between the mean obtained from the data and by the bootstrap method was under 10^{-4}. As we worked mostly with mean values, we also used the jackknife technique, re-calculating the results as many times as the number of images in our set, leaving out one image each time; all jackknife sets had the same distribution of values, with small numeric differences.

5   Conclusions

In this paper we detailed an fuzzy model of visual complexity. This model fits well with a subjective measure of complexity, based on perceived time. From a theoretical viewpoint we verified a strong correlation between spatial and temporal dimensions of complexity. Our results support the possibility to include the human information processing into the standard measure of visual complexity. Future work will be devoted to improve our experiment with more image categories.

References

1. M. Cardaci: The Mental Clock Model - Studies on the Structure of Time. Physics to Psycho(patho)logy, (L. Buccheri, V. Di Gesù, Saniga Eds.), Kluwer Academic/Plenum Publishers, New York, 2000.
2. R. E. Ornstein: The Psychology of Consciousness. Freeman and Company, San Francisco, 1972.
3. V. Di Gesù, S. Roy: Fuzzy measures for image distance. Advances in Fuzzy Systems and Intelligent Technologies (F. Masulli, R. Parenti, G. Pasi Eds.), Shaker Publishing (NL), 2000.
4. D. Marr, E. Hildreth: Theory of Edge Detection. Proc. R. Soc. Lond. B., Vol. 207, pp 187–217, 1980.

5. M. Petrou: The Differentiating Filter Approach to Edge Detection. Advances in Electronics and Electron Physics, Vol. 88, pp 297–345, 1994.
6. V. Di Gesù, C. Valenti, "Symmetry operators in computer vision", in Vistas in Astronomy, Pergamon, Vol. 40, No 4, pp 461–468, 1996.
7. A. De Luca, S. Termini: Information and Control. Vol. 20, p 301, 1972.
8. A. Oliva, M. L. Mack, M. Shrestha, A. Peeper: Identifying the perceptual dimensions of visual complexity of scenes. *Proc. of the 27th Annual Meeting of the Cognitive Science Society*, Chicago, 2004.

3D Brain Tumor Segmentation Using Fuzzy Classification and Deformable Models

Hassan Khotanlou[1], Jamal Atif[1], Olivier Colliot[2], and Isabelle Bloch[1]

[1] GET - Ecole Nationale Supérieure des Télécommunications, Dept TSI,
CNRS UMR 5141, 46 rue Barrault, 75634 Paris Cedex 13, France
[2] McConnell Brain Imaging Center, MNI, McGill University,
3801 University, Montréal, Québec, H3A2B4, Canada
{Hassan.Khotanlou, Jamal.Atif, Isabelle.Bloch}@enst.fr,
colliot@bic.mni.mcgill.ca

Abstract. A new method that automatically detects and segments brain tumors in 3D MR images is presented. An initial detection is performed by a fuzzy possibilistic clustering technique and morphological operations, while a deformable model is used to achieve a precise segmentation. This method has been successfully applied on five 3D images with tumors of different sizes and different locations, showing that the combination of region-based and contour-based methods improves the segmentation of brain tumors.

1 Introduction

Brain tumor segmentation from MR images is a challenging task that involves various disciplines including medicine, MRI physic, radiologist's perception, and image analysis based on intensity and shape. The literature is rich with techniques for segmenting normal brain structures, but many of these methods fail in the presence of a pathology. Actually the techniques that are intended for tumors leave significant room for increased automation, applicability and accuracy. In brain tumor studies, existence of abnormal tissues is most of the time easy to detect but accurate and reproducible segmentation and characterization of abnormalities still remain a challenging task.

Let us briefly summarize existing work, classically divided into region-based and contour-based methods. In the first class, a tumor segmentation method using knowledge based and fuzzy techniques was proposed by Clark et al. [2]. This method has two drawbacks. First, it requires multichannel images such as T1, T2 and PD. Furthermore a training phase prior to segmenting a set of images is necessary. Other methods are based on statistical pattern recognition techniques. Kaus et al. [6] have proposed a method for automatic segmentation of small brain tumors using a statistical classification method and atlas registration. Moon et al. [9] have also used the EM algorithm and atlas prior information for automatic tumor segmentation. These methods fail in the case of large deformations in the brain and they also require multichannel images (T1, T2, PD and contrast enhanced images) for classification. Prastawa et al. [12]

I. Bloch, A. Petrosino, and A.G.B. Tettamanzi (Eds.): WILF 2005, LNAI 3849, pp. 312–318, 2006.

consider the tumor as an outlier and use a statistical classification for rough segmentation and then geometric and spatial constraints for final segmentation. The fuzzy connectedness method was proposed by Moonis et al. [10] for tumor segmentation. In this semi-automatic method, the user must select the region of the tumor. The calculation of connectedness is achieved in this region and the tumor is delineated in 3D as a fuzzy connected object containing the seed points of the tumor that were selected by the user. Other methods such as data fusion [1], atlas based [4] and transformation [13] methods have been developed, with similar drawbacks.

In contour-based methods, Lefohn et al. [7] have proposed a semi-automatic method for tumor segmentation by level sets. The user selects the tumor region and after the deformation process, he adapts the level set parameters. Zhu and Yang [15] introduce an algorithm using neural networks and a deformable model. Their method processes each slice separately and is not a real 3D method. Ho et al. [5] have proposed level set evolution with region competition for tumor segmentation. Their algorithm uses two images (T1-weighted with and without contrast agents) and calculates a tumor probability map using classification, histogram analysis and the difference between the two images, and then this map is used as the zero level of the level set evolution. The deformable methods suffer from the difficulty of determining the initial contour, and tuning the parameters.

In this paper we propose a fully-automatic method that is a combination of region-based and contour-based methods. It does not require any user supervision, works in 3D and on standard routine T1 aquisitions. It combines a fuzzy classification method (FPCM) [11], morphological operations and a parametric deformable model, thus taking advantages of both approaches while cancelling their drawbacks. The method is detailed in Section 2, and results are presented in Section 3.

2   Tumor Segmentation Procedure

A preliminary stage consists of brain segmentation. For this purpose, a robust method using histogram scale-space analysis and morphological operations [8] is applied. This method first calculates statistical parameters of the main classes of tissues, which will be used in the classification procedure. After extracting the brain, the histogram based Fuzzy Possiblistic C-Mean method is used for rough segmentation of the tumor. This rough segmentation is used as the initial surface of a deformable model for the final precise tumor segmentation.

2.1   Classification Using FPCM and Morphological Operations

Fuzzy Possibilistic c-Means was introduced by Pal et al. [11] for classification. It is a combination of Fuzzy c-Means and Possibilistic c-Means algorithms. In data classification, both membership and typicality are mandatory for data structures interpretation. FPCM computes these two factors simultaneously. FPCM solves the noise sensitivity defect of FCM and also overcomes the problem of coincident clusters of PCM. The objective function of FPCM is:

$$J_{m,\eta}(U, T, V; X) = \sum_{i=1}^{c} \sum_{k=1}^{n} (u_{ik}^m + t_{ik}^\eta) \|X_k - V_i\|^2 \tag{1}$$

where $m > 1$, $\eta < 1$, $0 \leq u_{ik} \leq 1$, $0 \leq t_{ik} \leq 1$, $\sum_{i=1}^{c} u_{ik} = 1$, $\forall k$, $\sum_{k=1}^{n} t_{ik} = 1$, $\forall i$, X_k denotes the characteristics of a point to be classified (here we use grey levels), V_i is the class center, c the number of classes, n the number of points to be classified, u_{ik} the membership of point X_k to class i, and t_{ik} is the possibilistic typicality value of X_k associated with class i.

In order to detect and label the tumor we use a histogram based FPCM that is faster than the classical FPCM implementation. Since the process of the brain extraction provides an estimation of the cerebro spinal fluid (CSF), white matter (WM) and gray matter (GM) radiometric characteristics, we exploit this information to overcome the classifical initialization problem, i.e. the algorithm is initialized with class centers which are very close to the final ones. We classify the extracted brain into five classes, CSF, WM, GM, tumor and background (at this study stage we do not consider the edema). To obtain the initial value of the class centers, we use the results of histogram analysis in the extraction step. We have used the mean of the CSF, WM and GM ($\mathbf{m_G}$, $\mathbf{m_W}$ and $\mathbf{m_C}$) (calculated in brain extraction step) as the centers of their classes. For background, the value zero is used. To select the tumor class we assume that the tumor has the highest intensity among the five classes (this is the case in our study where we are interested in hyper-intensity pathologies such as full-enhancing tumors).

Several binary morphological operations (opening, erosion, largest component selection, etc.) are then applied to the tumor region in order to correct misclassification errors. The results of this step for two images are shown in Figure 1.

2.2   Refinement Using a 3D Deformable Model

To obtain an accurate segmentation, a parametric deformable method, that has been applied successfully in our previous work to segment internal brain structures [3], is used. The segmentation obtained from the previous processing is transformed into a triangulation using an isosurface algorithm based on tetrahedra and is decimated and converted into a simplex mesh $\mathbf{X}$. The evolution of the deformable surface $\mathbf{X}$ is described by the following dynamic force equation:

$$\gamma \frac{\partial \mathbf{X}}{\partial t} = \mathbf{F}_{int}(\mathbf{X}) + \mathbf{F}_{ext}(\mathbf{X}) \tag{2}$$

where $\mathbf{F}_{int}$ is the internal force that specifies the regularity of the surface and $\mathbf{F}_{ext}$ is the external force that drives the surface towards image edges. The chosen internal force is:

$$\mathbf{F}_{int} = \alpha \nabla^2 \mathbf{X} - \beta \nabla^2 (\nabla^2 \mathbf{X}) \tag{3}$$

where α and β respectively control the surface tension (prevent it from stretching) and rigidity (prevent it from bending) and ∇^2 is the Laplacian operator. It

(a)   (b)   (c)   (d)

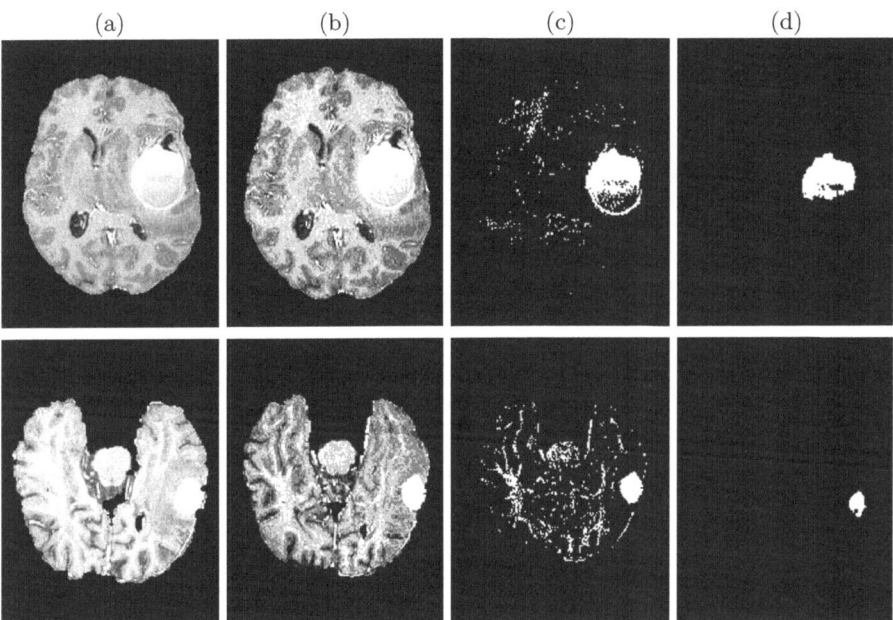

Fig. 1. Results obtained in the classification step for two 3D images. (a) One axial slice of extracted brain. (b) Result of FPCM classification. (c) Result of the thresholding. (d) Result after morphological operations.

is then discretized on the simplex mesh using the finite difference method [14]. In our case, the external force is derived from image edges. It can be written as:

$$\mathbf{F}_{ext} = v(x, y, z) \tag{4}$$

where v is a Generalized Gradient Vector Flow (GGVF) field introduced by Xu et al. [14]. A GGVF field v is computed by diffusion of gradient vector of a given edge map and is defined as the equilibrium solution of the following diffusion equation:

$$\frac{\partial v}{\partial t} = g(\|\nabla f\|)\nabla^2 v - h(\|\nabla f\|)(v - \nabla f) \tag{5}$$

$$v(x, y, z, 0) = \nabla f(x, y, z) \tag{6}$$

where f is an edge map and the functions g and h are weighting functions which can be chosen as follows:

$$\begin{cases} g(r) = e^{-\frac{r}{\kappa}^2} \\ h(r) = 1 - g(r) \end{cases} \tag{7}$$

To compute the edge map, a linear spatial filtering which is usually associated to Canny-Deriche edge detector is applied. Our experience shows that the GGVF is not sensitive to parameter k and we set it to $k = 0.05$ for all cases. The parameters involved in $\mathbf{F}_{int}$ were set to $\alpha = 0.25$ and $\beta = 0.0$. Again the same parameters were used for all tests.

(a)   (b)   (c)   (d)

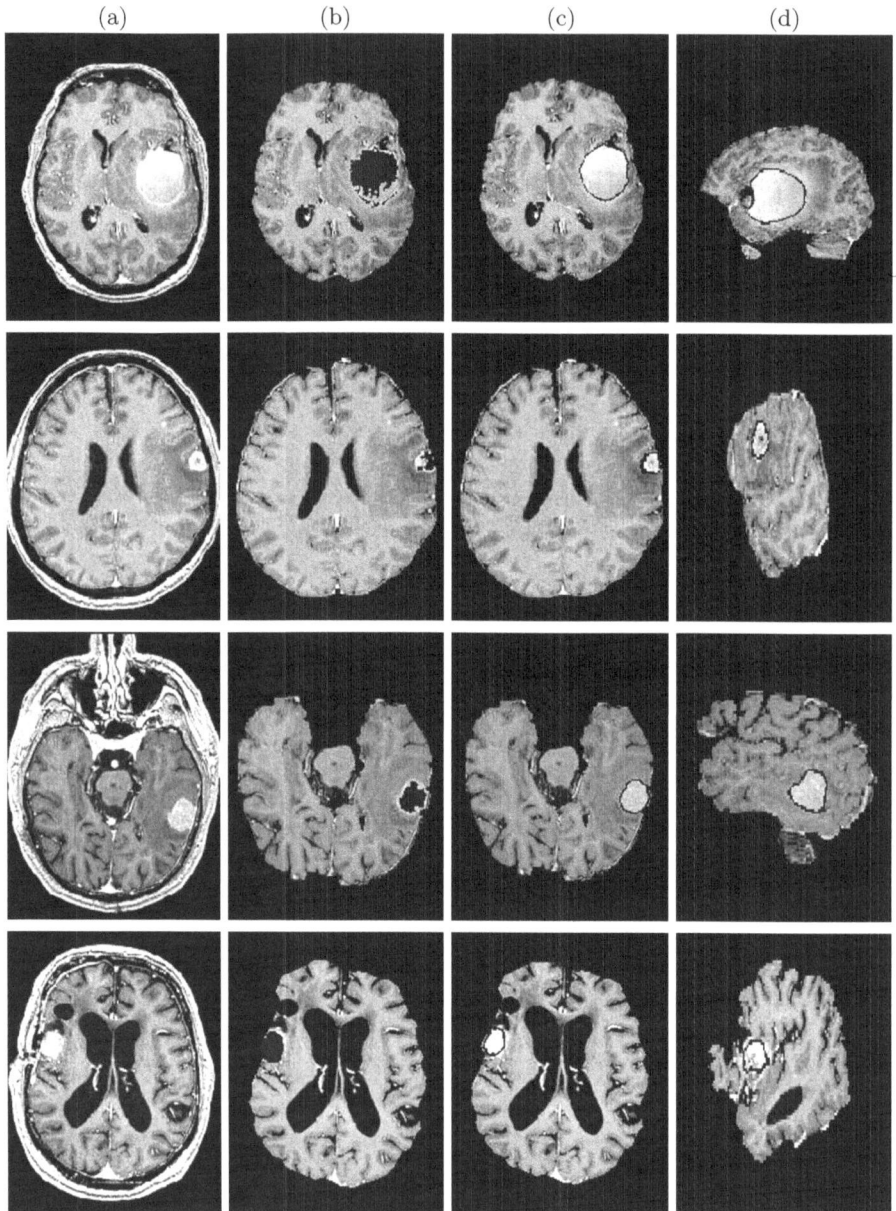

Fig. 2. Results obtained on four 3D images (obtained with the same parameters). (a) One axial slice of the original 3D image. (b) Result of the extracted brain and classification. (c) Final contour, superimposed on the axial slice. (d) Final contour, superimposed on a sagittal slice.

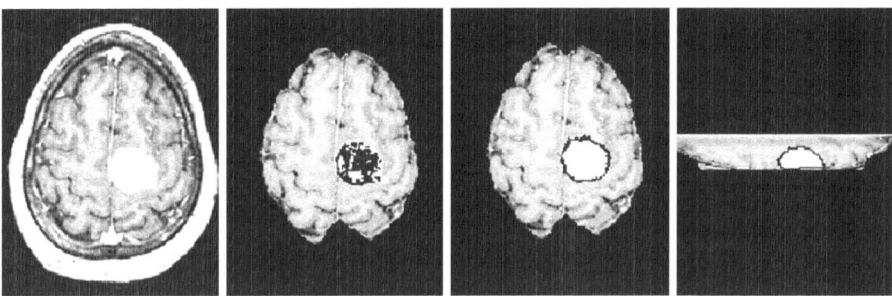

Fig. 3. Results obtained on one 3D image (256 × 256 × 22 voxels and obtained with the same parameters). (a) One axial slice of the original 3D image. (b) Result of the extracted brain and classification. (c) Final contour, superimposed on the axial slice. (d) Final contour, superimposed on a sagittal slice.

3   Results and Conclusion

We applied our algorithm to five different real 3D T1-weighted MR images (256× 256 × 124 voxels and 256 × 256 × 22 voxels). They exhibit tumors with different sizes and at different locations. We obtained good results for the five datasets without changing any parameter. The segmentation results of five datasets are shown in Figures 2 and 3.

We developed a hybrid algorithm using contour-based and region-based methods to segment brain tumors in 3D MR images. It exploits the advantages of fuzzy classification for automating the algorithm and the good quality segmentation result of deformable models to improve the segmentation. This is achieved by combining the FPCM classification method, morphological operations and a parametric 3D deformable model. Application on several datasets with different tumor sizes and different locations shows that this method works automatically with high quality of segmentation, and is robust to inter-individual variability for the all types of fully enhancing tumors. More tests are however necessary to further validate the approach. For quantitative evaluation of the results of segmentation, unfortunately there are not any standard images or methods. One way consists in comparing the results with manual segmentations. We are preparing these images for further evaluations with the help of medical experts.

Future work aims at assessing spatial relations to other structures around the tumor. Also we are extending this method for detecting several tumors in the brain and segmenting the edema.

Acknowledgments. We would like to thank Professor Desgeorges at Val-de-Grâce hospital for providing the images and his medical expertise. The image used in Figure 3 was provided by the Center for Morphometric Analysis at Massachusetts General Hospital and is available at http://www.cma.mgh.harvard.edu/ibsr/.

References

1. A.S Capelle, O.Colot, and C. Fernandez-Maloigne. Evidential segmentation scheme of multi-echo MR images for the detection of brain tumors using neighborhood information. *Information Fusion*, 5:203–216, 2004.
2. M.C. Clark, L.O. Lawrence, D.B. Golgof, R. Velthuizen, F.R. Murtagh, and M.S. Silbiger. Automatic tumor segmentation using knowledge-based techniques. *IEEE Transaction on Medical Imaging*, 17(2), April 1998.
3. O. Colliot, O. Camara, R. Dewynter, and I. Bloch. Description of brain internal structures by means of spatial relations for MR image segmentation. In *The International Society of Optical Engineering. SPIE 2004 Medical Imaging*, volume 5370, pages 444–455, 2004.
4. M.B. Cuadra, C. Pollo, A. Bardera, O. Cuisenaire, J. Villemure, and J. Thiran. Atlas-based segmentation of pathological MR brain images using a model of lesion growth. *IEEE Transactions on Medical Imaging*, 23(10):1301–1313, October 2004.
5. S. Ho, E. Bullitt, and G. Gerig. Level set evolution with region competition: Automatic 3D segmentation of brain tumors. In *International Conference on Pattern Recognition*, pages 532–535, 2002.
6. M.R. Kaus, S.K. Warfield, A. Nabavi, E. Chatzidakis, P.M. Black, F.A. Jolesz, and R. Kikinis. Segmentation of meningiomas and low grade gliomas in MRI. In *International Conference on Medical Image Computing and Computer-Assisted Intervention (MICCAI)*, volume LNCS 1679, pages 1–10, 1999.
7. A. Lefohn, J. Cates, and R. Whitaker. Interactive, GPU-based level sets for 3d brain tumor segmentation. Technical report, University of Utah, April 2003.
8. J.-F. Mangin, O. Coulon, and V. Frouin. Robust brain segmentation using histogram scale-space analysis and mathematical morphology. In *International Confrence on Medical Image Computing and Computer-Assisted Interventio (MICCAI)*, pages 1230–1241, 1998.
9. N. Moon, E. Bullitt, K.V. Leemput, and G. Gerig. Model-based brain and tumor segmentation. In *Intenational Conference on Pattern Recognition*, volume 1, pages 526–531, 2002.
10. G. Moonis, J. Liu, J.K. Udupa, and D.B. Hackney. Estimation of tumor volume with fuzzy-connectedness segmentation of MR images. *American Journal of Neuroradiology*, 23:352–363, 2002.
11. N.R. Pal, K. Pal, and J.C Bezdek. A mixed c-mean clustering model. In *IEEE International Conference on Fuzzy Systems*, pages 11–21, 1997.
12. M. Prastawa, E. Bullitt, S. Ho, and G. Gerig. A brain tumor segmentation framework based on outlier detection. *Medical Image Analysis*, 18(3):217–231, 2004.
13. H. Soltanian-Zadeh, M. Kharrat, and P.J. Donald. Polynomial transformation for MRI feature extraction. In *SPIE*, volume 4322, pages 1151–1161, 2001.
14. C. Xu and J.L. Prince. Snakes, shapes and gradient vector flow. *IEEE transaction on Image Processing*, 7:359–369, 1998.
15. Y. Zhu and H. Yang. Computerized tumor boundary detection using a Hopfield neural network. *IEEE Transactions on Medical Imaging*, 16(1):55–67, 1997.

A Hybrid Architecture for the Sensorimotor Exploration of Spatial Scenes

Kerstin Schill, Christoph Zetzsche, and Thusitha Parakrama

Kognitive Neuroinformatik, University of Bremen, Germany
kschill@informatik.uni-bremen.de

Abstract. Humans are very efficient in the analysis, exploration and representation of their environment. Based on the neurobiological and cognitive principles of human information processing, we develop a system for the automatic identification and exploration of spatial configurations. The system sequentially selects "informative" regions (regions of interest), identifies the local structure, and uses this information for drawing efficient conclusions about the current scene. The selection process involves low-level, bottom-up processes for sensory feature extraction, and cognitive top-down processes for the generation of active motor commands that control the positioning of the sensors towards the most informative regions. Both processing levels have to deal with uncertain data, and have to take into account previous knowledge from statistical properties and learning. We suggest that this can be achieved in a hybrid architecture which integrates a nonlinear filtering stage modelled after the neural computations performed in the early stages of the visual system, and a cognitive reasoning strategy that operates in an adaptive fashion on a belief distribution.

1 Introduction

In many application contexts, intelligent systems are confronted with problem of using their resources in the most efficient way for extracting a maximum amount of information from their current environment. Here we consider a system for the analysis, representation and exploration of spatial scenes. Biological vision systems have evolved an efficient design for solving this task, and this design has also been adapted for artificial vision systems in the context of "active vision" [1][2]. First, the pattern recognition capabilities are concentrated in a small region of the visual field, the central fovea, whereas the periphery has only limited optical resolution and processing power. With a static eye, one can hence only see a small spot of the environment with good quality, but this spot can be rapidly moved with fast saccadic eye movements of up to 700 deg/sec towards all the "relevant" regions of a scene. Second, there exists an efficient process which controls the selection of these informative regions. The selection is determined by both bottom-up processes, which extract the salient local features in a scene, and by cognitive top-down processes, which are determined by the memory, internal states and current tasks [3] [4]. Togther, this system allows for drawing efficient conclusions about a current scene and the environment.

I. Bloch, A. Petrosino, and A.G.B. Tettamanzi (Eds.): WILF 2005, LNAI 3849, pp. 319–325, 2006.

2 Maximizing the Information Gain

A basic principle in the development of our system is the maximization of the information that the system can extraxt from the environment. This is achieved on two interacting processing levels (Fig. 1). On the bottom-up level, sensory feature extraction is achieved by neural operators that are optimized with respect to the efficient exploitation of the statistical redundancies in natural scenes (for review see, e.g., [5],[6]). For the cognitive level, we have developed a top-down strategy which guides the system's explorative actions towards those regions of the environment which have the greatest potential for changing the current internal belief distribution of the system. The two processing levels interact in three parallel streams, two bottom-up streams and one top-down stream. In the first bottom-up stream, nonlinear neurons of the early sensory processing stage provide information about all salient locations that could possibly be of interest for the exploration. In the other bottom-up stream, those salient regions which are selected by the top-down strategy are analysed by a set of orientation-selective wavelet filters. This results in a feature vector for each fixated region, and a combination of these feature vectors and the associated motor commands yields the *sensorimotor features*, which are the basic elements of the scene representation. This is in contrast to classical computer vision approaches, even in active vision, which typically operate on purely sensory data.

The sensorimotor features induce a belief distribution on a hierarchically structured representation of scene hypotheses. A belief-based reasoning strategy operates on this hierarchical structure and selects the optimal top-down command, i.e., the motor action towards this salient position, which currently promises to maximize the information about the given environment.

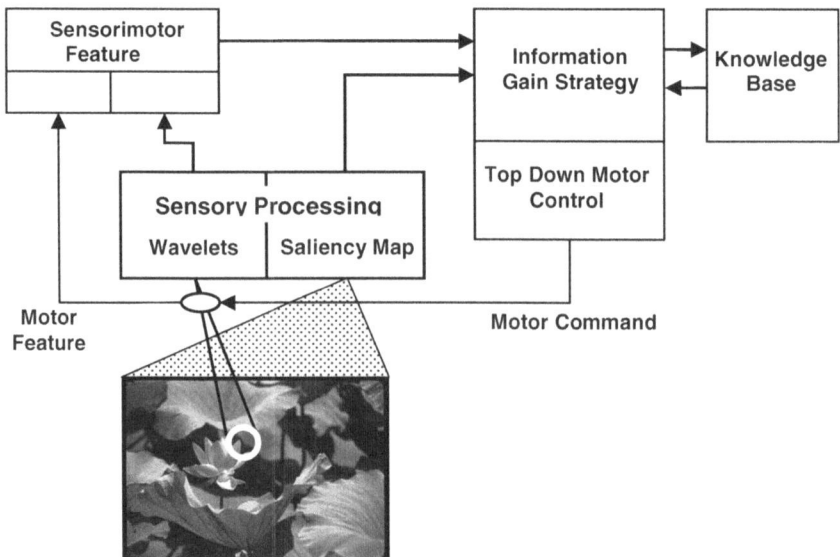

Fig. 1. Hybrid System Architecture

3 Nonlinear Neural Processing

The sensory processing stage of our system consists of a wavelet-like image decomposition by size- and orientation-specific filters and of nonlinear operators which model the extraclassical receptive field properties of cortical neurons. The spatial resolution is maximal in the centre and lower in the periphery. The sensory system has two functions: (i) identification of a-priori informative locations (salient features) within the scene, which can be a potential goal of saccadic fixations (cf. [7]), and (ii) provision of high-resolution information about the actually fixated local pattern.

Regarding the potential fixation candidates we conducted a statistical investigation of experiments on eye movements of human subjects on natural scenes. Analysis of the fixated regions in terms of polyspectra (a generalization of the power spectrum [8]) revealed a clear bias to fixate image regions with frequency components of multiple orientations (e.g. image regions with curved edges, junctions, or occlusion patterns) [9]. Formally, these highly informative local patterns can be defined as intrinsically two-dimensional (i2D-) signals, and the framework of Volterra-Wiener systems [10] can be employed to design i2D-selective operators [6]. The response of a quadratic Volterra system can be described as

$$u_2(\mathbf{x}) = \iint h_2(\mathbf{x}_1, \mathbf{x}_2) \cdot u_1(\mathbf{x} - \mathbf{x}_1) \cdot u_1(\mathbf{x} - \mathbf{x}_2) \, d\mathbf{x}_1 d\mathbf{x}_2 \qquad (1)$$

where $h_2(\mathbf{x})$ is the kernel of the system [10]. The critical condition for the selectivity to i2D signals can be imposed as condition on the spectral kernel $H_2(\mathbf{f})$ (the Fourier transform of $h_2(\mathbf{x})$) as

$$H_2(\mathbf{f}_1, \mathbf{f}_2) = H_2\left(\left[f_{x1}, f_{y1}\right], \left[f_{x2}, f_{y2}\right]\right) = 0 \quad \forall \, f_{x1} \cdot f_{y2} = f_{y1} \cdot f_{x2} \qquad (2)$$

The resulting operators mimic the nonlinear extraclassical receptive field effects of neurons in the visual cortex [6][11]. With respect to the image statistics, they perform a higher-order whitening strategy [6]. The susceptibility to noise is reduced by the band limitation (low spatial resolution). The result is similar to the saliency computation used in [7], although the latter is based on an iterative center-surround modification of linear filter outputs, whereas our operators provide a direct nonlinear processing. Two examples of the resulting nonlinear feature maps (saliency maps) are shown in Fig. 2.

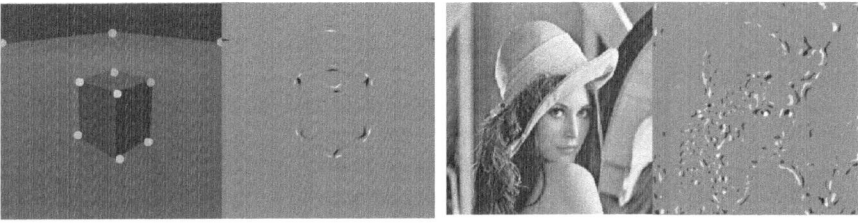

Fig. 2. Potential fixation candidates (salient features) as derived from non-linear i2D-selective neural operators. Left: application to a cubic object in a VR environment we use for spatial navigation tasks. Right: application to a more complex natural image.

The second function of the sensory processing stage is the provision of a sensory characterization of the fixated image regions. At each fixated location the outputs of a set of orientation-selective wavelet-like filters are used to compute a visual sensory feature vector $\mathbf{v}$. The sensory feature vectors $\mathbf{v}_1$ and $\mathbf{v}_2$ at the start and the end of the eye movement, together with the eye movement motor data $\mathbf{e}$ are combined into the sensorimotor feature vector $\mathbf{s}=[\mathbf{v}_1, \mathbf{e}, \mathbf{v}_2]$. Via a coarse coding this $\mathbf{s}$ is assigned to one of the sensorimotor feature classes S_i, which are the basic elements of the scene representation.

4   Uncertain Sensorimotor Features and Belief Combination

A sensorimotor feature S_i usually supports one or more scene hypotheses H_k with a certain degree of belief $m_i(H_k)$. In order to express this belief and to combine the belief induced by a number of sensorimotor features the "Belief Theory" of Shafer is used [12]. In this theory the frame of discernment θ is the set of all possible singleton hypotheses in the domain, in our system the set of all individual scenes that are considered. The resulting hypotheses space 2^θ comprises all possible subsets H_k of θ. The heart of the theory is the Dempster Rule of Combination. This rule can be used to update the current belief $m(H_k)$ in a hypothesis H_k by incoming evidence $m_i(H_l)$ about a hypothesis H_l as induced by a sensorimotor feature S_i as:

$$\forall H_k \in 2^\theta, H_k \neq \varnothing \; H_l, H_n \in 2^\theta : m(H_k) = \sum_{H_l \cap H_n = H_k} m\;(H_l) * m_i(H_n) * K^{-1} \tag{3}$$

with the normalization constant $K = 1 - \sum_{H_l \cap H_n = \varnothing} m(H_l) * m_i(H_n)$

In using this normalization K we assume a closed world in which all possible hypotheses are known. For the alternative case of an open world assumption Smets [13] has shown the advantages of omitting this normalization. The suggested information gain principle is independent of the respective normalization assumptions and can work with or without it.

In our case the hypotheses space 2^θ is reduced to a strict hierarchy T. Each non-leaf element corresponds to a scene class and each leaf element corresponds to an individual scene (or more precisely to the finest scene class). In order to combine the m-values in such a strict hierarchy we are using an approximation of the Dempster rule for tree-like knowledge structures [14]. The current belief $m_T(H_k)$ is the combination of the basic belief masses $m_{Ti}(H_k)$ [13] which correspond to those yet selected sensorimotor features S_i that induce belief for a hypothesis H_k in the hierarchy T. In this calculation one has to consider all those combinations of hypotheses H_l and H_n in the hierarchy which have H_k as intersection. It should be noted that one sensorimotor feature S_i usually induces belief m_{Ti} not only for one but for several different hypotheses H_k in T, i.e. a sensorimotor feature S_i induces a basic belief assignment m_{Ti} which supports a number of scene classes and/or individual scenes H_k to different degrees of belief $m_{Ti}(H_k)$.

The basic belief mass $m_{Ti}(H_k)$ is learned while the system generates fixation sequences on one or more training scenes. The "knowledge base" of the system, i.e. the information about all analyzed scenes, is continuously updated using all sensorimotor features S_i that result from the fixation sequences and the respective scene hypotheses H_k. The knowledge base consists of a matrix $M = \{x_{ik}\}$, where each row i is associated with a sensorimotor feature S_i and each column k is associated with a scene hypothesis (or scene class hypothesis) H_k. Each element of M relates sensorimotor features S_i and possible hypotheses H_k by the "relative frequency" of S_i selected in a scene H_k of the hierarchy:

$$\forall(H_k \in T): \quad r_i(H_k) = \frac{x_{ik}}{\sum_l x_{il}}$$

(4)

where x_{ik} is the frequency of a certain sensorimotor feature S_i selected on a scene H_k and Σx_{il} is the sum over all occurences of S_i in all yet analyzed scenes. We have shown that $r: T \rightarrow [0,1]$ corresponds to a basic belief assignment (bba) $m: 2^\theta \rightarrow [0,1]$ of the Dempster-Shafer theory [15].

5   A Hybrid Inference Strategy IBIG

The basic principle of our strategy IBIG (Inference by Information Gain) is to determine those data which, when collected next, would yield the largest information gain with respect to the actual belief distribution in the hypotheses space [16]. Applied to scene analysis the inference strategy determines the region of interest that promises the maximum information gain with respect to all activated scene hypotheses [15].

The information gain is calculated in the following way: a potential belief $\hat{m}_{Ti}(H_k)$ is calculated separately for each sensorimotor feature $\hat{S}_i$, i.e., for each $[\mathbf{v}_1, \mathbf{e}, \mathbf{v}_2]$ that is compatible with the currently fixated pattern $\mathbf{v}_1$, the saliency map of the potential targets $\mathbf{v}_2$, and the eye movement $\mathbf{e}$. For each such $\hat{S}_i$, (i) all hypotheses H_k that can be influenced by this $\hat{S}_i$ are determined, (ii) the potential beliefs $\hat{m}_{Ti}(H_K)$ are calculated according to the Dempster rule of combination for hierarchical hypotheses trees, and (iii) the potential information gains $I_i(H_k)$ are determined. by the absolute difference between the current belief $m_T(H_k)$ of a scene hypothesis H_k in the hierarchy and the potential belief $\hat{m}_{Ti}(H_k)$ that could be reached for this hypothesis due to $\hat{S}_i$: $I_i(H_k) = |m_T(H_k) - \hat{m}_{Ti}(H_k)|$. In a final step, the $\hat{S}_i$ with the maximum $I_i(H_k)$ is selected and determines the next eye movement.

6   Discussion

Let us describe one cycle of the resulting system behaviour (a screenshot of the system is shown in Fig. 3). Assume a region of interest in the scene has been determined. The fovea is then directed towards this location by a saccadic movement. The wavelet

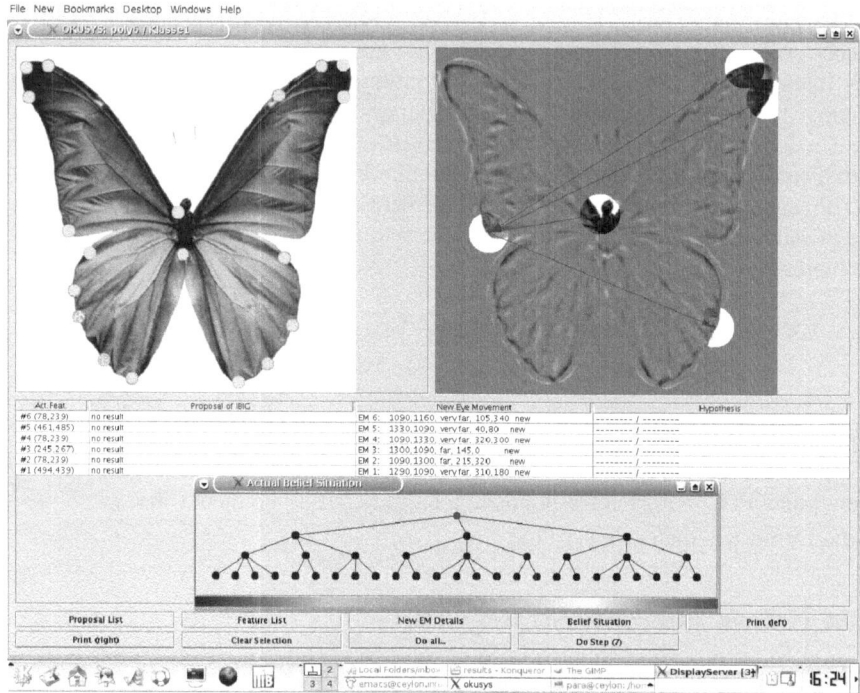

Fig. 3. Screen dump of the system analysing a butterfly from a hierarchical taxonomy of butter-flies which are categorized depending on the profile and pattern of the wings. On the left side the candidate fixations based on i2D-selective features are marked. The right side shows the output of the nonlinear operator with an overlaid fixation sequence. The tree of hierarchical representation is shown at the bottom.

vector at this new foveal position and the vector at the previous position are combined with the motor data into the sensorimotor feature S_i. This sensorimotor feature is used to update the current belief distribution in the knowledge base. IBIG then considers the remaining candidate targets, as derived from the nonlinear i2D-selective neural operator (only those sensorimotor features are considered whose stored pre-saccadic feature vector is consistent with the actual feature vector at the given foveal position, and whose relative target position is within the spatial range of one of the candidate locations.). From the set of possible targets it selects the one that promises the maximum information gain, i.e. the maximum change of the current belief situation, and a new saccade is initiated towards this location. This cycle is repeated until a sufficient belief threshold for one of the hypotheses is reached.

In conclusion, we have developed a system for the analysis of a spatial environment. It is a hybrid system that combines sensory filtering operations with an adaptive cognitive reasoning strategy. All modules are designed according to the principle of maximum information gain. The resulting architecture is able to cope with the typical incompleteness and uncertainty in sensory input data and provides an efficient representation and processing of spatial configurations.

Acknowledgment. Supported by DFG, SFB TR8 6023 Spatial Cognition.

References

1. Ballard, D.H.: Animate Vision. Artificial Intelligence, 48 (1991) 57–86
2. Fermüller, C., Aloimonos, Y.: Vision and Action. Image and Vision Computing 13 (1995) 725-744
3. Yarbus, A.L.: Eye Movements and Vision, Plenum Press. New York (1967)
4. Noton, D., Stark, L.: Scanpaths in Saccadic Eye Movements while Reviewing and Recognizing Patterns. Vision Res. 11 (1971) 929-942
5. Simoncelli, E., Olshausen, B.: Natural Image Statistics and Neural Representation. Annu. Rev. Neurosci. 24 (2001) 1193–1216
6. Zetzsche, C., Krieger, G.: Nonlinear Mechanisms and Higher-Order Statistics in Biological Vision and Electronic Image Processing: Review and Perspective. J. Electron. Imaging 10(1) (2001) 56-99
7. Itti, L., Koch, C.: Feature Combination Strategies for Saliency-Based Visual Attention Systems. J. Electron. Imaging, 10(1) (2001) 161-169
8. Nikias, C. L., Petropulu, A. P. : Higher-Order Spectral Analysis: A Nonlinear Signal Processing Framework. Prentice-Hall, Englewood Cliffs, NJ (1993)
9. Zetzsche, C., Schill, K., Deubel, H., Krieger, G., Umkehrer, E., Beinlich, S.: Investigation of a Sensorimotor System for Saccadic Scene Analysis: An Integrated Approach. In: Pfeifer, R., Blumenberg, B., Meyer, J.-A., Wilson, S.W. (eds): From Animal to Animats - Proc. 5th Intl. Conf. Soc. Adaptive Behavior, Vol. 5, MIT Press, Cambridge, MA (1998) 120-126
10. Schetzen, M.: The Volterra and Wiener Theories of Nonlinear Systems. Krieger's, Malabar, FL, updated edition (1989)
11. Shapley, R.: A New View of the Visual Cortex. Neural Networks 17 (2004) 615-623
12. Shafer, G.: A Mathematical Theory of Evidence, Princeton University Press, Princeton (1976)
13. Smets, P.: "Belief Functions", In: Smets, P., Mamdani, E.H., Dubois, D., Prade, H. (eds.): Non-Standard Logics for Automated Reasoning, Academic Press, London, UK, (1988) 253-286
14. Gordon, J., E. H. Shortliffe, E.H.: A Method for Managing Evidential Reasoning in a Hierarchical Hypothesis Space. Artif. Intell. 26 (1985) 323-357
15. Schill, K., Umkehrer, E., Beinlich, S., Krieger, G., Zetzsche, C.: Scene Analysis with Saccadic Eye Movements: Top-Down and Bottom-Up Modelling. J. Electron. Imaging 10(1) (2001) 152-160
16. Schill, K.: Decision Support Systems with Adaptive Reasoning Strategies. In: Freksa, C., Jantzen, M., Valk, R. (eds.): Foundations of Computer Sciences: Theory, Cognition, Application, Springer, Berlin Heidelberg New York (1997) 417-427

KANSEI-Based Image Retrieval Associated with Color

Sunkyoung Baek[1], Miyoung Cho[1], Myunggwon Hwang[1], and Pankoo Kim[2,*]

[1] Dept. of Computer Science, Chosun University, Gwangju 501-759, Korea
{zamilla100, irune, hmk2958}@chosun.ac.kr
[2] Dept. of CSE, Chosun University, Korea
pkkim@chosun.ac.kr

Abstract. Nowadays, the processing of KANSEI information is very important in intelligent computing field. Particularly, it is very interesting in image retrieval to deal with human's KANSEI. In this paper, we use natural language for the representation of KANSEI, including the image structure of Human's idea, which we can not observe. And then, a KANSEI-Adjective is used as a natural language querying method: In other words, this paper presents the image retrieval based on KANSEI. We propose the background image retrieval based on KAC (KANSEI-Adjective of Color) to represent the sensibility of color. Our method for processing of KANSEI information is the measure of similarity by using the adaptive Lesk algorithm in WordNet. In our experimental results, we are able to retrieve background images with the most appropriate color in term of the query's feeling. Furthermore, the method achieves an average rate of 63% user's satisfaction.

1 Introduction

KANSEI in Japanese means by sensibility that is to sense, recall, desire and think of the beauty in objects [1]. KANSEI is expressed usually with emotional words for example, beautiful, romantic, fantastic, comfortable etc [2]. The concept of KANSEI is strongly tied to the concept of personality and sensibility. KANSEI is an ability that allows humans to solve problems and process information in a faster and personal way.

The processing of KANSEI information is very important in intelligent computing field. Particularly, it is very interesting in image retrieval to deal with human's KANSEI. So many investigators give a trial of using KANSEI for image retrieval. However, the research in this area is still under its primary stage because it is difficult to process emotion or sensibility of human. KANSEI is the Knowledge based on individual experience and distinct from each person.

The existing image retrieval system is capable of understanding the semantics of visual information based on generic features such as color, texture and shape, and it is well within the realm of the technically possible [3]. However, we focus on image retrieval based on KANSEI which is more intelligent than that based on content.

In order to apply KANSEI methods to image retrieval, Hayashi et al. [4] attempted to train a neural network to predict human KANSEI with impression words that

---

* Corresponding author.

I. Bloch, A. Petrosino, and A.G.B. Tettamanzi (Eds.): WILF 2005, LNAI 3849, pp. 326–333, 2006.
© Springer-Verlag Berlin Heidelberg 2006

would be evoked by outdoor scenery images. Shibata et al. [5] defines KANSEI methods as "discriminated subjective interpretation which can be categorized as groups of adjectives." They emphasize the contrast between psychology and computer science, proposing KANSEI methods as the means to join the two. Their research goal is to develop an image retrieval system for street landscapes based on the KANSEI model. Kobayashi made a relation between color and language at the researching of color image standardization and Haruyoshi expresses a language that a lot of color included the image at the questionnaire in Japan [6] [7]. Color Wheel Pro explains meaning 9 based color including a Red and local color. And then, Hewlett-Packard defines the color meaning at 20 colors in the USA. In the Republic of Korea, IRI develop the I.R.I adjective image scale at a visual and symbol of Korean's KANSEI [8].

The image retrieval system based on KANSEI has emerged as a promising yet challenging research area in the past few years. However, as yet, no existing system is capable of completely understanding the KANSEI of color information. In this paper, we propose the image retrieval based on KANSEI of color. We experiment the background image retrieval based on KANSEI-Adjective to represent the sensibility of color.

2 The Proposed Architecture

KANSEI information of color. First of all, we need KANSEI information of color. Because the definition of color by human is subjective and ambiguous, most of the researchers used SD (Semantic Differential) technique through replication or statistics. In this paper, we use 20 colors that are defined "The meaning of color" by HP as KANSEI database of fig.1 [9]. The table lists vocabularies that express KANSEI about each of the color. In this paper, we named the meaning of color "KAC".

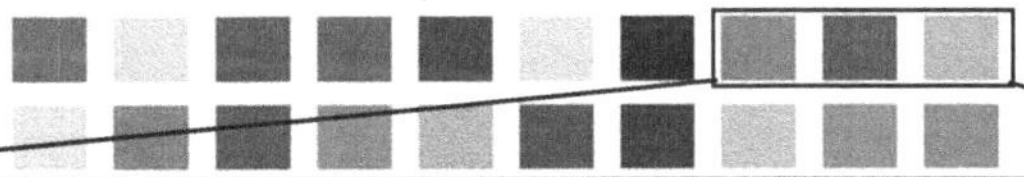

Color	KAC(KANSEI-Adjective of Color)
Lavender	enchanting, nostalgic, delicate, floral, sweet, fashionable
Blue	true, healing, tranquil, stabile, peaceful, harmonic, wise, trustable, calm, confidential, protective, secure, loyal
Fuchsia	hot, sensual, exciting, bright, funny, energetic, feminine

Fig. 1. "The Meaning of Color" of Hewlett-Packard Co

Relationship of adjectives for similarity measure. We can cover that various queries including KANSEI on the basis of the WordNet. We take advantage of WordNet that is a kind of lexical database [10]. In this paper, we put forward method of using Adjective in WordNet. We use 4 relations (attribute, also see, similar to, pertainym of) for similarity measure between KAC and query. We show the 4 relations as follows.

Table 1. Various relationships for similarity measure

Relationship	Adjectives
Attribute	650
Also see	2,714
Similar to	22,492
Pertainym of	4,433

Similarity Measure between Query and KAC. We use the adaptive Lesk algorithm put forward by Banerjee and Pedersen for query of extension. As Lesk's algorithm disambiguates a target word by selecting the sense whose dictionary gloss shares the largest number of words with the glosses of neighboring words, it only uses 4 relations for similarity measure between adjectives [11].

The original Lesk algorithm compares the glosses of a pair of concepts and computes a score by counting the number of words that are shared between them. This scoring mechanism does not differentiate between single word and phrasal overlaps and effectively treats each gloss as a bag of words. For example, it assigns a score of 3 to $bank_2$: (sloping **land** especially beside a **body** *of* **water**) and lake: (**body** *of* **water** surrounded by **land**), since there are 3 overlapping words: land, body, water. Note that stop words are removed, so *of* is not considered an overlap.

However, there is a Zipfian relationship between the lengths of phrases and their frequencies in a large corpus of text [12]. The longer the phrase is, the less likely it is to occur multiple times in a given corpus. A phrasal n–word overlap is a much rarer occurrence than a single word overlap. Therefore, we assign an n word overlap the score of n^2. This gives an n–word overlap a score that is greater than the sum of the scores assigned to those n words if they had occurred in two or more phrases, each less than n words long. This is true since the square of a sum of positive integers is strictly greater than the sum of their squares. That is, $(a_0 + a_1 + ... + a_n)^2 > a_0^2 + a_1^2 + ... + a_n^2$, where a_i is a positive integer. For the above gloss pair, we assign the overlap **land** a score of 1 and **body of water** a score of 9, leading to a total score of 10.

Table 2. Tracing example between words

Synset1: *friendly#a#1*
Synset2: *peaceful#a#1*
Functions: *also* glos - *also* glos : 47
Overlaps: 1 × "of" 1 × "characterized by" 1 × "characterized by friendship and good will" 2 × "to" 4 × "or"
Functions: *also* glos - glos : 3
Overlaps: 1 × "by" 2 × "or"
Functions: *also* glos - *sim* glos : 20
Overlaps: 1 × "of" 1 x "conducive to" 1 × "and" 1 × "characterized by" 1 × "nature" 2 × "to" 1 × "disposed to" 1 × "inclined" 2 × "or" ~
Overlaps: 1 × "of" 2 × "the" 1 × "by" 1 × "and" 1 × "not" 3 × "or"
Functions: *sim* glos - glos : 4
Overlaps: 1 × "by" 1 × "not" 2 × "or"
Functions: *sim* glos - *sim* glos : 10
Overlaps: 1 × "of" 1 × "by" 1 × "and" 3 × "a" 1 × "disposed" 1 × "not" 2 × "or"

Table 2 shows tracing the result of similarity measure between KAC and query using the adaptive Lesk algorithm. The following table shows similarity measure using semantic relation between *friendly#a#1* and *peaceful#a#1*. Here, *also* indicates **also see** relation between senses of word and then *sim* is **similar to** relation. For example, the overlap value is calculated using **also see** relation among grosses of synset1 and synset2. It is sum of square of overlapped words using the adaptive Lesk algorithm. It is calculated $1\times$"of", $1\times$"characterized by", $1\times$"characterized by friendship and good will", $2\times$"to", $4\times$"or". Result is $1^2+2^2+6^2+2+4=47$. When we calculate, we consider the words whose values less than 100 are insignificant words such as article, adjunction and so on. So we just select more than 100 values from calculated values.

The following table shows the result of similarity measure between KAC and Query.

Table 3. Similarity measure between KAC and Query

KAC	Similarity according to Query					
	Warm	Energetic	Alterative	Fortunate	Interest	Peaceful
Optimistic	28	38	7	21	26	27
Dynamic	49	**175**	20	49	55	50
Excite	68	25	11	18	54	26
Sexy	54	64	24	43	57	60
Intense	58	52	14	43	83	65
Aggressive	59	57	18	29	52	61
Powerful	69	23	15	46	57	38
Energetic	43	**673**	13	32	39	62
Vigorous	19	**122**	12	11	20	28
Elegant	45	63	21	35	44	61
Rich	24	25	10	21	17	26
Mature	**198**	17	12	28	57	34
Healing	23	11	**251**	9	32	23
Peaceful	49	62	22	31	57	**712**
Wise	43	44	11	37	42	43
Calm	27	20	14	14	24	**139**
Protective	49	39	27	35	66	44
Secure	47	40	15	51	**111**	56
Loyal	18	22	9	22	32	24
Natural	39	34	16	31	49	46
Lucky	25	12	14	80	46	17
Hopeful	30	23	13	**133**	16	23
Successful	65	37	18	**142**	71	48
Generous	23	20	15	8	29	30
Romantic	45	14	10	9	33	23
Soft	63	52	14	46	63	59
Delicate	51	40	23	24	54	48
Sweet	52	33	10	38	67	40
Friendly	92	62	23	58	67	**103**
Tender	**137**	44	21	53	**198**	84
Hot	**1284**	80	33	77	**115**	88
Sensual	21	16	10	16	21	18
Bright	52	41	12	39	55	39

3 Experimental Procedures and Results

We did the experiment for retrieval of background image based on KANSEI. The experiment performed to evaluate the matching of color through similarity measure between the query and KAC. We use backgrounds images of Microsoft Co, as database included 481 images.

In this experiment, we propose 1) this method to extract dominant color using color histogram of background images, then, 2) normalize the extracted color based on 20 colors "meaning of color" by Hewlett-Packard Co. Next phase is 3) the measure similarity between KAC and user's query using the adaptive Lesk algorithm. Using these results, we 4) match most adaptive color and retrieve the background images.

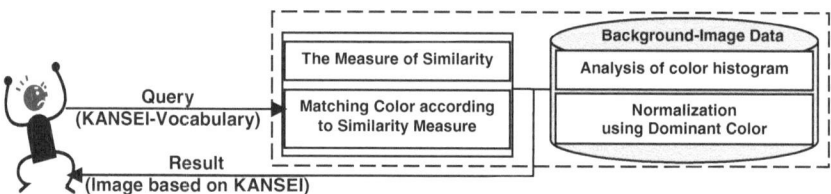

Fig. 2. Phases of experiment

We know the following result in a process of experiment. It indicates how the color is matching with KANSEI vocabulary by using result of similarity measure. Also, we know that selected KAC is the best matching with query. So, table 4 shows us the final results of matching adaptive colors to vocabulary's feeling in term of user's query.

Table 4. Result of matching colors

Query	Similarity	KAC	Color	Query	Similarity	KAC	Color
Warm	1284	Hot	Fuchsia	Interest	198	Tender	Light pink
	198	Mature	Burgundy		115	Hot	Fuchsia
	137	Tender	Light pink		111	Secure	Blue
Energetic	673	Energetic	Bright red	Peaceful	712	Peaceful	Blue
	673	Energetic	Fuchsia		139	Calm	Blue
	175	Dynamic	Bright red		120	Tranquil	Blue
Alterative	251	Healing	Blue	Fortunate	142	Successful	Green
	251	Healing	Green		133	Fortunate	Green

Fig. 3 shows the result of running the query with KANSEI relativity and color. It retrieves background images including suitable color which represents KANSEI about "energetic" query and "peaceful" query. And the O/X checkboxes showed in the figure are an implement for measure of user's satisfaction. Also we measure the satisfaction of antonym query for evaluation.

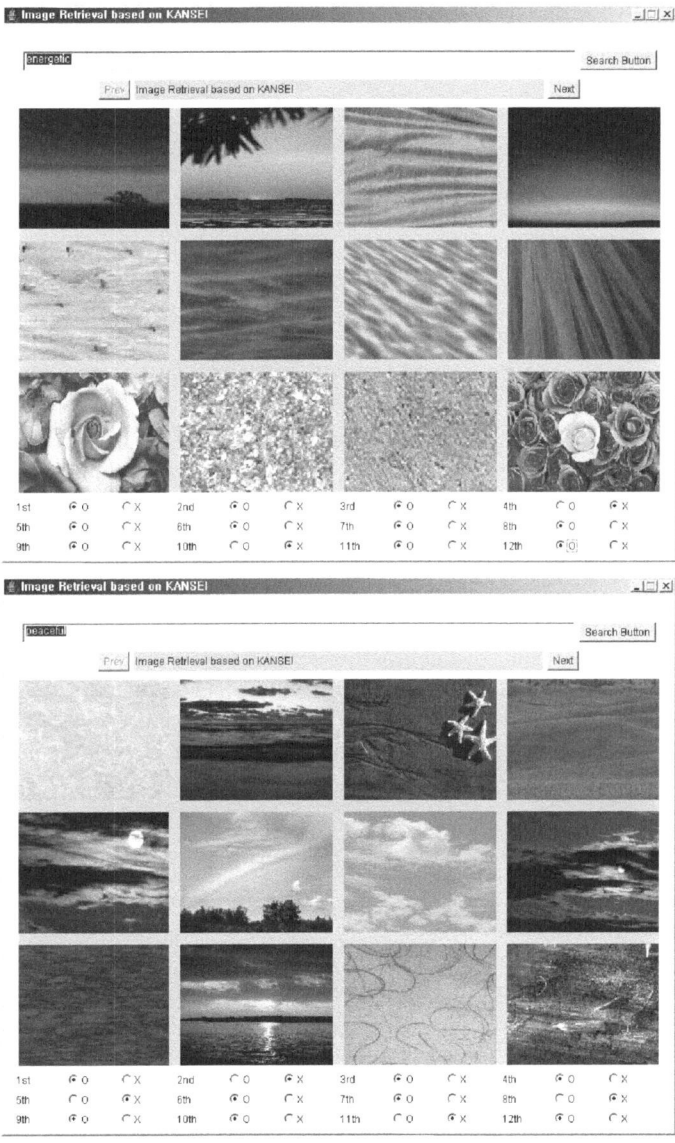

Fig. 3. Results of "energetic" query and "peaceful" query

We exclude KANSEI about texture, shape, pattern and so on as low-level features of visual information, and only concentrate on the processing method of KANSEI about color. We just measure user's satisfaction for evaluation of image retrieval based on KANSEI since KANSEI information retrieval is hard to measure accuracy. We experiment the image retrieval according to any 30 queries (KANSEI- Vocabulary)

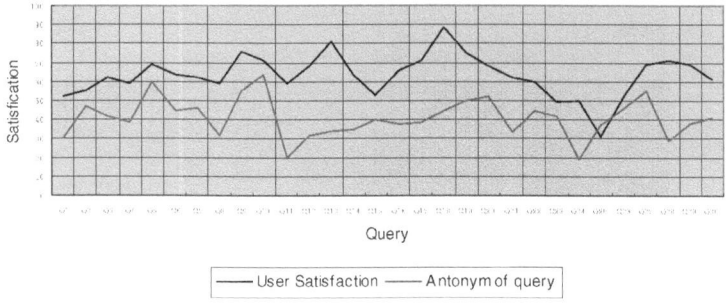

Fig. 4. Graph of evaluation in term of query

and then measure satisfaction to 50 people-oriented. In this experiment, user's satisfaction indicates much difference from 31% to 89% according to KANSEI. The reason of difference is that KANSEI strongly depends on personal disposition.

In this fig.4, the wave-blue line stands for the rate of user satisfaction when the users query the certain words. And the wave-red line indicates the rate for satisfaction of the antonymous meanings of the corresponding words. The later rate was also obtained by checking the O/X checkboxes based on the query results in the last step. We can see that there is the distinguish difference between these two rate. It manifests that our method proposed for query is more efficient and accurate.

In experimental result, our method achieves an average rate of 63% user's satisfaction. Our result is not so high because background image includes objects, shape, pattern and texture (not only color).

4 Conclusions

In this paper, we proposed a method of measuring similarity between KAC and user's query. On the basis of this method, we show how to retrieve background images based on KANSEI. In result, we retrieve most appropriate image by human's KANSEI and also achieves user's satisfaction. Our research means to use similarity of adjective on WordNet as lexical Ontology and applying KANSEI Information to image retrieval.

In our future works, we will expand the range of visual information (not only color) and will study the relation between visual information and KANSEI based intelligent computation methods.

Acknowledgement

This work was supported by the Korea Research Foundation Grant. (KRF-2004-042-D00171).

References

1. Hideki Yamazaki, Kunio Kondo, "A Method of Changing a Color Scheme with KANSEI Scales," Journal for Geometry and Graphics, vol. 3, no. 1, pp.77-84, 1999
2. Shunji Murai, Kunihiko Ono and Naoyuki Tanaka, "KANSEI-based Color Design for City Map," ARSRIN 2001, vol. 1, no. 3, 2001
3. A.Ono, M. Amano, M. Hakaridani, T. Satou, M. Sakauchi, "A fiexible Content-based Image Retrival System with Combined Scene Description Keyword," Proceeding of Multimedia 96, pp.201-208, 1996
4. Hayashi, T., Hagiwara, M, "An image retrieval system to estimate impression words from images using a neural network", 1997-IEEE International Conference on Systems, Man, and Cybernetics-Computational Cybernetics and Simulation, vol 1, 150-5, IEEE, New York, NY, 1997
5. Shibata, T., Kato, T, "KANSEI image retrieval system for street landscape discrimination and graphical parameters based on correlation of two images", IEEE-SMC'99 Conference Proceedings-1999 IEEE International Conference on Systems, Man, and Cybernetics, , vol 6, 247-52, IEEE, Piscataway, NJ, 1999
6. Kobayshi Singenobu, "Color Image Scale," Kodansha America, 1990
7. Haruyoshi Nagumo, "Color Image Chart," Chohyung Publishing Co., 2000
8. I.R.I, "Adjective Image Scale", http://www.iricolor.com/04_colorinfo/colorsystem.html
9. Hewlett-Packard, "The Meaning of Color", http://www.hp.com/united-states/public/color/meaning.html
10. George A. Miller, "WordNet: a lexical database for English," Communications on the ACM, 1995
11. S. Banerjee, T. Pedersen, "An adapted Lesk algorithm for word sense disambiguation using WordNet," In Proceedings of the Third International Conference on Intelligent Text Processing and Computational Linguistics, Mexico City, pp. 136–145, 2002
12. S. Banerjee, T. Pedersen, "Extended gloss overlaps as a measure of semantic relatedness," In Proceedings of the Eighteenth International Joint Conference on Artificial Intelligence, Acapulco, pp. 805–810, 2003

Mass Detection in Mammograms Using Gabor Filters and Fuzzy Clustering

M. Santoro, R. Prevete, L. Cavallo, and E. Catanzariti*

Department of Physical Sciences, University of Naples Federico II,
INFN, Section of Naples
{santoro, ezio, prevete}@na.infn.it

Abstract. In this paper we describe a new segmentation scheme to detect masses in breast radiographs.

Our segmentation method relies on the well known *fuzzy c-means* unsupervised clustering technique using an image representation scheme based on the local power spectrum obtained by a bank of Gabor filters.

We tested our method on 200 mammograms from the CALMA database. The detected regions have been validated by comparing them with the radiologists hand-sketched boundaries of real masses. The results, evaluated using ROC curve methodology, show that the greater flexibility and effectiveness provided by the fuzzy clustering approach benefit from an image representation that combine both intensity and local frequency information.

1 Introduction

Developing and testing new segmentation schemes that can be successfully applied to mammographic images is a big challenge motivated by the significant increase, during the last few years, of the demand for computerized medical imaging systems in radiology [1]. This is due mostly to the institution of radiological screening programs for early diagnosis of cancer in most industrialized countries.

At present, mammograms, i.e. X-ray images of the breast, are considered the best method for screening wide groups of asymptomatic women for an effective early detection of breast cancer. These screening programs are generating a large number of mammograms which must be interpreted by a limited number of expert radiologists. Therefore, computerized systems have been developed to aid radiologists working in mammography to assess correct diagnosis. The goal of these systems is to focus the radiologist's attention on suspicious areas.

Masses are one of the classic mammographic signs of malignancy. They are regions of the breast that appear "different" from other areas of the same (or opposing) breast. They generally exhibit poor image contrast and are largely

---

* The work described in this paper has been developed in the frame of a national collaboration between physicians and radiologists supported by the Italian National Institute for Nuclear Physics (MAGIC5 Project).

I. Bloch, A. Petrosino, and A.G.B. Tettamanzi (Eds.): WILF 2005, LNAI 3849, pp. 334–343, 2006.

similar to the surrounding breast tissue which is also not uniform. This fuzzy nature exhibited by mammographic images has led to the use of fuzzy image processing techniques for their interpretation. In this paper we propose an unsupervised fuzzy segmentation scheme to detect masses in mammograms based on single image decomposition.

2   Background

Any automatic system for the detection of malignant masses can be conceptually schematized as follows. After a preprocessing stage in which all the uninteresting areas of the mammogram, such as the identifying label, are filtered out from the image, a two step procedure is generally applied: first, the breast region is fully examined in search of suspected zones. At this early stage the aim is to detect solitary areas of the breast which appear different from other areas, i.e., areas where tissues seem to divert from normality. Since most masses show as regions with homogeneous gray level intensity, this first step is usually handled as a typical texture segmentation problem. Textural features are extracted from the breast region and different classes of texture patterns are detected on the basis of such features. The result of this first step is the segmentation of the breast image in the so called ROI's, regions of the image requiring further analysis. At this point, a strong sensitivity is requested but a weak specificity is tolerated.

At the second step, ROI's so detected are further processed to filter out those regions not likely to correspond to masses. False positives can be reduced by using computationally inexpensive criteria based, for example, on area, circularity or contrast. This phase is meant to increase the specificity, i.e., to lower the number of regions not likely to correspond to mass tissue.

A comparison can be made at this stage, in which corresponding views of the right and left breast are inspected in search of asymmetric densities which might indicate the possible presence of masses.

Many approaches skip either step 1 or step 2 or they do not explicitly distinguish them. Quite often, both steps are skipped by image analysis algorithms but the job of identifying ROI is left to an expert radiologist who manually draw mass boundaries.

3   The Algorithms

The first step in the mass detection process is the segmentation of the breast image in suspicious and normal regions. Once a suspicious region is found it is easier to proceed with the analysis because, as we pointed out in the previous paragraph, masses found within a mammogram often will have enough distinguishing characteristics to be classified as masses. However, the detection of suspicious regions is a harder task because most masses exhibit poor image contrast and may be highly connected to surrounding normal tissue. Therefore, in order to do that, we need powerful local textural descriptors capable of detecting the structural properties of the image at different levels of resolution both in the spatial

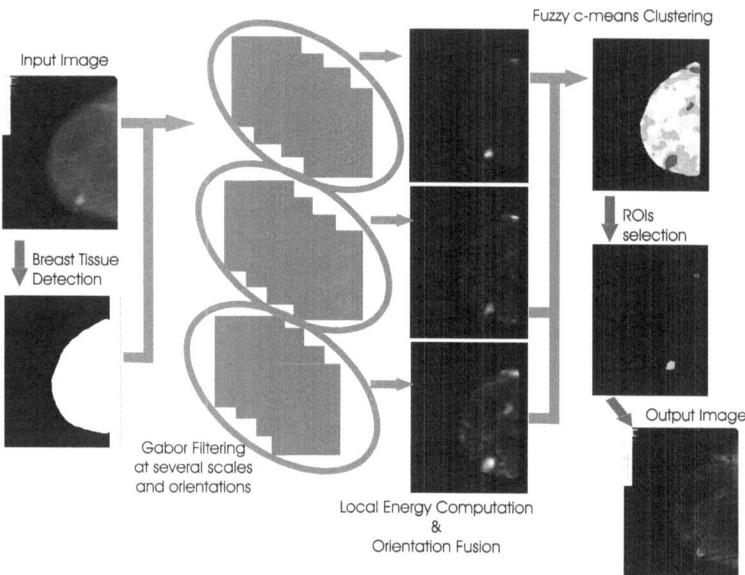

Fig. 1. The pipeline of our segmentation scheme

and in the spatial-frequency domain. Our approach employs the class of analytic functions known as Gabor Elementary Functions. Gabor filters have been shown to be very effective by several medical imaging algorithms. They were used, for example, by Ferrari et al. [2] to detect global signs of asymmetry in the fibro-glandular discs of the left and right mammograms of a given subject, by Cen et al. [3] for the registration of 3-D ultrasound images and by us [4] to detect clustered microcalcifications in mammograms. Due to the great variability in the textural appearance of both the pathologic tissue and the normal breast pattern no prior knowledge about the imaged domain is assumed, hence, an unsupervised scheme of processing must be used. In our approach, the Fuzzy C-means clustering algorithm [5] was employed. A fuzzy segmentation method retains more information from the original image than classical segmentation methods. As far as the number of classes is concerned, we experimentally increased the number of clusters until the class of candidate masses would stand-up quite clearly from the clusters representing different types of textural tissue. The overall work–flow of our segmentation scheme is shown in figure 1

3.1   Feature Extraction and Gabor Filters

In order to extract textural features we convolved the mammographic images with a 2–D version of Gabor filters. Such filters are linear and local and their convolution kernel is the product of a Gaussian with a plane–wave function. A 2–D Gabor filter acts as a local band–pass filter with optimal joint localization properties in the spatial and in the spatial–frequency domain. The analytic function $h(x, y)$ that represents these filters is:

$$h(x, y) = g((x - x_0)', (y - y_0)')e^{2\pi i(u_0(x-x_0)+v_0(y-y_0))} \tag{1}$$

where:

$$g(x, y) = \frac{1}{2\pi\sqrt{\sigma_x\sigma_y}}e^{-\frac{1}{2}\left[\frac{x}{\sigma_x}^2+\frac{y}{\sigma_y}^2\right]} \tag{2}$$

is a two-dimensional Gaussian centered on the origin and scaled by widths σ_x and σ_y, and $(x', y') = (xcos\varphi+ysin\varphi, -xsin\varphi+ycos\varphi)$ represent the coordinates in the rotated reference system $X'O'Y'$.

Typically, when a Gabor filtering approach is adopted, the image is filtered with a set of filters of different preferred orientations and spatial frequencies that cover appropriately the spatial frequency domain. The results of such convolutions are then joined to form a vector whose components are the pixel feature vector sets. In essence, we choose the bank of Gabor filters in such a way that their central frequency form a lattice sampling the spatial frequency plane [4]. A uniform covering of the frequency half-plane with minimal overlapping at the extremes of their bandwidths is obtained by a sampling rate that, in a polar reference system, is constant with respect to the orientation, and logarithmic with respect to the spatial frequency.

The main difficulty to directly use the feature vectors thus obtained is the high dimensionality of such vectors, which typically have from 12 to 40 components [6]. A clustering algorithm using the complete feature vectors will be very slow and could not be used effectively in a CAD system.

Several approaches have been proposed in literature to decrease the dimensionality of Gabor feature vector. We follow the approach proposed by Jain in [6] We compute the *texture local energy* for each pixel in the convolved images. The use of spatial context allows us to reduce the spatial resolution of the images which leads to a lesser number of feature vectors to be clustered.

An optional post–processing step, in which the responses of filters corresponding to the different orientations are merged together for each value of the spatial resolution, permits a further reduction of the number of features. In the case of mammograms this procedure proved effective in reducing computational overload.

3.2 Fuzzy C-Means Clustering

Fuzzy C-Means is an iterative clustering algorithm developed by Dunn in 1973 [5] and improved by Bezdek in 1981 [7].

It is considered a fuzzy version of the more famous k-means algorithm because it allows feature vectors to belong to multiple clusters, with varying degrees of membership. It represents a significant improvement with respect to the problem of erroneous classifications of vectors. Since each vector belongs to many classes, it does not introduce a large bias as is the case with k-means.

The main difference between k-means and fuzzy c-means is the objective function: $F = \sum_{i=1}^{k}\sum_{j=1}^{n}(w_{ij})^m\|\boldsymbol{x}_i^{(j)} - \boldsymbol{c}_j\|^2$.

The function contains the multiplicative terms $(w_{ij})^m$ corresponding to the membership values. Each $(w_{ij})^m$ can be considered an element of a membership matrix U and represents the degree of membership, in the range $[0; 1]$, of a feature vector xi to the fuzzy cluster c_j . The exponent m, called fuzzifier, determines the level of cluster fuzziness. A large m results in smaller memberships and hence, in fuzzier clusters. Usually good values of m are chosen on the base of some prior domain knowledge or on the base of experimental trials.

A problem with a pixel–based approach to image segmentation is the difficulty to introduce context dependent information into the segmentation scheme. Usually, post–processing procedures are separately applied to the segmented image. During the last few years, however, some attempts have been made to introduce contextual information in the point–by–point segmentation scheme. We studied two variations of the fuzzy clustering approach and we evaluated their effectiveness with respect to the standard FCM in the context of our image domain.

The first algorithm we considered was the *Spatially Guided Fuzzy C-Means* [8], where spatial information, in the form of geometrical shape description which can vary from local intensity neighborhood to a more extended shape model, is introduced during the construction of the cluster prototypes. In this way, both the spectral and spatial neighborhoods of a pixel determine the pixel contribution to a cluster prototype. Results reported in [8] with this augmented fuzzy clustering approach show more homogeneous regions and less spurious pixels.

The second alternative we explored was the algorithm proposed in [9], the *Bias Corrected Fuzzy C-Means*. In such algorithm the objective function of the standard FCM algorithm is modified in order to compensate for spatial inhomogeneities and to allow the labeling of a vector to be influenced by the labels in its immediate neighborhood. The neighborhood effect acts as a regularizer and biases the solution toward piecewise-homogeneous labeling. Such a regularization is reported to be effective in the segmentation of images corrupted by salt and pepper noise, as is the case in medical images.

A qualitative evaluation of preliminary results (not shown here) obtained by applying the proposed segmentation scheme to several images formed by combining different textures patches from Brodatz database showed that, among the three algorithms we implemented, *Fuzzy C-Means* performed the best.

The bad performance of *Spatially Guided Fuzzy C-means* can be schematically explained by the fact that, in our approach, each feature vector already contains information from the pixel spatial neighborhoods. Therefore, it can be argued that the further dependency from the spatial context introduced by the SG–FCM causes erroneous aggregations leading to oversized regions. *Bias–Corrected Fuzzy C-means* also wrongly segments the image. In fact, most of the noise in the original image has been removed by the smoothing filter we had applied when computing the *texture local energy*.

On the basis of the above sketched considerations, we decided to choose FCM for performing clustering.

4   Experimental Results

To test our algorithm we used the CALMA database of mammographic images [10] that was developed in the frame of a national collaboration, supported

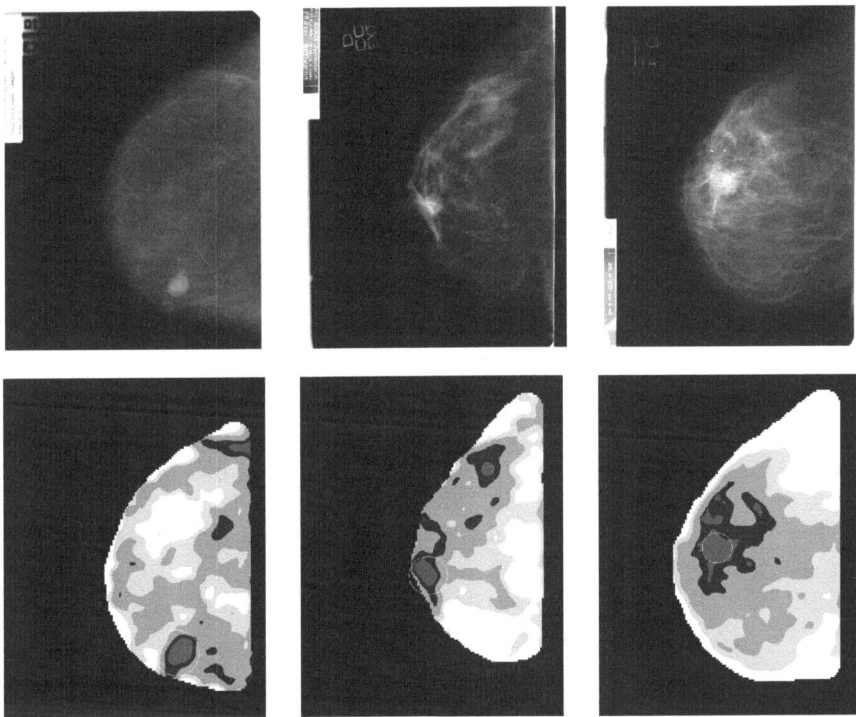

Fig. 2. On the top you can see three mammographic images. On the bottom there are the corresponding segmented images.

by the Italian National Institute for Nuclear Physics, between several Italian Universities and some hospitals. We selected 200 images corresponding to the cranio–caudal projection.

Due to the ill-defined textural characteristics of candidate mass regions and to the great variability in the textural appearance of the normal breast pattern, a segmentation of the breast image in two classes, normal and abnormal, is bound to fail. We experimentally set to six the number of clusters for the clustering algorithm. Figure 2 shows the results of the segmentation on two sample images. The detected ROIs, corresponding to the cluster with the highest response in the *texture local energy*, are shown in red.

Our system has detected 193 of 219 total masses, with an efficiency of 88%. The total number of detected ROIs is 966, 773 of which are false positive; and thus we obtained an average of less than 4 false positives per image.

4.1 Discussion

In order to reduce the number of ROIs, we classified them as either *mass* or *normal tissue* on the basis of four different global parameters: its *absolute dimension* (in pixels), its *relative dimension* with respect to the dimension of the

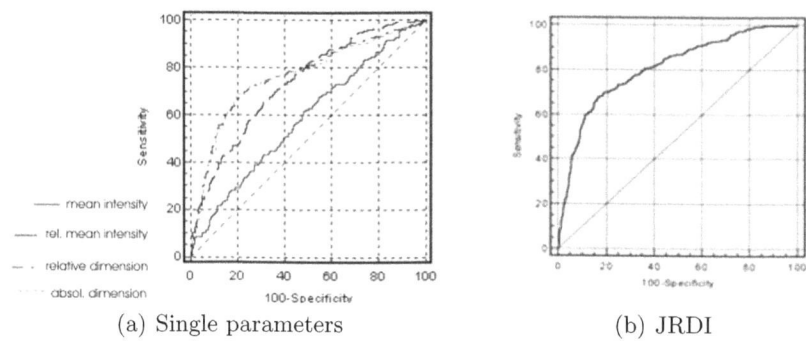

(a) Single parameters   (b) JRDI

Fig. 3. ROC curves measuring the performances of our segmentation scheme

breast (again in pixels), its *mean intensity* and, finally, its *relative intensity* with respect to the average intensity of the breast tissue. We let the threshold value for these parameters vary and then we calculated the sensibility $(se = \frac{TP}{TP+FN})$ and the specificity $(sp = \frac{TN}{TN+FP})$.

In order to classify a ROI as true/false positive or true/false negative, we check if the ROI drawn by the radiologist intersects the ROIs found by our algorithm, and also if the normalized area of the intersection is greater than a fixed threshold value.

Automatically detected ROIs that satisfy the above conditions are considered true positive, while the others are marked as false positive. We consider false negatives the ROIs detected by the radiologist but not detected by our system. Table 1 shows the areas under the ROC curves for the four above mentioned parameters. The ROC curves are shown and compared in fig. 3(a).

The best accuracy has been obtained using the *relative dimension* parameter. However, if we consider only the higher levels of sensitivity, the curve corresponding to the *relative mean intensity* parameter is more effective. We found best to combine them by defining a new paramenter, called *joint relative dimension and intensity* or (JRDI), that is a weighted sum of the two. In such a way, we obtained an area under the ROC curve equal to $0,808 \pm 0,018$ (see fig. 3(b) for the curve).

These results are very promising compared to some other recent results obtained in the same field [11, 10].

Table 1. Area under the ROC curves for each parameter

Parameter	Area
absolute dimension	$0,762 \pm 0,020$
relative dimension	$0,767 \pm 0,020$
mean intensity	$0,583 \pm 0,022$
relative mean intensity	$0,733 \pm 0,020$
JRDI	$0,808 \pm 0,018$

5 Conclusions

In this paper we have presented a new segmentation scheme to detect suspicious areas in breast radiographs. The scheme relies on the well known unsupervised clustering technique called *fuzzy c-means* that overcomes some of the main drawbacks of the standard *k-means* algorithm by allowing each sample point to belong to all clusters at the same time. Every single pixel has a degree of membership to each cluster: the objective function, to be minimized at each iteration of the algorithm, is the Euclidean distance between the sample points and the cluster centers, weighted by the corresponding membership values. Clustering is performed using image features based on the local power spectrum obtained by a bank of Gabor filters. We found that the greater flexibility and effectiveness provided by the fuzzy clustering approach benefited from a multiresolution image representation combining both local intensity and frequency information.

One of the main problems to cope with when using Gabor filters is the specification of an appropriate set of parameters. In fact, the number of filters can range from twelve to about forty, depending on the granularity of the sampling in the local-power spectrum. Indeed, the larger the number of filters the greater is the accuracy of the image representation. However, a larger number of convolution filters leads necessarily to an unacceptable computational effort that could make our system useless for the radiologist. Moreover, it could turn out that good parameters set for one dataset could prove less effective with a different one. We had in mind both problems while designing the algorithm.

As far as the computational complexity is concerned, it depends on the number and the size of the convolution kernels, which in turn depend on the linear dimensions and the modulating frequencies of Gabor functions. Therefore, one is confronted with the problem of assigning specific values to these parameters, with the goal of covering the spatial and spatial–frequency domains as uniformly as possible while keeping the computation manageable. As for the orientation ϑ, we considered Gabor filters at an angle $\Delta\vartheta = 45$ degree apart from each other. We found out that a finer resolution in the orientation ϑ does not lead to a significant refinement of segmentation results; rather, a separation among orientations smaller than 30 degrees produces more noise and worse results. Finally, although other choices are possible, in actual computations we decided to take the square module of the symmetrical and antisymmetrical response of each Gabor filter. Only four convolutions were thus required for each given spatial resolution, in fact, ϑ and $\vartheta + \pi$ differ only in sign. As for the spatial resolution, for each assigned orientation, the standard deviation σ assumes the values 1, 2 and 4. These values were selected on the basis of the size of the structures of interest in the mammograms. Given the average dimensions of breast masses, the program needs to detect intensity changes in the image plane occurring in the range of 0.5 to 4-5 cm. Each image in the CALMA dataset is 2656x2067 pixels large with a resolution of 0.085 cm, thus the size of the filters we used, combined with a down-sampling of the image by a factor of 10 in the pre-processing stage, provide us with such desired resolution.

By using the minimum set of filters the time necessary to segment each image in the CALMA database ranges approximately from 4 minutes up to a maximum of 8 minutes, depending on the ratio between the size of the breast region detected and the image size. This is a good performance if we consider that no code optimization has been made yet and that an average time of more than one patient per hour is quite good in the actual context of mammographic screening programs.

We would like to add a final consideration regarding the robustness of the proposed approach. Images contained in the CALMA database have been collected by different hospitals using different machines. As a result, CALMA images are quite heterogeneous in their visual characteristics and, therefore, they represent a good testing ground for CAD programs. However, we have not tested our program extensively on images from different databases. As for the dependence of the parameter values on image resolution, one could think of adding a module that automatically set both the proper range of filter sizes and the downsampling factor on the basis of the actual spatial resolution of the images being used and the approximate size of the objects to be detected. This information is always available to the radiologists that can easily set the needed parameters at the beginning of the analysis.

As we already mentioned, the results obtained with our approach have proved quite encouraging compared to other recent classification systems that use the same database [11, 10] .

References

1. Doi, K., MacMahon, H., Katsuragawa, S., Nishikawa, R., Jian, Y.: Computer-aided diagnosis in radiology: potential and pitfalls. Eur J Radiol **31** (1999) 97–109
2. Ferrari, R., Rangayyan, R., Desautels, J., Frre, A.: Analysis of asymmetry in mammograms via directional filtering with gabor wavelets. IEEE Transactions on Medical Imaging **20** (2001) 953–964
3. Cen, F., Jiang, Y., Zhang, Z., Tsui, H., Lau, T., Xie, H.: Robust registration of 3–d ultrasound images based on gabor filter and mean–shift method. In: ECCV-2004 (European Conference on Computer Vision). Computer Vision Approaches to Medical Image Analysis, Prague, Czech Republic (2004)
4. Catanzariti, E., Cimminello, M., Prevete, R.: Computer aided detection of clustered microcalcifications in digitized mammograms using gabor functions. In IEEE, ed.: Proceed. of 12th International Conference of Image Analysis and Processing, Mantova, Italy (2003) 266–270
5. Dunn, J.: A fuzzy relative of the isodata process and its use in detecting compact well-separated clusters. Journal of Cybernetics **3** (1973) 32–57
6. Jain, A., Farrokhnia, F.: Unsupervised texture segmentation using gabor filters. Pattern Recognition **24** (1991) 1167–1186
7. Bezdek, J.: Pattern Recognition with Fuzzy Objective Function Algorithms. Plenum Press, New York (1981)
8. J.C. Noordam, W.v.d.B., Buydens, L.M.C.: Unsupervised segmentation of predefined shapes in multivariate images. Journal of Chemometric (2003) 216–224

9. Ahmed, M.N., Yamany, S.M., Mohamed, N., Farag, A.A., Moriarty, T.: A modified fuzzy c-means algorithm for bias field estimation and segmentation of mri data. IEEE Trans. Med. Imaging **21** (2002) 193–199

10. Fauci, F., et Al.: A massive lesion detection algorithm in mammography. Pysica Medica **XXI** (2005) 21–28

11. Bellotti, R., et Al.: Massive lesion detection in mammographic images using haralik textural features. Submitted for revision to IEEE Trans. on Nuclear Physics (2005)

MRF Model-Based Approach for Image Segmentation Using a Chaotic MultiAgent System

Kamal E. Melkemi[1,2], Mohamed Batouche[2], and Sebti Foufou[3]

[1] University of Biskra, Computer Science Department, 07000 Biskra, Algeria
melkemi@mailcity.com
[2] University of Constantine, LIRE laboratory, 25000 Constantine, Algeria
batouche@wissal.dz
[3] University of Burgundy, LE2I laboratory, UFR sciences, BP 47870,
21078 Dijon Cedex, France
sfoufou@u-bourgogne.fr

Abstract. In this paper, we propose a new Chaotic MultiAgent System (CMAS) for image segmentation. This CMAS is a distributed system composed of a set of segmentation agents connected to a coordinator agent. Each segmentation agent performs Iterated Conditional Modes (ICM) starting from its own initial image created initially from the observed one by using a chaotic mapping. However, the coordinator agent receives and diversifies these images using a crossover and a chaotic mutation. A chaotic system is successfully used in order to benefit from the special chaotic characteristic features such as ergodic property, stochastic aspect and dependence on initialization. The efficiency of our approach is shown through experimental results.

Keywords: Image Segmentation, Markov Random Field, MultiAgent Systems, Genetic Algorithms, Chaotic System.

1 Introduction

In this work, we are interested in image segmentation based on Markov Random Field (MRF) model [11, 2, 7, 8, 15]. We cite the two main algorithms: the Besag's ICM [2] and the Simulated Annealing (SA) [11, 16]. Starting with a suboptimal configuration, the ICM maximizes the probability of the segmentation field by deterministically and iteratively changing pixel classifications. The ICM is computationally efficient [8], but it strongly depends on the initialization. Theoretically, SA always converges to the global optimum [11], but it remains a computationally intensive method for the image segmentation compared to ICM [8]. Other image segmentation approaches using Genetic Algorithms (GAs) are reported in [1, 3, 5, 18].

Richard et al. [24] have used a MultiAgent System (MAS) for image processing concerning the high level vision task which is Magnetic Resonance Imaging brain scans interpretation. Also, Duchesnay et al. [9] have proposed a (MAS) application for image segmentation.

I. Bloch, A. Petrosino, and A.G.B. Tettamanzi (Eds.): WILF 2005, LNAI 3849, pp. 344–353, 2006.

In this paper, we propose a new Chaotic MultiAgent System (CMAS) for image segmentation. This CMAS is a distributed system composed of a set of segmentation agents connected to a coordinator agent.

In the initialization of the CMAS, each segmentation agent creates an initial image from the observed one using K-means and a chaotic mapping. In the evolution cycle of the CMAS, each segmentation agent performs ICM starting from its own initial image, then transmits this image, the segmented one and the fitness function value to the coordinator agent. This latter, receives the messages, saves the best of segmentations, performs a crossover and a chaotic mutation, then retransmits the new initial images to the different agents for another segmentation cycle.

In chaotic systems and heuristics literature, we find only few papers dealing with hybrid heuristic in a chaotic system [6, 23]. In the paper [6], the authors apply the neural network heuristic. They have proposed the chaotic SA with the transiently chaotic neural network, as an approximation method, in order to cope with the combinatorial optimization problems. Recently, Ji Mingjun et al. [23] have proposed an application of chaotic attractor in SA. The characteristic features of chaotic systems allow without ambiguity the improvement of the SA efficiency [23].

In the CMAS, a chaotic mapping is introduced as a new agent behavior in order to improve the efficiency of the CMAS. Indeed, differing from classical probabilistic behaviors, chaos phenomena are a set of unpredictable behaviors. In fact, we benefit from the special chaotic characteristic features such as ergodic property, stochastic aspect and dependence on initialization. These features allow to this approach to escape the local optimum and converge to a global one.

The organization of this paper is as follows. After this brief introduction, the section 2 presents the related concepts. In the section 3, we describe the steps of the CMAS. Preliminary results are reported in the section 4. Several conclusions are drawn in the last section.

2   Related Concepts

An image $S = \{1, ..., t, .., MN\}$ specifies the grey levels for all pixels in an $MN-$lattice ($MN = M \times N$), where t is called a site. The true and the observed images are represented by the MN-random vectors: $X = (X_1, \ldots, X_t, \ldots, X_{MN})$, $X_t \in \{1, \ldots, C\}$, $Y = (Y_1, \ldots, Y_t, \ldots, Y_{MN}), Y_t \in \{0, \ldots, 255\}$. Let Ω be the set of all possible configurations. The observed image is obtained by adding Gaussian noise process to the ideal image [8]. A neighborhood system $NS = (N_i \subset S, i \in S)$ is a subset collection N_i of S according to: (1) $i \notin N_i$ and (2) $j \in N_i \Leftrightarrow i \in N_j$. A clique c is a set of points which are all neighbors to each other: $\forall r, t \in c, r \in N_t$. Let $X = (X_1, \ldots, X_{MN}) \in \Omega$. X is a MRF with respect to NS if:

1. $\forall x \in \Omega : P(X = x) > 0$
2. $\forall t \in S \quad x \in \Omega : P(x_i/x_j, j \in S - \{i\}) = P(x_i/x_j, j \in N_i)$

X is a MRF on S with respect to NS if and only if $P(X = x)$ is a Gibbs distribution defined by the a-priori probability $P(X = x) = e^{-U(x)}/Z$ where $Z = \sum_{x \in \Omega} e^{-U(x)}$ is the partition function and $U(x)$ is the energy function:

$$U(x) = \sum_{t=1}^{MN} \sum_{r \in N_t} \theta_r \delta(x_t, x_r) \tag{1}$$

where θ_r are the clique parameters, $\delta(a, b) = -1 \; if \quad a = b, 1 \; if \; a \neq b$.

The a-posteriori probability $P(x/y)$ is a Gibbs distribution given by: $P(x/y) = e^{-U(x/y)}/Z_y$ where Z_y is the normalization constant and $U(x/y)$ is the energy function [15] given in equation 2:

$$U(x/y) = \sum_{t=1}^{MN} [\ln(\sqrt{2\Pi}\sigma_{xt}) + \frac{(y_t - \mu_{xt})^2}{2\sigma_{xt}^2} + \sum_{r \in N_t} (\beta\delta(x_t, x_r))] \tag{2}$$

where β is a positive model parameter that controls the homogeneity of the image regions.

3   The Chaotic MAS for Image Segmentation

In this work, we define a new MRF model-based distributed system for image segmentation.

In fact, there is potential risk that distributed evolutionary design process can be attracted to a local minimum, especially when small sizes of subpopulations are used with good individuals as initial parents [20]. However, this danger can be avoided by using random initialization [20], which coincides exactly with our approach. Chaos is utilized as a random number generator, because it was realized that one could take advantage of the intrinsic features of a chaotic system and turn them in to an aperiodic sequence of random numbers.

3.1   The Chaotic System

Among the various non linear mappings considered by researchers, the simplest so-called logistic map, which exhibiting order-to-chaos transitions (see equation 3). So, the one-dimensional logistical map is introduced as follows:

$$z_{k+1} = f(\mu, z_k) = \mu z_k(1 - z_k), \quad z_k \in [0, 1], \quad k = 0, 1, \ldots \tag{3}$$

where z_k is the variable value z at the kth iteration and μ is so-called bifurcation parameter of the system.

In the equation 3 (see Fig. 1), the variable z_k represents the extinction rate where 0 represents extinction and 1 the maximum viable population. The bifurcation parameter μ represents the growth rate of the population.

According to the equation 3, we assume that the higher the scale of the growth rate, the higher the size of the population.

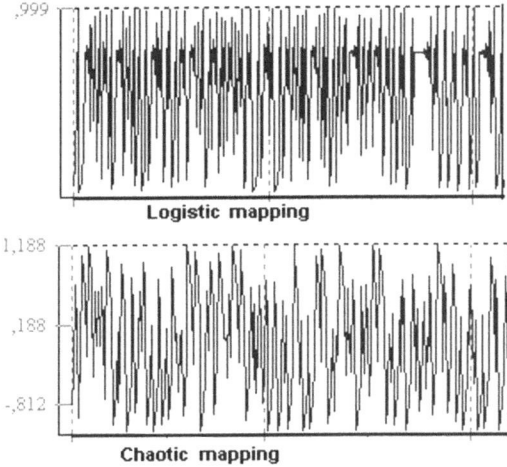

Fig. 1. The logistic mapping and the second new chaotic mapping from chaotic neuron. $\mu = 4$, $k = 300$, $z_0 = 0.01$, $\eta = 0.9$ and $\gamma = 5$.

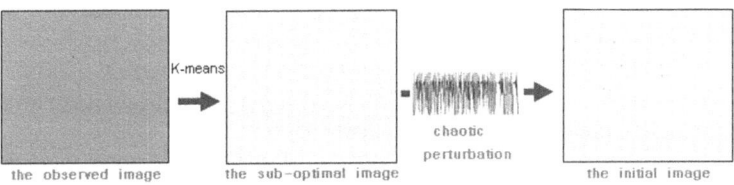

Fig. 2. The chaotic initialization of CMAS

The second chaotic system used in the chaos simulated annealing [23] is derived from chaotic neuron [23, 25] and produced by a new chaotic map defined in the equation 4.

$$z_{k+1} = \eta z_k - 2tanh(\gamma z_k)exp(-3z_k^3), \quad \eta \in [0,1], \quad k = 0, 1, \ldots \quad (4)$$

where z represents the internal state of the neuron, η is a damping factor of nerve membrane and the second term of the equation 4 given by $f(z_k) = 2tanh(\gamma z_k)exp(-3z_k^3)$ is a non-linear feedback. So, we generate the initial solution according to a chaotic mapping by using the equation of the second chaotic map (see equation 4) instead the equation 3.

We use the extreme sensitivity of chaos to the starting values [22] in the initialization of the CMAS. In the chaotic initialization (see Fig. 2), we create an initial configuration by using K-means and perturbing certain site labels according to a chaotic mapping.

The application of chaos in SA proved its superiority and obtained very good results in [23], by escaping local optima. In the SA using the new chaotic map (see equation 4) gains in precision, but consumes more time to converge compared to the SA using logistic map (see Fig. 1) which spends relatively less CPU time [23].

Let $x^0 = (x_1^0, \ldots, x_s^0, \ldots, x_{MN}^0)$ be an initial image created by using K-means. This initial image will undergo chaotic perturbation by applying the formula 5 as follows:

For a given site s selected randomly with a probability of 0.005 the site label $x_s^0 \in \{1, .., C\}$ is created by using the formula:

$$x_s^0 = \alpha \lfloor C * z_{k_s} \rfloor + (1 - \alpha) \lfloor C * w_{k_s} \rfloor \tag{5}$$

For a given z_0, the different chaotic variables $z_{k_s}, s = 1, 2, \ldots, MN$ are generated by the logistic map (see equation 3) and for a given w_0, the different chaotic variables $w_{k_s}, s = 1, 2, \ldots, MN$ are generated by the new chaotic map 4, where k_i is an integer created randomly in the set $\{1, \ldots, 400\}$ and α is a parameter in the interval $[0, 1]$.

In fact, the use of chaos as a random number generator was realized, so that one could take advantage of the intrinsic features of a chaotic system and turn them into an aperiodic sequence of random numbers [21].

3.2   The Genetic Operators

The population is a set of initial images created from the observed one. Let A be an individual of the population. A gene corresponds to a site label $A[i, j] \in \{1, 2, \ldots, C\}$, which the *alphabet* corresponds to the label set $\{1, 2, \ldots, C\}$. Each individual is encoded as a *chromosome* called a *genotype*, and each chromosome is evaluated with a measure of *fitness* via the energy function given in equation 2.

Let A be a chromosome and $i_1, i_2 \in \{1, \ldots, N\}$, $j_1, j_2 \in \{1, \ldots, M\}$, where $i_1 < i_2$ and $j_1 < j_2$. Let P be a rectangle of the image A limited by the points (i_1, i_2) and (j_1, j_2) defined by: $P = \{A_{i,j} \in A / i_1 \leq i \leq i_2, j_1 \leq j \leq j_2\}$. We can show that each part P of A is evaluated by the $U(P/y) = \sum_{i=i_1}^{i_2} \sum_{j=j_1}^{j_2} U(A_{i,j}/y)$, where y is the observed image. The genetic operation of reproduction is based on the Darwinian principle of reproduction and survival of the fittest [12, 14, 17].

The crossover combines, randomly with a probability of 0.9, the genetic material of two parent chromosomes as initial images to produce offsprings corresponding to new initial images, which will be used by the segmentation agents (see Fig. 3). For each mating, the crossover positions are selected randomly as numbers of lines and numbers of columns (see Fig. 3).

The mutation is a rare but extremely important event in GA. When a site label is mutated, it is randomly selected with a probability of 0.005 and replaced with another category from the alphabet by using the same process as that of initialization (see Fig. 4).

3.3   The CMAS Architecture

We consider k segmentation agents connected to a coordinator agent as presented in Figure 5. These cooperating agents contract in a common action after identifying a common goal.

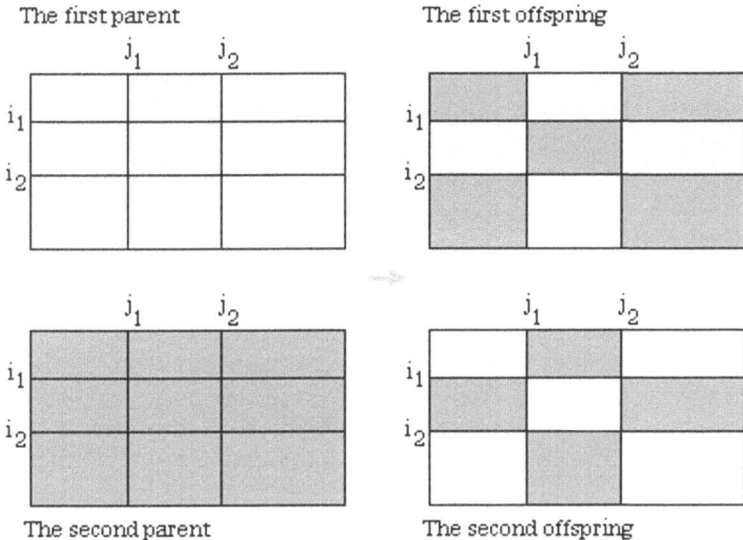

Fig. 3. The two parents represent the initial images and the two offspring are the new initial images. i_1, i_2 are the cross line points and j_1, j_2 are the cross column points.

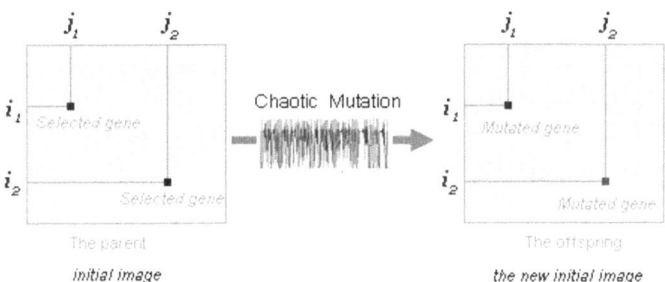

Fig. 4. Two sites mutation of an initial image. (i_1, j_1), (i_2, j_2) are the sites of mutation.

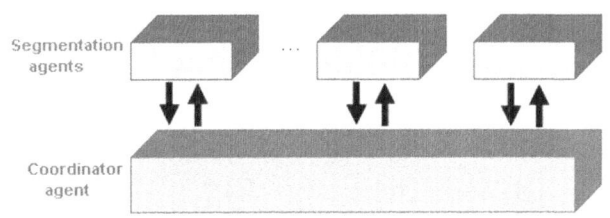

Fig. 5. Communication network of the CMAS

The CMAS intensively attempts to find the best segmentation by genetically breeding of a population of starting images.

The CMAS for image segmentation operates as follows:

- **1. Initialization:** Each segmentation agent creates an initial image from the observed one by using K-means and a chaotic perturbation.
- **2. Each segmentation agent:** Performs ICM starting from its own initial image which sent together with the segmented one and its fitness to the coordinator agent.
- **3. The coordinator agent:**
  - receives the messages from the agents,
  - compares the Best-Segmentation with the current one and saves the best one in the variable Best-Segmentation with its fitness in the variable U^*.
  - performs randomly a crossover on peers initial images, applies a chaotic mutation and retransmits the new offsprings to the different agents.
- **4. The process repeats:** The steps 2 and 3 until a stability of the system.

4   Experimental Results

We present both synthetic and real results of the CMAS compared with the ICM. We assume an isotropic second-order Ising model, so in equation 1, $\theta_1 = \theta_2 = \theta_3 = \theta_4 = \beta$. We have used one value of β which is kept constant through each segmentation. The segmentation is evaluated by both visual examination and energy function. The observed y is the same starting discrete data for all algorithms. These experiments are performed by using Builder C++ 6 on a Pentium 4, CPU 2.66 GHz with 256 MB.

Figure 6 shows two noisy synthetic images. In the up sub-figure, it can be seen that different regions are better segmented by CMAS than by ICM (see Table 1), despite the interference and the thinness of some regions. Also, in the down sub-figure 6, the CMAS extracts better the circles than ICM in spite of the degradation of the image.

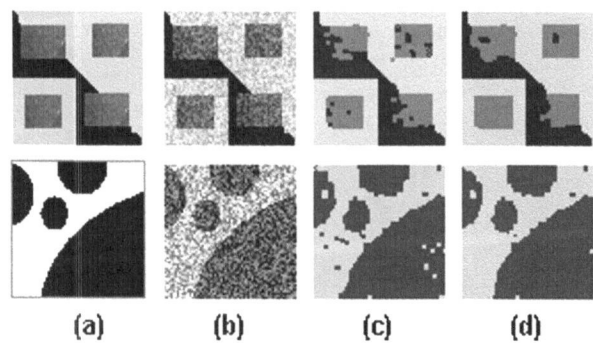

(a)   (b)   (c)   (d)

Fig. 6. Segmentation of 64×64 noisy synthetic scenes. (a) true images (b) noisy synthetic images, (c) ICM results, (d) CMAS results. CMAS Iterations=100 and $\alpha = 1$.

Table 1. Minimal values of energy functions and parameters

Experiment	Approach	$U(x^*/y)/MN$
the up sub Figure 6.(c)	ICM	−3.378
the up sub Figure 6.(d)	CMAS	−4.528
the down sub Figure 6.(c)	ICM	−3.6615
the down sub Figure 6.(d)	CMAS	−4.7351
the Figure 7.(b)	ICM	−3.9488
the Figure 7.(c)	CMAS	−4.3494
Figure 7.(e)	ICM	−2.5936
Figure 7.(f)	CMAS	−4.0297

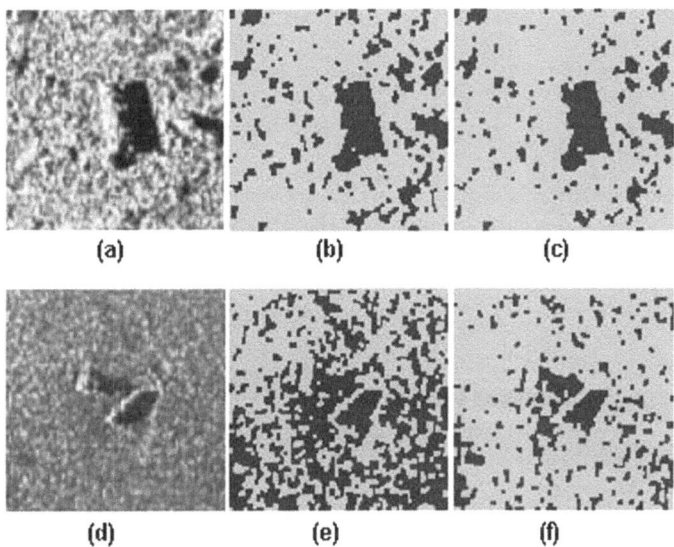

Fig. 7. A two-class segmentation of sonar images involving: (a) a rock shadows, and (b) a cylindrical object shadows. CMAS iterations=100 and $\alpha = 0$. (a) real sonar image, (b) ICM result of (a), (c) CMAS result of (a), (d) real sonar image, (e) ICM result of (d), (f) CMAS result of (d).

In figure 7.(a), the cast shadow of a manufactured object has a geometric shape. The CMAS result is better than the ICM one. In the two-class segmentation of a sonar image involving a rock shadows (see Fig. 7.(d)), dissimilar to the cast shadow of the cylinder (see Fig. 7 (a)), the ICM do not allow to eliminate the speckle noise effect.

5   Conclusion

We have introduced a cooperative chaotic approach for image segmentation. The competition/cooperation activity of the different agents is achieved to insure that this process accedes to the best segmentation. ICM is a deterministic optimization method based on a descent gradient strategy. In fact, the CMAS increases the possibilities to find good segmentations across a number of parallel ICM processes, each one starting from its own sub-optimal image.

The autonomous agents of the CMAS provide modularity in which one agent does not need to have a precise knowledge of the internal structure of others in order to add a new behavior to the system.

A new scheme for generating good pseudo-random numbers based on the composition of chaotic maps is introduced. In fact, we have applied a chaotic map in CMAS in order to benefit from chaotic characteristics such as ergodic property and sensitivity dependence on initialization phase.

Both synthetic and real images have been used to assess the validity and performance of the approach. Experimental results are very encouraging which demonstrate the feasibility, the convergence and the robustness of the method.

References

1. P. Andrey, Selectionist Relaxation: Genetic Algorithms applied to Image Segmentation, Image and Vision Computing, 17, pp. 175–187, 1999.
2. J. Besag, On the statistical analysis of dirty pictures, Jou. of the Royal Stat. Soc., Vol. 48, pp. 259–302, 1986.
3. B. Bhanu and S. Lee, Genetic Learning for Adaptive Image Segmentation, Kluwer Academic Press, 1994.
4. A. H. Bond and L. Gaser, Readings in Distributed Artificial Intelligence, Morgan Kaufmann, San Mateo, CA, 1988.
5. S. Cagnoni, A. B. Dobrzeniecki, R. Poli, and J. C. Yanch, Genetic Algorithm-based Interactive Segmentation of 3D Medical Images, Image and Vision Computing, 17, pp. 881–895, 1999.
6. L. N. Chen and K. Aihara, Chaotic simulated annealing by a neural network model with transient chaos, Neural Networks, vol.8, num. 6, pp.915–930, 1995.
7. H. Derin and H. Elliott, Modeling and segmentation of noisy and textured images using Gibbs random fields, IEEE Trans. on PAMI, 9, pp. 39–55, 1987.
8. R. C. Dubes, A. K. Jain, S. G. Nadabar, and C. C. Chen, MRF model-based algorithms for image segmentation, IEEE Trans. on PAMI, pp. 808–814, 1990.
9. E. Duchesnay, J. J. Montois and Y. Jacquelet, Cooperative agents society organized as an irregular pyramid: a mammography segmentation application, Pattern Recognition Letters, Elsevier Science Inc. Vol. 24, Issue 14, pp. 2435-2445, 2003.
10. J. Ferber, Multi-Agent System: An Introduction to Distributed Artificial Intelligence, Harlow: Addison Wesley Longman, 1999.
11. S. Geman and D. Geman, Stochastic Relaxation, Gibbs Distributions, and the Bayesian Restoration of Images, IEEE Trans. on PAMI, Vol. 6, No. 6, pp. 721–741, 1984.
12. D. E. Goldberg, Genetic algorithm in search, optimization and machine learning, Reading, MA: Addison-Wesley, pp. 28–60, 1989.
13. R. M. Haralick and L. G. Shapiro, Survey: Image Segmentation. Computer Vision, Graphics and Image Processing, pp. 100–132, 29: 1985.
14. J. H. Holland, Adaptation in natural and artificial system, Ann Arbor, MI: The University of Michigan press, 1975.
15. Z. Kato, T. Zerubia and M. Berthod, Satellite image classification using a Metropolis dynamics, in Proc. of ICQSSP, San Francisco, Vol. 3, pp. 573–576, 1992.

16. S. Kirkpatrick, C. D. Gelatt and M. P. Vecchi, Optimization by Simulated Annealing, in Science, Vol. 220, No. 4, pp. 671–680, 1983.
17. J. R. Koza, Survey of genetic algorithms and genetic programming, in Proc. of the Wescon95-Conf. Record: Microelectronics, Communications Technology, Producing Quality Products, Mobile and Portable Power, Emerging Technologies, San Francisco, CA, USA, 1995.
18. S. Lakshmanan and H. Derin, Simultaneous Parameter Estimation and Segmentation of Gibbs Random Fields Using Simulated Annealing, IEEE Trans. on PAMI, Vol. 11, pp. 799–813, 1989.
19. Adina Magda Florea, Introduction to Multi-Agent Systems, International Summer School on Multi-Agent Systems, Bucharest, 1998.
20. Rafal Kicinger, Tomasz Arciszewski and Kenneth De Jong, Distributed Evolutionary Design: Island-Model Based Optimization of Steel Skeleton Structures in Tall Buildings, IEEE CEC 2004, Portland, USA, 2004.
21. Po-Han Lee, Soo-Chang Pei and Yih-Yuh Chen, Generating Chaotic Stream Ciphers Using Chaotic Systems, in Chinese Journal of physics, Vol.41, No.6, 2003.
22. E . N. Lorenz, Deterministic non-periodic flow, J. Atmos. Sci., vol.20, pp.130, 1963.
23. Ji Mingjun and Tang Huanwen, Application of Chaos Simulated Annealing, Chaos, Solitons and Fractals, 21,pp. 933–941, 2004.
24. N. Richard, M. Dojat and C. Garbay, Multi-agent Approach for Image Processing: A Case Study for MRI Human Brain Scans Interpretation, AIME, pp. 91-100, 2003.
25. L. Yang and T. Chen, Application chaos in genetic algorithms. Commun Theor Phys, vol.38, pp.168–172, 2002.

Duality vs Adjunction and General Form for Fuzzy Mathematical Morphology

Isabelle Bloch

GET - Ecole Nationale Supérieure des Télécommunications,
Dept. TSI - CNRS UMR 5141 LTCI,
46 rue Barrault, 75013 Paris, France
Isabelle.Bloch@enst.fr

Abstract. We establish in this paper the link between the two main approaches for fuzzy mathematical morphology, based on duality with respect to complementation and on the adjunction property, respectively. We also prove that the corresponding definitions of fuzzy dilation and erosion are the most general ones if a set of classical properties is required.

1 Introduction

Extending mathematical morphology to fuzzy sets was addressed by several authors during the last years. Some definitions just consider grey levels as membership functions, or use binary structuring elements. Here we restrict ourselves to really fuzzy approaches, where fuzzy sets have to be transformed according to fuzzy structuring elements. Initial developments can be found in the definition of fuzzy Minkowski addition [1]. Then this problem has been addressed by several authors independently, e.g. [2,3,4,5,6,7,8,9]. These works can be divided into two main approaches. In the first one [2], an important property that is put to the fore is the duality between erosion and dilation. A second type of approach is based on the notions of adjunction and fuzzy implication, and was formalized in [8]. The aim of this paper is twofold. First, we will clarify the links between both approaches (which are summarized in Section 2) and establish the conditions of their equivalence (Section 3). Then, in Section 4, we will prove that the definitions of dilation and erosion in these approaches are the most general ones if we want them to share a set of classical properties with standard mathematical morphology.

2 Summary of the Two Main Approaches

Let us first briefly recall the two main approaches. Fuzzy sets are defined on a space $\mathcal{S}$, through their membership functions from $\mathcal{S}$ into $[0,1]$. The set of fuzzy sets on $\mathcal{S}$ is denoted by $\mathcal{F}$, and $\leq$ is the partial ordering defined by $\mu \leq \nu \Leftrightarrow \forall x \in \mathcal{S}, \mu(x) \leq \nu(x)$. This defines a lattice $(\mathcal{F}, \leq)$.

I. Bloch, A. Petrosino, and A.G.B. Tettamanzi (Eds.): WILF 2005, LNAI 3849, pp. 354–361, 2006.

2.1 Fuzzy Morphology by Formal Translation Based on t-Norms and t-Conorms

The first attempts to build fuzzy mathematical morphology were based on translating binary equations into fuzzy ones, as developed in [2]. This translation is done term by term, by substituting all crisp expressions by their fuzzy equivalents. For instance, intersection is replaced by a t-norm, union by a t-conorm, sets by fuzzy set membership functions, etc.

An important property that was put to the fore in this approach is the duality between erosion and dilation. We consider here morphological dilation and erosion, i.e. based on a structuring element.

Let $\varepsilon_B(X)$ denote the erosion of the set X by B, defined by $x \in \varepsilon_B(X) \Leftrightarrow B_x \subseteq X$, where B_x denotes B translated at point x. The translation of this expression into fuzzy terms leads to a natural way to define the erosion of a fuzzy set μ by a fuzzy structuring element ν, as:

$$\forall x \in \mathcal{S}, \ \varepsilon_\nu(\mu)(x) = \inf_{y \in \mathcal{S}} T[c(\nu(y-x)), \mu(y)], \tag{1}$$

were T is a t-conorm and c a complementation. This corresponds to a degree of inclusion of ν, translated at x, in μ. The dual of erosion in the crisp case is $\delta_B(X) = (\varepsilon_{\check{B}}(X^c))^c$, where $\check{B}$ denotes the symmetrical of B with respect to the origin. Accordingly, by duality with respect to the complementation c, fuzzy dilation is then defined by:

$$\forall x \in \mathcal{S}, \ \delta_\nu(\mu)(x) = \sup_{y \in \mathcal{S}} t[\nu(x-y), \mu(y)], \tag{2}$$

where t is the t-norm associated to the t-conorm T with respect to the complementation c. This definition of dilation corresponds to the translation of the following set equivalence: $x \in \delta_B(x) \Leftrightarrow \check{B}_x \cap X \neq \emptyset \Leftrightarrow \exists y \in \mathcal{S}, \ y \in \check{B}_x \cap X$. The fuzzy dilation at x is expressed as the degree of intersection of ν translated at x and μ, which is dual of the degree of inclusion used for the erosion. These forms of fuzzy dilation and fuzzy erosion are very general, and several definitions found in the literature appear as particular cases, such as [5, 3, 10] (see e.g. [2, 11] for a comparison).

Finally, fuzzy opening (respectively fuzzy closing) is simply defined as the combination of a fuzzy erosion followed by a fuzzy dilation (respectively a fuzzy dilation followed by a fuzzy erosion), by using dual t-norms and t-conorms.

The detail of properties of these definitions can be found in [2]. Most properties of classical morphology are satisfied whatever the choice of t and T. But in order to get true closing and opening, i.e. which are extensive (respectively anti-extensive) and idempotent, a necessary and sufficient condition on t and T is $t[b, T(c(b), a)] \leq a$, which is satisfied for Lukasiewicz t-norm and t-conorm for instance.

2.2 Fuzzy Morphology Using Adjunction and Residual Implications

A second type of approach is based on the notions of adjunction and fuzzy implication. Here the algebraic framework is the main guideline, which contrasts with the previous approach where duality was imposed in first place.

Fuzzy implication is often defined as [12]: $Imp(a, b) = T[c(a), b)]$. Fuzzy inclusion, as used in the previous approach, and therefore fuzzy erosion, is related to implication by the following equation: $\mathcal{I}(\nu, \mu) = \inf_{x \in \mathcal{S}} Imp[\nu(x), \mu(x)]$.

This suggests another way to define fuzzy erosion (and dilation), by using other forms of fuzzy implication. One interesting approach is to use residual implications: $Imp(a, b) = \sup\{\varepsilon \in [0, 1], t(a, \varepsilon) \leq b\}$. This provides the following expression for the degree of inclusion: $\mathcal{I}(\nu, \mu) = \inf_{x \in \mathcal{S}} \sup\{\varepsilon \in [0, 1], t(\nu(x), \varepsilon) \leq \mu(x)\}$. This definition coincides with the previous one for particular forms of t, typically Lukasiewicz t-norm.

The derivation of fuzzy morphological operators from residual implication has been proposed in [4], and then developed e.g. in [7]. One of its main advantages is that it leads to idempotent fuzzy closing and opening. This approach was formalized from the algebraic point of view of adjunction in [8]. It has then been used by other authors, e.g. [9]. This leads to general algebraic fuzzy erosion and dilation. Let us detail this approach. A fuzzy implication I is a mapping from $[0, 1] \times [0, 1]$ into $[0, 1]$ which is decreasing in the first argument, increasing in the second one and satisfies $I(0, 0) = I(0, 1) = I(1, 1) = 1$ and $I(1, 0) = 0$. A fuzzy conjunction is a mapping from $[0, 1] \times [0, 1]$ into $[0, 1]$ which is increasing in both arguments and satisfies $C(0, 0) = C(1, 0) = C(0, 1) = 0$ and $C(1, 1) = 1$. If C is also associative and commutative, it is a t-norm. A pair of operators (I, C) are said adjoint if:

$$C(a, b) \leq c \Leftrightarrow b \leq I(a, c). \tag{3}$$

The adjoint of a conjunction is a residual implication.

Fuzzy dilation and erosion are then defined as:

$$\forall x \in \mathcal{S}, \ \delta_\nu(\mu)(x) = \sup_{y \in \mathcal{S}} C(\nu(x - y), \mu(y)), \tag{4}$$

$$\forall x \in \mathcal{S}, \ \varepsilon_\nu(\mu)(x) = \inf_{y \in \mathcal{S}} I(\nu(y - x), \mu(y)). \tag{5}$$

Note that (I, C) is an adjunction if and only if $(\varepsilon_\nu, \delta_\nu)$ is an adjunction on the lattice $(\mathcal{F}, \leq)$ for an $y\nu$.

Opening and closing derived from these operations by combination have all required properties, whatever the choice of C and I. Some properties of dilation, such as iterativity, require C to be associative and commutative, i.e. a t-norm. This will be further investigated in Section 4.

3   Links Between Both Approaches

3.1   Dual vs Adjoint Operators

If C is a t-norm, then the dilation in the second approach is exactly the same as the one obtained in the first approach. To understand further the relation between both approaches for erosion, we define

$$\hat{I}(a, b) = I(c(a), b).$$

Then $\hat{I}$ is increasing in both arguments, and if I is further assumed satisfy $I(a,b) = I(c(b),c(a))$ and $I(c(I(a,b)),d) = I(a,I(c(b),d))$, then $\hat{I}$ is commutative and associative, hence a t-conorm. In the following, in order to simplify notations we simply take $c(a) = 1 - a$ which is the most usual complementation, but the derivations and results hold for any c.

Equation 5 can be rewritten as:

$$\varepsilon_\nu(\mu)(x) = \inf_{y \in S} \hat{I}(1 - \nu(y - x), \mu(y)),$$

which corresponds to the fuzzy erosion of the first approach. The adjunction property can also be written as:

$$C(a,b) \leq c \Leftrightarrow b \leq \hat{I}(1 - a, c).$$

However, pairs of dual t-norms and t-conorms are not identical to pairs of adjoint operators. Let us take a few examples. For $C = \min$, its adjoint is $I(a,b) = b$ if $b < a$, and 1 otherwise (known as Gödel implication). But the derived $\hat{I}$ is the dual of the conjunction defined as $C(a,b) = 0$ if $b \leq 1 - a$ and b otherwise. Conversely, the adjoint of this conjunction is $I(a,b) = \max(1 - a, b)$ (Kleene-Dienes implication), the dual of which is the minimum conjunction. Lukasiewicz operators $C(a,b) = \max(0, a+b-1)$ and $\hat{I}(a,b) = \min(1, a+b)$ are both adjoint and dual, which explains the exact correspondence between both approaches for these operators. Table 1 summarizes the differences between dual and adjoint operators.

Table 1. A few dual and adjoint operators: dual and adjoint are generally not identical, except in the case of Lukasiewicz operators (among these examples)

conjunction	dual t-conorm	adjoint implication I	$\hat{I}$
$\min(a,b)$	$\max(a,b)$	$\begin{cases} b \text{ if } b < a \\ 1 \text{ otherwise} \end{cases}$ (Gödel)	$\begin{cases} b \text{ if } b < 1 - a \\ 1 \text{ otherwise} \end{cases}$
$\begin{cases} 0 \text{ if } b \leq 1 - a \\ b \text{ otherwise} \end{cases}$	$\begin{cases} b \text{ if } b < 1 - a \\ 1 \text{ otherwise} \end{cases}$	$\max(1 - a, b)$ (Kleene-Dienes)	$\max(a,b)$
$\max(0, a+b-1)$	$\min(1, a+b)$	$\min(1, 1 - a + b)$ (Lukasiewicz)	$\min(1, a+b)$

3.2   Equivalence Condition

The first main result of this paper is expressed in the following theorem.

Theorem 1. *The condition for dual t-norms and t-conorms leading to idempotent opening and closing (i.e. $t(b, T(1-b,a)) \leq a$) is equivalent to the adjunction property between C and I for $t = C$ and $T = \hat{I}$.*

Proof. Let us assume that the adjunction property is satisfied for $t = C$ and $T = \hat{I}$, i.e.

$$t(a, b) \le c \Leftrightarrow b \le T(1 - a, c). \tag{6}$$

Applying this property to the tautology $T(1 - b, a) \le T(1 - b, a)$ leads directly to:

$$t(b, T(1 - b, a)) \le a. \tag{7}$$

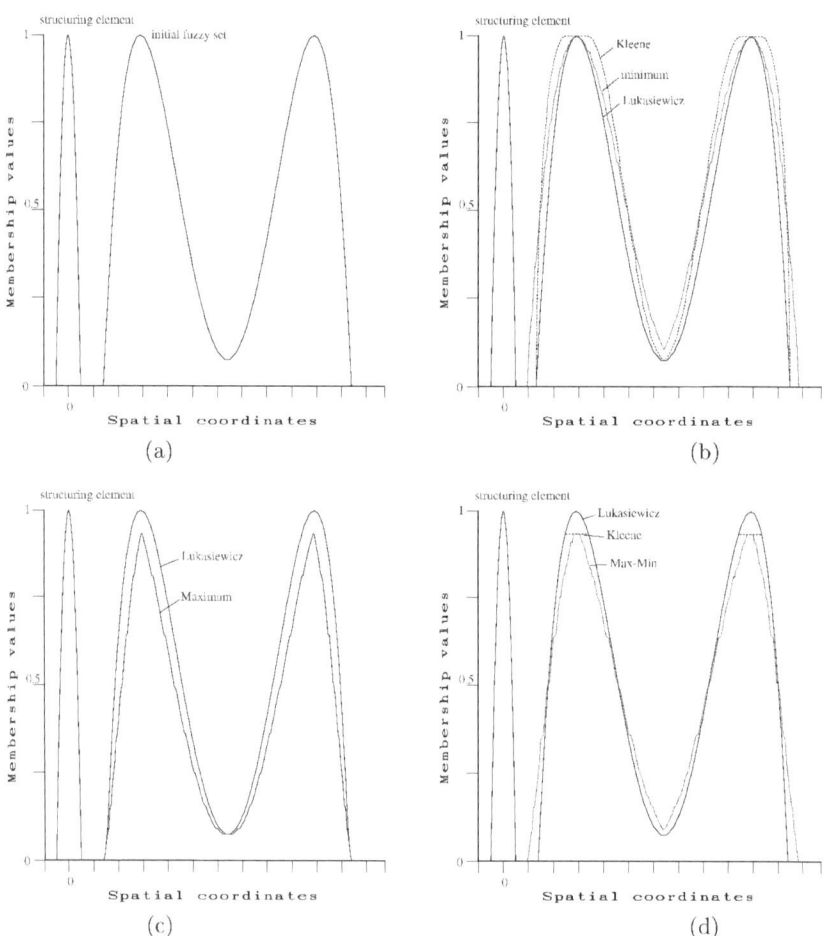

Fig. 1. Illustration of some morphological operations on a one-dimensional example. (a) Initial fuzzy set and fuzzy structuring element. (b) Dilations using the minimum, Lukasiewicz and Kleene-Dienes conjunctions. (c) Erosions using the maximum and Lukasiewicz t-conorms. (d) Opening using max-min, Lukasiewicz and Kleene-Dienes operators.

Let us now assume that we have the property expressed by Equation 7 for dual operators. If $b \leq T(1 - a, c)$, then since t is increasing, we have $t(a, b) \leq t(a, T(1 - a, c))$ which is less than c by Eq. 7. This implies $t(a, b) \leq c$.

Since t and T are dual, Eq. 7 is equivalent to $1 - T(1 - b, 1 - T(1 - b, a)) \leq a$, and, by exchanging the roles of $1 - a$ and a and then of a and b, to $T(1 - a, t(a, b)) \geq b$.

Now, if $t(a, b) \leq c$, since T is increasing, we have $T(1 - a, t(a, b)) \leq T(1 - a, c)$. Since the first term is greater than b, this implies $b \leq T(1 - a, c)$.   ■

This new result completes the link between both approaches by showing that duality and adjunction are generally not compatible, and that in case dual operators lead to true opening and closing, the condition on these operators is equivalent to the adjunction property. This means that in case duality and adjunction are compatible, the two approaches lead exactly to the same definitions.

3.3 Illustrative Example

In order to show the influence of the choice of the conjunctions, t-conorms, implications, we illustrate a few operations on a one-dimensional example in Figure 1. Dilation, erosion and opening are performed using different operators. When using adjoint operators, opening is a "true" opening (i.e. increasing, anti-extensive and idempotent). It is clear in this figure that when using min and max for instance, which are dual but not adjoint, opening is not anti-extensive (it is not idempotent either, but it is increasing). On the contrary, using Kleene-Dienes adjoint operators, the opening is anti-extensive (Figure 1 d). However, other properties of erosion and dilation are lost, due to the weaker properties of the conjunction with respect to the ones of t-norms. These aspects will be further investigated in Section 4. The results obtained with Lukasiewicz operators are in this case very close to the original fuzzy set. However, all properties of all operations hold when using these operators.

4 General Forms of Fuzzy Morphological Dilation and Erosion

The second main result of this paper establishes the general form of fuzzy dilation and erosion, in order to satisfy a set of properties. Let $\delta_\nu(\mu)$ be a morphological dilation. Let us consider the following general form of δ:

$$\delta_\nu(\mu)(x) = g(f(\nu(x - y), \mu(y)), y \in \mathcal{S}), \tag{8}$$

where f is a mapping from $[0, 1] \times [0, 1]$ in $[0, 1]$ and g is a mapping from $[0, 1]^{\mathcal{S}}$ into $[0, 1]$ (the result is then a fuzzy set).

Theorem 2. *The compatibility of fuzzy dilation with classical dilation in case ν is crisp, its increasingness, and the commutativity with the supremum lead to the only possible form of δ:*

$$\delta_\nu(\mu)(x) = \sup_{y \in \mathcal{S}} t(\nu(x - y), \mu(y))$$

where t is a conjunction. If the commutativity $(\delta_\nu(\mu) = \delta_\mu(\nu))$ and iterativity $(\delta_\nu \delta_{\nu'}(\mu) = \delta_{\delta_\nu(\nu')}(\mu))$ properties are also required, then t has to be a t-norm.

From this dilation, a unique erosion such that $(\varepsilon_\nu, \delta_\nu)$ is an adjunction is derived:

$$\varepsilon_\nu(\mu)(x) = \inf_{y \in S} I(\nu(y - x), \mu(y)),$$

where I is the adjoint of t.

If duality is required, $\hat{I}$ has to be the dual of t.

Proof. Let g_1 be the version of g applying on one variable only. Most results are derived by considering constant membership functions. Increasingness of δ implies that $g_1.f$ should be increasing in μ and ν. If ν is crisp, the compatibility with classical dilation implies that $\forall a \in [0, 1], g_1.f(1, a) = a$. Therefore $g_1.f$ is a conjunction.

Further properties such as commutativity and iterativity imply $g_1.f$ be commutative and associative, respectively, i.e. it should be a t-norm.

It is easy to prove that g_1 has to be a bijection (one to one mapping). It follows that $f(1, a) = g_1^{-1}(a)$. Let $\mu'(y) = g_1^{-1}(\mu(y))$. The compatibility with classical morphology implies $\sup_{y \in S} \mu(y) = g(g_1^{-1}(\mu(y)), y \in S)$, i.e. $\sup_{y \in S} g_1(\mu'(y)) = g(\mu'(y), y \in S)$. Therefore $\delta_\nu(\mu)(x) = \sup_{y \in S} g_1.f(\nu(x - y), \mu(y))$. From the properties of t-norms, this form commutes with the supremum.

From a dilation δ_ν, a general result on adjunctions guarantees that there exists a unique erosion ε_ν such that $(\varepsilon_\nu, \delta_\nu)$ is an adjunction, and it is given by:

$$\varepsilon_\nu(\mu) = \bigvee \{\mu', \delta_\nu(\mu') \leq \mu\}.$$

We have the following equivalences, by denoting $g_1.f = t$ and I the adjoint of t:

$$\delta_\nu(\mu') \leq \mu \Leftrightarrow \forall x \in S, \delta_\nu(\mu')(x) \leq \mu(x)$$
$$\Leftrightarrow \forall x, y \in S, t(\nu(x - y), \mu'(y)) \leq \mu(x)$$
$$\Leftrightarrow \forall x, y \in S, \mu'(y) \leq I(\nu(x - y), \mu(x))$$
$$\Leftrightarrow \forall y \in S, \mu'(y) \leq \inf_{x \in S} I(\nu(x - y), \mu(x))$$

Since ε_ν is the supremum of μ' verifying this equation, we have: $\varepsilon_\nu(\mu)(y) = \inf_{x \in S} I(\nu(x - y), \mu(x))$.

Now, if duality is required between ε_ν and δ_ν with respect to complementation, it is straightforward to show that t and $\hat{I}$ have to be dual operators.

Having both duality and adjunction is possible under the conditions expressed in Theorem 1. ∎

In [3], a similar approach was developed for deriving a general form of fuzzy inclusion (from which fuzzy erosion is derived). Since weaker properties are required, this approach leads to the use of weak t-norms and t-conorms (they are not associative and do not admit 1 (respectively 0) as unit element, in general). Properties of morphological operators are then weaker (no iterativity can be expected, no compatibility with classical morphology), and this is therefore somewhat less interesting from a morphological point of view. Our approach overcomes these drawbacks.

5   Conclusion

This paper exhibits the exact conditions to have a convergence between the two main approaches for fuzzy morphology. Although the underlying principles are not compatible in general, it is interesting to note that in case they are consistent, then both approaches are equivalent. Furthermore, they provide the most general forms in order to satisfy a set of reasonable properties as in classical morphology. These two new results clarify the status of different forms of mathematical morphology.

References

1. Dubois, D., Prade, H.: Inverse Operations for Fuzzy Numbers. In Sanchez, E., Gupta, M., eds.: Fuzzy Information, Knowledge Representation and Decision Analysis, IFAC Symposium, Marseille, France (1983) 391–396
2. Bloch, I., Maître, H.: Fuzzy Mathematical Morphologies: A Comparative Study. Pattern Recognition **28** (1995) 1341–1387
3. Sinha, D., Dougherty, E.R.: Fuzzification of Set Inclusion: Theory and Applications. Fuzzy Sets and Systems **55** (1993) 15–42
4. Baets, B.D.: Idempotent Closing and Opening Operations in Fuzzy Mathematical Morphology. In: ISUMA-NAFIPS'95, College Park, MD (1995) 228–233
5. Bandemer, H., Näther, W.: Fuzzy Data Analysis. Theory and Decision Library, Serie B: Mathematical and Statistical Methods. Kluwer Academic Publisher, Dordrecht (1992)
6. Popov, A.T.: Morphological Operations on Fuzzy Sets. In: IEE Image Processing and its Applications, Edinburgh, UK (1995) 837–840
7. Nachtegael, M., Kerre, E.E.: Classical and Fuzzy Approaches towards Mathematical Morphology. In Kerre, E.E., Nachtegael, M., eds.: Fuzzy Techniques in Image Processing. Studies in Fuzziness and Soft Computing. Physica-Verlag, Springer (2000) 3–57
8. Deng, T.Q., Heijmans, H.: Grey-Scale Morphology Based on Fuzzy Logic. Journal of Mathematical Imaging and Vision **16** (2002) 155–171
9. Maragos, P.: Lattice Image Processing: A Unification of Morphological and Fuzzy Algebraic Systems. Journal of Mathematical Imaging and Vision **22** (2005) 333–353
10. Rosenfeld, A.: The Fuzzy Geometry of Image Subsets. Pattern Recognition Letters **2** (1984) 311–317
11. Bloch, I.: Fuzzy Mathematical Morphology and Derived Spatial Relationships. In Kerre, E., Nachtegael, N., eds.: Fuzzy Techniques in Image Processing. Springer Verlag (2000) 101–134
12. Dubois, D., Prade, H.: Fuzzy Sets and Systems: Theory and Applications. Academic Press, New-York (1980)

A Fuzzy Mathematical Morphology Approach to Multiseeded Image Segmentation

Isabelle Bloch[1], Gabriele Martino[2], and Alfredo Petrosino[2]

[1] Ecole Nationale Supérieure des Télécommunications,
Dept. TSI - CNRS UMR 5141 LTCI
`Isabelle.Bloch@enst.fr`
[2] Department of Applied Science,
University of Naples "Parthenope"
`alfredo.petrosino@uniparthenope.it`

Abstract. We propose an innovative segmentation algorithm based on mathematical morphology operators. This definition is based on a morphological and fuzzy pattern-matching approach, and consists in comparing an object to a fuzzy landscape representing the degree of satisfaction of an affinity relationship. It has good formal properties, it is flexible, it fits the intuition, and it can be used for structural pattern recognition under imprecision. Moreover, it also applies in 3D and for fuzzy objects issued from images.

1 Introduction

The spatial arrangement of objects in images provides important information for recognition and interpretation tasks, in particular when the objects are embedded in a complex environment like in medical images. Relationships between objects, in particular for image segmentation purposes, can be described in terms of affinity between them, and it is the aim of this paper to address the problem of defining such relationships. From our every day experience, it is clear that any all-or-nothing definition leads to unsatisfactory results in several situations, even of moderate complexity. Fuzzy approaches are all the most interesting when imprecision in images has to be taken into account. Indeed, the representation of image regions as spatial fuzzy sets is useful to take into account the imprecision inherent to images. Several different image segmentation methods have been proposed in literature in the past [6]. In this paper we propose a new approach based on mathematical morphology [3].

In the present work, we show how the shape of objects in the image can be utilized to define a new segmentation algorithm. This method has been inspired on a work about fuzzy relative position between objects according to morphological operators [1]. Indeed, this algorithm is based on the concept of affinity. Affinity is a fuzzy relation defined between pixels of the image and its goal is to capture the grade of their "hanging togetherness".

The paper is organized as follows: Section 2 is dedicated to the basic concepts, specifically inherently to the fuzzy affinity; Section 3 presents how affinity can

I. Bloch, A. Petrosino, and A.G.B. Tettamanzi (Eds.): WILF 2005, LNAI 3849, pp. 362–368, 2006.
© Springer-Verlag Berlin Heidelberg 2006

be used with some morphological operators for the formulation of a segmentation algorithm; Section 4 describes the proposed algorithm, finally Section 5 the performances evaluation.

2 Fuzzy Affinity

Let us represent the image domain by S and let $\mathcal{S} = (S, f)$ be a fuzzy scene, where f is the pixel intensity function. We define a fuzzy relation κ in S, with its membership function μ_κ; we would want κ to be such that $\mu_\kappa(c, d)$ is a function of $V(c, d)$, that is the neighbours of both c and d and of $f(c)$ and $f(d)$, that is the intensity features. We can use the following functional form for μ_κ,

$$\mu_\kappa(c, d) = g(\mu_\psi(c, d), \mu_\phi(c, d)). \tag{1}$$

ψ and ϕ respectively represent the *homogeneity-based* and the *object-feature-based* components of affinity and μ_ψ and μ_ϕ the respective membership functions. The strenght of relation ψ indicates the degree of local hanging togetherness of spels because of their intensity similarities. The strenght of relation ϕ indicates the degree of local hanging togetherness of pixels because of the similarity of their features values to some(specified) object feature. The function g can be considered as a fusion operator; we have chosen an average type fusion operator, which achieves a compromise between both pieces of information:

$$g = \sqrt{\mu_\phi \mu_\psi}. \tag{2}$$

For the homogeneity-based affinity, we assume the following expression:

$$\mu_\psi(c, d) = W_\psi(|f(c) - f(d)|). \tag{3}$$

In our work, μ_ψ is assumed as a Gaussian function with zero mean:

$$W_\psi(x) = e^{\frac{x^2}{2k_\psi^2}}, \text{with} x = |f(c) - f(d)|.$$

The treatment of μ_ϕ is somewhat different from that of μ_ψ. We consider the object feature as well as background feature to formulate μ_ϕ. We use an object membership function W_o, as well as background membership function W_b, to capture the idea of membership of any pixel to the respective regions and then combine them to obtain μ_ϕ.

For our purpose, we choose them as Gaussian functions. Namely, we set
$$W_o(x) = e^{\frac{(x - m_o)^2}{2k_o^2}} \text{ and } W_b(x) = e^{\frac{(x - m_b)^2}{2k_b^2}}.$$
Finally, we consider any points c and d to have a high *object-feature-based* affinity only if both c and d have high object membership (i.e., the value of W_o) and both have low background membership (i.e., the value of W_b). The functional form chosen to reflect this strategy is as follows:

$$\mu_\phi(c, d) = \begin{cases} 1, & \text{if } c = d \\ \frac{\mathcal{W}_o(c, d)}{\mathcal{W}_o(c, d) + \mathcal{W}_b(c, d)}, & \text{otherwise} \end{cases} \tag{4}$$

where

$$\mathcal{W}_o(c,d) = \min[W_o(f(c)), W_o(f(d))], \tag{5}$$

and

$$\mathcal{W}_b(c,d) = \max[W_b(f(c)), W_b(f(d))]. \tag{6}$$

3 Segmentation Algorithm

Let us consider an image, whose domain is indicated by S, and a set of reference objects in it O_i, for $i = 1, 2, ..., n$ where n is the number of desired objects. These objects could be considered as binary or fuzzy. In the last case a specific fuzzy relation o_i [1] is defined on each of them, as indicated in Section 3.1; the related membership function $\mu_{o_i}(c)$ indicates, for some pixel $c \in S$, the degree of satisfaction of the specific fuzzy relation o_i.

In order to establish the relationships between the binary or fuzzy objects O_i and S for segmentation purpose, we choose the following approach:

1. We first define a fuzzy "landscape" μ_{f_i}, around the reference objects O_i, as fuzzy sets such that the membership values of each point, that is μ_{f_i}, corresponds to the degree of satisfaction of the fuzzy relation κ defined before.
2. We then compare S to the fuzzy landscapes μ_{f_i} in order to evaluate how well a specific object O_i matches S. This is done using a fuzzy pattern-matching approach.

3.1 Definition of the Fuzzy Landscape

The goal of the definition of the fuzzy landscape is to point out the relations of affinities between pixels of the image.

The definition of the fuzzy landscape can be adapted to binary and fuzzy objects; for our purpose, we have only considered fuzzy objects. The fuzzy objects have been defined by means of a specific fuzzy relation o_i, "**degree of membership to a specific manually selected object O_i**". The membership function μ_{o_i}, for each of the selected objects O_i, has been defined by means of a Gaussian funtion whose mean and variance are related to the mean and variance of the pixels of the selected regions of the image. So, given an object O_i [2], each point $c \in S$ is characterized by its "**degree of affinity with object O_i**". The affinity relation is the one defined before, indicated by κ. As pointed in Section 4, we compute several different affinity relations κ_i, one for each of the object O_i we want to segment.

---

[1] O_i and o_i are two different notations: the first one indicates the reference objects while the last one the fuzzy relations defined on them.

[2] In our application, an operator selects, on a display of a slice of the scene, for each object O_i, a region of the object and a region of the background using a mouse-controlled brush.

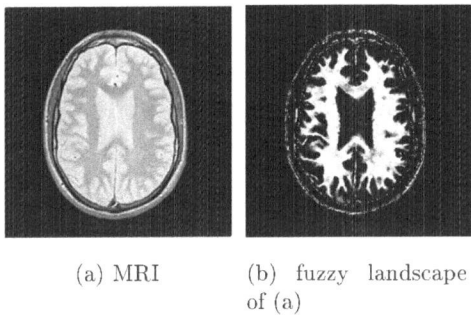

(a) MRI   (b) fuzzy landscape
                 of (a)

Fig. 1. Illustration of the concept of the fuzzy landscape of a MRI image: (a) original image, (b) fuzzy landscape of (a)

In establishing the fuzzy landscape μ_{f_i} of a specific object O_i, in fuzzy case, we choose a method that combines directly μ_{o_i}, describing the membership to some O_i, with the strenght of affinity μ_κ [1]. In fuzzy terms the following holds:

$$\mu_{f_i}(c) = \max_{d \in Supp(O_i)} t[\mu_{o_i}(c), \mu_{\kappa_i}(c,d)], \qquad (7)$$

where t is a t-norm. This definition can be adapted to the case of multiple objects, in that case $\mu_\kappa(c,d)$ is simply replaced by $\mu_\kappa^{(i)}(c,d)$, for $i = 1, 2, ...n$.

In Fig.1(a) and Fig.1(b) is illustrated the concept of fuzzy landscape. It is an interesting rapresentation of the fuzzy landscape for MRI image. Fig.1(a) shows an original a MRI image, Fig.1(b) represents the fuzzy landscape according to the fuzzy relation "**degree of affinity to the gray matter**". It's clear that points of Fig.1(b) that are embedded in the gray matter have high membership values (white pixels of the images).

3.2 Fuzzy Pattern Matching

The fuzzy landscape defined before allows us to define objects in the image based on some specific characteristic of affinity. The process of objects extraction, for segmentation purpose, is determined by means of the evaluation of the degree of matching between S and the fuzzy landscape of the objects O_i obtained from the previous step.

Let us denote by μ_S the membership function of S, which is a function of S in $[0,1]$, where S is the image domain. An appropriate tool for defining the degree of matching of S with respect to each fuzzy landscape μ_{f_i} is the fuzzy pattern-matching approach [2]. Following this approach, the evaluation of the matching between two possibility distributions consists of two numbers, a necessity degree Π and a possibility degree N. Π and N are computed for each object O_i, $i = 1, \ldots, n$, according to the following expressions:

$$\Pi_i(x) = \sup_{y \in S} t[\mu_{f_i}(y - x), \mu_S(y)] \quad \forall x \in S \qquad (8)$$

$$N_i(x) = \inf_{y \in S} T[\mu_{f_i}(y - x), 1 - \mu_S(y)] \quad \forall x \in S \qquad (9)$$

where t is a t-norm (fuzzy intersection) and T a t-conorm (fuzzy union) [3]. In the crisp case, these equations reduce to:

$$\Pi_i(x) = \sup_{y \in S} \mu_{f_i}(y) \quad \forall x \in S \tag{10}$$

$$N_i(x) = \inf_{y \in S} \mu_{f_i}(y) \quad \forall x \in S \tag{11}$$

The possibility and necessity can be interpreted in terms of fuzzy matheatical morphology, since the possibility is equal to the dilation of μ_S by μ_{f_i}, while the necessity is equal to the erosion [3].

4   The Proposed Algorithm

In this section we review the main steps of the proposed segmentation algorithm.

As first step, the user must select on the image, for each of the objects O_i he desires to segment, an object and a background region. These regions are used as statistical samples of training for the computation of the parameters of the affinity, that is m_o and m_b for the mean and κ_o, κ_b and κ_ψ for the standard deviation.

The second step is represented by the computation of the fuzzy landscapes μ_{f_i} for each of the the desired objects O_i. We propose here a fast algorithm for computing them.

The algorithm consists in performing two passes on the image, one in the conventional sense, and one in the opposite sense. For each point c, we store the point $Q = O(c)$ from which the maximum affinity is obtained. For a point c, we don't consider all points in O_i as for exhaustive method, but only those of neighbourhood of c. Specifically, we compute the fuzzy landscape as:

$$\mu_{f_i}(c) = \max_{d \in V(c)} t[\mu_{f_i}(O(d)), \mu_S(O(d))],$$

where $V(c)$ denotes the neighbourhood of c. Let d_c be the point d for which the maximum affinity value is obtained

$$d_c = argmax_{d \in V(c)} t[\mu_{f_i}(O(d)), \mu_S(O(d))]$$

Then, we set: $O(c) = O(d_c)$.

As final result of the second step, each point c of the image is characterized by its membership, its degree of affinity with each of the manually selected objects O_i, that is $\mu_{f_i}(c)$, $\forall c \in S$ and $\forall i = 1, 2, ...n$. Then, as pointed before in Section 3, the fuzzy landscapes of the objects are matched with S. This is realized by means of the computation of the values Π_i and N_i. As in the computation of the fuzzy landscape, this is realized $\forall c \in S$ and $\forall i = 1, 2, ...n$. The computation of $\Pi_i(c)$ and $N_i(c)$, as regards fuzzy case, is realized as indicated in (10) and (11). The final decision is to assign a point $c \in S$ to an object for which it has the maximum degree of matching.

The computational time of the algorithm is $O(cn_c n)$, where c is the number of pixels, $n_c = |V(c)|$, that is the number of neighbours of c, and n the number of objects being extracted. This algorithm is quite fast with respect to the one of the Fuzzy Connectednees in [7]. Even if the computational time is quite similar $(O(c(n+1)))$ the differences are in operational costs, that is in terms of ms.

Below we propose the final image segmentation algorithm:

1. Estimate parameters for affinity.
2. **for** each object i **do**
3.   **for** each pixels c **do**
4.   compute $\mu_{f_i}(c)$ according to (7);
5. **for** each object i **do**
6.   **for** each pixels c **do**
7.   compute $g_i(c) = \frac{\Pi_i(c)+N_i(c)}{2}$ according to (10) and (11);
8. **for** each pixels c **find**
9. $p_i = arg\max g_i(c)$;
10. Output p_i.

5   Experimental Results

In order to evaluate the experimental results of the algorithm, we have tested it segmenting MRI images. Based on [4], we have used 100 MRI images of the IBSR (http://www.cma.mgh.harvard.edu/ibsr/). The results obtained by this algorithm have been compared to the results of the Fuzzy Connectedness segmentation algorithm [7]. We have used the SSIM-INDEX algorithm in order to evaluate the degree of similarity between the original signal (groundtruth) and the distorted signal (segmented image), [9].

In Fig.2(b) is shown an example of segmentation. The groundtruth is shown in Fig.2(c), while the original images in Fig.2(a).

In Table 1(a) are shown the results obtained by the proposed algorithm on the 100 MRI images of the IBSR. Each row of the tables reports the results of 5 images. We obtain a 90% of similarity that is comparable with results achieved on the same data set by similar algorithms based on fuzzy connectivity [5].

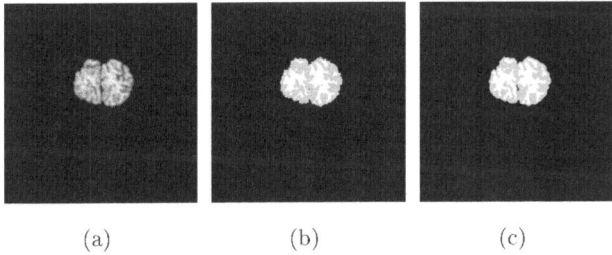

(a)   (b)   (c)

Fig. 2. Segmentation obtained by the proposed algorithm: (a) original image , (b) segmented image, (c) groundtruth

Table 1. Performances evaluation

Images - SSIM-INDEX				
1-5	0.9691	0.9621	0.9616	0.9474	0.9227
6-10	0.9130	0.8992	0.9079	0.8905	0.8882
11-15	0.8806	0.8790	0.8718	0.8675	0.8522
16-20	0.8564	0.8493	0.8524	0.8541	0.8698
21-25	0.8781	0.8610	0.8676	0.8598	0.8465
26-30	0.8572	0.8481	0.8515	0.8595	0.8605
31-35	0.8606	0.8673	0.8768	0.8709	0.8787
36-40	0.8743	0.8908	0.8979	0.9082	0.9147
41-45	0.9263	0.9238	0.9318	0.9330	0.9379
46-50	0.9335	0.9425	0.9453	0.9482	0.9593
Mean 0.8942				

(a) SSIM-INDEX of the proposed algorithm for the first 50 images.

References

1. I. Bloch. *Fuzzy Relative Position Between Objects in Image Processing: A Morphological Approach.* IEEE Transaction On Pattern Analisys and Machine Intelligence, vol. 21, n.7, July 1999.
2. I. Bloch, H. Maitre. *Fuzzy Adjacency between Image Objects.* International Journal of Uncertainty, Fuzzyness and Knowledge-Based Systems, 5(6):615-653, 1997.
3. I. Bloch, H. Maitre. *Fuzzy mathematical morphologies: a comparative study.* Pattern Recognition vol.28, no.9, pp 1341-1387, 1995.
4. J.-M. Lee, U. Yoon, S.-H. Nam, J.-H. Kim, I.-Y. Kim, S.-I.Kim. *Evaluation of automated and semi-automated skull-stripping algorithms using similarity index and segmentation error.* Computers in Biology and Medicine vol.33, pp 495-507, 2003.
5. G. Martino, A. Petrosino, *Fuzzy Connectivity and its Application to Image Segmentation,* in B. Apolloni, R. Tagliaferri eds., *WIRN05 - Lecture Notes in Computer Science,* Springer Verlag, 2005.
6. D.-L.Pham, C. Xu, J.- L.Prince. *A survey of current methods in medical image segmentation.* Annual Review of Biomedical Engineering, January 1998.
7. J.K. Udupa, P.K Saha. *Fuzzy Connectedness and Image segmentation,* in Proc. of the IEEE, vol.91, no.10 , October 2003.
8. J.K. Udupa, P.K Saha, D. Odhner. *Scale-based fuzzy connectedness image segmentation: Theory, algorithms and applications in image segmentation.* Comput. Vision Image Understanding, vol.77, pp 145-174, 2000.
9. Z. Wang, L. Lu, A.C Bovik. *Video quality assessment based on structural distortion measurement.* Signal Processing: image communication, vol.19, no.2, pp 121-132. February 2004.

Neuro-fuzzy Analysis of Document Images by the KERNEL System

Ciro Castiello, Przemysław Górecki, and Laura Caponetti

Dipartimento di Informatica,
Università degli Studi di Bari,
Via E. Orabona, 4 - 70126 Bari, Italy
{castiello, przemyslaw, laura}@di.uniba.it

Abstract. Document image analysis represents one of the most relevant topics in the field of image processing: many research efforts have been devoted to devising automatic strategies for document region classification. In this paper, we present a peculiar strategy to extract numerical features from segmented image regions, and their employment for classification purposes by means of the KERNEL system, a particular neuro-fuzzy framework suitable for application in predictive tasks. The knowledge discovery process performed by KERNEL proved to be effective in solving the problem of distinguishing between textual and graphical components of a document image. The information embedded into sample data is organised in form of a fuzzy rule base, which results to be accurate and comprehensible for human users.

1 Introduction

The goal of document processing is to convert document into more usable form, composed of text and graphics. Proper detection and recognition of text in the document images allows for better compression of documents, querying for specified words, or reediting the document. A document image, usually obtained by scanning or photographing the original document, is first divided into regions which are then classified as text or graphics. To classify document regions correctly, a set of features must be firstly extracted from each region. Successively, a classification mechanism has to be applied in order to distinguish between text and graphics, on the basis of the examination of the extracted features [1].

Many different methods for document image classification have been proposed throughout the last few years, which use different feature sets for classification tasks: spline wavelet decomposition [2], classification of features such as shape and area of connected components [3], or Delaunay triangulation [4] to mention only a few.

In this paper, starting from the investigation of segmented document images, we propose a neuro-fuzzy approach for image region classification. In particular, Hough transform is applied for region skew detection and a Power Spectral Density (PSD) analysis of a projection of a region along skew angle is performed. The classification is based on the shape and location of the prominent peaks in

I. Bloch, A. Petrosino, and A.G.B. Tettamanzi (Eds.): WILF 2005, LNAI 3849, pp. 369–374, 2006.

PSD spectrum vector. Additionally, connected components are extracted from a region and their shape properties such as area, perimeter, first and second order moments are calculated. Weighted averages of those values extend the feature vector to increase the classification accuracy.

In order to accomplish the region classification process, the extracted information is involved in a knowledge discovery process, based on the employment of KERNEL, a particular neuro-fuzzy system developed for dealing with classification and regression tasks. The idea is to present a set of heterogeneous features to the system, in order to obtain a fuzzy rule base which could provide accurate knowledge organisation and readable information for human understanding. As we are going to detail in the following, the knowledge discovery process performed by KERNEL is based on successive learning steps of a particular neural network, the neuro-fuzzy network, which realises a suitable hybridisation between connectionist learning and fuzzy logic. The application of KERNEL to the region classification problem, based on the analysis of properly identified input features, produced satisfactory results both in terms of accuracy and comprehensibility of the obtained base of knowledge.

The paper is organised as follows. In section 2 we detail the feature extraction process we adopted in our research; section 3 briefly presents the KERNEL system and finally section 4 is devoted to the illustration of experimental results and conclusive remarks.

2   Feature Extraction for Region Description

The general process of document image analysis can be decomposed into three distinct phases. Firstly, a document image should be segmented into regions, then a feature extraction process should be performed to successively classify each obtained region as text or graphics. The research activity described in this paper assumes that a database of segmented images is available, from which a set of numerical features has to be extracted. They are employed to classify each region of the document image by means of the KERNEL system.

The first step of the feature extraction process consists in detecting the region skew angle, which is important for correct classification of text regions. A text line is a group of characters, symbols and words, that are adjacent and through which a straight line can be drawn. The dominant orientation of the text lines determines the skew angle. The region skew angle ϕ is determined by means of the Hough transform: it can be identified as the angle for which the Hough transform of an image region has the maximum value [5,6,7]. Successively, an evaluation of the projection of the document region along ϕ is performed, in order to obtain the projection vector v_p. The elements of v_p codify the information deriving from the analysed region. Particularly, for a text region, v_p should exhibit regular, high frequency sinusoidal-like shape, with peaks and valleys corresponding to the text lines and the interline spacings, respectively. In contrast, when a graphic region is considered, such regularities cannot be observed.

To measure the periodicity of the vector v_p, a Power Spectral Density (PSD) analysis is performed. In fact, for large text regions, the PSD coefficients show a significant peak around the frequency value corresponding approximately to the number of text lines. For graphic regions, the spectrum presents only 1 or 2 peaks around the lowest frequency values. As a result, a vector v_{psd} is calculated as PSD of the Fourier Transform of the vector v_p [1]. In most cases, the number of the components of the PSD spectrum vector v_{psd} appears to be too high for a direct use in classification tasks. Therefore, we resolved to extract information from v_{psd} by reducing its dimensionality. Particularly, we divide v_{psd} into 7 intervals, whose lengths correspond to a scaled Fibonacci sequence, with multiplying factor two (i.e., 2, 4, 6, 10, 16, 26, 42). In this way, we are able to exploit most of information, since it is accumulated in the first part of the spectrum. For each interval, the maximum value of v_{psd} is derived: all these values (normalised with respect to the highest one) represent the first 7 components of the feature vector v_f, which will be employed for the successive region classification stage.

To increase classification accuracy, a further step of additional feature extraction is devoted to evaluate some statistical information, concerning the connected components in the analysed region. In practice, a set of connected components (i.e., single characters, words, or graphics) is obtained from a binarisation of the considered region. Then, for each component, the averages of the shape ratios (namely, the ratio between the squared perimeter and the area of the component) and the averages of the second order momenta, along horizontal and vertical axes, are evaluated. In this way, the information embedded into the feature vector v_f can be extended with 3 additional values, corresponding to each computed average.

At the end of the overall feature extraction process, all the regions of the segmented document image can be represented as 10-components feature vectors. They are employed as input data of the KERNEL system, for classification purposes, as detailed in the next section.

3   Classification of Image Regions by the KERNEL System

In this section we are going to briefly overview the peculiar system employed to perform the classification of the image regions, namely the KERNEL system. KERNEL is a particular neuro-fuzzy system, designed to derive from input data a form of structured knowledge, expressed in terms of fuzzy rules, which can be employed to tackle classification or regression problems. A summary description of the system is presented, addressing the reader to some previous publications of ours for further details [8,9].

Combining the neural learning capabilities with the readability of fuzzy rules, the system performs a knowledge discovery process to generate a fuzzy inference model adopting fuzzy rules of the following type:

IF x_1 is A_1^k **AND** ... **AND** x_n is A_n^k **THEN** y_1 is b_{1k} **AND** ... **AND** y_m is b_{mk},

where $\mathbf{x} = (x_1, \ldots, x_n)$ is the input variable vector, $\mathbf{y} = (y_1, \ldots, y_m)$ is the output variable vector, A_i^k are fuzzy sets and b_{jk} are fuzzy singletons, with reference to the k-th rule ($k = 1, \ldots, K$). Fuzzy sets A_i^k are defined by Gaussian membership functions $\mu_{ik}(x_i) = \exp(-(x_i - c_{ik})^2/\sigma_{ik}^2)$, where c_{ik}, σ_{ik} are the centres and the widths of the Gaussian functions, respectively. Indicating by $\mu_k(\mathbf{x}) = \prod_{i=1}^{n} \mu_{ik} x_i$ the activation strength of the k-th rule, the final output of the predictive model can be obtained by: $y_j = (\sum_{k=1}^{K} \mu_k(\mathbf{x}) b_{jk})/(\sum_{k=1}^{K} \mu_k(\mathbf{x}))$, ($j = 1, \ldots, m$).

This fuzzy inference model is encoded into a specific neural network, whose topology reflects the parameters and the organisation of the fuzzy rule base. The neuro-fuzzy network is composed of three layers: in the first layer nodes are collected in K groups (corresponding to the K rules), each composed of n units (corresponding to the n fuzzy sets of every rule). These nodes estimate the values of the Gaussian membership functions. The second layer is composed of K units (corresponding to the K rules of the fuzzy model). Each node evaluates the fulfilment degree of every rule, and the nodes of the third layer supply the final output of the system.

The neuro-fuzzy network described above constitutes the core of the KERNEL system, which is organised into three distinct components. The first component extracts knowledge from data in the form of a fuzzy rule base. The basis for this process of knowledge extraction consists in a clustering of input data, performed by an unsupervised learning of the neuro-fuzzy network. The second component improves the accuracy of the fuzzy rule base, performing a supervised learning of the neuro-fuzzy network. The third component enhances the comprehensibility of the fuzzy rule base, reducing the number of rules (through a structure reduction of the neural network) and adjusting the configuration of the input membership functions via a genetic algorithm.

In the course of the experimental session concerning the image region classification, the input vector $\mathbf{x}$, involved in the fuzzy inference model, corresponds to the ten-dimensional feature vector v_f, derived during the feature extraction process. The output vector $\mathbf{y}$ is related to the classes of the classification task (textual and graphical regions).

4   Experimental Results and Conclusive Remarks

To test the effectiveness of the presented methodology, we employed the Document Image Database available from the University of Oulu [10]. This database includes 233 images of articles, scanned from magazines and newspapers, books and manuals. The images vary both in quality and contents: some of them contain text paragraphs only (with Latin and Cyrillic fonts of different sizes), while others contain mixtures of text, pictures, photos, graphs and charts. Moreover, only the minority of the documents is characterised by Manhattan page layout.

From those document images, 306 graphic regions and 894 text regions have been extracted and automatically labelled, according to the information included in the database. After applying the feature extraction process to the entire set of document regions, we built up a dataset comprising 1200 samples that was successively partitioned into a training and test set, each one of 600 samples.

The proportion between graphic and text regions into the training and test sets reflects the original ratio characterising the overall dataset. The employment of the KERNEL system, for executing the classification phase, has been realised on the basis of the training process of a neuro-fuzzy network with 10 inputs (corresponding to the components of the input vector v_f) and 2 outputs (related to the two classes involved in the tasks). During the first step, the unsupervised learning stage produced an initial fuzzy rule base comprising 12 rules. They were successively refined in terms of accuracy during the second learning step. Finally, the overall comprehensibility of the knowledge has been enhanced by reducing the complexity of the fuzzy rule base (namely, the number of involved fuzzy rules which decreased from 12 to 8) and adjusting the Gaussian membership function configurations. In table 1 the classification results are reported for each of the derived fuzzy models, both for training and test set, together with the number of

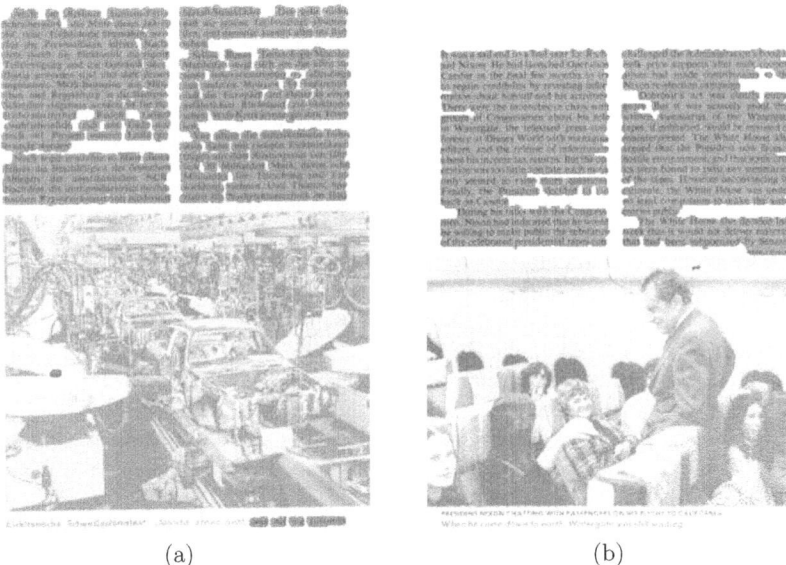

(a)   (b)

Fig. 1. Classification results obtained for two sample images. Dark regions have been classified as text, while light regions have been classified as graphics.

Table 1. Classification results provided by each one of the derived fuzzy predictive models

	number of	% classification	
	rules	Training set	Test set
Initial fuzzy rule base	12	95.71	93.53
Refined fuzzy rule base	12	95.80	93.60
Final fuzzy rule base	8	95.48	92.72

involved rules. As it can be observed, the obtained results are quite satisfactory in terms of accuracy and the final knowledge optimisation step contributed to determine a simpler and more comprehensible fuzzy rule base.

As guidelines for future work, the overall methodology could be further enhanced by improving the angle detection step during the feature extraction process (introducing more appropriate interpretations of Hough transform). Moreover, the KERNEL system could be able to produce more accurate results, by introducing some additional features (related to the texture properties of a region).

References

1. William K. Pratt, *Digital Image Processing*, PIKS Inside, 3rd Edition, 2001.
2. Shulan Deng, Shahram Lati, Emma Regentova, *Document segmentation using polynomial spline wavelets*, Pattern Recognition 34, p.2533-2545, 2001.
3. Oleg Okun, Matti Pietikainen, Jaakko Sauvola, *Document skew estimation without angle range restriction*, Int. J. Document Analysis and Recognition, 1999.
4. Yi Xiao, Hong Yan, *Text region extraction in a document image based on the Delaunay tessellation*, Pattern Recognition 36, 799-809, 2003.
5. Jain, A. K., Yu, B., *Document representation and its application to page decomposition*, IEEE Trans. on PAMI, 20 (1998) 254–308 .
6. Hinds S. C., Fisher J. L., D'Amato D. P.: A document skew detection method using run-length encoding and Hough transform. Proc. of the 10th Int. Conference on Pattern Recognition (ICPR), (1990) 464–468.
7. Srihari S. N., Govindaraju V.: Analysis of textual images using the Hough transform. Machine Vision Applications, 2 (1989) 141–153.
8. Castellano G. , Castiello C., Fanelli A. M.: KERNEL: A Matlab toolbox for Knowledge Extraction and Refinement by NEural Learning. Lecture Notes in Computer Science, 2329(1), Springer Verlag Berlin Heidelberg, (2002) 970–979.
9. Castellano G. , Castiello C., Fanelli A. M.: KERNEL: A system for Knowledge Extraction and Refinement by NEural Learning. Proc. of KES 2002, IOS Press (2002) 443–447.
10. University of Oulu, Finland, *Document Image Database*, http://www.ee.oulu.fi/research/imag/document/

Intelligent Knowledge Capsule Design for Associative Priming Knowledge Extraction

JeongYon Shim

Division of General Studies, Computer Science, Kangnam University,
San 6-2, Kugal-ri, Kihung-up,YongIn Si, KyeongKi Do, Korea
Tel.: +82 31 2803 736
mariashim@kangnam.ac.kr

Abstract. Intelligent Knowledge Capsule was designed for the functions of knowledge acquisition,Memory retention and Knowledgeretrieval. Specially in this paper, focusing on Knowledge Retrieval process Associative Priming Knowledge Extraction mechanism is designed. A hierarchical associative memory including long term memory, short term memory and synonym net for keyword is established and using this structure Associative Priming Knowledge Extraction mechanism is processed. We apply this mechanism to virtual memory and test the retrieving process.

1 Introduction

As a result of many brain researches, it is known that human being does not see the real images in the world but sees the virtual images made by selected input data and previous stored knowledge in his brain. Most of living things as well as human being have similar mechanisms. Living things have evolved this ability to survive because it is only way for small brain to process huge information of the outside. For this reason, various talents appear in the functions of living things. The capability of a living thing depends on how well the memory is structured and has a good mechanism for knowledge acquisition, retention and retrieval.

Generally living things reacts on the familiar stimulus very easily. Especially in the case of data which is in enclosing circumstance, it is easy to activate the associative priming knowledge stored in the memory.

As the information technologies are developing quickly,information circumstance is getting more complex and data are piling up in a huge scale. Facing this situation, the importance of developing more intelligent system is becoming high. However there is a difficulty for processing the synonyms or similar concepts to keyword used for knowledge retrieval.It is not an efficient way to include all the synonym of keywords in the memory and cause waste of memory. In this case it is desirable to manage the synonym net in a separate way.

For this reason, as one of efforts for developing an intelligent system Intelligent Knowledge Capsule was designed for processing the functions of knowledge acquisition, memory retention and knowledge retrieval. In this paper, we describe the structure of Intelligent Knowledge Capsule and its main functions shortly

I. Bloch, A. Petrosino, and A.G.B. Tettamanzi (Eds.): WILF 2005, LNAI 3849, pp. 375–384, 2006.
© Springer-Verlag Berlin Heidelberg 2006

because it is introduced in the previous papers. Focusing on Knowledge retrieval we design the flexible structure of associative memory including short term /long term memory for efficient knowledge extraction and propose Associative Priming Knowledge Extraction mechanism considering synonym net. In the last step, We apply this mechanism to virtual memory and test the retrieving process.

2   Intelligent Knowledge Capsule Design

2.1   The Structure of Intelligent Knowledge Capsule

Intelligent knowledge capsule is designed for data acquisition, selection, storing and extraction.

As shown in Fig. 1, This system consists of Learning memory, Rule base, and Episode memory. They are related to the others according to their association. The memory is constructed by information acquisition process. The obtained data from the knowledge environment come into Input Interface and are temporarily stored in Temporary memory. They are selected and distributed by the basic mechanism. For autonomous learning mechanism. Learning engine receives the training data of the special domain and process its learning mechanism. Episode memory stores the event oriented facts with the information of time and location and memorize them according to the flow of events sequentially. Memory Index which composed of Short term memory Index and Long term memory Index is used for efficient knowledge retrieval.

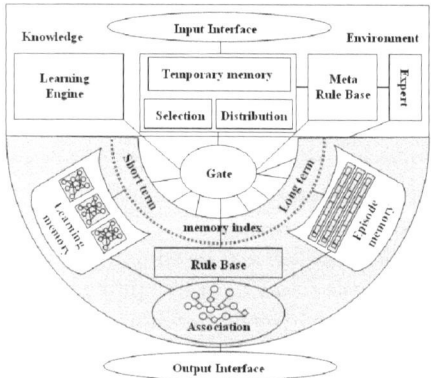

Fig. 1. The structure of Knowledge capsule

2.2   Knowledge Acquisition: Learning in KLN

Knowledge Learning Frame has the Hierarchical structure and also has KLN (Knowledge Learning Net) module which consists of modular Neural networks representing the domain knowledge for the autonomous learning process as

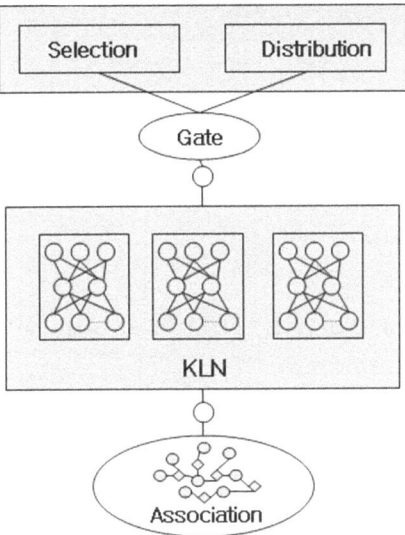

Fig. 2. Knowledge learning frame

shown in Figure 2[1]. These domains are connected to the corresponding areas in the other layers vertically and related with the associative relation in the association module.The observed mixed input data are selected and distributed to the corresponding NN(Neural Network) in the Selection/Distribution module by filtering function F_i, which is the criteria for determining the state of firing.

$$F_i = P(C_i|e_1, e_2, \ldots, e_n) = \prod_{k=1}^{n}(C_i|e_k) \tag{1}$$

where C_i denotes a hypothesis for disease class and $e_k = e_1, \ldots, e_n$ denotes a sequence of observed data. F_i can be obtained by calculating the belief in C_i.
If filtering factor F_i is over the threshold, q_i, $(F_i \geq q_i)$, the corresponding class is fired. The corresponding KLN of the fired class starts the learning or perception mechanism and produces the output. The values of the cells that don't belong to the activated class, are filtered and cleared. The structure of NN is three layered neural network trained by BP learning algorithm[3].

2.3 Association

Association level consists of nodes and arcs. one node have knowledge and connected to other nodes with an arc of associative relation. Knowledge nodes are connected to their neighbors according to their associative relations horizontally and connected to NN of the previous layer vertically. Their relations are represented by the relational graph as shown in Figure 3.

Association are formed in the classical paradigm. System have a bias for associating stimuli that are likely to be related in the environment. The system is also

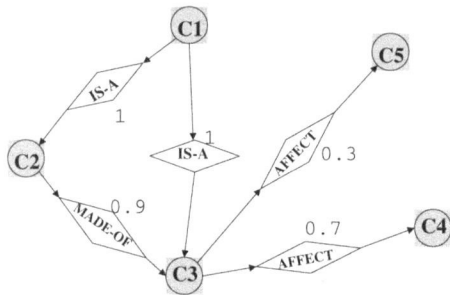

Fig. 3. Knowledge graph in Association level

sensitive to the likelihood of an association between the two stimuli(associative bias). It is might be characterized as forming a statistical inference. Bayesian statistical models for inference about probabilistic relationships consider not only the data but also prior beliefs about what relationships are likely. The fundamental relationship in Bayesian statistics is

$$PosteriorBelief = PriorBelief * Evidence$$

That is,

$$P(h|e) = \frac{P(e|h)P(h)}{p(e)}$$

Associative strength,R_{ij}, between C_i and C_j in the knowledge graph is represented by Bayesian Posterior belief and calculated by equation (2).

$$R_{ij} = P(a_i|a_j)D \tag{2}$$

where D is the direction arrow, $D = 1 or -1$, $i = 1, \ldots, n$, $j = 1, \ldots, n$.

The relation is characterized by the associative strength of relation which is not fixed but changed by incoming evidence.

Table 1. Associative list

node1	Relation	strength	node2
C_1	IS-A	1.0	C_2
C_1	IS-A	1.0	C_3
C_2	MADE-OF	0.9	C_3
C_3	AFFECT	0.7	C_4
C_4	SELF	0.0	NULL
C_3	AFFECT	0.3	C_5
C_5	SELF	0.0	NULL

The knowledge graph of Figure 2. is transformed to Associative list of Table 1 for the efficient memory management.

3   Associative Priming Knowledge Extraction Mechanism

3.1   Associative Memory Structuring for Efficient Retrieving

In this section, Associative memory in Intelligent knowledge capsule for efficient knowledge retrieving is designed. Figure 4 shows modules take charge of structuring associative memory and its knowledge extraction mechanism. After keyword coming from Input interface is checked in synonym net of temporary memory, representative keyword(R-keyword) substitutes for the coming keyword and relation between coming keyword and R-keyword is checked. Then R-keyword is propagated to Associative memory through Gate and used for knowledge extraction mechanism.

As shown in Figure 5, Associative memory has a hierarchical structure which consists of short-term memory, long-term memory and forgetting pool. The associative knowledge is represented as a knowledge graph associated with nodes and relations. The knowledge activated very frequently or knowledge regarded as an important one is stored in short-term memory and old knowledge which does not be frequently used is stored in long-term memory. The knowledge which is not useful any more is discarded in Forgetting pool. These three types of memory are connected to vertically and has a flexible characteristics that knowledge nodes and their association moves and are changed periodically. The node which does not be activated frequently in short-term memory can move to long-term

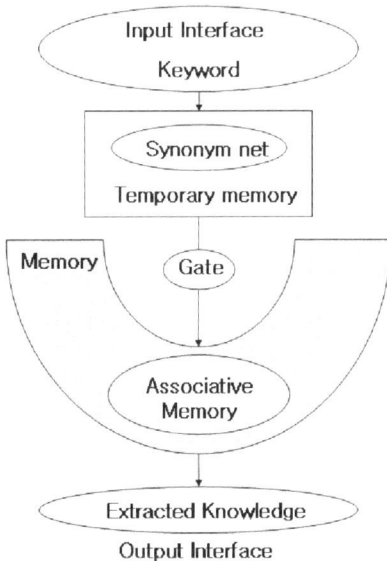

Fig. 4. Modules for knowledge extraction in Intelligent knowledge capsule

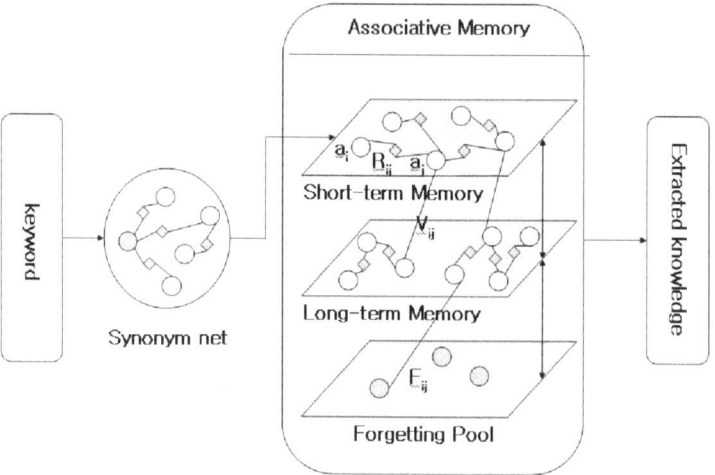

Fig. 5. The structure of associative memory including synonym net

memory and the node activated very frequently can move to short-term memory. Some node in short-term memory is connected to node in long-term memory with the activating degree of long-term memory, V_{ij} and some node in long-term memory is connected to the node with forgetting degree,F_{ij}, in Forgetting pool.

3.2 Synonym Net

One of big trouble things in information processing is the problem of synonyms. There can be many expression called as "synonyms" for representing same meaning. One keyword has not only synonyms but also words which is not exactly matched and has very similar meaning. If memory contains all the words including synonyms and similar expressions as a keyword, there may occur complex problem and waste of memory. For preventing this problem and developing the efficient mechanism, it is necessary to figure out efficient mechanism for memory maintenance.

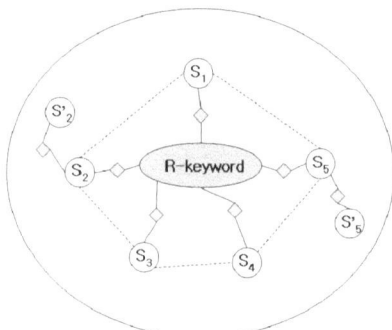

Fig. 6. Synonym net

So, in this section 'synonym net' which represents the group of synonyms is proposed. Synonym net has one representative keyword(R-keyword),its synonyms and its similar word in group. On the basis of R-keyword, synonyms and similar words are connected to R-keyword with relational strength which represents how far it is from R-keyword as shown in Figure 6.

If one keyword comes through Input Interface, it is checked in synonym net how much related to R-keyword and the value of relation is stored in Temporary memory for Associative Priming Knowledge Extraction mechanism. After checked, R-keyword substitutes for coming keyword and is propagated to short-term memory in Associative memory for Knowledge Retrieval.

3.3   Associative Priming Knowledge Extraction Mechanism

It is known that memory records become more available when associated information is in the environment. The ability of a cue to make associated information more available is referred to as associative priming. The associative priming knowledge in memory reacts on the coming keyword very easily.

In this section this system performs Associative Priming Knowledge Extraction mechanism using the structure of associative memory. It consists of four steps, i.e, Synonym net search, Short-term memory search, Long-tern memory search and Forgetting pool search.

Taking one keyword, this system processes Associative Priming Knowledge Extraction mechanism and produces the related extracted knowledge using the following algorithm.

Algorithm 1. Associative Priming Knowledge Extraction algorithm

STEP 1: Input the Keyword.

Synonym net search:
STEP 2: Search the corresponding keyword in synonym net.
STEP 3 : If (not found)
    Print ("There is no corresponding keyword!")
    Else (found)
    calculate the relational strength to R-Keyword;
    R-keyword = Keyword;

Short-term memory Search:
STEP 4: Search the corresponding R-keyword in the Associative list of short
    term memory
STEP 5 : If (not found)
    Print ("There is no corresponding keyword!")
    Else (found)
    associative-knowledge-extraction;
    If(longterm-wanted != n)
    goto STEP 8 ;
    Else(longterm-wanted)
    Long-term memory Search;

Long-term memory Search:
STEP 6: Search the corresponding R-keyword in the Associative list of short
    term memory
STEP 7: If (not found)
    Print ("There is no corresponding keyword!")
    Else (found)
    associative-knowledge-extraction;
    If(longterm-wanted != n)
    goto STEP 8;
    Else(longterm-wanted)
    Long-term memory Search;

associative-knowledge-extraction:
    next=associative-list;
    i = found-i;
    while((next.node2[i]!= null) and (i !=N))
    { Put next.node1[i] to F-list;
    Put next.R[i] to F-list;
    Put next.node2[i] to F-list ;
    i=i+1; }

STEP 8: Stop.

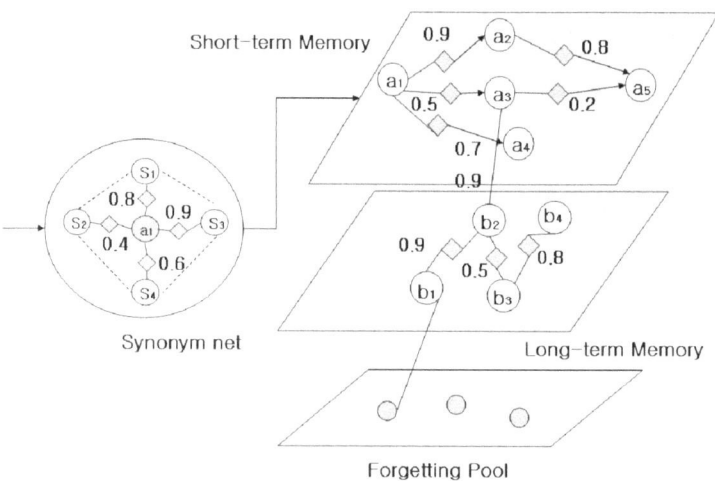

Fig. 7. Example of virtual memory

4   Experiments

This system is applied to the virtual memory as shown in Figure 7 and tested the
process of Associative Priming Knowledge Extraction. Figure 8 shows the results

of Knowledge retrieval considering synonym search, short-term memory search and Long-term memory search from Associative Priming Knowledge Extraction mechanism. As shown in the results, It is found that this extraction algorithm works successfully and can retrieve the correct data.

As it is shown in the experiment output, the structure of memory is very flexibly designed and data can be extracted by the reactive level.

```
Associative Priming Knowledge Extraction mechanism starts...
Enter the keyword? s1
... R-keword is a1
... s1-s 0.8
...
Do you want short term memory search (y/n)?y
short term memory search starts...
a1 0.9 a2 0.8 a5
a1 0.5 a3 0.2 a5
a1 0.7 a4
...
Do you want long term memory search (y/n)?y
long term memory search starts...
a3 0.9 b2 0.9 b1
...
Do you want Forgetting pool search(y/n)?n
...finished.
```

Fig. 8. The retrieved knowledge extracted by Associative Priming Extraction algorithm

5   Conclusion

We proposed knowledge capsule which has a hierarchical structure and many functions of selection, learning,storing and data extraction. Based on this knowledge structure, it has a flexible memory and Associative Priming Extraction mechanism. We designed the concept of Synonym net and the structure of associative memory including Short-term memory, Long-term memory and Forgetting pool and proposed Associative Priming Knowledge Extraction algorithm. We tested associative knowledge retrieval function with virtual memory. As a result of experiments, we could find this system has a flexible memory and can produce the extracted data according to the reactive level. This function is applicable to build the flexible memory in intelligent brain system.

References

1. Jeong-Yon Shim, Knowledge Retrieval Using Bayesian Associative Relation in the Three Dimensional Modular System,pp630-635, *Lecture Notes in Computer Science(3007), 2004.*
2. John R. Anderson Learning and Memory *Prentice Hall.*

3. Laurene Fausett: Fundamentals of Neural Networks,Prentice Hall
4. Simon Haykin: Neural Networks,Prentice Hall
5. Jeong-Yon Shim, Chong-Sun Hwang,Data Extraction from Associative Matrix based on Selective learning system,IJCNN'99, Washongton D.C
6. John R. Anderson,Learning and Memory,Prentice Hall

A Flexible Intelligent Associative Knowledge Structure of Reticular Activating System: Positive/Negative Masking

JeongYon Shim

Division of General Studies, Computer Science, Kangnam University,
San 6-2, Kugal-ri, Kihung-up,YongIn Si, KyeongKi Do, Korea
Tel.: +82 31 2803 736
mariashim@kangnam.ac.kr

Abstract. As the information circumstance is getting more complicated, the requirements for implementing the efficient intelligent system adopting human brain functions is getting high. We focus on the
function of Reticular Activating System which takes charge of information selection.In this paper we designed Reticular Activating System
with Positive/Negative masking in the associative memory and Thinking chain extraction mechanism specially implemented for flexible memory structure. The proposed Reticular Activating system has Knowledge
acquisition, selection, storing, reconfiguration and retrieving part. P/N
masking mechanism for flexible memory is specially designed and tested
with virtual memory.

1 Introduction

The greatest ability of human brain consists not in simply storing the information
but in thinking something from his memory. Every memory groups together with
other memory and is configured harmoniously. It depends on integrated activity
of whole memory. It is known that human being does not see real images of the
world but actually perceives virtual images made in his brain. The reason why
human being should see the virtual image in the brain is that perceiving virtual
images made of selected data is one of the efficient ways for processing huge data
in the world. The best way of putting the huge world in a small human brain
is information selection, forming the virtual images with selected information
and efficient processing. Reticular Activating System in the brain is known as a
system which selects the information coming from the outside. As the computer
technologies are developing very quickly, the information society is getting more
complicated and enclosed by huge data environment.

On the other hands, human thinking way depends on the deferent viewpoints.
A person who has positive viewpoints has a positive thinking way and a person who has negative viewpoints has a negative thinking way. The aspects of
viewpoints are very important in the information processing because the results
of thinking and their actions are different. It means that thinking chain and

I. Bloch, A. Petrosino, and A.G.B. Tettamanzi (Eds.): WILF 2005, LNAI 3849, pp. 385–394, 2006.
© Springer-Verlag Berlin Heidelberg 2006

direction of inference are changed by the viewpoint. The thinking way has a key of acting characteristics in human beings. These processing ability may be an efficient way for human being to survive in a difficult complex environments.

As the computer technologies are developing very quickly, the information society is getting more complicated and enclosed by huge data environment. In this huge data circumstance, the requirement of efficient intelligent systems adopting human brain functions is getting high. For making more flexible intelligent system, we consider the factors of different viewpoints because ,in the case of human thinking way, the results of thinking depends on personal characteristics and thinking view points, i.e. positive thinking way or negative thinking way.

For the purpose of implementing these functions , Reticular Activating System was designed in the previous paper[8]. In this paper, we introduce the main functions of designed Reticular Activating system first, focus on the design of associative memory configuration and describe Thinking chain extraction retrieved by Positive/Negative masking in detail. Positive/Negative(P/N) masking is specially designed in this system. This system is implemented by following three step design.

Reticular Activating System: The structure and main function of Reticular Activating System is described.

Reconfiguration of associative memory: knowledge is stored in associative memory according to the associative relations.

Positive/Negative Masking: In this step, the associative knowledge net masked by Positive/Negative masking vector is produced.

Positive/Negative thinking chain retrieval: thinking chain is retrieved from the associative knowledge net activated by P/N signal.

Inference/ Decision making: Using the retrieved thinking chain, Inference or Decision making step are made.

In the last step, Thinking chain retrieval process from the associative memory is experimented with virtual memory design . We tested the variation of thinking chain extraction in two aspects of P/N switching and level extraction.

2   Reticular Activating System

2.1   The Structure of Reticular Activating System

In this section Reticular Activating System which can select and store the information was designed. As shown in Fig.1, this system has a hierarchical structure and it consists of Knowledge acquisition, Selection and Storing to Memory.

First,Knowledge acquisition part has multi modular NN(neural Network)s and perform the learning process with the training data according to the categories. It uses BP(Back Propagation) algorithm. The output nodes of Modular

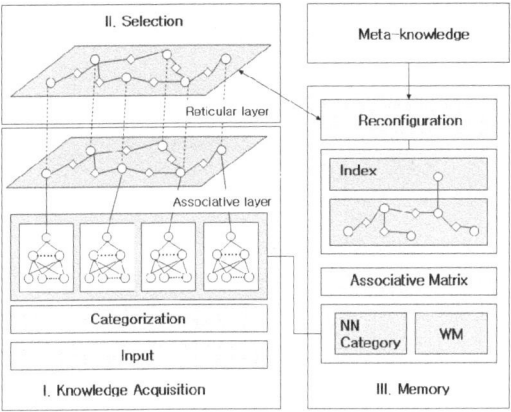

Fig. 1. Associative memory frame

NN are connected to nodes in Associative layer which has logical network connected by associative relations. Its learning mechanism is described in the paper [1] in detail.

Second, Reticular Activating layer has a knowledge net which consists of nodes and their associative relations. The nodes in knowledge net are connected to the nodes of Associative layer vertically. The importance value is assigned to the connection weight of this vertical relation. Selection module performs selecting process with these values of associative relations and vertical relations using the criteria given by Meta Knowledge.

Third, Storing to Memory consists of two part of Knowledge Reconfiguration and storing the values for NN. In Reconfiguration, the selected nodes and relations are reconfigured and stored in memory. The knowledge net is performed by attaching nodes centering around common node. After reconfiguration the centering node is connected to index which is used in searching process. In the case of polysemy, the common node is connected by On/Off switching to multiple knowledge net. The another part of memory is storing the values for NN. After finishing the learning process of modular NN, this system stores the values of category, parameters and weight matrix. These stored values are used for perception, inference and knowledge retrieval.

Reticular Activating System performs the functions of Learning, Selection , memory reconfiguration and Knowledge retrieval as these three parts collaborates on a work interactively.

2.2 Structuring Associative Memory

Structuring Associative memory is made by configuring the knowledge net which consists of nodes and arcs. Each knowledge node is connected to others with associative relational strength as shown in Figure 2. For efficient information processing,the graph of Knowledge net is represented as associative list of Table 1. Associative list is used for retrieving the Thinking chain.

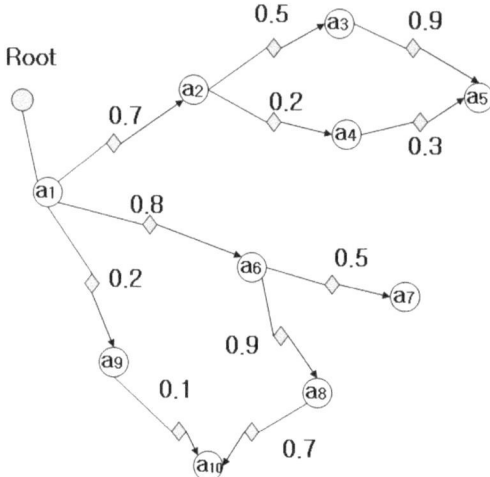

Fig. 2. Knowledge net in associative memory

Table 1. Associative list

node1	Relation	node2
Root	1.0	a_1
a_1	0.7	a_2
a_1	0.8	a_6
a_1	0.2	a_9
a_2	0.5	a_3
a_2	0.2	a_4
a_3	0.9	a_5
a_4	0.3	a_5
a_5	0.0	null
a_6	0.5	a_7
a_6	0.9	a_8
a_7	0.0	null
a_8	0.7	$a_1 0$
a_9	0.1	$a_1 0$
$a_1 0$	0.0	null

2.3 Reconfiguration of Associative Memory

In this section ,Selection, Storing and Knowledge Retrieval of Reticular Activating System are described. This system has a same structure of multi modular NN ,learning and functions of Associative layer as explained in the paper [1]. As shown in figure 3, the nodes of reticular layer are connected each other with Associative relation , R_{ij} horizontally and are also connected to the nodes in Associative layer with connection weight R_{ij} vertically.

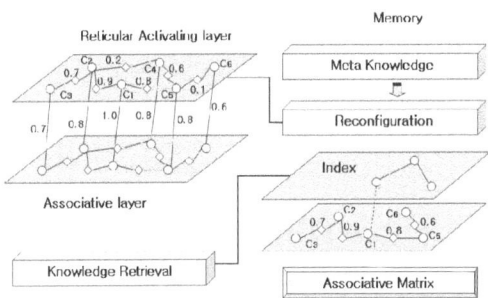

Fig. 3. Reticular Activating Layer

The connection weights R_{ij} and S_{ij} are used for selecting when Meta knowledge gives a criteria.

Storing Criteria:
$V_i \geq 0.5, R_{ij} \geq 0.5$

If Meta knowledge gives a following storing criteria, the nodes and relations satisfying this criteria are selected. The selected nodes and relations are reconfigured by attaching the related nodes centering common node [7]. This centering common node is connected to the index node in Index layer and used for searching as a keyword.

3 Positive/Negative Masking

3.1 The System Flow of P/N Masking Mechanism

As one of implementing Positive thinking and Negative thinking way, we designed P/N masking mechanism. First we define " P/N masking" as a frame for retrieving Positive/Negative Thinking chain.

Figure 4 shows the system flow of P/N mechanism. The knowledge coming from I/O interface is connected to other knowledge by its associative strength and stored in associative memory. Associative memory provides an important basic frame for extracting the thinking chains. If control signal arrives in P/N switch, P/N switch works. P/N switch takes part in controlling the masking type, That is, Positive masking or Negative masking. If one masking type is selected, the related knowledge net is activated according to P/N signal and the associative Thinking chains are retrieved. The extracted Thinking chains are used for inference and decision making process.

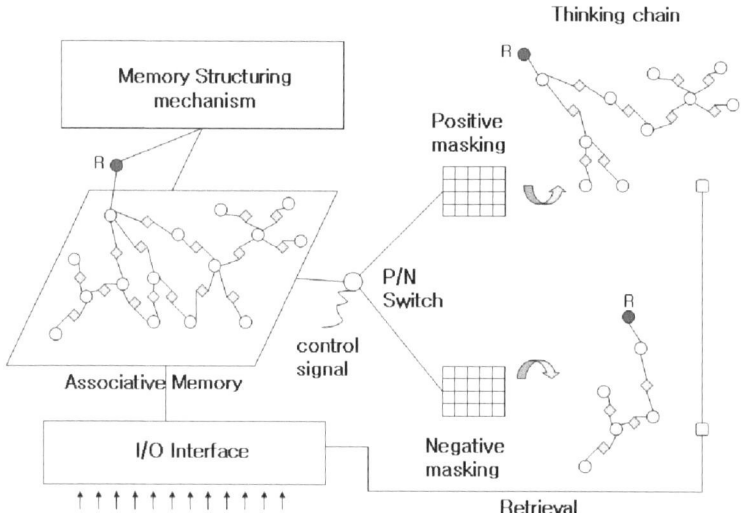

Fig. 4. The system flow of P/N masking mechanism

3.2   P/N Masking

The main function of P/N masking is retrieving Thinking chains related to different thinking points. For implementing this mechanism, we designed P/N masking mechanism. P/N masking is made by masking P/N vector to the connected associative strength in Knowledge net. Due to this function, the interpretation of associative memory can be always changing and associative memory maintains a flexible storage.

P vector/N vector is defined as a vector for P/N masking. P vector and N vector are represented as:

$$P = [p_{ij}], N = [n_{ij}] \tag{1}$$

After P/N masking, associative strengths are changed by equation (2).

$$R'_{ij} = R_{ij} * p_{ij}..If\,P\,switch, R'_{ij} = R_{ij} * n_{ij}..If\,N\,switch \tag{2}$$

4   Positive/Negative Thinking Chain Retrieval

After P/N masking finished,Thinking chain can be extracted. Thinking chain is defined as thinking flows retrieved from the memory in this paper.

Thinking chain extraction mechanism is : System takes one keyword and searches the corresponding data in Associative list. If it is found,Forward-chaining and Backward-chaining process are started from the found keyword in Associative list. The related Thinking chains are extracted using Forward/ Backward chaining process.

Thinking chain extraction algorithm is as follows.

Algorithm 1 : Thinking chain extraction algorithm

STEP 1: Input the Keyword.
STEP 2: Search the corresponding keyword
   in the Associative list
STEP 3 : If (not found)
   Print ("There is no corresponding keyword!")
   Else (found)
   forward-chaining;
   backward-chaining;
STEP 4: Output the retrieved Thinking chain.
   Print F-list
forward-chaining;
   next=associative-list;
   i = found-i;
   while((next.node2[i]!= null) and (i !=N))
   { Put next.node1[i] to F-list;
   Put next.R[i] to F-list;
   Put next.node2[i] to F-list ;
   i=i+1; }
backward-chaining;
   next=associative-list;
   i = found-i;
   while((next.node2[i]!= null) and (i !=0))
   {Put next.node1[i] to B-list;
   Put next.R[i] to B-list;
   Put next.node2[i] to B-list ;
   i=i-1; }
STEP 5: Stop.

4.1   Inference and Decision Making

Using the retrieved Thinking chains, Inference and Decision making process can be made in an appropriate ways. These Thinking chains can produce the answer for query.

5   Experiments

P/N masking mechanism is applied to virtual memory. The knowledge net shown in Figure 2 was used for testing P/N Masking process and Thinking Chain extraction algorithm. P vector and N vector for P/N masking is as follows.

$$P = [1.0, 1.0, 0.0, 0.7, 1.0, 0.8, 1.0, 1.0, 1.0, 0.0, 0.0, 0.0, 0.0, 0.0, 0.0]$$

$$N = [1.0, 0.5, 1.0, 0.0, 0.0, 0.0, 0.0, 0.0, 0.0, 1.0, 1.0, 1.0, 0.9, 0.0, 1.0]$$

After P/N masking was applied, knowledge net masked by P/N vector in Figure 5 was obtained.

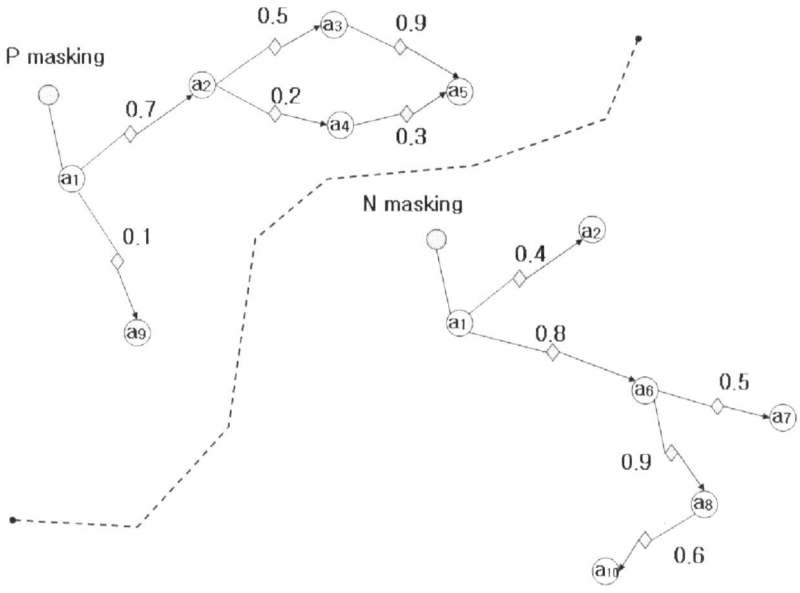

Fig. 5. The changed knowledge net masked by P/N vector

Table 2 shows changed values of associative strength of Knowledge net and the retrieved results of Thinking chains from the knowledge net masked by P/N masking and Figure 6 depicts the result of Thinking chain extraction. As a result,it was found that Thinking chains controlled by Positive/ negative thinking way were extracted successfully.

6   Conclusion

In this paper, we propose Reticular Activating system which has functions of selective reaction, learning and inference. This system consists of Knowledge acquisition, selection , storing and retrieving part. Reticular Activating layer is

Table 2. The changed associative strengths by P/N masking

a_i - a_j	R_{ij}	P	R_{ij}^P	N	R_{ij}^N
0-1	1.0	1.0	1.0	1.0	1.0
1-2	0.7	1.0	0.7	0.5	0.4
1-6	0.8	0.0	0.0	1.0	0.8
1-9	0.2	0.7	0.1	0.0	0.0
2-3	0.5	1.0	0.5	0.0	0.5
2-4	0.2	0.8	0.2	0.0	0.0
3-5	0.9	1.0	0.9	0.0	0.0
4-5	0.3	1.0	0.3	0.0	0.0
5-5	0.0	1.0	0.0	0.0	0.0
6-7	0.5	0.0	0.0	1.0	0.5
6-8	0.9	0.0	0.0	1.0	0.9
7-7	0.0	0.0	0.0	1.0	0.0
8-10	0.7	0.0	0.0	0.9	0.6
9-10	0.1	0.0	0.0	0.0	0.0
10-10	0.0	0.0	0.0	1.0	0.0

```
... Thinking Chain extraction
Type of masking?(P/N)P

... P masking
... The extracted Thinking Chains
... a1 0.7 a2 0.5 a3 0.9 a5 0.0 null
... a1 0.7 a2 0.2 a4 0.3 a5 0.0 null
... a1 0.1 a9 0.0 null

Type of masking?(P/N)N

... N masking
... The extracted Thinking Chains
... a1 0.4 a2 0.0 null
... a1 0.8 a6 0.5 a7 0.0 null
... a1 0.8 a6 0.9 a8 0.6 a10 0.0 null
```

Fig. 6. Retrieved thinking chain by P/N masking

connected to Meta knowledge in the high level of this system and takes part in Data Selection. Reconfiguration and Positive/Negative(P/N) masking Switching mechanism in knowledge net are specially designed. We applied this system to the virtual memory and tested P/N masking mechanism.

As a result of testing, we could find that it can extract the related thinking chains efficiently. It is expected that Reticular Activating system , the concept of its P/N masking and Thinking chain extraction can contribute to implement flexible associative memory and efficient retrieval mechanism.

References

1. Jeong-Yon Shim:Knowledge Retrieval Using Bayesian Associative Relation in the Three Dimensional ModularSystem,Lecture Notes in Computer Science, Vol.3007,Springer-Verlag,(2004)630-635
2. E. Bruce Goldstein,Sensation and Perception,BROOKS/COLE
3. Judea Pearl : Probabilistic reasoning in intelligent systems, networks plausible inference,Morgan kaufman Publishers (1988)
4. Laurene Fausett: Fundamentals of Neural Networks,Prentice Hall
5. Simon Haykin: Neural Networks,Prentice Hall
6. Jeong-Yon Shim, Chong-Sun Hwang,Data Extraction from Associative Matrix based on Selective learning system,IJCNN'99, Washongton D.C
7. John R. Anderson,Learning and Memory,Prentice Hall
8. Jeong-Yon Shim, "Automatic Knowledge Configuration by Reticular Activating System" Lecture Notes in Artificial Intelligence, Springer-Verlag, (2005)

Selective Immunity-Based Model Considering Filtering Information by Automatic Generated Positive/Negative Cells

JeongYon Shim

Division of General Studies, Computer Science, Kangnam University,
San 6-2, Kugal-ri, Kihung-up, YongIn Si, KyeongKi Do, Korea
Tel.: +82 31 2803 736
mariashim@kangnam.ac.kr

Abstract. Biological system has a very efficient immunity system which selects important signals and protects its body. The functions of immunity system can be successfully adopted to design an intelligent system in the information society. Accordingly in this paper Immunity based system which can select the important data from a large amount data is proposed . we define filtering factor as a criterion for reacting and selecting the data. This system is designed to have learning, perception & inference and Data extraction and to have an additive learning mechanism for the new obtained important information. This system is applied to the area for the analysis of customer's tastes and its performance is analyzed and compared

1 Introduction

The Biological system has a very efficient immunity system which can select an important data and protect its body. It does not make all the input data from the outside an object of information processing. If one should process all the obtained information in the internal part, one could not hold out any longer because of the overload for processing As the internet environment is developing, the requirement of information filtering is getting high. Immunity based systems are self-maintenance systems learned from and inspired by the immune system. It deals with information related to the system itself, and not with data from outside the system as typically exemplified by a pattern classifier that processes data not relevant to the pattern classifier itself. Thus Immunity based system deals with the self-related data or more specifically, challenges to the survivability of the system such as faults of the system, noise in the control signal, malicious attacks against the system and so on [1]. For making more intelligent system in the information society it is necessary to implement the automatic filtering system where concept of immunity system is adopted. Accordingly in this paper, we propose the modeling of immunity based system considering filtering factor. This system was designed to have learning, perception & inference and Data extraction and to have an additive learning mechanism for including the new important information.

I. Bloch, A. Petrosino, and A.G.B. Tettamanzi (Eds.): WILF 2005, LNAI 3849, pp. 395–403, 2006.

2 Immunity-Based Systems

An immunity based system is one that involves a self-maintenance system. This property is placed in an axiomatic position in defining Immunity based system. In addressing first point, it is obvious that the system must be distributed into components capable of evaluating each other. Another important property of Immunity based system is that the system, that is the self incorporates components capable of evaluating or interacting with each other. Because of their mutual evaluation and interaction characteristics, normal components of the system(the self) and faulty ones(the nonself) constitute an ad hoc network formed on the spot depending on the situations in which the system is applied. To overcome the second point, immunity based system incorporates an adaptive system by diversity and selection. This may not provide an optimal solution, but instead a feasible one.

The immune system uses similar solutions, suitably amplified and modified, for similar problems. Immunity based system has the following three properties:

o self maintenance system with monitoring not only of the nonself but also of the self
o distributed system with autonomous components capable of mutual evaluation on adaptive system with diversity and selection

In selecting tasks for Immunity system, the following are worth mentioning:

o Immunity based system is meant for a specific task : the self-nonself recognition. The task of self-nonself discrimination is neither a kind of pattern recognition nor classification. Both pattern recognition and classification deal with data not related system itself. Further, both pattern and classes must be beforehand.
o In self-nonself discrimination, there is a trade-off between misidentifying self as nonself and vice versa (i.e., false positive and false negative in terms of detection theory).

The ultimate goal is Immunity based system based on the organic view of the immune system pioneered by Metchinikoff, Jerne and Burnet. That is, Immunity system is formalized as a system whose interdependency is so strong that any other entities not in harmony with the system will be eliminated by the self-organizing and maintenance process. However , this goal has not been attained yet.

2.1 Self-maintenance System

For Immunity based system to be self-maintenance systems, two remarks are in order. First, the self-maintenance property of Immunity based system comes from the self-nonself discrimination problem that the Immunity based system faces. Immunity based system must deal with challenges that affect the system itself, not with data that can be defined without referring to the system. This Immunity based system characteristics is consistent with Tauber's view on nonself from the viewpoint of immunological self. ' Most saliently, the network 'knows' only itself, and it is in the perturbation of the system that reaction occurs ,i.e. not to the 'foreign' but to the 'disturbance' of the system itself. Second, the problems addressed by the Immunity based system are not a pattern classification where the matter is only mapping to a number of classes, but problems related to a specific dichotomy between 'the self' and 'others'. This dichotomy between the self and others is not simply a classification with two classes, but is qualitative at a different level (meta level) from classes captured at the same level.[1]

3   The Modeling of Immunity-Based System Considering Filtering Factor

Biological system takes an important input signals selected by filtering. During the process only information reacted by the corresponding receptive cell can be passed and the other things are discarded. But we can not exclude the possibility that new important information which has an important effect on the perception and inference may be included in the discarded data. To be an efficient system, it should have a flexible function adaptive to a new environment. The new mechanisms for reaction made by receptive cell and increment learning for new obtained data are proposed in this paper.

3.1   Receptive Field for Information Filtering Personal Preference

Fig. 1 shows the structure of proposed system. This system consists of Receptive field, Learning module and Associative Map. Receptive field is described in detail. Receptive field has reactive cells, the area of new cells and Filtering gate.

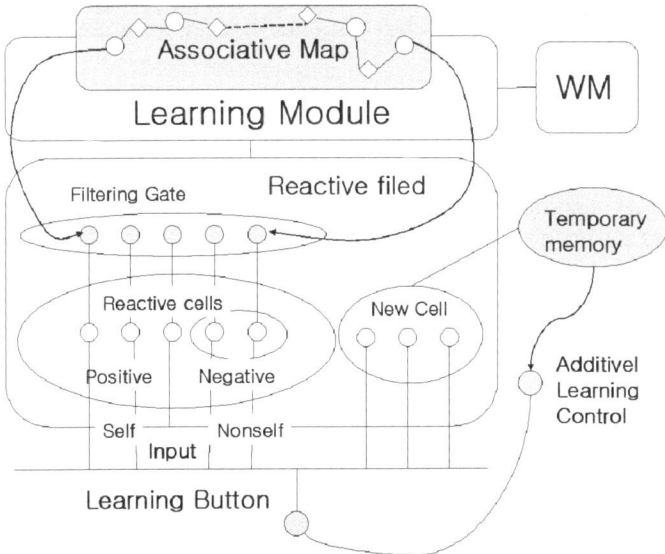

Fig. 1. Receptive Field

The filtering gate is connected to Associative Map and controlled by system and regulates the reactive degree. Input data propagated from the input layer react on the receptive cells and are filtered by filtering gate connected to the receptive cells.

We define reactive degree for selecting the outside data as filtering factor. In the first step, the system takes the values for filtering factor from experts and allocate the readjusted values to the nodes in Associative map and connected filtering gate. The readjusted value of filtering factor, Fi, calculated by the following equation 1.

$$F_i = \frac{R_i}{Max(R_i)} \tag{1}$$

Where R_i is

$$0 \leq R_i \leq 1$$

3.2   Additive Learning

In information filtering mechanism, only data filtered through receptive cell are accepted and other data are discarded. But important information may be included in the discarded data. It is important to find these data and to include them in the learning process. As shown in fig.1, new cells are included in Receptive field. They take part in selecting the new important data by estimating their appearing frequency, Ai (eq. 2).

$$A_i = P(N_i|E) \tag{2}$$

Where Ni is the number of appearing and E is the evidence of data in the new cell. This system takes the selected data in the new cells and starts additive learning mechanism.

3.3   Additive Learning Mechanism

The proposed mechanism for additive learning does not start the learning process again from the beginning point but performs the learning algorithm using the weight values set in the previous learning step. It allocates the random values to the new connection formed by additional new nodes. The following Algorithm 1 shows additive learning mechanism.

Algorithm 1: Additive Learning

STEP 1: Prepare the training data in forms of input-output pairs.
　　Selection:
STEP 2: Select the new important data by estimating their appearing frequency,
$$A_i = P(N_i|E)$$
STEP 3: Connect the new nodes to the previous structure.
STEP 4: Assign the weight values stored in WM(Weight Matrix).

STEP 5: Assign the random values to the new connections. **Learning:**
STEP 6: Get the input data, Ii.

Filtering:
STEP 7: Get the values of R_i for personal preference and calculate F_i.

$$F_i = \frac{R_i}{Max(R_i)}$$

STEP 8 : Calculate the reactive value in the receptive field.

$$S_i = \sum_{i=1}^{n} I_i \cdot Fi$$

Learning:
STEP 9: Calculate the output H_j in hidden layer.

$$H_j = \frac{1}{1 + e^{-\sum_{i=1}^{n} W1_{ij}S_i}}$$

STEP 10: Calculate the actual output, $y_1', y_2', \cdots, y_n'$

$$y_j = \frac{1}{1 + e^{-\sum_{i=1}^{n} W2_{ij}S_i}}$$

STEP 11:Calculate the error between the desired output,$d_1, d_2, \cdots, d_n$ and actual output,$y_1', y_2', \cdots, y_n'$.

$$E = \frac{(d - y)^2}{2}$$

STEP 12: IF $(E > \epsilon)$ THEN goto Learning ELSE goto STEP 14
STEP 13: Adapt the weights propagating the error backward to Inference layer
and input layer. Goto STEP 9.

$$\Delta W2_{ij} = \eta \delta 2_{ij} H_i$$

$$\Delta W1_{ij} = \eta \delta 1_{ij} H_i$$

STEP 14: stop

4   Perception & Inference and Data extraction

4.1   Associative Map

Associative Map consists of nodes and their associative relations shown in Fig.2.
Each node has a value of filtering factor for reactivating the data from outside
in receptive field. The filtering factors are store in filtering matrix B as following
Table 1. They are transformed to the Associative Matrix (AM) in Table 2.

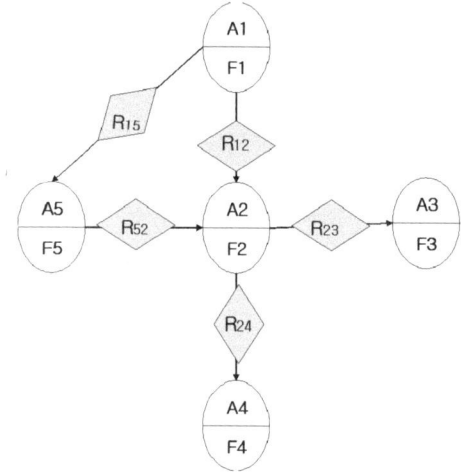

Fig. 2. Relational Graph

Table 1. Filtering Value

A_1	A_2	A_3	A_4	A_5
F_1	F_2	F_3	F_4	F_5

Table 2. Associative Matrix(AM)

	A_1	A_2	A_3	A_4	A_5
A_1	1	R_{12}	0	0	R_{15}
A_2	$-R_{12}$	1	R_{23}	R_{24}	$-R_{52}$
A_3	0	$-R_{23}$	1	0	0
A_4	0	$-R_{24}$	0	1	0
A_5	$-R_{51}$	R_{52}	0	0	1

The value of associative relation is calculated by equation 3.

$$R_{ij} = P(A_j|A_i) \cdot D \qquad (3)$$

Where D is the direction of an arrow,-1,1, and $i - 1, 2, \cdots, n, j = 1, 2, \cdots, n$.
AM relations are converted to the following numerical terms

$$A = \begin{bmatrix} 1.0 & 1.0 & 1.0 & 0.0 & 0.0 \\ -1.0 & 1.0 & 0.7 & 0.0 & 0.0 \\ -1.0 & -0.7 & 1.0 & 0.6 & 0.3 \\ 0.0 & 0.0 & -0.6 & 1.0 & 0.0 \\ 0.0 & 0.0 & -0.3 & 0.0 & 1.0 \end{bmatrix}$$

Filtering Matrix, B , is as follows :

$$B = \begin{bmatrix} 0.5 & 0.6 & 0.7 & 0.1 & 0.0 \end{bmatrix}$$

4.2 Data Extraction

When Data extraction process is chosen in selection mode, the system starts to extract the related knowledge distributed in the neural network and Associative Map. The knowledge for perception is stored in the connection weights and the facts are connected by their associative relations in Associative Map. Data extraction mechanism extracts the related facts using this Associative Matrix. The algorithm of Data extraction is described in Algorithm 2.

Algorithm 2: Data extraction mechanism in AM

STEP 1: Search for associated nodes in the row of the activated node in AM.
STEP 2: IF((not found) AND (found the initial activated node)), Goto STEP 3.
ELSE . Output the found fact.
. Add the found fact to the list of inference paths. Goto STEP 1
STEP 3: STOP

As a result of performing this Data extraction mechanism, inferential paths are produced :

A1 (R13 1.0) A3 (R34 0.6) A4 Null A1 (R12 1.0) A2 (R23 0.7) A3 Null A2 (R23 0.7) A3 (R35 0.3) A5 Null A3 (R35 0.3) A5 Null

This mechanism can elicit the related facts referring the inferential paths. More detailed description of this mechanism is provided in the paper[1].

5   Experiments

The proposed system applied to the area for the analysis of customer's tastes. We tested this system with 20 input factors and analyzed its results for decision-making(three types). The raw data was filtered by filtering function as shown in Fig. 3. Fig. 4 shows the comparison of the case considering filtering factor and the case without considering filtering. The result of the case considering filtering is different from the other one. We also tested the performance of additive learning. In this experiment, the error converging curves are compared. As shown in Figure 5 and Figure 6, the case of using the previous weight values(Case1) is more efficient than the case of relearning from the beginning point(Case2). In Case 1, the error was converged after 753 training iteration. Compared to this test , only 421 iteration was needed for error conversion. Figure 6 shows the result of Data extraction mechanism using the following Associative Matrix A.

$$A = \begin{bmatrix} 1.0 & 1.0 & 1.0 & 0.0 & 0.0 \\ -1.0 & 1.0 & 0.7 & 0.0 & 0.0 \\ -1.0 & -0.7 & 1.0 & 0.6 & 0.3 \\ 0.0 & 0.0 & -0.6 & 1.0 & 0.0 \\ 0.0 & 0.0 & -0.3 & 0.0 & 1.0 \end{bmatrix}$$

$$B = \begin{bmatrix} 0.5 & 0.6 & 0.7 & 0.1 & 0.0 \end{bmatrix}$$

After perception mode was performed, the node A1 was activated and the inferential paths of the associative knowledge were extracted by Data extraction mechanism.

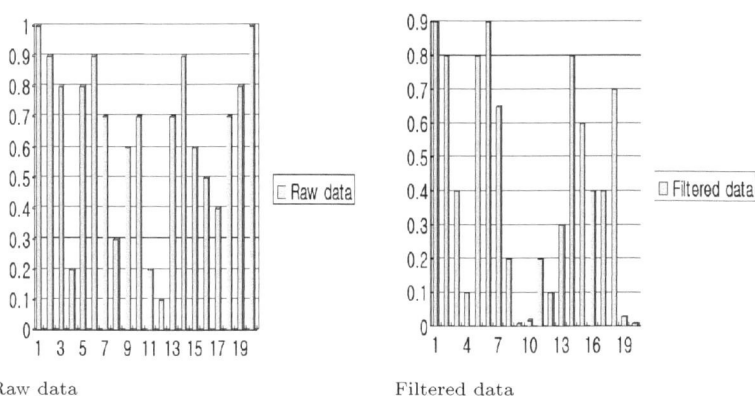

Raw data   Filtered data

Fig. 3.

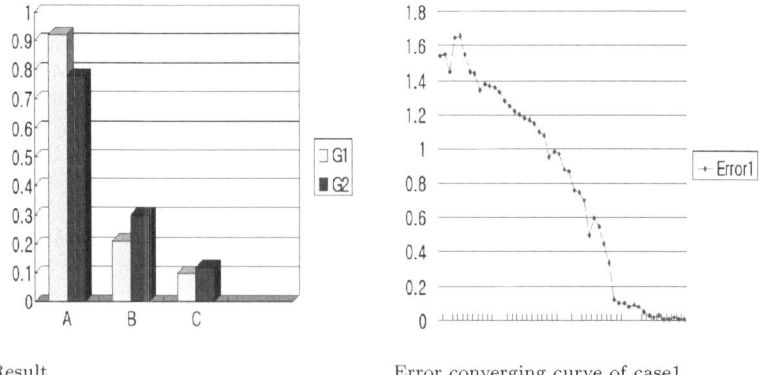

Result   Error converging curve of case1

Fig. 4.   **Fig. 5.**

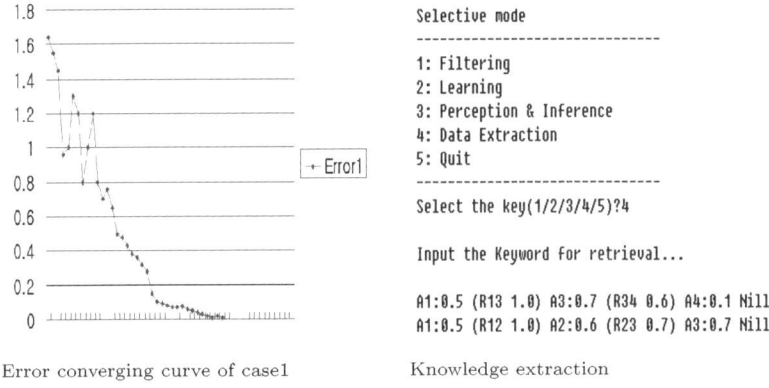

Error converging curve of case1   Knowledge extraction

Fig. 6.

6   Conclusion

Immunity based system which can select the important data from a large amount data is proposed. we define filtering factor as a criterion for reacting and selecting the data. This system is designed to have learning, perception & inference and Data extraction and to have an additive learning mechanism for the new obtained important information. The proposed system is applied to the area for the analysis of customer's tastes and its performance is analyzed and compared. As a result of testing , we could find that the input data was reacted by filtering factor and showed the different output. The proposed system was also can be usefully applied to many areas in the internet environment where the abilities for obtaining the new knowledge, filtering the data automatically and extracting the related information are required.

References

1. Jeong-Yon Shim, Knowledge Retrieval Using Bayesian Associative Relation in the Three Dimensional ModularSystem,pp630-635, *Lecture Notes in Computer Science(3007), 2004.*
2. John R. Anderson Learning and Memory *Prentice Hall.*
3. Laurene Fausett Fundamentals of Neural Networks *Prentice Hall.*

Exploring the Way for Meta-learning with the MINDFUL System

Ciro Castiello, Giovanna Castellano, and Anna Maria Fanelli

CILab - Computational Intelligence Laboratory,
Dipartimento di Informatica, Università degli Studi di Bari,
Via E. Orabona, 4 - 70126 Bari Italy
{castiello, castellano, fanelli}@di.uniba.it

Abstract. Meta-learning practices concern the dynamical search of the bias presiding over the behaviour of artificial learning systems. In this paper we present an original meta-learning framework, namely the MINDFUL (Meta INDuctive neuro-FUzzy Learning) system. MINDFUL is based on a neuro-fuzzy learning strategy providing for the inductive processes applicable both to ordinary base-level tasks and to more general cross-task applications. The peculiar organisation of the system allows a suitable meta-knowledge management, in order to carry on meta-learning investigations and to develop life-long learning strategies.

1 Introduction

The applied research in the field of artificial intelligent systems often deals with empirical evaluations of machine learning algorithms to illustrate the selective superiority of a particular model. This kind of approach is characterised by a "case study" formulation that has been recognised and criticised in literature [1,2]. The selective superiority demonstrated by a learner in a case study application reflects the inherent nature of the so-called *base-learning* strategies, where data-based models exhibit generalisation capabilities when tackling a particular task. Precisely, base-learning approaches are characterised by the employment of a fixed bias, that is the ensemble of all the assumptions, restrictions and preferences presiding over the learner behaviour. This means a restricted domain of expertise for each learning model, and a reduction in its overall scope of application. The limitations of base-learning strategies can be theoretically established: the no free lunch theorems express the fundamental performance equality of any chosen couple of learners (when averaged on every task), and deny the superiority of specific learning models outside the case study dimension [12].

Obviously, if we want to perform pragmatic investigations of particular domains, base-learning approaches represent a quite satisfactory way of proceeding to obtain adequate results. Whenever we are interested in following a line of research with a broader scope, involving some kind of cross-domain applications, the resort to somewhat different methodologies is advisable. By focusing the attention on the role of bias, we characterise the *meta-learning* approach as

I. Bloch, A. Petrosino, and A.G.B. Tettamanzi (Eds.): WILF 2005, LNAI 3849, pp. 404–409, 2006.

a dynamical search of a proper bias, that should be able to adapt the learner behaviour to the particular task at hand. The research field of meta-learning represents a novel approach aiming at designing artificial learners with enhanced capabilities, possibly capable of profiting from accumulated past experience [10,11]. In this way, the formulation of the model evaluation could overcome the case study dimension and the limitations of the base-learning strategies.

The research activity described in this paper takes part in the investigation on meta-learning, presenting a particular framework based on neuro-fuzzy integration, namely the MINDFUL (Meta-INDuctive neuro-FUzzy Learning) system. The proposed meta-learning framework is organised in order to employ a single learning scheme, working both as base- and meta-learner. The combination of neural learning and fuzzy logic allows to derive from data a base of interpretable knowledge, codified in form of fuzzy rules, which could prove to be effective in solving any base-level task at hand. Besides the usual base-learning process, the MINDFUL system brings forward also a meta-learning activity, where the same knowledge-based methodology is adopted to examine a set of meta-features describing the properties of specific tasks. In this way, the meta-learner provides an explicit meta-knowledge, in terms of fuzzy rules, representing a significant form of bias to direct the learning process of the base-learner. The MINDFUL system is able to prove its appropriateness in retaining the knowledge accumulated during learning, showing improved performances when tackling new tasks.

The paper is organised as follows. In the next section we are going to introduce the MINDFUL system, presenting the underlying hybrid learning framework. The nature of the involved meta-knowledge and different directions to be followed for the employment of the system are presented in section 3. Finally, some conclusive applicability remarks are drawn in section 4.

2   MINDFUL: A Framework for Meta-INDuctive Neuro-FUzzy Learning

The central tenet in meta-learning research consists in devising suitable techniques for allowing an artificial learner to modify its bias, in order to better tackle the tasks under analysis. Therefore, when designing a meta-learning strategy, a proper identification of the involved bias represents a key point, which is important also for determining the suitable mechanism to direct the learner behaviour. Our peculiar idea for a meta-learning framework is centred on the working scheme of an intelligent system that realises a neuro-fuzzy integration. It represents our proposal for a prototype tool, the MINDFUL system, which stands as a candidate for fulfilling the meta-learning requirements.

Our approach differs to some extent from the majority of the meta-learning schemes proposed in literature. Most of them, in fact, perform a dynamic search of the learning bias by assimilating the concept of bias with the specific learner, properly chosen for the task at hand. In this way, several strategies of model combination and model selection have been developed, involving different learners which constitute a pool of bias-specific candidates. A number of research projects produced results in this context; prominent examples include the METAL and Statlog projects [6,7]. On our part, we aim at employing a single learning scheme, endowed with the capability of improving its performance: a neuro-fuzzy model plays the twofold role of base-learner (to tackle ordinary predictive learning tasks) and meta-learner (to produce some form of meta-knowledge). By doing so, we characterise our strategy on the basis of a number of key points. Firstly, the idea of meta-learning is translated to a more qualified level, since it is not intended as simply picking a learning procedure among a pool of candidates, but it focuses on a deeper analysis of a learning model behaviour, in order to understand and possibly to improve it. Moreover, the choice for a single learning model should be suitable to preserve the uniformity of the whole system and to reduce its complexity, even in terms of comprehensibility. Finally, the neuro-fuzzy strategy, applied both at base and meta-level, endows also the meta-learning procedure with the benefits deriving from the integration of the connectionist paradigm with fuzzy logic.

In the following, we are going to trace a broad outline of the working scheme underlying the MINDFUL system, highlighting the particular nature of the meta-knowledge involved in the meta-level activity. We shall indicate also different directions that can be followed to bring the meta-learning system into action.

2.1   The Working Scheme of the MINDFUL System

The core of the MINDFUL system is represented by a neuro-fuzzy strategy: it provides for the inductive process applicable both to the ordinary base-learning level and to a more general cross-task level. The combination of connectionist learning with fuzzy logic permits to organise the inductive process in such a way that, starting from the analysis of observational data, the learning model is able to produce a fuzzy rule base which eventually codifies the processed information in a linguistically comprehensible fashion. This is accomplished by postulating the formal equivalence between the fuzzy rule base and a particular neural network (the *neuro-fuzzy network*), reflecting in its topology the structure of the fuzzy inference system. The learning scheme is articulated in two successive steps, intended to firstly initialise a knowledge structure and then to refine the obtained fuzzy rule base. During the first step, a clustering of the input data is performed by means of an unsupervised learning process of the neuro-fuzzy network. A rival penalised mechanism is employed to adaptively determine a proper structure of the network. In this way, an initial knowledge is extracted from data and expressed in form of fuzzy rules. The base of knowledge is successively refined during the second step, where a supervised learning process of the neuro-fuzzy network (based on a gradient descent technique) is accomplished, in order to attune the parameters of the fuzzy rule base to the numerical data. The fuzzy inference system codifies the knowledge in fuzzy rules of the form:

$$\text{IF } x_1 \text{ is } A_1^r \text{ AND} \ldots \text{AND } x_m \text{ is } A_m^r \text{ THEN } y_1 \text{ is } b_1^r \text{ AND} \ldots \text{AND } y_n \text{ is } b_n^r, \quad (1)$$

where the index $r = 1, \ldots, R$ indicates the r-th rule among the R comprised into the rule base; A_i^r are fuzzy sets (defined in terms of Gaussian membership

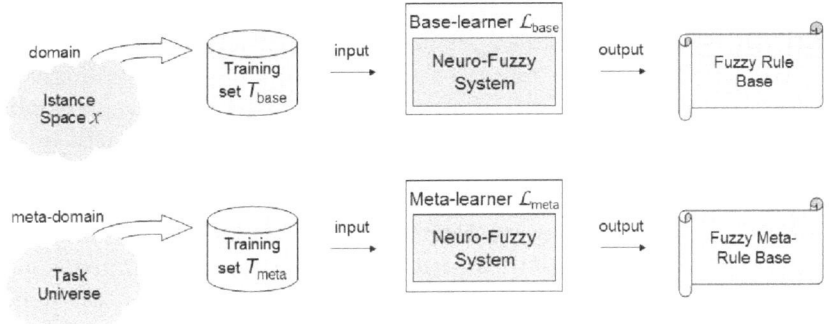

Fig. 1. The working scheme proposed for the base- and the meta-learner

functions over the input components x_i, $i = 1, \ldots, m$); b_j^r are fuzzy singletons (defined over the output components y_j, $j = 1, \ldots, n$). Here we do not provide further details concerning the formalisation of the fuzzy inference system (whose general scheme is comparable to the TSK method of fuzzy inference [9]) and the learning algorithms, addressing the reader to some other previous works of ours [3,4]. Instead, in the following section we would like to underline how the distinct base- and meta-learning practices performed by the MINDFUL system are articulated on the basis of the common working scheme described above.

3   Meta-knowledge Characterisation and Directions for Meta-learning

A couple of learners is involved in the overall activity of the framework, namely a base-learner $\mathcal{L}_{base}$, working on the basis of a training set T_{base}, and a meta-learner $\mathcal{L}_{meta}$, employing a training set T_{meta} (as depicted in fig. 1). Indeed, while the learning procedure yielding to the generation of knowledge in form of fuzzy rules is the same for both learners, the information embedded in the training sets is quite different. In fact, $T_{base} = \{(\mathbf{x}_k, \mathbf{y}_k)\}_{k=1}^{K}$ is composed by an ensemble of K instances, arranged as a pair of input-output vectors, which describe a given base-level task. The meta-training set $T_{meta} = \{(\mathbf{z}_k, \mathbf{b}_k)\}_{k=1}^{K'}$ is similarly formalised, but the input vector $\mathbf{z}_k$ expresses a kind of higher-level task knowledge. Particularly, $\mathbf{z}_k$ comprises a number of meta-features (general, statistical and information theoretical measures) which characterise the base-level dataset describing a particular task. (Actually, these kinds of meta-features appear to be widely exploited in meta-learning contexts [6]; in [5] we engaged a deeper analysis oriented to define the most relevant meta-features for task discrimination.) The output vector $\mathbf{b}_k$ represents the bias yielding the best learning performances when applied to $\mathcal{L}_{base}$, during the base-level activity addressing the solution of the task connected with $\mathbf{z}_k$. Practically, the bias refers to the configurations of parameters presiding over the overall learning process; the best configuration is determined by experimenting at base-learning level. Therefore, while T_{base} is established on the basis of the problem at hand which $\mathcal{L}_{base}$ ultimately aims at solving, the arrangement of T_{meta} is empirically performed. However, both the base- and the meta-learner, being characterised by the same neuro-fuzzy learning procedure, end up by originating a base of knowledge with fuzzy rules in the form described in (1). In this way, the meta-knowledge produced by $\mathcal{L}_{meta}$ supplies the proper bias for future base-level task analyses.

The introduced working scheme could be differently exploited to give rise to meta-learning practices: here we delineate a pair of directions which could be followed to bring the MINDFUL system into action. Actually, a first approach consists in employing the system to tackle a number of specific task domains, retaining the experience accumulated during base-learning applications. This kind of situation relies on the compilation of a meta-training set which is built up once and for all at the end of a base-level session of experiments. While the meta-training set is experimentally derived from base-level task analyses, the meta-knowledge obtainable from the meta-learning process can be exploited to determine the proper bias in novel circumstances. In this way, the meta-knowledge serves as a resource for improving future learning performances in the same task domains.

This preliminary evaluation should pave the way for a broader approach based on a *life-long learning* strategy. In this case, the framework should be capable of dealing with an incremental meta-training set, possibly enlarging as time goes by, in order to fit novel task information and to attune the meta-knowledge to different contexts of real-world problems. Obviously, suitable mechanisms can be devised for allowing the meta-training set to enlarge whenever the available meta-knowledge appears to be inadequate for improving the learning performances on unseen tasks. In particular, a resort to base-learning activity could be renewed to update the applicable bias information. By doing so, the MIND-FUL system can act as a self-adaptive learner, even working without compiling an initial training set. Beginning with no experience at all, the system would be initially constrained to make use of the few pieces of information it is able to gather (building up the meta-knowledge little by little, when facing its first tasks). In other words, MINDFUL would employ an almost fixed form of bias at start. As more tasks are observed, however, the system would be able to use the accumulated meta-knowledge to change its own bias, according to the characteristics of each task.

4   Applicability Remarks

The realisation of the MINDFUL system offers the opportunity for exploring the way for meta-learning practices, with the possibility of following different directions, in the way they have been delineated in the previous section. The original aspects characterising the framework among the meta-learning approaches proposed in literature, basically concern the employment of a single learning model (which has to be directed in its work by performing a dynamic bias search) and the adoption of a hybrid procedure representing the core of the system (based on a neuro-fuzzy approach). In this way, MINDFUL differentiates from the common model selection or model combination strategies, and benefits both at base- and meta-level of activity from the fruitful integration of neural learning capabilities with fuzzy knowledge representation.

Actually, here we would like to underline how the meta-learning prototype system has been preliminary tested, involving a number of synthetic task domains [4]. Following the first previously described direction, founded on the realisation of a particular meta-training set, a preliminary session of base-learning experiments helped to identify the best bias configurations applicable in every task domain. By extracting the meta-feature vectors from the analysed datasets, a meta-training set has been established, properly correlating the dataset characterisations with the identified biases. The hybrid learning procedure has been successively replied over the meta-training set, to produce a base of meta-knowledge, in form of fuzzy rules, suitable for use in future task analyses. To evaluate the meta-knowledge obtained at the end of the meta-learning process, novel tasks have been considered from the synthetic domains and they have been tackled employing the bias configuration suggested by the fuzzy (meta-)rules. The final performance results proved the effectiveness of the derived meta-knowledge, thus encouraging a more detailed application of the MINDFUL system in real-world contexts of life-long learning.

References

1. Aha, D. W. (1992). Generalizing from case studies: a case study. *Proc. of the 9th Int. Conf. on Machine Learning*.
2. Brodley, C. (1993). Addressing the selective superiority problem: automatic algorithm/model class selection. *Proc. of the 10th Int. Conf. on Machine Learning*.
3. Castellano, G., Castiello, C., Fanelli, A. M., Mencar, C. (2005). Knowledge discovery by a neuro-fuzzy modeling framework. *Fuzzy Sets and Systems*, 149:187–207.
4. Castiello, C. (2004). Meta-Learning: a Concern for Epistemology and Computational Intelligence. PhD Thesis. University of Bari - Italy.
5. Castiello, C., Castellano, G., Fanelli, A. M. (2005). Meta-data: characterization of input features for meta-learning. Lecture Notes Artificial Intelligence, 3558:457-468.
6. Kalousis, A., Hilario, M. (2000). Model selection via meta-learning: a comparative study. *Proc. of the 12th Int. IEEE Conference on Tools with AI*.
7. Michie, D., Spiegelhalter, D. J., Taylor, C. (1994). *Machine learning, neural and statistical classification*. Ellis Horwood Series in Artificial Intelligence.
8. Ortega, J., Koppel, M., Argamon, S. (2001). Arbitrating among competing classifiers using learned referees. *Knowledge and Information Systems*, 3:470-490.
9. Sugeno, M., Kang, G. T. (1988). Structure identification of fuzzy model. *Fuzzy sets and systems*, 28:15–33.
10. Thrun, S., Pratt, L., eds (1998). *Learning to Learn*. Kluwer Academic Publisher.
11. Vilalta, R., Drissi, Y. (2002). A perspective view and survey of meta-learning. *Artificial Intelligence Review*, 18:77–95.
12. Wolpert, D. H. and Macready, W. G. (1997). No free lunch theorems for optimization. *IEEE Transactions on Evolutionary Computation*, 1(1):67–82.

Using Fuzzy Logic to Generate the Mesh for the Finite Element Method

Guido Sangiovanni

Department of Aerospace Engineering, Politecnico di Milano,
via La Masa, 34 – 20156 – Milano, Italy
sangiovanni@aero.polimi.it

Abstract. The aim of this work is to prove the efficacy of a Soft Computing approach to the problem of generating the best suited mesh for solving a differential problem with the Finite Element Method. Using Fuzzy Logic, it is possible to introduce a set of linguistic *if-then* rules reproducing the human expert reasoning used for creating the mesh.

1 Introduction

The Finite Element Method (FEM) is one of the most used techniques for solving Partial Differential Problems. The idea of FEM is to divide the domain of definition of the problem into small regions, called *elements* of the *mesh*, where an approximation of the solution is searched. This approximation converges to the exact solution as elements dimensions tend toward zero; obviously, the smaller the elements are, the higher their number is. Experience, competence, and high level knowledge, guide humans in finding the optimal mesh spacing that brings the approximation error to an acceptable level using the smallest number of elements [1], [2], [3].

The current numerical approach to the problem of finding the best grid is the *mesh adaptation strategy*. An initial solution is calculated using a first guess grid and the approximation error is estimated; reducing the dimensions of the elements with larger error values, decreases the overall error in a very efficient way, by modifying the grid only where it is needed. This process is iterated until the estimated error is below a given threshold or a maximum number of elements is reached [3], [4].

In this paper, an alternative solution to this problem is obtained using Soft Computing methods, following the works of Manevitz and Givoli. In [1] and [2] they tackle the problem of automating the FEM using Expert Systems and Self-Organizing Neural Networks respectively for node numbering and mesh placement. In the present work, Fuzzy Logic is used for the mesh generation process because it allows reproducing the qualitative reasoning typical of humans, by translating numerical inputs into linguistic values (such as '*good*', '*near*', '*high*') and by evaluating some *if-then* rules in parallel [5]. The translation of FEM expert's reasoning and rules for solving mesh generation problem will be described. The result is the so called *metric*, i.e. a map $h : \Omega \rightarrow \Re^+$ which specify the dimension of the element in each point of the domain Ω.

I. Bloch, A. Petrosino, and A.G.B. Tettamanzi (Eds.): WILF 2005, LNAI 3849, pp. 410–419, 2006.

2 Problem Formalization and Algorithm

Let us consider the Poisson problem with mixed boundary conditions:

$$\begin{cases} -div(v \cdot \nabla(u)) = f & in\ \Omega \subset \Re^2 \\ u = g & on\ \Gamma_D \\ \partial_n u = h & on\ \Gamma_N \end{cases} \tag{1}$$

where u is the unknown solution, v is the viscosity function, Γ_D and Γ_N are the part of the border where, respectively, the Dirichlet boundary condition and the Neuman boundary condition are applied [4].

Solving this problem with the FEM in an '*intelligent*' way requires having an idea of the general behaviour of the solution over the domain. Where it will be smooth, large elements will be required, while elements will be smaller where the solution exhibits great changes. An expert of FEM is able to identify *a priori* the regions where it is necessary to have small elements, by simply looking at the parameters of the problem and by combining this information with a high-level reasoning.

FEM experts state that elements must be very small where:

− there is a singularity in the boundary conditions;
− a point is present where B.C. types are different;
− the border presents edges with angles lower than 90° and higher than 270°;
− the forcing function f is high (because it causes high variations in u);
− the gradient of the viscosity v has a high absolute value.

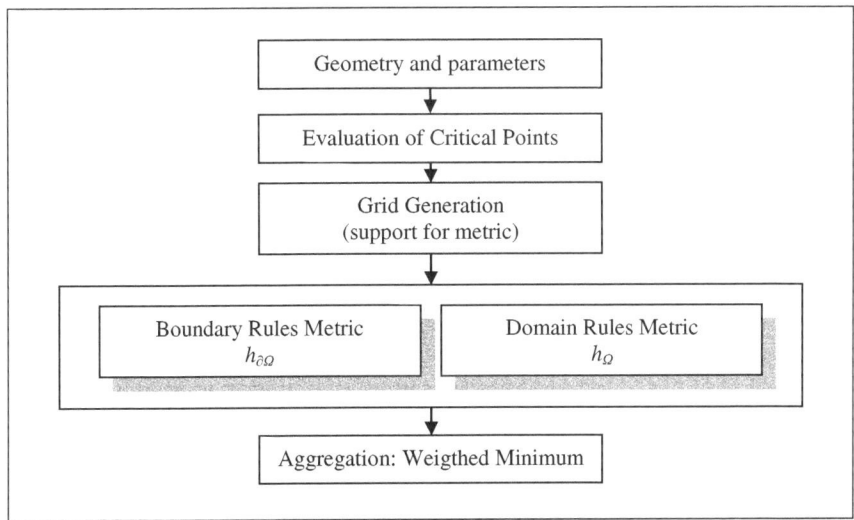

Fig. 1. Architecture of the proposed algorithm; the output of the last step is the metric

The first step of the algorithm is the creation of differential problem using the *pdetool* toolbox of *Matlab©*. The next step is the evaluation of some *Critical Points*, i.e. points where the types of B.C. are different, B.C. functions have discontinuities or the angle of the border of the domain is very big or very small. It is also necessary to calculate the gradient of the B.C. functions (g) along the borders, using the local tangent vector. With such set of information and some fuzzy rules, the *Boundary Rules' Metric* $h_{\partial\Omega}$ is generated. The following step is the generation of the *Domain Rules' Metric* h_Ω, which requires the evaluation of the modulus of the forcing function f and of the gradient of the viscosity. The two metrics are combined with the *Weighted Minimum* operator, a T-norm that allows mixing the two maps according to the distance from the borders (Fig. 1).

3 Fuzzy Rules and Membership Functions

There are four sets of fuzzy rules, based on the considerations of FEM experts:

1. Type rules – for the difference of the type of B.C. in a critical point;
2. Discontinuity rules – for a discontinuity in B.C. functions;
3. Connectivity rules – taking into account the angle of the boundary at the segments intersections;
4. Domain rules – for the absolute value of the functions f and $|\nabla(v)|$.

For the first three types of rules, the variables used are the distance from a Critical Point, the value of the discontinuity, and the angle (in degrees) as inputs, and the

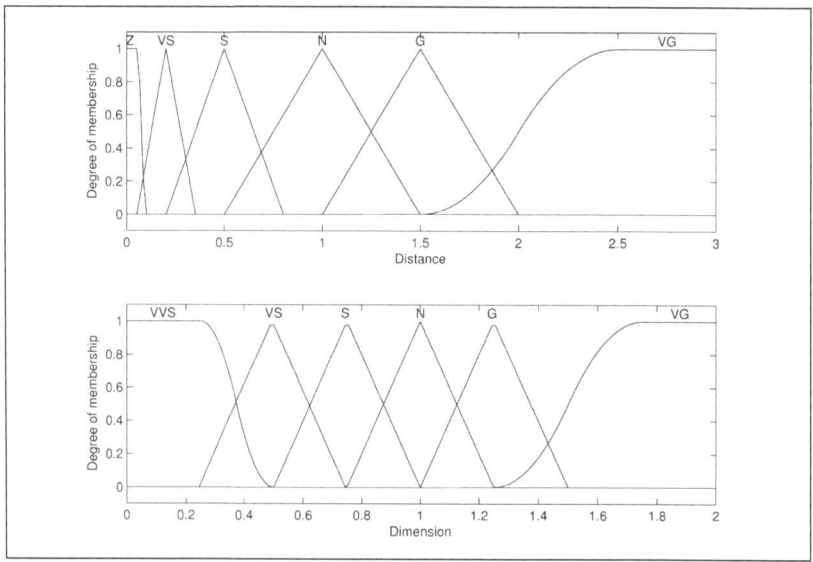

Fig. 2. Membership functions for *Type rules*. Distance and Dimension are respectively the input and the output of these set of linguistic rules.

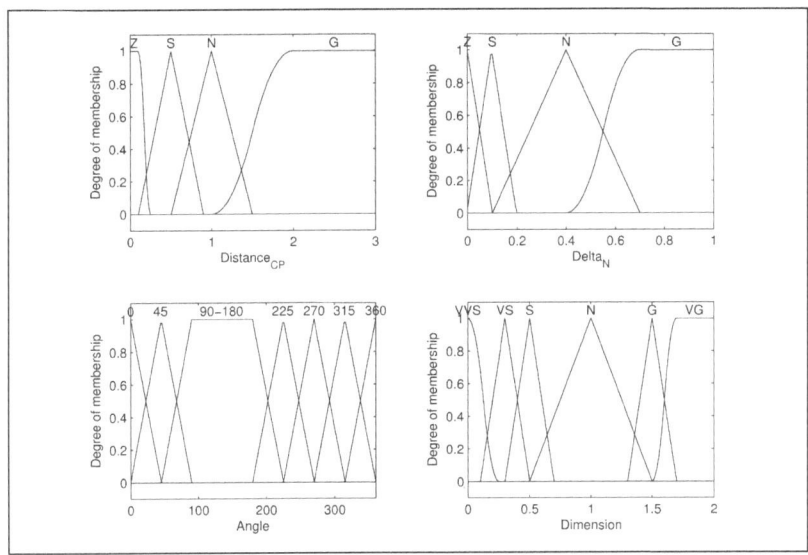

Fig. 3. Membership functions for *Discontinuity and Connectivity rules*. Distance$_{CP}$, Delta$_N$, and Angle are the input variables, whereas Dimension is the output of these sets of rules.

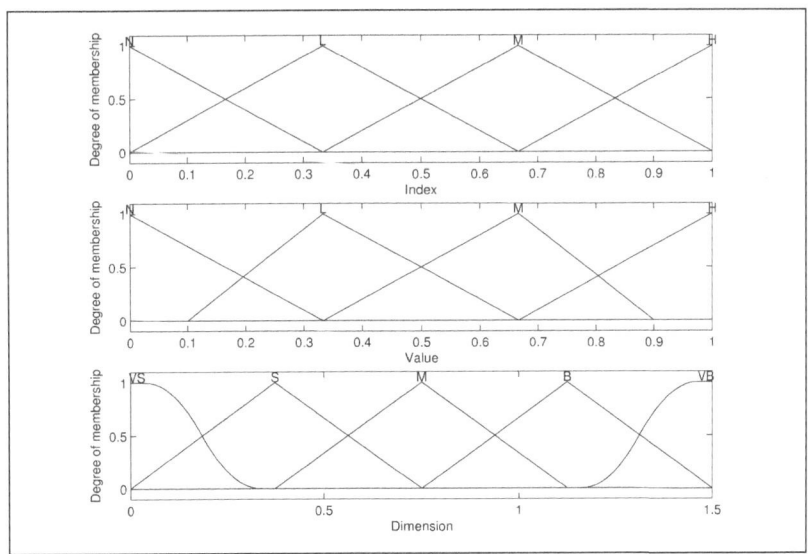

Fig. 4. Membership functions for *Domain rules*, with Index and Value as input variables

dimension of the element as output (Fig. 2 and Fig. 3). For Domain rules, input variables are the normalized value of the function respect to its maximum and its mean value and an user-defined value called index that states the importance of the maximum of the function (Fig. 4). It is possible to notice that membership functions are not regular triangular sets, but they are designed according to FEM experts.

In Fig. 5 the four rules sets are reported in the usual table format. Fuzzy sets are indicated with the capital letters associated to linguistic labels: Z stands for *Zero*, VS for *Very Small*, G for *Great*, N for *Normal*, L for *Low*, M for *Medium*, H for *High*, B for *Big* and so on; in case of angular values, they are fuzzy numbers [5]. It has to be noticed than each set of rules is evaluated in parallel and results are combined with a T-norm that is the usual product with a linear interpolation, necessary for keeping the correct maximum and minimum values of the whole metrics.

d	Z	VS	S	N	G	VG
h	VVS	VS	S	N	G	VG

(a)

d_{CP}	Δ_N			
	Z	S	N	G
Z	N	VS	VVS	VVS
S	N	S	S	VS
N	N	N	N	S
G	N	N	N	N

(b)

Index	Value			
	N	L	M	H
N	VB	VB	VB	VB
L	VB	VB	B	M
M	VB	B	M	S
H	B	M	S	VS

(c)

d_{CP}	Angle [°]						
	0	45	90-180	225	270	315	360
Z	VVS	S	N	S	VS	VVS	VVS
S	VS	N	N	N	S	VS	VS
N	S	N	N	N	N	S	S
G	N	N	N	N	N	N	N

(d)

Fig. 5. The four rules sets in the table format. (a) Type Fuzzy Rules (b) Discontinuity Fuzzy Rules (c) Domain Fuzzy Rules; (d) Connectivity Fuzzy Rules. As an example, in (c) if the function in a point of the domain has a *medium* value and the importance of the maximum of the function is *low*, the dimension of the element is *big*.

4 Numerical Tests

In order to judge the performances of the algorithm, a session of tests has been realized on a simple geometry (a circle) with a lot of different features. A qualitative

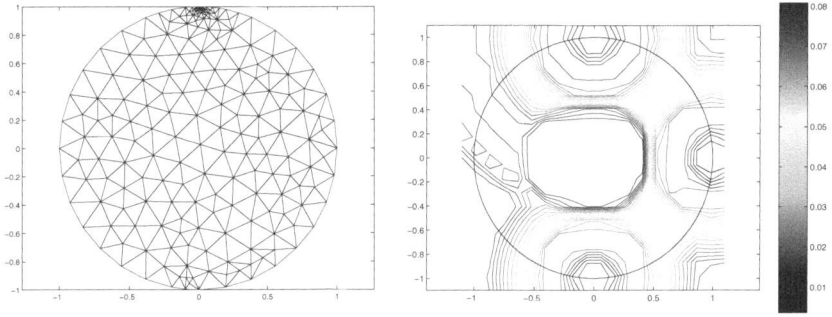

Fig. 6. Discontinuity in Dirichlet B.C. in (0, ±1) and (1,0)

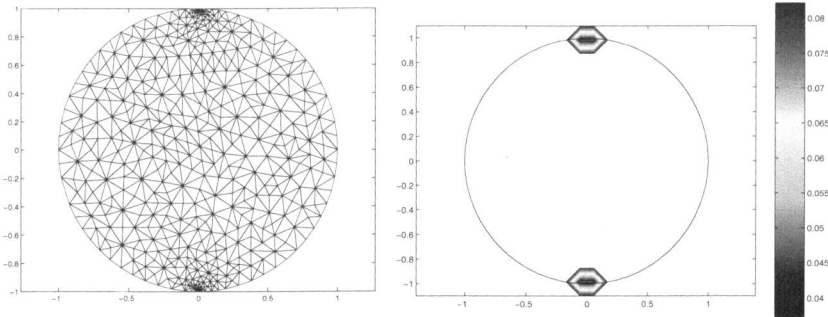

Fig. 7. Different type of B.C. in (0, ±1)

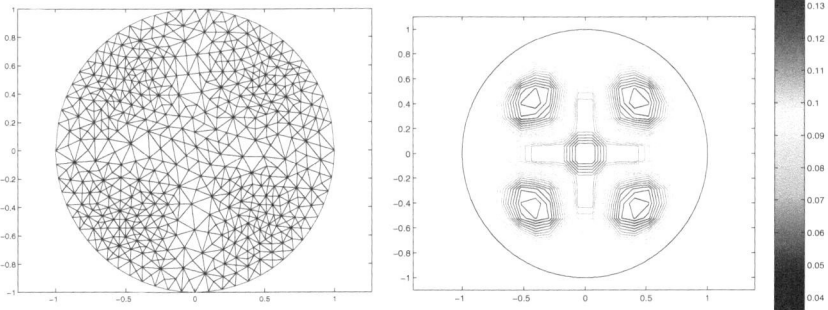

Fig. 8. Forcing function $f = sin(\pi \cdot x) \cdot sin(\pi \cdot y)$

comparison could be made with the grid produced by a *Matlab©* function that performs adaptive triangular mesh generation for the PDE problem, until the minimum error value or the maximum number of elements is reached (from Fig. 6 to Fig. 11). In each case the metric created (on the right column of figures) is consistent with the grid produced by the optimization method (on the left). Where the optimization procedure has produced concentrations of small elements, the same result is obtained via a soft computing approach, and viceversa.

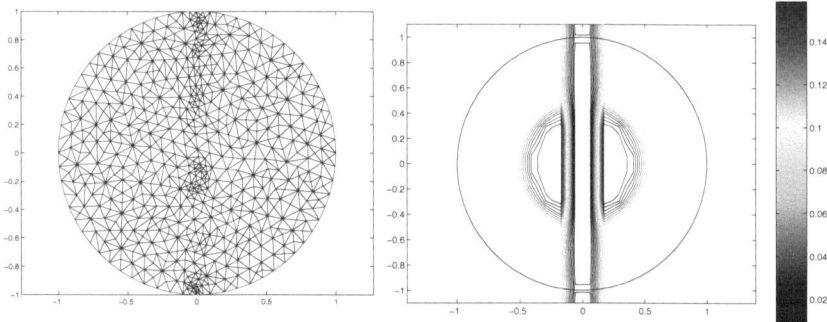

Fig. 9. Viscosity function is $v = 0.1$ if $x < 0$ and $v = 10$ if $x > 0$

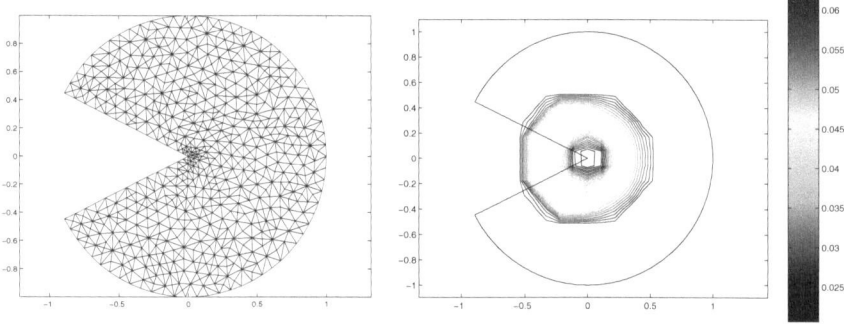

Fig. 10. Presence of a convexity in $(0, 0)$

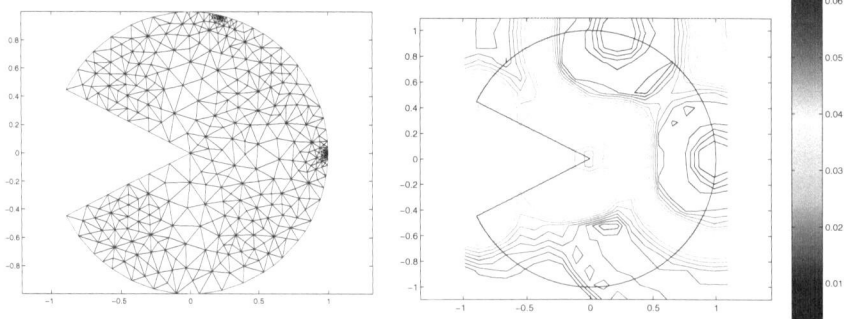

Fig. 11. Mixing all the conditions

In Fig. 6 it is possible to notice that since the three discontinuities in the Dirichlet B.C. have different absolute values, the grid is denser where there is the biggest one; on the contrary, the metric produced with the algorithm, has nearly the same values around the three Critical Points, thus showing less sensitivity.

If the geometry has convexities but the other parameters are regular, it is possible to show the ability of the algorithm to catch them (Fig. 10). On the same geometry the

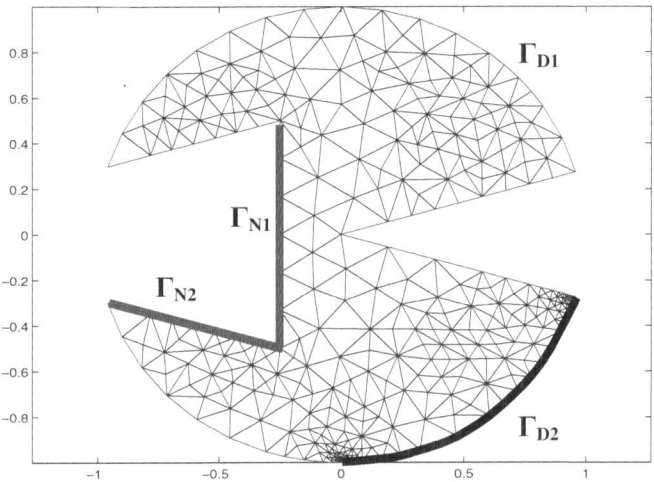

Fig. 12. Test on a general geometry with mixed boundary conditions (N for Neuman and D for Dirichlet conditions); mesh obtained via adaptation methodology

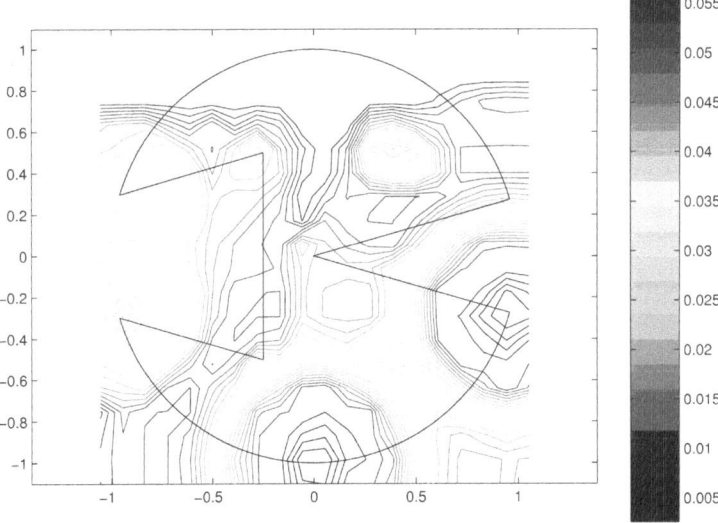

Fig. 13. Metric produced for the general test case with the proposed Fuzzy Logic approach

algorithm is evaluated when all the conditions are present: the metric produced is consistent with the grid realized using the mesh adaptation method (Fig. 11); other tests have been realized, obtaining very good results, but they are not here reported.

In a more general PDE problem it is possible to judge the rightness of the proposed algorithm. In Fig. 12 the domain and the different boundaries for the following test case are reported:

$$\begin{cases} -\Delta(u)=\sin(\pi \cdot x)\cdot \sin(\pi \cdot y) & in\ \Omega \subset \Re^2 \\ u=0 & on\ \Gamma_{D1} \\ u=0.1 & on\ \Gamma_{D2} \\ \partial_n u=0 & on\ \Gamma_{N1} \\ \partial_n u=0.1 & on\ \Gamma_{N2} \end{cases} \qquad (2)$$

In Fig. 13 it is possible to recognize the performance of the proposed algorithm. The two most important critical points (0, -1) and ~(0.97,-0.23) are identified and the value of the metric is very small all around that region. On the other part of the domain, the metric is wider, except where there is the maximum value of the forcing function (near the point (0.5,0.5)) and where there are other critical points. These critical points are localized in the region where there are different types of boundary conditions and where convexities of the border are presents.

5   Conclusions

In this work the efficacy of the fuzzy approach to the problem of generating a mesh for solving a differential problem using the Finite Element Method has been proved, comparing the metric produced with grids realized using a standard optimization technique.

The most relevant characteristic of the algorithm is its ability to reproduce the human expert reasoning by using fuzzy logic. Whereas the adaptation strategy solves the problem iteratively, until a good solution is reached, this soft-computing approach evaluates some features of the differential problem and finds a good solution: time and computation could be saved because no iterations are required, thus speeding up the FEM solving process.

It has to be noticed that a direct and more effective comparison between a traditional mesh adaptation strategy and the fuzzy reasoning approach here described will be possible after the conversion of the metric into a mesh, by using a triangle meshing algorithm (see [6] for a general survey).

The main weakness of the method is its dependence on some normalization constants derived via heuristics; these constants are necessary to use general membership functions, but they are not here reported because of lack of space.

Another advantage of the algorithm is the possibility to modify the rules by using some learning algorithms, widely tested in the Soft Computing field [1], [2], [5]. Learning the rules and adapting them to a specific PDE problem would be very useful in the first phase of a project. With the classical mesh adaptive methods, even if the geometry changes in a little detail, it is necessary to start from the beginning to define the grid, trashing all the work done on a similar problem: learning would allow maintaining the knowledge acquired in the past into fuzzy rules.

Acknowledgment

The author wish to thank Professor Luca Formaggia of the Department of Mathematics of Politecnico di Milano for his technical support on the definition of the criteria required to create the fuzzy rules, and for his careful revision of this paper.

References

1. L. Manevitz, D. Givoli, Automating the Finite Element Method: A Test-Bed for Soft Computing, Applied Soft Computing Journal, July 2003, vol. 3, no. 1, pp. 37-51(15)
2. L. Manevitz, D. Givoli, The Finite Element Method and Soft Computing, in Advances in Soft Computing – Engineering Design and Manufacturing, R. Roy, T. Furuhashi and P.K. Chawdhry, eds., Springer-Verlag, London, 1999
3. A.Quarteroni, Modellistica Numerica per Problemi Differenziali, Springer-Verlag Italia, Milan, 2000 (in Italian). (290 p.); 2nd Edition, 2003 (330 p.)
4. L. Formaggia, S. Perotto, Error Estimation for Finite Element Methods, in 31st Computational Fluid Dynamics Lecture Series, Von Karman Institute, Belgium, 2000
5. G. J. Klir, B. Yuan, Fuzzy Sets and Fuzzy Logic - Theory and Applications, Prentice Hall PTR, New Jersey, 1995
6. S. J. Owen, A Survey of Unstructured Mesh Generation Technology, Proceedings 7th International Meshing Roundtable, Dearborn, MI, October 1998

Unidirectional Two Dimensional Systolic Array for Multiplication in $GF(2^m)$ Using LSB First Algorithm

Soonhak Kwon[1], Chang Hoon Kim[2], and Chun Pyo Hong[3]

[1] Inst. of Basic Science and Dept. of Mathematics,
Sungkyunkwan University, Suwon 440-746, Korea
shkwon@skku.edu
[2] Dept. of Computer and Information Engineering,
Daegu University, Kyungsan 712-714, Korea
chkim@dsp.taegu.ac.kr
[3] Dept. of Computer and Communication Engineering,
Daegu University, Kyungsan 712-714, Korea
cphong@daegu.ac.kr

Abstract. The two dimensional systolic array for multiplication in binary field $GF(2^m)$ with LSB (Least Significant Bit) first algorithm proposed by Yeh et al. has the unfavorable property of bidirectional data flows compared with that of Wang and Lin which use MSB (Most Significant Bit) first algorithm. In this paper, by using a polynomial basis with LSB first algorithm, we present an improved bit parallel systolic array over $GF(2^m)$. Our two dimensional systolic array has unidirectional data flows with 7 latches in each basic cell. Therefore our systolic array has a shorter critical path delay and has the same unidirectional data flows to the multipliers with MSB first scheme.

Keywords: Systolic array, VLSI, fault tolerant architecture, LSB first algorithm, finite field, data flow.

1 Introduction

Arithmetic of finite fields, especially finite field multiplication, found various applications in many VLSI architectures. Moreover, arithmetic of $GF(2^m)$ is easily realized in a circuitry using a few logical gates. Some popular multiplication arrays are Berlekamp's bit serial multipliers [9,10] which use a dual basis, and bit parallel multipliers of Massey-Omura type [11,12,13] which use a normal basis. Above mentioned arrays and other traditional multiplication arrays have some unappealing characteristics. For example, they have irregular circuit designs. In other words, their hardware structures may be quite different for varying choices of m for $GF(2^m)$, though the multiplication algorithm is basically same for each m. Moreover as m gets large, the propagation delay also increases. So deterioration of the performance is inevitable.

I. Bloch, A. Petrosino, and A.G.B. Tettamanzi (Eds.): WILF 2005, LNAI 3849, pp. 420–426, 2006.
© Springer-Verlag Berlin Heidelberg 2006

A systolic array does not suffer from the above problems. It has a regular structure consisting of a number of replicated basic cells, each of which has the same circuit design. So overall structures of systolic arrays are same and not depending on a particular choice of m for $GF(2^m)$. Furthermore since each basic cell is only connected with its neighboring cells, signals can be propagated at a high clock speed. Accordingly, the computational delay of the non systolic arrays in [9,10,11,12,13] is very long when compared with systolic arrays if m is large. There are systolic arrays using a polynomial basis [1,2,6,8], a dual basis [3,4,7] and a normal basis [5]. However a standard polynomial basis is preferred if one considers an array applicable to all finite fields and if one does not want a basis conversion process. When one uses a polynomial basis to multiply two elements in a finite field, there are basically two types of multiplication algorithms, namely, LSB (least significant bit) first scheme and MSB (most significant bit) first scheme.

To find a systolic arrangement, MSB first scheme is used by Wang and Lin in [1] and LSB first scheme is used by Yeh et al. in [2]. A design in [2] has a shorter critical path delay than that of [1] due to increased parallelism among internal computations. On the other hand, the two dimensional array in [2] has bidirectional data flows whereas [1] has unidirectional data flows. Note that a system with unidirectional data flows gains advantages in terms of chip cascadability, fault tolerance, and wafer-scale integration compared with the system with bidirectional data flows. In this paper, we present a two dimensional systolic array with unidirectional data flows and show that our array has a shorter critical path delay while achieving the same unidirectional data flows compared with that of MSB first scheme. Thus our construction of systolic arrays provides possible applications for designing fault tolerant architectures [14,15] computing a power sum, an exponentiation and an inversion, so far most of which have been designed by using MSB first scheme.

2   LSB First Algorithm

Let $GF(2^m)$ be a finite field of 2^m elements. Let

$$F(x) = f_0 + f_1 x + \cdots + f_{m-1} x^{m-1} + x^m \in GF(2)[x]$$

be an irreducible polynomial over $GF(2)$ and let α be any root of $F(x)$. Then $\alpha \in GF(2^m)$ and $\{1, \alpha, \alpha^2, \cdots, \alpha^{m-1}\}$ is a standard polynomial basis over $GF(2)$. An element $A \in GF(2^m)$ is uniquely represented by $A = a_0 + a_1\alpha + a_2\alpha^2 + \cdots + a_{m-1}\alpha^{m-1}$ for some $a_0, a_1, \cdots, a_{m-1} \in GF(2)$. Now let $B = \sum_{i=0}^{m-1} b_i\alpha^i$ and $C = \sum_{i=0}^{m-1} c_i\alpha^i$ be other elements in $GF(2^m)$. We want to compute the product sum $AB + C$ by LSB first scheme,

$$AB + C = C + A \sum_{i=0}^{m-1} b_i\alpha^i = C + \sum_{i=0}^{m-1} b_i A\alpha^i.$$

For each i, let

$$A\alpha^i = \sum_{j=0}^{m-1} u_j^i \alpha^j.$$

Then we have

$$\sum_{j=0}^{m-1} u_j^{i+1} \alpha^j = A\alpha^{i+1} = \sum_{j=0}^{m-1} u_j^i \alpha^{j+1}$$

$$= \sum_{j=1}^{m-1} u_{j-1}^i \alpha^j + u_{m-1}^i \alpha^m = \sum_{j=1}^{m-1} u_{j-1}^i \alpha^j + u_{m-1}^i \sum_{j=0}^{m-1} f_j \alpha^j \qquad (1)$$

$$= u_{m-1}^i f_0 + \sum_{j=1}^{m-1} (u_{j-1}^i + u_{m-1}^i f_j)\alpha^j,$$

Therefore we have

$$u_j^{i+1} = u_{j-1}^i + u_{m-1}^i f_j, \;\; \text{with } u_{-1}^i = 0, \;\; 0 \leq i,j \leq m-1, \qquad (2)$$

which implies that $A\alpha^i$ can be recursively computed by the above relation. Moreover for each i, let

$$\sum_{j=0}^{m-1} s_j^i \alpha^j = C + \sum_{j=0}^{i-1} b_j A\alpha^j.$$

Then we find

$$\sum_{j=0}^{m-1} s_j^{i+1} \alpha^j = C + \sum_{j=0}^{i} b_j A\alpha^j = C + \sum_{j=0}^{i-1} b_j A\alpha^j + b_i A\alpha^i$$

$$= \sum_{j=0}^{m-1} s_j^i \alpha^j + b_i \sum_{j=0}^{m-1} u_j^i \alpha^j \qquad (3)$$

$$= \sum_{j=0}^{m-1} (s_j^i + b_i u_j^i)\alpha^j.$$

Therefore $C + \sum_{j=0}^{i} b_j A\alpha^j$ can also be recursively calculated by the relation,

$$s_j^{i+1} = s_j^i + b_i u_j^i, \;\; 0 \leq i,j \leq m-1. \qquad (4)$$

From the above two observations regarding s_j^i and u_j^i, we deduce that $AB + C = \sum_{j=0}^{m-1} s_j^m \alpha^j$ can be computed by the following algorithm.

Table 1. Bit-level LSB first algorithm

INPUT: $A = \sum_{j=0}^{m-1} a_j \alpha^j$, $B = \sum_{j=0}^{m-1} b_j \alpha^j$, $C = \sum_{j=0}^{m-1} c_j \alpha^j$

OUTPUT: $S = \sum_{j=0}^{m-1} s_j^m \alpha^j$ /* The result of $AB + C$. */

$(u_0^0, u_1^0, \cdots, u_{m-1}^0) \leftarrow (a_0, a_1, \cdots, a_{m-1})$

$(s_0^0, s_1^0, \cdots, s_{m-1}^0) \leftarrow (c_0, c_1, \cdots, c_{m-1})$ /* Initialize. */

for $i = 0$ to $m - 1$ do

    for $j = m - 1$ down to 0 do

    $s_j^{i+1} \leftarrow s_j^i + b_i u_j^i$.

    $u_j^{i+1} \leftarrow u_{j-1}^i + u_{m-1}^i f_j$ with $u_{-1}^i = 0$.

    end

end

---

3 Two Dimensional Systolic Array with Unidirectional Data Flows

The two dimensional systolic array in [2] needs 7 latches in each basic cell. Therefore it has the same hardware complexity to the array in [1] with MSB first scheme. On the other hand, [2] has bidirectional data flows whereas [1] has unidirectional data flows. Since unidirectional data flows is desirable if one wants a fault tolerant system, most of the systolic arrays [14,15] computing arithmetic operations such as a power sum, a division, an inversion and an exponentiation are based on the MSB first scheme in [1]. However, we may construct a unidirectional systolic array using LSB first scheme without sacrificing the hardware complexity of the basic cell. In

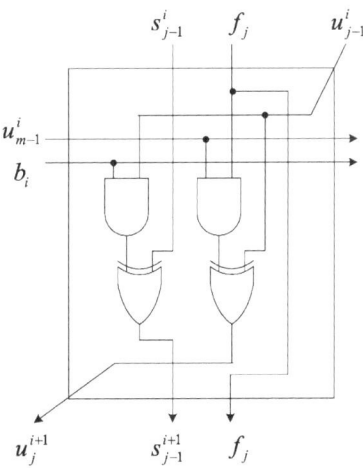

Fig. 1. A circuit of $(i, m - j - 1)$ basic cell where $0 \leq i \leq m - 1$ and $1 \leq j \leq m - 1$

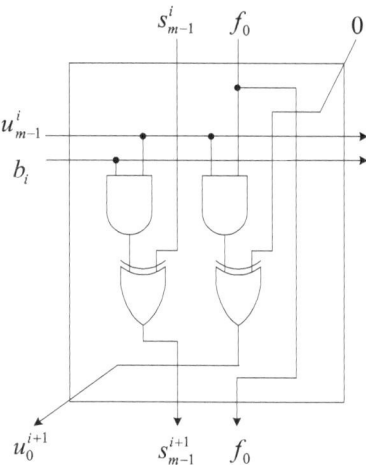

Fig. 2. A circuit of $(i, m-1)$ basic cell where $0 \le i \le m-1$

fact, we will present a bit parallel systolic array using LSB first scheme which has unidirectional data flows and has 7 latches in each basic cell.

In each $(i, m-j-1)$ basic cell, where $0 \le i \le m-1$ and $1 \le j \le m-1$, the operations

$$s_{j-1}^{i+1} \leftarrow s_{j-1}^i + b_i u_{j-1}^i \quad \text{and} \quad u_j^{i+1} \leftarrow u_{j-1}^i + u_{m-1}^i f_j$$

are simultaneously computed. Also in each $(i, m-1)$ basic cells, i.e. the cells in the rightmost column, the operations

$$s_{m-1}^{i+1} \leftarrow s_{m-1}^i + b_i u_{m-1}^i \quad \text{and} \quad u_0^{i+1} \leftarrow u_{m-1}^i f_0$$

are simultaneously computed. The resulting basic cells are shown in Fig. 1 and 2. Notice that we compute two outputs, s_{j-1}^{i+1} and u_j^{i+1}, simultaneously which have slightly different orderings from Table 1, where s_j^{i+1}, u_j^{i+1} are computed. This is because we want to optimize the architecture by using only the signal u_{j-1}^i instead of using both of the u_{j-1}^i, u_j^i in the computation.

The corresponding systolic array is shown in Fig. 3 for the case $m = 4$, where • is a latch (one bit delay element). Our multiplication array supports pipelined operation with latency $3m$ and throughput rate 1. After $2m$ clock cycles, we have the sequence of outputs,

$$s_{m-2}^m, s_{m-1}^m, \cdots, s_0^m \quad \text{and} \quad s_{m-1}^m,$$

where $AB + C = \sum_{j=0}^{m-1} s_j^m \alpha^j$. We compare our systolic array with other bit parallel arrays in Table 2. When compared with [1], we have a shorter critical path delay. Compared with [2], we have unidirectional data flows. Finally, compared with [3], we use a standard polynomial basis. Notice that the dual basis multiplier in [3] requires extra gates and wiring because of the basis conversion

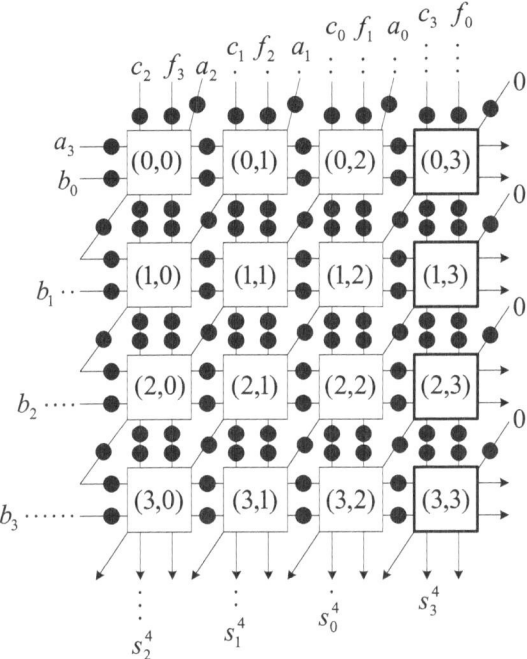

Fig. 3. A systolic array computing $AB+C$ using LSB first algorithm with unidirectional data flows

Table 2. Comparison with other two dimensional systolic arrays

	basis	AND	XOR	3XOR	Latch	Data Flows	Critical Path Delay
[1]	polynomial	2	0	1	7	unidirectional	$D_A+D_{3X}+D_L$
[2]	polynomial	2	2	0	7	bidirectional	$D_A+D_X+D_L$
[3]	dual	2	2	0	7	unidirectional	$D_A+D_X+D_L$
Fig. 3	polynomial	2	2	0	7	unidirectional	$D_A+D_X+D_L$

process. In Table 2, AND, XOR denote 2-input AND, XOR gates and 3XOR denotes a 3-input XOR gate. Also D_A, D_X and D_{3X} mean the delay time of an AND gate, a XOR gate and a 3XOR gate respectively.

4   Conclusions

In this paper, we proposed a two dimensional systolic array using LSB first algorithm with unidirectional data flows. Compared with the architectures with bidirectional data flows, our architecture is more suitable for fault tolerant systems which require fine wafer-scale integration. Since LSB first algorithm is always faster than MSB first algorithm due to the reduced critical path delay, our construction of the systolic array gives possible applications for designing other fault tolerant circuits computing an inversion, a division and an exponentiation, so far most of which have been designed by using MSB first algorithm.

Finally, it should be mentioned that the analogue for the bit serial case with LSB first algorithm is presented in [16], where the proposed architecture has a shorter critical path delay and comparable area complexity to the bit serial systolic array [1] with MSB first algorithm.

Acknowledgements. This work was supported by grant No. R01-2005-000-11261-0 from Korea Science and Engineering Foundation in Ministry of Science & Technology.

References

1. C.L. Wang and J.L. Lin, "Systolic array implementation of multipliers for finite fields $GF(2^m)$," *IEEE Trans. Circuits Syst.*, **38**, pp. 796–800, 1991.
2. C.S. Yeh, I.S. Reed and T.K. Troung, "Systolic multipliers for finite fields $GF(2^m)$," *IEEE Trans. Computers*, **C-33**, pp. 357–360, 1984.
3. S.T.J. Fenn, M. Benaissa and D. Taylor, "Dual basis systolic multipliers for $GF(2^m)$," *IEE Proc. Comput. Digit. Tech.*, **144**, pp. 43–46, 1997.
4. J.J. Wozniak, "Systolic dual basis serial multiplier", *IEE Proc. Comput. Digit. Tech.*, **145**, pp. 237–241, 1998.
5. C.Y. Lee, E.H. Lu and J.Y. Lee, "Bit parallel systolic multipliers for $GF(2^m)$ fields defined by all one and equally spaced polynomials," *IEEE Trans. Computers*, **50**, pp. 385–393, 2001.
6. B.B Zhou, "A new bit serial systolic multiplier over $GF(2^m)$," *IEEE Trans. Computers*, **C-37**, pp. 749–751, 1988.
7. M. Diab and A. Poli, "New bit serial systolic multiplier for $GF(2^m)$ using irreducible trinomials," *Electronics Letters*, **27**, pp. 1183–1184, 1991.
8. S.K. Jain, L. Song and K.K. Parhi, "Efficient semisystolic architectures for finite field arithmetic," *IEEE Trans. VLSI Syst.*, **6**, pp. 101–113, 1998.
9. E.R. Berlekamp, "Bit-serial Reed-Solomon encoders," *IEEE Trans. Inform. Theory*, **28**, pp. 869–874, 1982.
10. M. Wang and I.F. Blake, "Bit serial multiplication in finite fields," *SIAM J. Disc. Math.*, **3**, pp. 140–148, 1990.
11. M.A. Hasan, M.Z. Wang and V.K. Bhargava, "A modified Massey-Omura parallel multiplier for a class of finite fields," *IEEE Trans. Computers*, **42**, pp. 1278–1280, 1993.
12. C. Paar, P. Fleischmann and P. Roelse, "Efficient multiplier architectures for Galois fields $GF(2^m)$," *IEEE Trans. Computers*, **47**, pp. 162–170, 1998.
13. T. Itoh and S. Tsujii, "Structure of parallel multipliers for a class of finite fields $GF(2^m)$," *Information and computation*, **83**, pp. 21–40, 1989.
14. C.L. Wang, "Bit-level systolic array for fast exponentiation in $GF(2^m)$," *IEEE Trans. Computers*, **43**, pp. 838-841, 1994.
15. C.L. Wang and J.H. Guo, "New systolic array for $C + AB^2$, inversion, and division in $GF(2^m)$," *IEEE Trans. Computers*, **49**, pp. 1120–1125, 2000.
16. S. Kwon, C. H. Kim, and C. P. Hong, "Compact linear systolic arrays for multiplication using a trinomial basis in $GF(2^m)$ for high speed cryptographic processors," *ICCSA 2005, Lecture Notes in Computer Science*, **3480**, pp. 508–518, 2005.

Efficient Linear Array for Multiplication over NIST Recommended Binary Fields

Soonhak Kwon[1], Taekyoung Kwon[2], and Young-Ho Park[3]

[1] Inst. of Basic Science and Dept. of Mathematics,
Sungkyunkwan University, Suwon 440-746, Korea
shkwon@skku.edu
[2] School of Computer Engineering,
Sejong University, Seoul 143-747, Korea
tkwon@sejong.ac.kr
[3] Dept. of Information Security,
Sejong Cyber University, Seoul 143-747, Korea
youngho@cybersejong.ac.kr

Abstract. We propose a new linear array for multiplication in $GF(2^m)$ which outperforms most of the existing linear multipliers in terms of the area and time complexity. Moreover we will give a very detailed comparison of our array with other existing architectures for the five binary fields $GF(2^m), m = 163, 233, 283, 409, 571$, recommended by NIST for elliptic curve cryptography.

Keywords: linear array, VLSI architecture, NIST, elliptic curve cryptography.

1 Introduction

Efficient arithmetic of finite field is a critical factor for a low cost VLSI implementation in many applications such as coding theory and cryptography. In these applications, a polynomial basis [10,11] is usually preferred over a normal or a dual basis because of its simple field arithmetic and flexibility of the many available algorithms. Massy-Omura multiplier [1,2,3] is also one of the most popular multipliers these days because of simple squaring operation. There is another type of multiplier called Berlekamp dual basis bit serial multiplier [4]. Though considerable improvements have been made [5,6,7,8] on this multiplier, it does not seem to get much attention these days for cryptographic purposes because of long critical path delay and inconvenient basis conversion process. The multiplication arrays in [5,6,7] do not consider basis conversion and therefore have different bases for input and output values, which make it complicated for practical applications.

Our aim in this paper is to show that a suitably modified linear array from [4] is excellent also for cryptographic purposes. Our proposed array uses single basis consistently for input and output values and has a sequential structure (i.e. parallel in parallel out), which are our salient features that distinguish from previous results in [5,6,7,8]. Consequently our multiplier has a significantly reduced

I. Bloch, A. Petrosino, and A.G.B. Tettamanzi (Eds.): WILF 2005, LNAI 3849, pp. 427–436, 2006.
© Springer-Verlag Berlin Heidelberg 2006

critical path delay and a low area complexity which are superior or comparable to those of normal or polynomial basis multipliers. We will give a very detailed comparison of our multiplier with other proposed linear arrays for the five binary fields $GF(2^m)$, $m = 163, 233, 283, 409, 571$, recommended by NIST (National Institute of Standards and Technology) [9] for elliptic curve cryptography.

2   New Linear Array for Multiplication

Let $GF(2^m)$ be a finite field with characteristic two. There exists a basis for $GF(2^m)$ which is a vector space of dimension m over $GF(2)$. Two bases $\{\alpha_0, \alpha_1, \cdots, \alpha_{m-1}\}$ and $\{\beta_0, \beta_1, \cdots, \beta_{m-1}\}$ of $GF(2^m)$ are said to be dual if the trace map,

$$Tr : GF(2^m) \to GF(2) \quad \text{with} \quad Tr(\alpha) = \alpha + \alpha^2 + \cdots + \alpha^{2^{m-1}}, \quad (1)$$

satisfies $Tr(\alpha_i \beta_j) = \delta_{ij}$ for all $0 \leq i, j \leq m - 1$, where $\delta_{ij} = 1$ if $i = j$, and zero if $i \neq j$. It is easy [4] to see that a unique dual basis exists for a given basis. Also if $\{\beta_0, \beta_1, \cdots, \beta_{m-1}\}$ is the dual basis of $\{\alpha_0, \alpha_1, \cdots, \alpha_{m-1}\}$, then for any nonzero $\beta \in GF(2^m)$,

$$\{\beta^{-1}\beta_0, \beta^{-1}\beta_1, \cdots, \beta^{-1}\beta_{m-1}\} \quad \text{and} \quad \{\beta\alpha_0, \beta\alpha_1, \cdots, \beta\alpha_{m-1}\} \quad (2)$$

are dual to each other. Let $\{1, \alpha, \alpha^2, \cdots, \alpha^{m-1}\}$ be a polynomial basis for $GF(2^m)$ and let

$$\begin{aligned} f(X) &= f_0 + f_1 X + f_2 X^2 + \cdots + f_{m-1} X^{m-1} + X^m \\ &= (X - \alpha)(g_0 + g_1 X + \cdots + g_{m-1} X^{m-1}) \end{aligned} \quad (3)$$

be the unique irreducible polynomial of α over $GF(2)$, where f_i is in $GF(2)$ and g_i is in $GF(2^m)$ for all $0 \leq i \leq m-1$. Then the dual basis of $\{1, \alpha, \alpha^2, \cdots, \alpha^{m-1}\}$ is expressed [5] as

$$\left\{ \frac{g_0}{f'(\alpha)}, \frac{g_1}{f'(\alpha)}, \cdots, \frac{g_{m-1}}{f'(\alpha)} \right\}. \quad (4)$$

The dual basis bit serial multiplier [4] has two serious problems compared with most of the other multipliers with a normal or a polynomial basis. First it needs a basis conversion which requires extra circuitry .Second it has a long critical path delay which is a crucial drawback for cryptographic purposes, especially for elliptic curve cryptography where one should choose m large. We will show that these two problems can be solved using suitable data rearrangement and basis change technique.

Reducing the critical path delay: Since a bit serial architecture usually has a long critical path delay, we want to modify the architecture to the sequential architecture using the symmetry of the multiplication table $Tr(\beta\alpha^{i+k}x)$ in [5,6]. Let $\{\gamma_0, \gamma_1, \cdots, \gamma_{m-1}\}$ be the dual basis of $\{\beta, \beta\alpha, \beta\alpha^2, \cdots, \beta\alpha^{m-1}\}$ where β will be determined later. Write $x, y \in GF(2^m)$ as

$$x = \sum_{i=0}^{m-1} [x]_i \gamma_i \quad \text{and} \quad y = \sum_{i=0}^{m-1} y_i \alpha^i. \quad (5)$$

Letting $xy = \sum_{k=0}^{m-1}[xy]_k\gamma_k$, the dual basis multiplication formula in [5,6] sa ys that $[xy]_k = \sum_{i=0}^{m-1} y_i Tr(\beta\alpha^{i+k}x)$. That is,

$$[xy]_0 = y_0 Tr(\beta x) + y_1 Tr(\beta\alpha x) + \cdots + y_{m-1}Tr(\beta\alpha^{m-1}x),$$
$$[xy]_1 = y_0 Tr(\beta\alpha x) + y_1 Tr(\beta\alpha^2 x) + \cdots + y_{m-1}Tr(\beta\alpha^m x),$$
$$\cdots \tag{6}$$
$$\cdots$$
$$[xy]_{m-1} = y_0 Tr(\beta\alpha^{m-1}x) + y_1 Tr(\beta\alpha^m x) + \cdots + y_{m-1}Tr(\beta\alpha^{2m-2}x).$$

By defining the column vectors

$$Y = (y_0, y_1, \cdots, y_{m-1})^T \quad \text{and} \quad Z = ([xy]_0, [xy]_1, \cdots, [xy]_{m-1})^T, \tag{7}$$

as the transposition of the row vectors $(y_0, y_1, \cdots, y_{m-1})$ and $([xy]_0, [xy]_1, \cdots, [xy]_{m-1})$ respectively, w e have the matrix multiplication $Z = \mathcal{A}Y$ where the m by m matrix $\mathcal{A} = (a_{ij})$ is defined as $a_{ij} = Tr(\beta\alpha^{i+j}x), 0 \leq i, j \leq m-1$. The crucial property of the matrix $\mathcal{A}$ is that it is symmetric. Note that, in the bit serial construction of Berlekamp [4], each row vector of $\mathcal{A}$ is computed by a feedback shift register. Since $\mathcal{A}$ is symmetric, the column vectors of $\mathcal{A}$ are generated by the same shift register. Therefore we may compute the product xy sequentially. In other words, letting

$$A_j = (Tr(\beta\alpha^j x), Tr(\beta\alpha^{j+1} x), \cdots, Tr(\beta\alpha^{j+m-1} x))^T \tag{8}$$

be the jth column vector of $\mathcal{A}$ with $0 \leq j \leq m-1$, we compute the multiplication as follows;

$$Z = (\cdots(((A_0y_0) + A_1y_1) + A_2y_2) + \cdots) + A_{m-1}y_{m-1}. \tag{9}$$

Note that at the jth clock cycle ($0 \leq j \leq m-1$), A_j is multiplied to the constant y_j and the value A_jy_j is added to the partial sum $A_0y_0 + \cdots + A_{j-1}y_{j-1}$ to get the result $A_0y_0 + \cdots + A_jy_j$ which is stored in the register D_i, $0 \leq i \leq m-1$, for a partial summation.

T ec hniques of basis con ersion: Let $f(X) = 1 + X^{n_1} + X^{n_2} + \cdots + X^{n_t} + X^m$ be the irreducible polynomial of $\alpha \in GF(2^m)$ with $0 = n_0 < n_1 < n_2 < \cdots < n_t < m$. Recall that, from the previous expression of $f(X)$ in (3), we have

$$f(X) = (X - \alpha) \sum_{i=0}^{m-1} g_i X^i = \sum_{i=0}^{m-1} g_i X^{i+1} - \sum_{i=0}^{m-1} \alpha g_i X^i$$
$$= \sum_{i=0}^{m-2} g_i X^{i+1} - \sum_{i=0}^{m-2} \alpha g_{i+1} X^{i+1} + g_{m-1}X^m - \alpha g_0 \tag{10}$$
$$= \sum_{i=0}^{m-2} (g_i - \alpha g_{i+1}) X^{i+1} + g_{m-1}X^m - \alpha g_0.$$

From the above equations, it is straightforward to see that $g_i - \alpha g_{i+1}$ is 1 if there exists n_j such that $n_j = i+1$ and is zero if there is no such n_j, since $g_i - \alpha g_{i+1}$ is the coefficient of X^{i+1} of the polynomial $f(X) = 1 + X^{n_1} + X^{n_2} + \cdots + X^{n_t} + X^m$. Therefore using $g_i - \alpha g_{i+1} = \alpha^{-i-1} \sum_{i < n_j \leq i+1} \alpha^{n_j}$, we deduce

$$g_i = \alpha^{-i-1} \sum_{n_j \leq i} \alpha^{n_j} = \alpha^{m-i-1} + \alpha^{-i-1} \sum_{n_j > i} \alpha^{n_j}. \tag{11}$$

Note that, in [4,5,6,7], two values x and xy are expressed in terms of the basis $\{\gamma_0, \gamma_1, \cdots, \gamma_{m-1}\}$ and y is expressed with respect to $\{1, \alpha, \alpha^2, \cdots, \alpha^{m-1}\}$. However, in our paper, we will stick to the basis $\{\gamma_0, \gamma_1, \cdots, \gamma_{m-1}\}$ consistently. Therefore we need a basis conversion from $\{\gamma_0, \gamma_1, \cdots, \gamma_{m-1}\}$ to $\{1, \alpha, \alpha^2, \cdots, \alpha^{m-1}\}$ to express y in terms of $\{1, \alpha, \alpha^2, \cdots, \alpha^{m-1}\}$ from our initial choice of the basis $\{\gamma_0, \gamma_1, \cdots, \gamma_{m-1}\}$. Let $\gamma_i = \sum_{j=0}^{m-1} c_{ij} \alpha^j$ for all $0 \leq i \leq m-1$ and let $y = \sum_{i=0}^{m-1} y_i \alpha^i = \sum_{i=0}^{m-1} [y]_i \gamma_i$ be the expression of y with respect to $\{1, \alpha, \alpha^2, \cdots, \alpha^{m-1}\}$ and $\{\gamma_0, \gamma_1, \cdots, \gamma_{m-1}\}$, respectively. Then

$$\sum_{i=0}^{m-1} [y]_i \gamma_i = ([y]_0, [y]_1, \cdots, [y]_{m-1})(\gamma_0, \gamma_1, \cdots, \gamma_{m-1})^T$$

$$= ([y]_0, [y]_1, \cdots, [y]_{m-1})(c_{ij})(1, \alpha, \alpha^2, \cdots, \alpha^{m-1})^T \tag{12}$$

$$= (y_0, y_1, \cdots, y_{m-1})(1, \alpha, \alpha^2, \cdots, \alpha^{m-1})^T = \sum_{i=0}^{m-1} y_i \alpha^i,$$

where (c_{ij}) is a m by m matrix and $(\gamma_0, \gamma_1, \cdots, \gamma_{m-1})^T$ (resp. $(1, \alpha, \alpha^2, \cdots, \alpha^{m-1})^T$) is the transposition of the row vector $(\gamma_0, \gamma_1, \cdots, \gamma_{m-1})$ (resp. $(1, \alpha, \alpha^2, \cdots, \alpha^{m-1})$). Thus we have $(y_0, y_1, \cdots, y_{m-1}) = ([y]_0, [y]_1, \cdots, [y]_{m-1})(c_{ij})$ and (c_{ij}) can be regarded as the basis conversion matrix from $\{\gamma_0, \gamma_1, \cdots, \gamma_{m-1}\}$ to $\{1, \alpha, \alpha^2, \cdots, \alpha^{m-1}\}$. We define the excess number of the basis conversion to be 'the number of nonzero entries of the matrix (c_{ij}) minus m'. As long as the excess number is small, we can get a fairly simple basis conversion. For example, if the excess number is *zero*, then the rows of (c_{ij}) are the permutated ones of the identity matrix so that the basis conversion is just a permutation. Now following the approach in [8], we choose the mysterious constant $\beta \in GF(2^m)$ as

$$\beta = (\alpha^{n_s} f'(\alpha))^{-1} \tag{13}$$

where $s = \frac{t+1}{2}$, i.e. n_s is the power of the exact middle term of the irreducible $f(X) = 1 + X^{n_1} + \cdots + X^{n_s} + \cdots + X^{n_t} + X^m$. Then using the equation (11), one can show [8] that the excess number is $\sum_{i=s+1}^{t} n_i - \sum_{i=1}^{s-1} n_i$. For example, as is showed in [5,8], when $f(X) = X^m + X^k + 1$ is an irreducible trinomial, then the excess number is *zero* and the basis conversion from $\{\gamma_0, \gamma_1, \cdots, \gamma_{m-1}\}$ to $\{1, \alpha, \alpha^2, \cdots, \alpha^{m-1}\}$ is just a permutation. Thus no extra circuitry is needed in this case. Also when irreducible $f(X)$ is the following special pentanomial $X^m + X^{k+1} + X^k + X^{k-1} + 1$, then we have $t = 3, s = \frac{t+1}{2} = 2$ and thus the excess number is $2 = (k+1) - (k-1)$ which w as already sho wn in [6]. Howev er for a general irreducible polynomial, the excess n umber may not be small and in that case, for each y_i which is a linear combination of the signals $[y]_0, [y]_1, \cdots, [y]_{m-1}$, the depth and the number of necessary XOR gates of the X OR tree with respect to y_i should be determined precisely. A detailed analysis leads to the following lemma which explains the basis conversion of y terms at each coefficients.

Lemma 1. *Let* $\gamma_i = \alpha^{n_s} g_i$ *for all* $0 \le i \le m-1$. *Let* $y = \sum_{i=0}^{m-1} [y]_i \gamma_i = \sum_{i=0}^{m-1} y_i \alpha^i$ *be the expr ession of* y *with respect to the bases* $\{\gamma_0, \gamma_1, \cdots, \gamma_{m-1}\}$ *and* $\{1, \alpha, \alpha^2, \cdots, \alpha^{m-1}\}$, *resp ctively. Then for each* $0 \le i \le m-1$, y_i *is the sum of at most* s *elements of* $[y]_j$ *with* $0 \le j \le m-1$. *Thus each coefficient* y_i *is obtained by using an XOR tree of depth at most* $\lceil \log_2 s \rceil$ *with at most* $s-1$ *XOR gates. Also the total number of necessary XOR gates to generate* $y_0, y_1, \cdots, y_{m-1}$ *using the signals* $[y]_0, [y]_1, \cdots, [y]_{m-1}$ *is exactly* $\sum_{i=s+1}^{t} n_i - \sum_{i=1}^{s-1} n_i$.

Recall that we hav e used the basis $\{\gamma_0, \gamma_1, \cdots, \gamma_{m-1}\}$ which is dual to the basis $\{\beta, \beta\alpha, \beta\alpha^2, \cdots, \beta\alpha^{m-1}\}$ with $\beta = (\alpha^{n_s} f'(\alpha))^{-1}$. From the relations (2,4) and (13), we easily find that $\{\gamma_0, \gamma_1, \cdots, \gamma_{m-1}\} = \{\alpha^{n_s} g_0, \cdots, \alpha^{n_s} g_{m-1}\}$, which justifies the expression of γ_i in Lemma 1. Because of Lemma 1, we can now precisely describe the basis conversion using XOR trees. The necessary XOR gates for each tree can also be determined easily. For example, when one has an irreducible pentanomial $f(X) = X^m + X^{n_3} + X^{n_2} + X^{n_1} + 1$ with $0 = n_0 < n_1 < n_2 < n_3 < m$, the total number of necessary XOR gates for the basis con ersion is $n_3 - n_1$ and each X OR tree, if it exists, is just a single XOR gate with depth one because $t = 3$ and $s = \frac{t+1}{2} = 2$. That is, among all coefficients $y_0, y_1, \cdots, y_{m-1}$ of $y = \sum_{i=0}^{m-1} y_i \alpha^i = \sum_{i=0}^{m-1} [y]_i \gamma_i$, exactly $n_3 - n_1$ number of y_i are the sum of tw o elements from the set $\{[y]_0, [y]_1, \cdots, [y]_{m-1}\}$ and exactly $m - n_3 + n_1$ number of y_i are same to some members in $\{[y]_0, [y]_1, \cdots, [y]_{m-1}\}$.

3   Main Result with VLSI Realization

All the explanations in the previous section are realized in the circuit arrangement shown in Fig. 1. In Fig. 1, ev ery input of our array is expressed with respect to the basis $\{\gamma_0, \gamma_1, \cdots, \gamma_{m-1}\} = \{\alpha^{n_s} g_0, \cdots, \alpha^{n_s} g_{m-1}\}$. The input $y = \sum_{i=0}^{m-1} [y]_i \gamma_i$ is loaded via XOR trees to the lo wershift register so that the vector in the register has the value $(y_0, \cdots, y_{m-1})$, and this vector is cyclically shifted to the left by one position at every cycle. The upper shift register is loaded with the v ector A_0 and it is feedback shifted via the relations (6,8), which is the usual case of Berlekamp multipliers. The area complexity and the critical path delay of the proposed array are explained in Theorem 2.

Theorem 2. *Let* $f(X) = X^m + \sum_{i=0}^{t} X^{n_i}$ *be an irreducible p olynomial over* $GF(2)$ *with* $0 = n_0 < n_1 < \cdots < n_t < m$ *and let* α *be a zero of* $f(X)$. *Write* $s = \frac{t+1}{2}, \beta = (\alpha^{n_s} f'(\alpha))^{-1}$ *and* $f(X) = (X - \alpha)(\sum_{i=0}^{m-1} g_i X^i)$. *Then, by using the basis* $\{\alpha^{n_s} g_0, \alpha^{n_s} g_1, \cdots, \alpha^{n_s} g_{m-1}\}$, *we can construct a linear array for the*

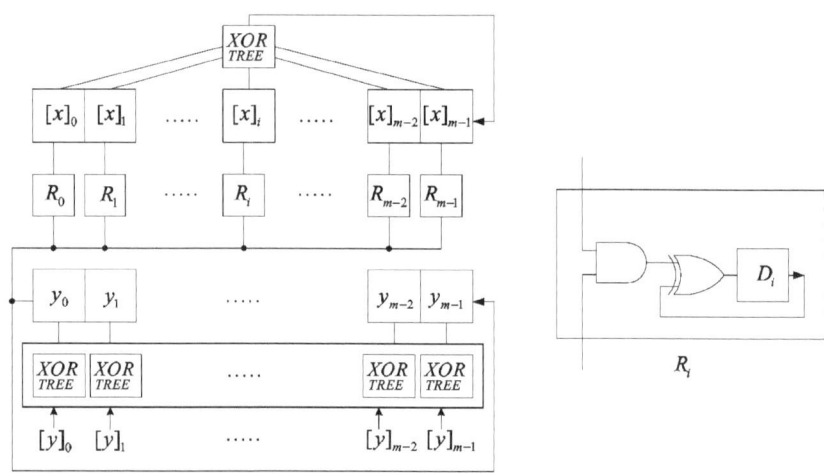

Fig. 1. A new linear array for multiplication in $GF(2^m)$ using the basis $\{\alpha^{n_s}g_0, \cdots ,$ $\alpha^{n_s}g_{m-1}\}$ with $s = \frac{t+1}{2}$ and $f(X) = X^m + \sum_{i=0}^{t} X^{n_i} = (X - \alpha)(\sum_{i=0}^{m-1} g_i X^i)$

multiplication of x and y in $GF(2^m)$ with $3m$ flip-flops, m AND gates, and $m + t + \sum_{i=s+1}^{t} n_i - \sum_{i=1}^{s-1} n_i$ XOR gates such that

1. *An XOR tree for the feed back shift register for the input x needs t XOR gates with depth of the tree $\lceil \log_2(t + 1) \rceil$. Also the basis conversion of the input y costs $\sum_{i=s+1}^{t} n_i - \sum_{i=1}^{s-1} n_i$ XOR gates with m XOR trees such that each XOR tree consists of at most $\frac{t-1}{2}$ XOR gates with depth $\lceil \log_2(t + 1) \rceil - 1$.*

2. *Our array has a parallel-in parallel-out structure and produces an output of xy at a rate of one every m clock cycle. The critical path delay of our array is $T_A + T_X$ if $t = 1$ (i.e. trinomial case) and it is $\lceil \log_2(t + 1) \rceil T_X$ if $t > 1$, where T_A and T_X are the delay time of a two input AND gate and a two input XOR gate, respectively.*

4   Comparison with Other Linear Arrays

In view of the result in Theorem 2, we may now claim that our proposed linear array is, in terms of the time and area complexity, superior to most of normal and polynomial basis multipliers currently used. Detailed comparisons are given in Table 1 and 2 for the five binary fields [9] recommended by NIST (National Institute of Standards and Technology) for elliptic curve cryptography. Also in Fig. 2 and 3, we show explicit circuits for VLSI implementation for the cases $GF(2^{163})$ and $GF(2^{233})$ as examples. Note that the circuits for other fields can be constructed similarly following Theorem 2.

The critical path delay of [2,3] using a Gaussian normal basis of type k is $T_A + (1 + \lceil \log_2 k \rceil)T_X$. Also the critical path delay of the polynomial basis multipliers is $T_A + T_X$ with LSB first scheme and is $T_A + 2T_X$ with MSB first scheme. On the

Table 1. Comparison of the critical path delay for the five binary fields

$GF(2^m)$	basis	$GF(2^{163})$	$GF(2^{233})$	$GF(2^{283})$	$GF(2^{409})$	$GF(2^{571})$
[1]	normal	$T_A + 10T_X$	$T_A + 9T_X$	$T_A + 11T_X$	$T_A + 11T_X$	$T_A + 13T_X$
[2]	normal	$T_A + 3T_X$	$T_A + 2T_X$	$T_A + 4T_X$	$T_A + 3T_X$	$T_A + 5T_X$
[3]	normal	$T_A + 3T_X$	$T_A + 2T_X$	$T_A + 4T_X$	$T_A + 3T_X$	$T_A + 5T_X$
[4,5]	dual	$T_A + 8T_X$	$T_A + 8T_X$	$T_A + 9T_X$	$T_A + 9T_X$	$T_A + 10T_X$
[10,11] LSB	polynomial	$T_A + T_X$	$T_A + T_X$	$T_A + T_X$	$T_A + T_X$	$T_A + T_X$
[10] MSB	polynomial	$T_A + 2T_X$	$T_A + 2T_X$	$T_A + 2T_X$	$T_A + 2T_X$	$T_A + 2T_X$
This paper	$\{\alpha^{n_s} g_i\}$	$2T_X$	$T_A + T_X$	$2T_X$	$T_A + T_X$	$2T_X$

other hand, our multiplier has a critical path delay $T_A + T_X$ if $f(X)$ is a trinomial and $2T_X$ if $f(X)$ is a pentanomial. This implies that our multiplier has a significantly shorter critical path delay compared with normal basis multipliers and has almost the same critical path delay compared with polynomial basis multipliers. See the difference of the critical path delay for $m = 163, 233, 283, 409, 571$ shown in Table 1. It should be mentioned that, since we are dealing with a linear multiplier, even a small increment of the critical path delay such as T_X results in a total delay of mT_X where m is the size of a field.

Remark: For irreducible polynomials, NIST [9] recommends to use $X^{163} + X^7 + X^6 + X^3 + 1, X^{233} + X^{74} + 1, X^{283} + X^{12} + X^7 + X^5 + 1, X^{409} + X^{87} + 1, X^{571} + X^{10} + X^5 + X^2 + 1$, respectively. NIST also recommends to use the lowest complexity Gaussian normal basis for the above fields and they are of type $4, 2, 6, 4, 10$, respectively.

The area complexity of our multiplier is also far lower than that of Massey-Omura multipliers [1,2,3] and is comparable to that of polynomial basis multipliers. For a binary field $GF(2^m)$ with odd m, the result in [3] shows that the type k Gaussian normal basis multipliers need $\frac{k+1}{2}m - \frac{k-1}{2}$ XOR gates, which is $\frac{3}{2}m - \frac{1}{2}$ for a type II ONB (optimal normal basis) and $\frac{k+1}{2}m - \frac{k-1}{2} > \frac{5}{2}m - 5$ for other Gaussian normal bases of type $k = 4, 6, 10$. On the other hand, our multiplier needs $m + 1$ XOR gates when trinomial is used and needs $m + 3 + n_3 - n_1$ ($< 2m + 3$) XOR gates when pentanomial is used. Note that the difference of the number of necessary XOR gates between our multiplier and the polynomial basis multipliers comes from the base change of our multiplier and is practically negligible. We omit the number of necessary AND gates in Table 2 because all the architectures in the table have almost the same number of AND gates.

Table 2. Comparison of the number of necessary XOR gates for the five binary fields

$GF(2^m)$	$GF(2^{163})$	$GF(2^{233})$	$GF(2^{283})$	$GF(2^{409})$	$GF(2^{571})$
[1]	648	464	1692	1632	5700
[2]	649	465	1693	1633	5701
[3]	406	349	988	1021	3136
[4,5]	165	233	285	409	573
[10,11]	166	234	286	410	574
This paper	170	234	293	410	582

5   Circuits for NIST Recommended Binary Fields

Let us explain how to construct a multiplication circuit for each binary field $GF(2^m)$, $m = 163, 233, 283, 409, 57$ 1. The recommended irreducible polynomials [9] for these fields are $X^{163} + X^7 + X^6 + X^3 + 1$, $X^{233} + X^{74} + 1$, $X^{283} + X^{12} + X^7 + X^5 + 1$, $X^{409} + X^{87} + 1$, $X^{571} + X^{10} + X^5 + X^2 + 1$, respectively. Since these polynomials are either trinomials or pentanomials, we will discuss basis conversion of these polynomials and derive explicit multiplication architectures. First let us consider the trinomial $f(X) = X^m + X^k + 1$. Then we get the exact middle term $\alpha^{n_s} = \alpha^k$ and from the equation (11), we have

$$\alpha^k g_i = \alpha^{k-i-1} \sum_{n_j \le i} \alpha^{n_j} = \alpha^{k-i-1} \qquad \text{if } 0 \le i < k$$

$$= \alpha^{k-i-1}(1 + \alpha^k) = \alpha^{k-i-1+m} \quad \text{if } k \le i < m. \tag{14}$$

Therefore the basis of our multiplier for the trinomial $f(X) = X^m + X^k + 1$ is

$$\{\alpha^k g_0, \cdots, \alpha^k g_{m-1}\} = \{\alpha^{k-1}, \alpha^{k-2}, \cdots, \alpha^0, \alpha^{m-1}, \alpha^{m-2}, \cdots, \alpha^k\}. \tag{15}$$

Next let us consider the pentanomial $f(X) = X^m + X^{n_3} + X^{n_2} + X^{n_1} + 1$. Then $\alpha^{n_s} = \alpha^{n_2}$ and again using the equation (11),

$$\alpha^{n_2} g_i = \alpha^{n_2-i-1} \qquad\qquad\qquad\qquad\qquad\quad \text{if } 0 \le i < n_1$$

$$= \alpha^{n_2-i-1}(1 + \alpha^{n_1}) = \alpha^{n_2-i-1} + \alpha^{n_2-i-1+n_1} \qquad \text{if } n_1 \le i < n_2$$

$$= \alpha^{n_2-i-1}(\alpha^{n_3} + \alpha^m) = \alpha^{n_2-i-1+n_3} + \alpha^{n_2-i-1+m} \quad \text{if } n_2 \le i < n_3$$

$$= \alpha^{n_2-i-1}\alpha^m = \alpha^{n_2-i-1+m} \qquad\qquad\qquad\quad \text{if } n_3 \le i < m.$$

Thus the basis $\{\gamma_0, \cdots, \gamma_{m-1}\} = \{\alpha^{n_2} g_0, \cdots, \alpha^{n_2} g_{m-1}\}$ of our multiplier for the pentanomial $f(X) = X^m + X^{n_3} + X^{n_2} + X^{n_1} + 1$ is

$$\{\alpha^{n_2-1}, \cdots, \alpha^{n_2-n_1}, \alpha^{n_2-n_1-1} + \alpha^{n_2-1}, \cdots, \alpha^0 + \alpha^{n_1},$$
$$\alpha^{n_3-1} + \alpha^{m-1}, \cdots, \alpha^{n_2} + \alpha^{n_2-n_3+m}, \alpha^{n_2-n_3+m-1}, \cdots, \alpha^{n_2}\}. \tag{16}$$

Example 1. Multiplication in $GF(2^{163})$ using $f(X) = X^{163} + X^7 + X^6 + X^3 + 1$: From the equations (16), we have the following basis $\{\gamma_0, \cdots, \gamma_{162}\}$ for the field $GF(2^{163})$,

$$\{\alpha^5, \alpha^4, \alpha^3, \alpha^2 + \alpha^5, \alpha + \alpha^4, \alpha^0 + \alpha^3, \alpha^6 + \alpha^{162}, \alpha^{161}, \alpha^{160}, \cdots, \alpha^6\}. \tag{17}$$

Therefore letting $y = \sum_{i=0}^{162} [y]_i \gamma_i = \sum_{i=0}^{162} y_i \alpha^i$, the basis conversion is

$$y = [y]_0 \alpha^5 + [y]_1 \alpha^4 + [y]_2 \alpha^3 + [y]_3(\alpha^2 + \alpha^5) + [y]_4(\alpha + \alpha^4) + [y]_5(\alpha^0 + \alpha^3)$$
$$+ [y]_6(\alpha^6 + \alpha^{162}) + [y]_7 \alpha^{161} + [y]_8 \alpha^{160} + \cdots + [y]_{162} \alpha^6$$
$$= [y]_5 \alpha^0 + [y]_4 \alpha + [y]_3 \alpha^2 + ([y]_2 + [y]_5)\alpha^3 + ([y]_1 + [y]_4)\alpha^4 + ([y]_0 + [y]_3)\alpha^5$$
$$+ ([y]_6 + [y]_{162})\alpha^6 + [y]_{161} \alpha^7 + [y]_{160} \alpha^8 + \cdots + [y]_6 \alpha^{162},$$

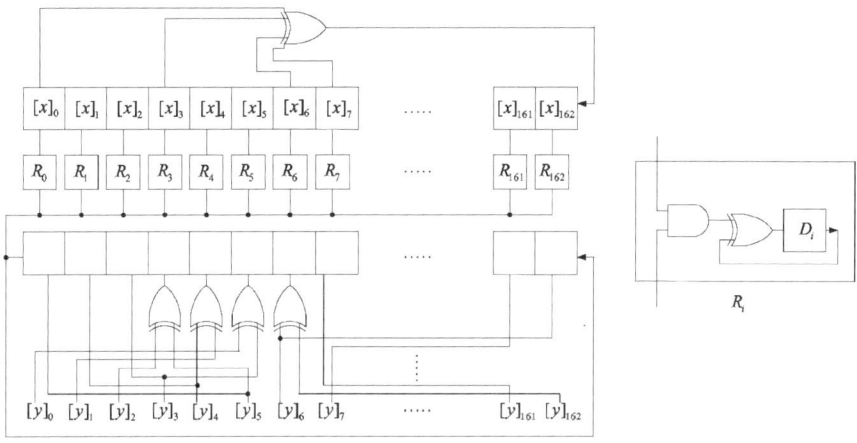

Fig. 2. A multiplication circuit using the basis $\{\gamma_0, \cdots, \gamma_{m-1}\} = \{\alpha^6 g_0, \cdots, \alpha^6 g_{162}\}$ in $GF(2^{163})$ with $f(X) = X^{163} + X^7 + X^6 + X^3 + 1 = (X - \alpha)(\sum_{i=0}^{162} g_i X^i)$

and the corresponding circuit for the multiplication is shown in Fig. 2. It should be mentioned that the same technique can be applied to other fields $GF(2^{283})$ and $GF(2^{571})$ having irreducible pentanomials.

Example 2. Multiplication in $GF(2^{233})$ using $f(X) = X^{233} + X^{74} + 1$: From the equation (15), we have the following basis $\{\gamma_0, \cdots, \gamma_{232}\}$ for the field $GF(2^{233})$,

$$\{\alpha^{73}, \alpha^{72}, \cdots, \alpha^0, \alpha^{232}, \alpha^{231}, \cdots, \alpha^{74}\}. \tag{18}$$

Thus the multiplication is easily realized in the shift register arrangement shown in Fig. 3. Note that one can construct a similar circuit for the case $GF(2^{409})$ with NIST recommended trinomial $f(X) = X^{409} + X^{87} + 1$.

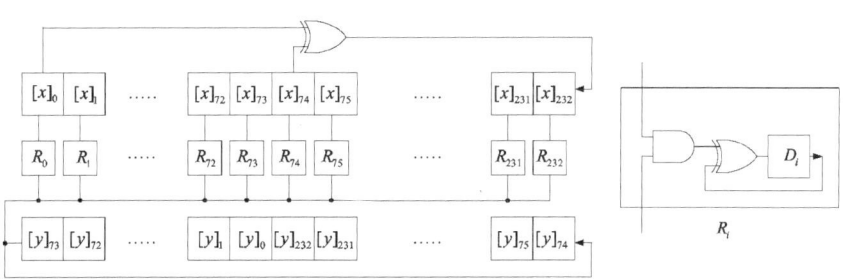

Fig. 3. A multiplication circuit using the basis $\{\alpha^{74} g_0, \cdots, \alpha^{74} g_{232}\}$ in $GF(2^{233})$ with $f(X) = X^{233} + X^{74} + 1 = (X - \alpha)(\sum_{i=0}^{232} g_i X^i)$

6   Conclusions

In this paper, w e proposed a new linear array for multiplication and show ed how to construct explicit multiplication circuits for the five NIST recommended binary fields $GF(2^m)$, $m = 163, 233, 283, 409, 571$. Table 1 and 2 show that our linear arra y has a significantly reduced critical path delay and a low erhardware complexity compared with other existing arra ys with normal or polynomial bases. Therefore our proposed linear array can be used in many applications where VLSI implementation of a low cost and high speed arithmetic unit is needed.

Acknowledgements. This work was supported b y grant No. R01-2005-000-11261-0 from Korea Science and Engineering Foundation in Ministry of Science & Technology.

References

1. J.L. Massy and J.K. Omura, "Computational method and apparatus for finite field arithmetic," *US Patent No. 4587627*, 1986.
2. G.B. Agnew, R.C. Mullin, I. Onyszch uk, and S.A. Vanstone, "An implementation for a fast public key cryptosystem," *J. Cryptology*, Vol. 3, pp. 63–79, 1991.
3. S. Kwon, K. Gaj, C. Kim, and C. Hong, "Efficient linear array for multiplication in $GF(2^m)$ using a normal basis for elliptic curve cryptography," CHES 2004: *Cryptographic Hardware and Embedded Systems, L ecture Notes in Computer Scienc e* V ol. 3156, pp. 76 91, 2004.
4. E.R. Berlekamp, "Bit-serial Reed-Solomon encoders," *IEEE Trans. Inform. Theory*, Vol. 28, pp. 869–874, 1982.
5. M. Wang and I.F. Blake, "Bit serial multiplication in finite fields," *SIAM J. Disc. Math.*, Vol. 3, pp. 140–148, 1990.
6. M. Morii, M. Kasahara, and D.L. Whiting, "Efficient bit-serial multiplication and the discrete-time Wiener-Hopf equation o ver finite fields," *IEEE T nns. Inform. Theory*, Vol. 35, pp. 1177–1183, 1989.
7. C.Y. Lee, Y.C. Lu, and E.H. Lu, "Low complexity systolic multiplier o ver $GF(2^m)$ using weakly dual basis," APCCAS 2002: *Asia Pacific Conference on Circuits and Systems*, Vol. 1, pp. 367–372, 2002.
8. D.R. Stinson, "On bit-serial m ultiplication and dual bases in $GF(2^m)$," *IEEE T nns. Inform. Theory*, Vol. 37, pp. 1733–1736, 1991.
9. NIST, "Digital Signature Standard," *FIPS Publication*, 186-2, February, 2000.
10. D. Hankerson, A.J. Menezes, and S.A. Vanstone, *Guide to Elliptic Curve Cryptography*, Springer–Verlag, 2004.
11. L. Song and K.K. Parhi, "Efficient finite field serial/parallel multiplication," *International Conference on Application Specific Systems, Architectures and Processors*, pp. 19–21, 1996.

Author Index

Lecture Notes in Artificial Intelligence (LNAI)

Vol. 3641: D. Ślęzak, G. Wang, M. Szczuka, I. Düntsch, Y. Yao (Eds.), Rough Sets, Fuzzy Sets, Data Mining, and Granular Computing, Part I. XXIV, 742 pages. 2005.

Vol. 3635: J.R. Winkler, M. Niranjan, N.D. Lawrence (Eds.), Deterministic and Statistical Methods in Machine Learning. VIII, 341 pages. 2005.

Vol. 3632: R. Nieuwenhuis (Ed.), Automated Deduction – CADE-20. XIII, 459 pages. 2005.

Vol. 3630: M.S. Capcarrère, A.A. Freitas, P.J. Bentley, C.G. Johnson, J. Timmis (Eds.), Advances in Artificial Life. XIX, 949 pages. 2005.

Vol. 3626: B. Ganter, G. Stumme, R. Wille (Eds.), Formal Concept Analysis. X, 349 pages. 2005.

Vol. 3625: S. Kramer, B. Pfahringer (Eds.), Inductive Logic Programming. XIII, 427 pages. 2005.

Vol. 3620: H. Muñoz-Ávila, F. Ricci (Eds.), Case-Based Reasoning Research and Development. XV, 654 pages. 2005.

Vol. 3614: L. Wang, Y. Jin (Eds.), Fuzzy Systems and Knowledge Discovery, Part II. XLI, 1314 pages. 2005.

Vol. 3613: L. Wang, Y. Jin (Eds.), Fuzzy Systems and Knowledge Discovery, Part I. XLI, 1334 pages. 2005.

Vol. 3607: J.-D. Zucker, L. Saitta (Eds.), Abstraction, Reformulation and Approximation. XII, 376 pages. 2005.

Vol. 3601: G. Moro, S. Bergamaschi, K. Aberer (Eds.), Agents and Peer-to-Peer Computing. XII, 245 pages. 2005.

Vol. 3600: F. Wiedijk (Ed.), The Seventeen Provers of the World. XVI, 159 pages. 2006.

Vol. 3596: F. Dau, M.-L. Mugnier, G. Stumme (Eds.), Conceptual Structures: Common Semantics for Sharing Knowledge. XI, 467 pages. 2005.

Vol. 3593: V. Mařík, R. W. Brennan, M. Pěchouček (Eds.), Holonic and Multi-Agent Systems for Manufacturing. XI, 269 pages. 2005.

Vol. 3587: P. Perner, A. Imiya (Eds.), Machine Learning and Data Mining in Pattern Recognition. XVII, 695 pages. 2005.

Vol. 3584: X. Li, S. Wang, Z.Y. Dong (Eds.), Advanced Data Mining and Applications. XIX, 835 pages. 2005.

Vol. 3581: S. Miksch, J. Hunter, E.T. Keravnou (Eds.), Artificial Intelligence in Medicine. XVII, 547 pages. 2005.

Vol. 3577: R. Falcone, S. Barber, J. Sabater-Mir, M.P. Singh (Eds.), Trusting Agents for Trusting Electronic Societies. VIII, 235 pages. 2005.

Vol. 3575: S. Wermter, G. Palm, M. Elshaw (Eds.), Biomimetic Neural Learning for Intelligent Robots. IX, 383 pages. 2005.

Vol. 3571: L. Godo (Ed.), Symbolic and Quantitative Approaches to Reasoning with Uncertainty. XVI, 1028 pages. 2005.

Vol. 3559: P. Auer, R. Meir (Eds.), Learning Theory. XI, 692 pages. 2005.

Vol. 3558: V. Torra, Y. Narukawa, S. Miyamoto (Eds.), Modeling Decisions for Artificial Intelligence. XII, 470 pages. 2005.

Vol. 3554: A.K. Dey, B. Kokinov, D.B. Leake, R. Turner (Eds.), Modeling and Using Context. XIV, 572 pages. 2005.

Vol. 3550: T. Eymann, F. Klügl, W. Lamersdorf, M. Klusch, M.N. Huhns (Eds.), Multiagent System Technologies. XI, 246 pages. 2005.

Vol. 3539: K. Morik, J.-F. Boulicaut, A. Siebes (Eds.), Local Pattern Detection. XI, 233 pages. 2005.

Vol. 3538: L. Ardissono, P. Brna, A. Mitrović (Eds.), User Modeling 2005. XVI, 533 pages. 2005.

Vol. 3533: M. Ali, F. Esposito (Eds.), Innovations in Applied Artificial Intelligence. XX, 858 pages. 2005.

Vol. 3528: P.S. Szczepaniak, J. Kacprzyk, A. Niewiadomski (Eds.), Advances in Web Intelligence. XVII, 513 pages. 2005.

Vol. 3518: T.-B. Ho, D. Cheung, H. Liu (Eds.), Advances in Knowledge Discovery and Data Mining. XXI, 864 pages. 2005.

Vol. 3508: P. Bresciani, P. Giorgini, B. Henderson-Sellers, G. Low, M. Winikoff (Eds.), Agent-Oriented Information Systems II. X, 227 pages. 2005.

Vol. 3505: V. Gorodetsky, J. Liu, V.A. Skormin (Eds.), Autonomous Intelligent Systems: Agents and Data Mining. XIII, 303 pages. 2005.

Vol. 3501: B. Kégl, G. Lapalme (Eds.), Advances in Artificial Intelligence. XV, 458 pages. 2005.

Vol. 3492: P. Blache, E.P. Stabler, J.V. Busquets, R. Moot (Eds.), Logical Aspects of Computational Linguistics. X, 363 pages. 2005.

Vol. 3490: L. Bolc, Z. Michalewicz, T. Nishida (Eds.), Intelligent Media Technology for Communicative Intelligence. X, 259 pages. 2005.

Vol. 3488: M.-S. Hacid, N.V. Murray, Z.W. Raś, S. Tsumoto (Eds.), Foundations of Intelligent Systems. XIII, 700 pages. 2005.

Vol. 3487: J.A. Leite, P. Torroni (Eds.), Computational Logic in Multi-Agent Systems. XII, 281 pages. 2005.

Vol. 3476: J.A. Leite, A. Omicini, P. Torroni, P. Yolum (Eds.), Declarative Agent Languages and Technologies II. XII, 289 pages. 2005.

Vol. 3464: S.A. Brueckner, G.D.M. Serugendo, A. Karageorgos, R. Nagpal (Eds.), Engineering Self-Organising Systems. XIII, 299 pages. 2005.

Vol. 3452: F. Baader, A. Voronkov (Eds.), Logic for Programming, Artificial Intelligence, and Reasoning. XI, 562 pages. 2005.

Vol. 3451: M.-P. Gleizes, A. Omicini, F. Zambonelli (Eds.), Engineering Societies in the Agents World V. XIII, 349 pages. 2005.

Vol. 3446: T. Ishida, L. Gasser, H. Nakashima (Eds.), Massively Multi-Agent Systems I. XI, 349 pages. 2005.

Vol. 3445: G. Chollet, A. Esposito, M. Faúndez-Zanuy, M. Marinaro (Eds.), Nonlinear Speech Modeling and Applications. XIII, 433 pages. 2005.

Vol. 3438: H. Christiansen, P.R. Skadhauge, J. Villadsen (Eds.), Constraint Solving and Language Processing. VIII, 205 pages. 2005.

Vol. 3435: P. Faratin, J.A. Rodríguez-Aguilar (Eds.), Agent-Mediated Electronic Commerce VI. XII, 215 pages. 2006.

Vol. 3430: S. Tsumoto, T. Yamaguchi, M. Numao, H. Motoda (Eds.), Active Mining. XII, 349 pages. 2005.